IET TELECOMMUNICATIONS SERIES 88

The Nine Pillars of Technologies for Industry 4.0

Other volumes in this series:

Volume 9	**Phase Noise in Signal Sources** W.P. Robins
Volume 12	**Spread Spectrum in Communications** R. Skaug and J.F. Hjelmstad
Volume 13	**Advanced Signal Processing** D.J. Creasey (Editor)
Volume 19	**Telecommunications Traffic, Tariffs and Costs** R.E. Farr
Volume 20	**An Introduction to Satellite Communications** D.I. Dalgleish
Volume 26	**Common-Channel Signalling** R.J. Manterfield
Volume 28	**Very Small Aperture Terminals (VSATs)** J.L. Everett (Editor)
Volume 29	**ATM: The broadband telecommunications solution** L.G. Cuthbert and J.C. Sapanel
Volume 31	**Data Communications and Networks, 3rd Edition** R.L. Brewster (Editor)
Volume 32	**Analogue Optical Fibre Communications** B. Wilson, Z. Ghassemlooy and I.Z. Darwazeh (Editors)
Volume 33	**Modern Personal Radio Systems** R.C.V. Macario (Editor)
Volume 34	**Digital Broadcasting** P. Dambacher
Volume 35	**Principles of Performance Engineering for Telecommunication and Information Systems** M. Ghanbari, C.J. Hughes, M.C. Sinclair and J.P. Eade
Volume 36	**Telecommunication Networks, 2nd Edition** J.E. Flood (Editor)
Volume 37	**Optical Communication Receiver Design** S.B. Alexander
Volume 38	**Satellite Communication Systems, 3rd Edition** B.G. Evans (Editor)
Volume 40	**Spread Spectrum in Mobile Communication** O. Berg, T. Berg, J.F. Hjelmstad, S. Haavik and R. Skaug
Volume 41	**World Telecommunications Economics** J.J. Wheatley
Volume 43	**Telecommunications Signalling** R.J. Manterfield
Volume 44	**Digital Signal Filtering, Analysis and Restoration** J. Jan
Volume 45	**Radio Spectrum Management, 2nd Edition** D.J. Withers
Volume 46	**Intelligent Networks: Principles and applications** J.R. Anderson
Volume 47	**Local Access Network Technologies** P. France
Volume 48	**Telecommunications Quality of Service Management** A.P. Oodan (Editor)
Volume 49	**Standard Codecs: Image compression to advanced video coding** M. Ghanbari
Volume 50	**Telecommunications Regulation** J. Buckley
Volume 51	**Security for Mobility** C. Mitchell (Editor)
Volume 52	**Understanding Telecommunications Networks** A. Valdar
Volume 53	**Video Compression Systems: From first principles to concatenated codecs** A. Bock
Volume 54	**Standard Codecs: Image compression to advanced video coding, 3rd Edition** M. Ghanbari
Volume 59	**Dynamic Ad Hoc Networks** H. Rashvand and H. Chao (Editors)
Volume 60	**Understanding Telecommunications Business** A. Valdar and I. Morfett
Volume 65	**Advances in Body-Centric Wireless Communication: Applications and state-of-the-art** Q.H. Abbasi, M.U. Rehman, K. Qaraqe and A. Alomainy (Editors)
Volume 67	**Managing the Internet of Things: Architectures, theories and applications** J. Huang and K. Hua (Editors)
Volume 68	**Advanced Relay Technologies in Next Generation Wireless Communications** I. Krikidis and G. Zheng
Volume 69	**5G Wireless Technologies** A. Alexiou (Editor)
Volume 70	**Cloud and Fog Computing in 5G Mobile Networks** E. Markakis, G. Mastorakis, C.X. Mavromoustakis and E. Pallis (Editors)
Volume 71	**Understanding Telecommunications Networks, 2nd Edition** A. Valdar
Volume 72	**Introduction to Digital Wireless Communications** Hong-Chuan Yang
Volume 73	**Network as a Service for Next Generation Internet** Q. Duan and S. Wang (Editors)
Volume 74	**Access, Fronthaul and Backhaul Networks for 5G & Beyond** M.A. Imran, S.A.R. Zaidi and M.Z. Shakir (Editors)
Volume 76	**Trusted Communications with Physical Layer Security for 5G and Beyond** T.Q. Duong, X. Zhou and H.V. Poor (Editors)

Volume 77	**Network Design, Modelling and Performance Evaluation** Q. Vien
Volume 78	**Principles and Applications of Free Space Optical Communications** A.K. Majumdar, Z. Ghassemlooy, A.A.B. Raj (Editors)
Volume 79	**Satellite Communications in the 5G Era** S.K. Sharma, S. Chatzinotas and D. Arapoglou
Volume 80	**Transceiver and System Design for Digital Communications, 5th Edition** Scott R. Bullock
Volume 81	**Applications of Machine Learning in Wireless Communications** R. He and Z. Ding (Editors)
Volume 83	**Microstrip and Printed Antenna Design, 3rd Edition** R. Bancroft
Volume 84	**Low Electromagnetic Emission Wireless Network Technologies: 5G and beyond** M.A. Imran, F. Héliot and Y.A. Sambo (Editors)
Volume 86	**Advances in Communications Satellite Systems Proceedings of the 36th International Communications Satellite Systems Conference (ICSSC-2018)** I. Otung, T. Butash and P. Garland (Editors)
Volume 89	**Information and Communication Technologies for Humanitarian Services** M.N. Islam (Editor)
Volume 92	**Flexible and Cognitive Radio Access Technologies for 5G and Beyond** H. Arslan and E. Başar (Editors)
Volume 93	**Antennas and Propagation for 5G and Beyond** Q. Abbasi, S.F. Jilani, A. Alomainy and M.A. Imran (Editors)
Volume 95	**ISDN applications in Education and Training** R. Mason and P.D. Bacsich

The Nine Pillars of Technologies for Industry 4.0

Edited by
Wai Yie Leong, Joon Huang Chuah and Boon Tuan Tee

The Institution of Engineering and Technology

Published by The Institution of Engineering and Technology, London, United Kingdom

The Institution of Engineering and Technology is registered as a Charity in England & Wales (no. 211014) and Scotland (no. SC038698).

First published 2020

The Institution of Engineering and Technology
Michael Faraday House
Six Hills Way, Stevenage
Herts, SG1 2AY, United Kingdom

www.theiet.org

British Library Cataloguing in Publication Data
A catalogue record for this product is available from the British Library

ISBN 978-1-83953-005-0 (hardback)
ISBN 978-1-83953-006-7 (PDF)

Typeset in India by MPS Limited
Printed in the UK by CPI Group (UK) Ltd, Croydon

Contents

About the editors

Wai Yie Leong is the Chairperson of The Institution of Engineering and Technology (Malaysia Local Network) and Vice President of The Institution of Engineers, Malaysia (IEM). She specializes in sensing and wireless communications and medical signal processing research including wireless sensor networks, wireless communications, and brain signal processing for signal conditioning and classification in various EEG-based mental tasks. She received the Women Engineer of the Year award in 2018, the IEM Presidential of Excellence Award in 2016, and ASEAN Meritorious Award in Science and Technology 2017. She holds a Ph.D. in Electrical Engineering from The University of Queensland, Australia.

Joon Huang Chuah lectures at the Department of Electrical Engineering, University of Malaya, Malaysia and is the Head of VLSI and Image Processing (VIP) Research Group. He specialises in artificial intelligence, image processing and integrated circuit design. He is the Principal Journal Editor of the Institution of Engineers, Malaysia. He received his B.Eng. from the Universiti Teknologi Malaysia, M.Eng. from the National University of Singapore, MPhil in Technology Policy from the Cambridge Judge Business School, and PhD from the University of Cambridge.

Boon Tuan Tee is an Associate Professor at the Faculty of Mechanical Engineering, Universiti Teknikal Malaysia Melaka, Malaysia. His research focuses on energy management system and green technology. He is a Professional Technologist (Green Technology) registered under Malaysia Board of Technologists. His other professional certifications include Certified Energy Manager under ASEAN Energy Management System (AEMAS) and Certified Professional in Measurement & Verification from Malaysia Greentech. He has been a member of the IET since 2015. He holds a Ph.D. degree in Engineering from University of Cambridge, UK.

Foreword

Industry 4.0 refers to the current trend of automation and data exchange in manufacturing technologies, also called Intelligent or Smart Manufacturing. There is an increasing number of organizations and countries where Industry 4.0 is becoming adopted including the United Kingdom (Industry 4.0 and the work around 4IR), the United States, Japan, China and the European Union. Several research organizations, with a leading role for the Fraunhofer Institute, are pushing a reference architecture model for secure data sharing based on standardized communication interfaces. The nine pillars of technology that are supporting this transition include: the internet of things (IoT), cloud computing, autonomous and robotics systems, big data analytics, augmented reality, cybersecurity, simulation, system integration and additive manufacturing. A key role is played by Industrial IoT with its many components (platforms, gateways, devices) but many more technologies play a role in this process including cloud, fog and edge computing, advanced data analytics, innovative data exchange models, artificial intelligence, machine learning, mobile and data communication and network technologies, as well as robotics, sensors and actuators. Over the IoT, cyber-physical systems communicate and cooperate with each other and with humans in real-time both internally and across organizational services in the value chain. Within smart factories, cyber-physical systems monitor physical processes, create a virtual copy of the physical world and make decentralized decisions. The aim of this edited book is to focus on the nine pillars of technology including innovative research, challenges, strategies and case studies.

Advances in science and technology continuously support the development of industrialization all over the world. The First Industrial Revolution used water and steam power to mechanize production, the Second used electric power to create mass production and the Third used automation for the manufacturing line. Now a Fourth Industrial Revolution, the digital revolution, is characterized by a fusion of technologies that is blurring the lines between the physical, digital and biological spheres. There are three reasons why today's transformations represent not merely a prolongation of the Third Industrial Revolution but rather the arrival of a Fourth and distinct one: velocity, scope and systems impact. The speed of current breakthroughs has no historical precedent. When compared with previous industrial revolutions, the Fourth is evolving at an exponential rather than a linear pace. Moreover, it is disrupting almost every industry in every country. And, the breadth and depth of these changes herald the transformation of entire systems of production, management and governance. The design principles of Industry 4.0 such as virtualization, interoperability, decentralization, service-oriented approaches, real-time capabilities and modularity all play a key role in the radical changes facing industry. In this book series, we describe the advantages of intelligent manufacturing systems and discuss how they will benefit the manufacturing industry by increasing productivity, competitiveness and profitability. (Benefits: enhanced productivity through optimization and automation, real-time, better quality products, sustainability, personalization and customization for consumers, and improved scalability and agility.)

In this book series, the nine pillars of technology that are supporting this transformation have been introduced: the IoT, cloud computing, autonomous and robotics systems, big data analytics, augmented reality, cybersecurity, simulation, system integration and additive manufacturing.

Technologies and Industry 4.0 are about reducing complexity and processes and add value. However, while business processes are changing rapidly, industries and manufacturers are struggling to exploit the full potential of digitization. From both a strategic and technological perspective, the Industry 4.0 roadmap visualizes every further step on the route towards an entirely digital enterprise. Many case studies have been discussed in this book series, which aims to benefit and create impact to industry players, scientists, academicians, researchers and manufacturers.

Chapter 1

The Nine Pillars of technology for Industry 4.0

Joon Huang Chuah[1]

1.1 Introduction

Industry 4.0 is a strategic initiative introduced by the German government during early 2010s to transform industrial manufacturing through digitalisation and exploitation of the potentials of new technologies. It is an effort to increase productivity and efficiency mainly in the manufacturing sector. Industry 4.0 production system aims to be highly flexible and should be able to produce individualised and customised products. In fact, it is an exciting employment of automation within manufacturing, covering the use of robotics, data management, cloud computing and the internet of things (IoT). It has started to show that artificial intelligence, robotics, smart sensors and integrated systems are an important part of a normal manufacturing process. In interaction with machines, it needs horizontal integration at every step in the production process. The Americans have the same concept for Industry 4.0 but prefer to call it Smart Factory.

Industry 4.0 has presented a lot of excitement and many in the various industries are presently deciding how to get involved. Many people consider that Industry 4.0 is in fact the repackaging and combination of different key technologies and technologies that have already existed. It has a refreshed acceptance by providing an overall wrapper that enables total operability, which essentially covers the tasks of collecting big data, processing and analysing the data, and improving the process through positive feedback for enhanced functionally and efficiency.

In the First Industrial Revolution, there was a drastic transition from manual labour to steam engine. It was the era where the mechanical advantage of steam engines was employed to lessen the dependence on the use of human labour. It was followed by the Second Industrial Revolution, which capitalised on the power of electricity. Steam engines were phased out by the introduction of electrical motors and analogue systems. Mass production assembly lines were ubiquitously built during this period. Meanwhile, the Third Industrial Revolution focused on automation of work. Computers and electronic systems were widely employed in factories and beyond. The Fourth Industrial Revolution, commonly abbreviated and called as IR 4.0, mainly relies on integration of

[1]Department of Electrical Engineering, Faculty of Engineering, University of Malaya, Kuala Lumpur, Malaysia

data, artificial intelligence, industrial machinery and communication to form a highly efficient, automated and intelligent industrial ecosystem. Industry 4.0 is the subset of IR4.0, which covers on the future state of industry characterised by complete digitisation of production flows.

Industry 4.0 encompasses six main principles, i.e. interoperability, virtualisation, decentralisation, real-time capability, service orientation and modularity. Interoperability allows the cyber-physical systems, humans and factories to connect and communicate effectively with each other via the IoT and the internet of services (IoS). Virtualisation provides a virtual copy of the factory, which is formed by linking sensor data with virtual plant models and simulation models. Decentralisation is the ability of cyber-physical systems with the factories to make optimal decisions on their own. Real-time capability is highly crucial as it helps to collect and analyse collected data and generate the insights instantly. Apart from that, to fulfil the principle of service orientation, it must offer services of cyber-physical systems, human and factories via the IoS. With the principle of modularity, it possesses a flexibility in adapting smart factories for changing requirements of individual modules.

Industry 4.0 is a disruptive concept that revolutionises how products are manufactured efficiently and creates a new generation of industrial manufacturing systems that will be totally different from the existing ones. Implementing all these capabilities in the factories is only just the beginning. They need to be able to work seamlessly in an interconnected way within the manufacturing environment to be counted as Industry 4.0. Importantly, Industry 4.0 determines the future state of industry characterised by digitisation of production flows and it is an exciting breakthrough as innovation comes with sustainable processes. As industrial production is highly driven by global competition and fast-changing market requests, extreme advances in the current manufacturing technologies can only fulfil them. The introduction of Industry 4.0 is therefore timely and important for ensuring the sustainability of a manufacturing business. It is a promising concept based on integration of manufacturing and business processes, and integration of all business stakeholders in the value chain, i.e. suppliers and customers.

1.2 The Nine Pillars

The Nine Pillars of Industry 4.0, as depicted in Figure 1.1, are the main constituents of technology that transforms isolated cell production into a fully optimised, integrated and automated production flow. This technological blending will essentially lead to a higher production efficiency among suppliers, manufacturers and customers, as well as between machines and machines, and between machines and humans. The framework of Industry 4.0 is formed with the constituting pillars and their main role as shown in Table 1.1.

Various international organisations have been influencing the trend and conducting impactful research and development to support the exciting rollout of Industry 4.0. Table 1.2 shows the international organisations which are the main contributors and supporters to the progress of Industry 4.0.

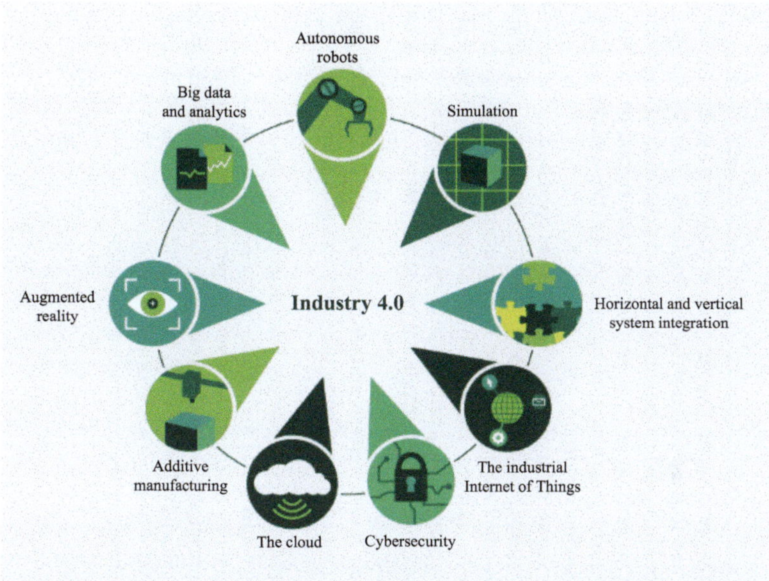

Figure 1.1 Nine Pillars of Industry 4.0 (Source: BCG analysis)

Table 1.1 The Nine Pillars and their main role

Pillar	Main role
Autonomous Robots	Performing tasks with minimal or no human intervention
Simulation	Creating in the virtual world and predicting the outcomes of an action
Horizontal and Vertical System Integration	Linking, collaborating and co-operating horizontally and vertically
Industrial Internet of Things	Making objects communicate with each other and with humans
Cybersecurity	Protecting computing systems and networks
Cloud Computing	Accessing information from anywhere using the Internet
Additive Manufacturing	Creating 3D objects by adding material layer upon layer
Augmented Reality	Carrying out tasks in real-world environment interactively with virtual objects
Big Data and Data Analytics	Analysing and extracting useful information from large datasets

1.2.1 Autonomous robots

Robots have been commonly used to carry out repetitive, tedious and dangerous tasks. In recent days, robots are able to provide an ever-wider range of services while becoming more flexible, cooperative as well as autonomous. These robots

Table 1.2 International organisations involving in Industry 4.0

Organisation	Description
Acatech	An autonomous, independent and a non-profit organisation that supports policymakers and society, providing qualified technical evaluations and forward-looking recommendations. Also supporting knowledge transfer between science and industry, and encouraging the next generation of engineers.
Fraunhofer IAO	Helping shape the Industry 4.0 project since as early as 2011 as part of activities carried out by the Industry-Science Research Alliance; also developing new applications and business models for Industry 4.0 with industry and trade association partners.
German Research Center for Artificial Intelligence (DFKI)	On the forefront of Industry 4.0 research; the SmartFactory Living Lab performs operation and testing of the latest technologies in process engineering and piece goods under industrial conditions. The project 'RES-COM' examines the vision of an automatised conservation of resources through highly interconnected and integrated sensor-actuator systems. 'SmartF-IT' is looking at cyber-physical IT systems to master complexness of a new generation of multi-adaptive factories due to the intensive use of high-networked sensors and actuators, overcoming traditional production hierarchies of central control towards decentralised self-organisation.
Industrial Internet Consortium (IIC)	Formed to accelerate the development, adoption and widespread use of interconnected machines and devices and intelligent analytics through technology test-beds, use cases and requirements development.
Industry-Science Research Alliance	An advisory group, which brings together leading representatives from science and industry to accompany the High-Tech Strategy of inter-ministerial innovation policy initiatives
OPC Foundation	An industry consortium that creates and maintains standards for open connectivity of industrial automation devices and systems. OPC Foundation's Unified Architecture (OPC UA) is the backbone of interconnected communication in production, warehouse and transport logistics. The Reference Architecture Model for Industry 4.0 (RAMI 4.0) recommended only IEC standard 62541 OPC Unified Architecture (OPC UA) for implementing the communication layer.

(Continues)

Platform Industry 4.0	A platform for development of technologies, standards, business and organisational models and their practical implementation.
SmartFactoryKL	A pioneer for the technology transfer of key aspects of Industry 4.0 into practice; supporting the development, application and propagation of innovative automation technologies in different sectors as well as providing a basis for their extensive usage in science and industry.
W3C	An international community that develops open standards to ensure the long-term growth of the World Wide Web; defining Web standards to unlock the potential for open markets for suppliers and consumers of services.

eract with one another and able designed in a way to work safely with humans. Formed by two individual words 'collaborative' and 'robotics', Cobotics is a popular term used to describe robots assisting operators performing their daily operational tasks. Robots are becoming more intelligent and are now able to learn from human beings to perform various complex and challenging tasks.

Autonomous robots are being used in factories to take up a number of roles. Fully autonomous robots are usually employed for high-volume and repetitive processes as the accuracy, speed and durability of a robot present considerable advantages. Many manufacturing floors use robots to assist in performing tasks that are more intricate. Robots are commonly used to execute tasks such as lifting, holding and shifting heavy or bulky components. Robots are designed to be more flexible, cooperative and autonomous and they can interact with each other and operate safely with human beings [1]. Maintaining good safety, flexibility and versatility, autonomous robots can execute a given task with high precision and efficiency [2].

The introduction of autonomous robots plays a critical role in modern manufacturing industry. Robots can complete tasks intelligently, with consideration of versatility, safety and flexibility. With recent technological innovation, robots that are more autonomous are produced to support industrial revolution. Robots and humans are working hand in hand on interlinking tasks with the help of human–machine interface. The implementation of robots in manufacturing has been widespread and encouraging as it covers various functions, i.e. logistics, production, office management, etc. Since 2004, the number of multipurpose industrial robots developed by companies in Europe has increased by two times [3].

The inclusion of robots for automation in manufacturing drives companies to stay competitive internationally as this presents an efficient solution to meeting the skills gap in areas where suitably qualified workers are difficult to employ. With the implementation of autonomous robots, it allows employees to allocate more time on design, innovation, planning and strategy, which are equally critical for growth and success. The automation in manufacturing with robots will culminate in better employee safety and satisfaction, increased productivity and eventually higher profitability.

There are reasonable justifications to pull the factory owners to use autonomous robots in their respective entities. First, as there are many manufacturing processes involved in producing a product, robots may be employed to automate every task all the way from raw material handling towards finished product packaging. Second, robots do not need rest as humans do and can work 24 h a day to get task completed continuously. Third, the modern robots are developed in a way that they are highly flexible and can be conveniently customised to perform various functions including complex ones. Fourth, using robot for automation in manufacturing is cost-effective and will improve a company's bottom line. Finally, automation with robots will help in achieving high-speed manufacturing and faster delivery of products to customers.

Some robots are designed to be mobile in carrying out their tasks. Motion planning is a challenging aspect in the field of autonomous mobile robots, which allows them to travel from one position to another in various environments, which may include both static and dynamic obstacles [4]. Intelligent and optimised navigation solution with the

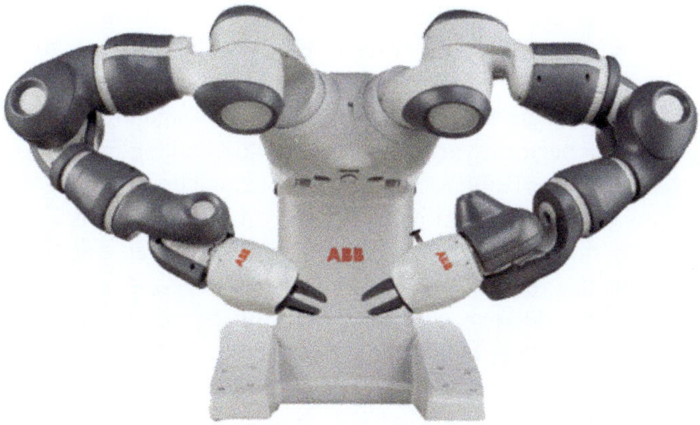

Figure 1.2 YUMI from ABB (Source: ABB)

implementation of artificial intelligence technique has been proposed to ensure maximum safety as well as a shorter and better routing option [5].

Autonomous robots from different suppliers such as ABB (shown in Figure 1.2), Bionic Robotics, Fanuc, Gomtec, Kuka, etc., are currently widely employed for various industrial uses. Various models of robots are developed by different key players to support the technological trend of Industry 4.0, as shown Table 1.3.

Increasingly, more countries are stepping up efforts of employing industrial robots in their manufacturing industry. Republic of Korea topped the list in 2016 with 631 industrial robots per 10,000 employees as shown in Figure 1.3.

The recent development of autonomous robots has been exceptional and encouraging. In many factories, autonomous robots are able to transfer raw materials, semi-finished and completed goods in a more efficient, faster and smarter way. As the robots operate based on a complex logic algorithm, they do not require any pre-set path to carry out their tasks. The use of robots has in fact made the manufacturing process more efficient as well as more cost-effective.

1.2.2 Simulation

Simulation is defined as an approximate imitation of the operation of a process or system. It is employed in many functions such as safety engineering, performance optimisation, training, testing, designing, etc. Simulation plays a critical role in Industry 4.0 as this technology helps to achieve better results in many ways. It essentially reduces unnecessary waste in time and resources while increasing efficiency in manufacturing. In addition, it significantly boosts the productivity and revenue of a manufacturing business. Apart from that, simulation is highly important during the design stage of a product because it allows one to duly evaluate the outcomes of the product and to make necessary changes if the product does not meet specifications. With the assistance of simulation steps, the quality of decision-making can be greatly enhanced and ascertained [6]. It is noteworthy that

Table 1.3 Industrial robots developed by key players of Industry 4.0

Company	Model	Description
ABB	Yumi	First collaborative robot from ABB. Features an advanced vision system, flexible hands, parts feeding systems, sensitive force control feedback and state-of-the-art robot control software that allows for programming through teaching.
Bionic Robotics	BioRob	A lightweight robot with advantages of flexibility, mobility and inherent safety properties required to implement effective and safe service robotics applications. Certified safe for use in close proximity with humans without needing further protective measures.
Epson Robots	Flexion	Able to fold through themselves, resulting in a much smaller footprint than typical six-axis robots. Vibration-reduction technology for performing precise tasks and handling delicate objects.
Fanuc	CR-35iA	A highly advanced collaborative robot that can work without safety fences; is relatively heavy and has a pedestal that is designed to be fixed to the ground.
F&P Personal Robotics	P-Rob	A six-axis robotic arm which is collaborative and safe, and powered by artificial intelligence.
Gomtec	Roberta	A lightweight, adaptable and inexpensive six-axis industrial robot. Specifically developed for small and medium-size businesses, which require flexible and efficient automation.
Honyen	King Kong	An articulated six-axis transfer robot. Consisting of a body and seven reducers, 12 servo motors and one controller, with excellent precision and control flexibility.
Kuka	KR 1000 Titan	A six-axis robot moving parts and components safely and precisely, even over long distances; powerful robot for heavy loads.
MABI Robotic	Speedy-10	A lightweight six-axis kinematics system with standard wrist with excellent damping characteristics; high positioning precision for high-speed applications, due to its high-resolution absolute feedback encoder.
Precise Automation	PF3400	World's first collaborative four-axis SCARA robot; achieving speeds and accelerations much faster than any other collaborative robot while still limiting forces to ISO collaborative robot standards; safe design.
Rethink Robotics	Sawyer	A high-performance collaborative robot designed to execute machine tending, circuit board testing and other precise tasks that are impractical to automate with conventional robots.
Universal Robots	UR series	A highly flexible robot, which performs well for optimising lightweight collaborative processes, e.g. moving, picking and placing.

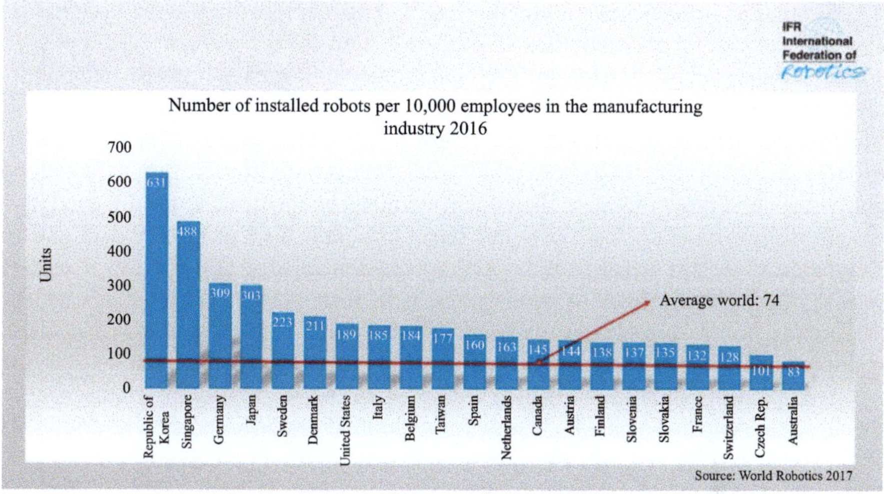

Figure 1.3 Countries with the most industrial robots in manufacturing (Source: International Federation of Robotics (IFR))

3D simulation of product development, material development and production processes has become a normal practice in many companies as a necessary step. It uses real-time data to mirror the physical world in a virtual model embodying humans, machines and products. Simulation also allows operators to test and optimise on machine settings for the next product in line in the virtual world prior to the actual physical reconfiguration, therefore keeping the machine downtime for setup to a minimum while increasing overall quality. The use of simulation tools is seen more commonly in manufacturing facilities to mirror the physical world in virtual model mainly to reduce machine setup time and to increase overall quality [1]. The uses of simulations for production processes will not only reduce the downtime but also bring down the production failures during start-up phase [7].

Simulation with computer systems provides a modelling and evaluation tool for analysing complex systems. Simulation has been used to enhance the design of complex systems as early as 1950s. Initially, simulations were conducted to verify system designs to find out if they fulfil certain predefined objectives. With runs of simulations, they essentially save time and resources by establishing proof of concept before building a physical system. Simulations are useful for evaluation of different options of system designs, which subsequently helps the designer to select the best performance or most optimised design. Furthermore, simulations can be used to help the designers to identify the strengths and weaknesses of a particular design. Significantly, a simulation model is able of leveraging specificity to enhance fidelity to the system that is being modelled. Hence, outputs from the simulation model can offer a trustworthy prediction of system performance. With the simulation results, the designer can then decide on whether to build a physical system or otherwise.

Simulation models are normally required to capture important systems and operations with a lot of details. Thus, simulation models are usually very complex and computationally expensive to run. The computation cost is even higher when multiple replications of simulations are necessary to control the noise in stochastic simulations. In addition, sufficient high-quality data must be provided to the simulation modellers in order for them to estimate probability distribution models for various sources of uncertainly and randomness in the system. Simulation models are usually used in the system design stage so that designers have sufficient time to collect data and run simulation to obtain statistically backed conclusion for decision-making. Explosive growth of computing power in recent years, in the forms of cloud computing and high-performing computing clusters, has encouraged further simulation-based decision-making, which was previously limited by computational cost concerns.

Simulation optimisation enables the system designer to search a large decision space, systematically and methodically, for an optimal design without being confined to only several specific options. This optimisation capability tremendously increases the scope simulation tool for complex design. It is currently an active research area and there are many commercial simulation software products coming with the integration of simulation optimisation solvers. Simulation optimisation is essentially a useful tool to find the optimal design of a system, based on a computer simulation model. It is employed to predict and evaluate performance of complex stochastic systems. Continuous efforts for simulation optimisation and exponential growth in computing power have made it more attractive to use simulations to optimise design and operations of systems directly [8].

1.2.3 Horizontal and vertical system integration

A system can be integrated horizontally and vertically. Horizontal integration can lead networking among cyber-physical systems to an unprecedented level to achieve greater efficiency. Every device and system of the same level of manufacturing in the same facility is connected with each other. There is also data communication between systems in different facilities, allowing tasks to be planned and distributed among themselves. With that in place, downtime at a particular facility can be covered or compensated by machines from another facility with little or no human intervention. Meanwhile, vertical integration is relatively a more challenging endeavour. Every system at different levels of the hierarchy has access to all the data collected or generated. The main challenge now is that different communication protocols are used at different levels. Thus, the systems are having a difficult time communicating and exchanging data with each other. However, if vertical integration in a facility is conducted correctly, the efficiency will be immensely improved. This can be resolved by using proper and suitable interfaces. The complete integration and automation of manufacturing processes vertically and horizontally shows automation of communications and cooperation particularly along standardised processes [9].

For the horizontal integration, it attempts to realise connected networks of cyber-physical and enterprise systems that achieve new levels of flexibility, automation and operational efficiency into production processes. This sort of integration may happen at different levels. On the level of production floor, internet-connected machine and production units become an object with well-defined properties within the production network. They constantly communicate their performance status and respond autonomously to changing production requirements. This results in better cost-effectiveness in producing items and lower machine downtime with predictive maintenance. Horizontal integration may occur across multiple production facilities, where production facility data are shared across the whole enterprise and this enables production tasks to be intelligently allocated among facilities to react efficiently and swiftly to changes in production variables. Furthermore, horizontal integration may take place across the entire supply chain where data transparency and high levels of automated collaboration across (i) the upstream supply and logistics chain that handles the production processes themselves and (ii) the downstream chain that delivers the finished products to market. On the other hand, the vertical integration helps to tie together all logical layers within an organisation such as production, research and development, quality assurance, information technology, sales and marketing, human resources, etc. This sort of integration allows data to flow freely among these layers in order for facilitating tactical and strategic decision to be made. Vertical integration essentially creates a competitive advantage as it enables a company to respond quickly and appropriately to changing market signals as well as new opportunities [10].

At present, it is discovered that many information systems are still not fully integrated. Most companies are seldom linked to their suppliers and customers with internet connection. Even within the own company, engineering design department is not connected directly to its production floor. Thus, with the introduction of Industry 4.0, it is hoped that most manufacturing company will take steps to be closely interlinked or interconnected between departments, and between its suppliers and customers. With cross-company universal data integration networks that enables completely automated value chain, the horizontal and vertical system integration among companies, departments, functions and capabilities will be more cohesive and efficient.

1.2.4 Industrial internet of things

The IoT is a huge number of devices, machines or systems that are connected through a network or the Internet for data sharing and manipulation. It is an ecosystem in which all the sensors and actuators with the ability to function separately and communicate with each other. The IoT is also sometimes being referred to internet of everything (IoE), covering the IoS, internet of manufacturing services (IoMs), internet of people (IoP), and integration of information and communication technology (IICT) [11]. Kelvin Ashton, a British technology pioneer, first started the concept of IoT in 1999. As can be seen in automotive industry, there are communications among different devices and systems in a car. The introduction of connected car enables real-time communications, e.g. navigational systems in combination with vehicle logs and setting of emergency calls in the event of an accident [12].

Industrial IoT (IIoT), first mentioned by General Electric, is a robust version of the IoT, which possesses increased ruggedness to survive the harsh environments of the industry. As a subcategory of the IoT, the IIoT is important as it uses smart sensors and actuators to enhance manufacturing processes. It is a network of intelligent devices connected to form a complete system that collect, exchange and analyse data. In general, an IIoT system consists of (i) intelligent devices that can collect, store and communicate data, (ii) private or public data communications infrastructure and (iii) data analytical systems that generate useful business information. It also covers the use of data collection and analytics in various industries such as manufacturing, energy, agriculture, transportation, healthcare, etc. IIoT devices range widely from small environmental sensors to complex industrial autonomous robots.

The IIoT systems are normally made up of a layered modular architecture of digital technology. These layers are device layer, network layer, service layer and content layer. The device layer normally consists of physical components such as sensors, machines, cyber physical systems (CPS), etc. Meanwhile, the network layer is commonly made up of network buses, communication protocols, cloud computing, etc. Service layers consist of software or applications that analyse data and eventually transform them into useful information. At the top is the content layer, which is a user-interface device like a monitor, display, tablet, smart glass, etc.

The IIoT is a game-changer for superior operational efficiency and presents an entirely new business model that benefits not only a certain company but also the society in general. Presently, IIoTs are employed in many industries to further their daily operation. The IIoT technology is aggressively implemented in manufacturing sector to improve production efficiency. An IIoT-enabled machine can have smart monitoring functions and predict potential problems, resulting in lower downtime and better overall efficiency. In the sector of supply chain, the IIoT technology is employed to handle supply ordering before they are running out. This helps keep necessary items in stock and reduces the amount of waste produced. In the retail line, the IIoT assists in terms of intelligent and fast decision-making for individual stores. As a business strategy with the implementation of IIoT technologies, display panels, which automatically refresh according to customer interest and their ability to put together smart promotion, could help a store to gain significant advantage over other companies. In healthcare, the IIoT is driving the industry to become more responsive, more precise and safer. It introduces devices that monitor patients remotely and inform the medical officers when there are irregularities or unusual patterns in the patients' status or monitoring data. The IIoT technology is also penetrating the sector of building management where the use of this technology could make building management tasks more secure and efficient. Sensor-driven climate control is employed and this will remove the inaccurate guesswork involved in manually changing a building's climate. In addition, with the IIoT technology, installation of networked devices to monitor the entrance of a building can enhance the security and allow quick response to be taken if there is a potential threat [13].

One of the main challenges for IoT in Industry 4.0 is the lack of common standards. Having devices connected together for data sharing is good; however, when all of them collect data in different formats and have different protocols, integrating them into a fully automated factory will be difficult and costly.

Big companies such as Bosch, Eclipse Foundation and others have been working in standard communications architectures and protocols such as production performance management protocol (PPMP), message queuing telemetry transport (MQTT) and OPC UA. With these in place, connected devices, including those in the production floor, can communicate with each other seamlessly. Another main challenge is concerning internet security. As devices are connected to the Internet, they are vulnerable and prone to hacking and cyber-abuse by hackers. Security experts are moving rapidly to tackle the cybersecurity concerns on the IoT, combining new technologies with standard IT security [14].

1.2.5 Cybersecurity

Cybersecurity is an extremely crucial component of Industry 4.0 as it protects computer systems, networks and data from malicious activities such as network attack, unauthorised access, data theft, disruption, damage, etc. It is becoming more important than ever due to the increased numbers of connected devices and systems. Safeguarding of data while maintaining the performance of systems is the main purpose of cybersecurity. It is an alarming fact that IT systems of many institutions are being attacked and intruded every day. It is highly vital that the factories must be aware of their potential weaknesses and well prepared against any possible incoming threats. It is very important to have robust cybersecurity in place as it ensures that the daily running of manufacturing activities is not severely affected, which may cost immense losses to the company.

For Industry 4.0, having an advanced identity and access management of machines and users and trustworthy communications systems is of utmost importance as the problem of cybersecurity threats becomes more serious with the increased connectivity and wider use of standard communications protocols. Increase of data density and fusion of information technology and operational technology bring a great challenge to cybersecurity. In recent years, many governments have treated cybersecurity as a main national issue with the highest level of importance. It is important to protect business information in digital shape against unauthorised access, theft and abuse. With expanding network connections, cyberattacks become more rampant as the data stolen can be used for gaining certain advantages in the financial and strategic forms.

Cyberattacks and internet threats have become increasingly serious in the past decade. Users of IoT systems, in particular, are directly and indirectly affected by these problems. Large companies or businesses are commonly exposed to malicious attacks, which cause immense financial losses in additions to other inconvenience such as system crashes, leak of data, privacy breaches, data corruption, slowness in systems, etc. The widespread use of connected devices and services has created immense need and spurred new forms of powerful cyber defence to combat cyberattack issue. In many companies, cybersecurity is identified as a main technology issue. Most big companies have strengthened the cyber defence and capability of their IT systems in order prevent the attacks. Millions of dollars have been allocated and spend to procure advanced systems and to develop new strategies with investment in IT security to bring down the risk of cyber threats.

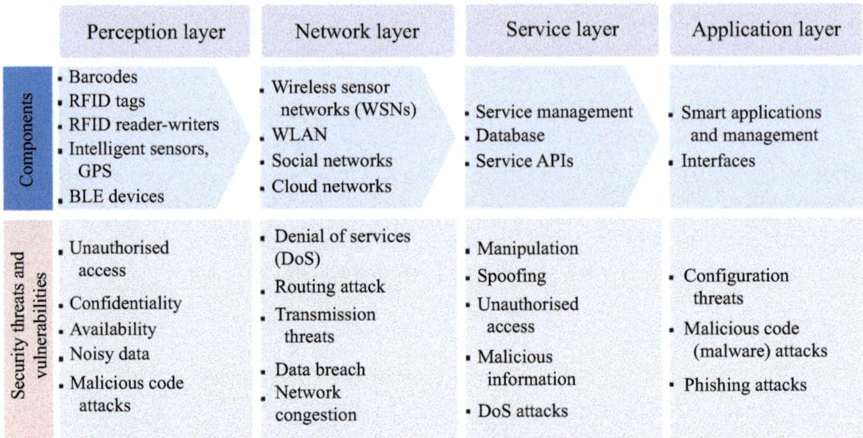

Figure 1.4 Layers of IoT system and their security threat [15]

In an IoT system, it can generally be divided into four main levels, i.e. perception layer, network layer, service layer and application. A system may be vulnerable and a cyberattack can occur at any level of the system. Figure 1.4 shows the four layers with their components, and the possible cyber threat and vulnerabilities.

To tackle the cybersecurity issues, various parties need to work together closely to bring the impact to the minimal. These stakeholders are IT security experts, manufacturers, regulators, standardisation community and academia. To ensure success in handing these issues, each of them should take up their roles proactively as shown in Figure 1.5. It covers important roles such as (i) promoting cross-functional knowledge on IT and OT security, (ii) clarifying liability among Industry 4.0 actors, (iii) fostering economic and administrative incentive for Industry 4.0 security, (iv) securing supply chain management processes, (v) harmonising efforts on Industry 4.0 security standards, (vi) establishing Industry 4.0 baselines for security inter-operability and (vii) applying technical measure to ensure Industry 4.0 security [16].

IT security experts should promote cross-functional knowledge in IT and OT security and secure supply chain management processes. This group of experts can help establish Industry 4.0 baselines for security interoperability besides applying technical measures to ensure Industry 4.0 security. Also, secure and reliable communications coupled with sophisticated identity and access management of machines and users are of high importance [1].

1.2.6 Cloud computing

Cloud computing refers to a remote system where information can be accessed remotely from anywhere using the Internet. With the implementation of cloud computing, a manufacturing system enables delivery of difference services through the Internet with resources including tools and applications such as databases, data storage, servers, networking, software, etc. Instead of keeping data on a proprietary

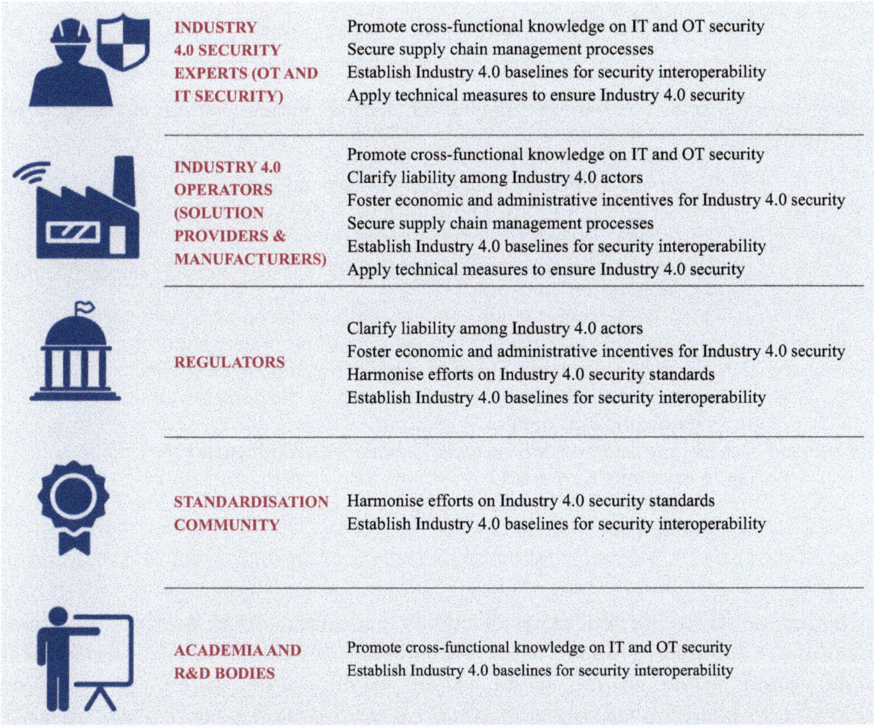

Figure 1.5 Recommended efforts to ensure Industry 4.0 security [16]

local storage device or hard disk, cloud computing stores the information to a remote database. When a computing device has a connection to the Internet, it may conveniently access the data on a remote server and run certain software to manipulate or process the data. To achieve faster response time even at the level of sub-second, an organisation requires data sharing across sites and companies as provided by cloud computing [1]. It has started to be a popular support for individual users as well as businesses as it has key advantages in terms of higher cost efficiency, better security, increased productivity, faster processing and superior efficiency.

Cloud computing is an important technology that helps people to work together. It is a collaborative tool that revolutionises the working culture of data management. Companies or businesses nowadays are more open to information sharing instead of keeping it to themselves. The practice of opening up will benefit the company as overall, achieving better financial results. With cloud computing, it has virtually infinite storage capabilities for the users. There is a need for a company to make the information accessible and actionable when more information is generated and collected. This will benefit every user who is accessing to the computing system for completing their tasks as well as the company as whole.

Table 1.4 Types of cloud

Cloud type	Description
Public cloud	Does not operate under IT of a company Usually offered by IT service operators Open to any person or company or general public
Community cloud	Only accessible to those companies or institutions that have joined a respective group with it specific requirements Shared by several organisations that have shared concerns (e.g. security, mission, compliance, etc.) May be managed by the organisations involved or a third party
Private cloud	Only available exclusively for one individual organisation or company Normally, physically located within the company network May be owned, managed and operated by the company, a third party or a combination of both
Hybrid cloud	A composition of two or more distinct cloud infrastructure (public, community or private) Bound together by standardised or proprietary – for data and application portability

In general, there are four types of clouds, i.e. public cloud, community cloud, private cloud and hybrid cloud [17]. The characteristics of them are as summarised in Table 1.4.

Cloud computing essentially facilitates real-time exchange of data, creating and promoting a sphere of digital integration and collaboration. A company or business that employs the services of cloud computing will better connect with its key stakeholders, allowing proactive management of supply chain, providing real-time visibility, achieving superior efficiency and enhancing risk management.

1.2.7 Additive manufacturing

Additive manufacturing is widely known as a process of joining materials together to form a physical object, with reference to a set of 3D model data, normally layer upon layer. This technology is commonly used to fabricate small batches of customised products that offer construction advantages, i.e. complex but lightweight designs. Meanwhile, it is as opposed to subtractive manufacturing, which is a conventional process by which a physical 3D object is created by successively removing material away from a solid block of material. Subtractive manufacturing is usually performed by manually cutting the material or mechanically by using a computerised numerical control (CNC) machine.

The physical part of the implementation of Industry 4.0 is limited by the functions and capabilities of the current manufacturing systems and this makes additive manufacturing a highly important component of Industry 4.0. Challenges of increasing individualisation of products and reducing time to market are seriously encountered by many companies to fulfil the needs of customers [18]. Non-traditional manufacturing methods are required to be redeveloped to achieve the mass customisation capability in Industry 4.0. Additive manufacturing has become a key technology for creating customised product due to its ability to fabricate items with advanced attributes in terms of shape, material, etc. The production has become faster and more cost-effective with the implementation of additive manufacturing technologies such as selective laser melting (SLM), fused deposition method (FDM), selective laser sintering (SLS), etc. [19]. As the quality of products fabricated with the additive manufacturing technologies has improved immensely in recent years, they are now being employed in various industries, e.g. manufacturing, construction, biomedical, aerospace and many more.

Despite the arising doubts about its applicability in mass production, the implementation of additive manufacturing in various industries is fast increasing due to the new technological advancements. Serving as an advanced technology to manufacture accurate and complex products, additive manufacturing is on its way to replace conventional manufacturing techniques [20].

1.2.8 Augmented reality

Augmented reality refers to a digital technology where the users have an interactive experience of real-world environment with the virtual objects enhanced by computer-generated perceptual information. Tools for augmented reality are mostly in their infancy but they have already started to create new waves for services. Previously, augmented reality found its applications only in certain important or dangerous tasks such as flight simulators. Recently, it has penetrated into the areas of repairs and maintenance. Remote repair instructions can be delivered to any part of the world as long as there is an internet connection. With this technology, it allows the technicians to acquire their skills by practising the maintenance steps repetitively until they are sufficiently competent.

Augmented reality has been introduced to many applications in the industry. It is now used in human robot collaboration (HRC), which is an area attempting to understand how to enhance the collaboration between human and robot using innovative interfaces. In fact, forming a trustworthy and safe human-robot system is undoubtedly a highly challenging task. Augmented reality is used to show information contextualised in the real environment, helping the operators to have better awareness of the movements of as well as forces applied by a robot.

Apart from that, augmented reality is used for maintenance, repair and assembly tasks. Employing augmented reality for these tasks can help to reduce overall costs. However, implementing augmented reality for these tasks may have its complexity, i.e. technicians might need to refer to instruction manual to complete the procedure. The continuous switch of attention between a system and manual may impose additional cognitive load the technicians. Augmented reality applications for maintenance and repair consist of a set of virtual assets, which give aid, indication or suggestion to the technicians. These virtual assets include animated 3D model describing the task to be performed, audio track providing instructions, textual label explaining steps, etc. With graphical assets overlaid and aligned with the machine to be repaired or maintained, this technology allows technicians to be properly guided to perform certain tasks, which could be dangerous in nature. Despite its usefulness, it is facing an implementation challenge, i.e. it might take a long time to create, change and improve the augmented reality procedures. Another great challenge faced is that it lacks a clear an accessible workflow to design and develop augmented reality applications for the industry [21]. Systems with the implementation with augmented reality also support many other services, e.g. selecting parts in a warehouse and sending repair instructions via mobile devices. Augmented reality can be used to provide workers with real-time information to assist in decision-making and work procedures. While the workers are checking an actual system calling for repair, they may receive repair instructions on how to conduct the work [1].

Augmented reality is also used for product quality control. As the variety of products grows rapidly in industry, the inspection task becomes increasingly complex. The checking process could become less effective due to the cognitive limitation of the workers. Implementing augmented reality can help in enhancing the inspection process as it enables a direct comparison between the real product and the ideal one. By wearing a dedicated AR device, the worker can effectively inspect the manufactured product by visualising a 3D representation of the ideal product superimposed on it [21].

Augmented reality is indeed a very powerful component for Industry 4.0. It has created a manufacturing floor where not only everything is connected but also viewable and interactive. The effectiveness of augmented reality does depend on the visualisation process itself, but on how data are visualised instead. Its ability to enhance the real space has been widely proven in many applications. It can essentially help a factory to improve manufacturing productivity significantly. In addition, augmented reality also enhances the reliability and safety of robotic systems. It also results in cost reduction and performance improvement of maintenance systems. Augmented reality is definitely one of the key technologies of Industry 4.0 enriching the roles of managers as well as workers.

1.2.9 Big data and data analytics

Big data analytics is commonly referred to as a common process of examining huge and varied datasets to extract crucial information (e.g. hidden pattern, correlation, irregularity, trend and preference) usually for decision-making. It is a systematic way of uncover important information or clues, which could not be easily done using traditional data-processing methods. Data analytics, once very popular among IT applications, is currently penetrating into the supply chain and manufacturing industry. The power of big data and data analytics can assist the manufacturing industry in lessening wastage and reducing downtime. In some factories, data are collected at different levels of manufacturing processes. When a product from the factories is identified to be defective, its manufacturing data could then be accessed, processed and analysed to arrive at a certain pattern. The manufacturing process step or steps that cause the formation of the pattern can be redesigned or readjusted to rectify the faulty issue. Predictive maintenance is used to estimate when the maintenance routine should be carried out and can be based on manufacturing data collected. It is considered to be more cost efficient as well as safer than the traditional maintenance practice.

Big data analytics is often considered as one of the key parts of Industry 4.0. Industrial big data analytics has in fact attracted both research and application interests from both industry and academia. The effective use of this technology will deliver a new wave of production growth and ultimately transform economies. The requirements of industrial big data must be fulfilled in order for companies to achieve operational efficiency in a cost-effective manner. The industrial big data analytics covers cloud-based data storage, operational data management system and hybrid service platform. Industrial big data analytics is also employed for manufacturing maintenance and service innovation focusing on automated data processing, health assessment and prognostics [22]. The collected data are analysed in order to identify the issues happened in different production processes and to predict the recurrences during operation [23]. The collection and analysis of data from various sources of manufacturing systems have become a norm for supporting real-time decision-making [1].

With the accelerated implementation of IoT and various sensors for data collection, the volume and velocity of data are growing in an exponential curve especially in the industrial manufacturing sector. The manufacturing sector needs to embrace state-of-the-art technologies to extract and analyse useful information from the big data and this is paving for the widespread of big data analytics. The obvious advantages or return of investment the manufacturers enjoy are in terms of superior product quality, higher operational efficiency, better flexibility and optimised cost efficiency.

1.3 Conclusions

This chapter mainly focuses on the importance of the Nine Pillars in ensuring the complete implementation of Industry 4.0 resulting in manufacturing with intelligent, efficient, cost-effective, individualised and customised production. The successful implementation is with the help of faster computers, smarter machines, smaller sensors, lower-cost storage and reliable data transmission. As the implementation of Industry 4.0 increases its intensity, new research related to these Nine Pillars are also producing exponential activities and discoveries. Industry 5.0, with the emphasis on human-machine interaction and collaboration for personalisation, has recently been introduced to take over and is considered as the next industrial revolution imminently. The subsequent industrial revolution will include impact on humanity, civil society, country governance, etc., apart from purely positive manufacturing outcomes.

The conceptual Industry 4.0 possesses a high impact and causes a wide range of changes to manufacturing processes and subsequently business models. It leads to mass customisation, improvement of quality of product, better productivity and faster speed in production. The mass customisation enables production of small volume even as small as single unique item, due to the capability of flexible machine configuration to fulfil customer specifications as well additive manufacturing. The flexibility encourages product innovation as new products can be manufactured rapidly without the complication of readjustment or setting up of a new production line. Therefore, with Industry 4.0, it allows the manufacturing of unique variants of a particular product and this may help in better control of inventory. Apart from that, the time for a product to be produced can be reduced as the technologies of digital design and virtual modelling help reduce the time between product design and its delivery.

The successful roll out of Industry 4.0 will provide countless unprecedented opportunities for paradigm improvement. It is there to fulfil the vision of intelligent automation while enabling more efficient, faster and more flexible processes. This helps to improve profitability, enhance goods quality and boost customer satisfaction. It has fundamental impact on the reputation and competitiveness of a business. In fact, all stakeholders are looking for smarter manufacturing, connected supply chain and value-added products and services, which are efficiently delivered through Industry 4.0. Ultimately, it is the society that reaps and enjoys the benefits from better manufacturing productivity and higher industrial growth.

References

[1] M. Rüßmann, M. Lorenz, P. Gerbert, *et al.*, Industry 4.0: The Future of Productivity and Growth in Manufacturing Industries, The Boston Consulting Group, pp. 1–16, 2015.

[2] M.A.K. Bahrin, M.F. Othman, N.H.N. Azli, M.F. Talib, Industry 4.0: A Review on Industrial Automation and Robotic, Jurnal Teknologi (Sciences & Engineering), 78(6–13), pp. 137–143, 2016.

[3] Roland Berger Strategy Consultants, Industry 4.0, The New Industrial Revolution: How Europe Will Succeed. pp. 1–24, 2014.

[4] A. Nasrinahar and J.H. Chuah, Effective Route Planning of a Mobile Robot for Static and Dynamic Obstacles with Fuzzy Logic, IEEE International Conference on Control System, Computing and Engineering (ICCSCE2016), Penang, Malaysia, pp. 34–38, 25–27 Nov 2016.

[5] A. Nasrinahar and J.H. Chuah, Intelligent Motion Planning of a Mobile Robot with Dynamic Obstacle Avoidance, Journal on Vehicle Routing Algorithms, vol. 1, pp. 89–104, 2018.

[6] G. Schuh, T. Potente, C. Wesch-Potente, A.R. Weber, J. Prote, Collaboration Mechanisms to Increase Productivity in the Context of Industrie 4.0, Robust Manufacturing Conference (RoMaC 2014), Procedia CIRP 19, pp. 51–56, 2014.

[7] S. Simons, P. Abé, S. Neser, Learning in the AutFab – the Fully Automated Industrie 4.0 Learning Factory of the University of Applied Sciences Darmstadt, 7th Conference on Learning Factories, CLF 2017, Procedia Manufacturing 9, pp. 81–88, 2017.

[8] J. Xu, E. Huang, L. Hsieh, L.H. Lee, Q. Jia, C. Chen, Simulation Optimization in the Era of Industrial 4.0 and the Industrial Internet, Journal of Simulation, 10(4), pp. 310–320, 2016.

[9] Status Report: Reference Architecture Model Industrie 4.0 (RAMI4.0), VDI, 2015.

[10] M. Schuldenfrei, Horizontal and Vertical Integration in Industry 4.0, https://www.mbtmag.com/business-intelligence/article/13251083/horizontal-and-vertical-integration-in-industry-40, 2019.
[11] R. Neugebauer, S. Hippmann, M. Leis, M. Landherr, Industrie 4.0 - Form the Perspective of Applied Research, 49th CIRP Conference on Manufacturing Systems (CIRP-CMS 2016), Procedia CIRP 57, pp. 2–7, 2016.
[12] P. Werr, How Industry 4.0 and the Internet of Things are Connected, https://www.iotevolutionworld.com/m2m/articles/401292-how-industry-40-the-internet-things-connected.htm, 2015.
[13] Hewlett Packard Enterprise, What Is Industrial Intenet of Things (IIoT)?, https://www.hpe.com/my/en/what-is/industrial-iot.html, 2020.
[14] A.C. Oliver, How IoT & Industry 4.0 Relate – and Why Manufacturers Should Care, https://lucidworks.com/post/how-are-iot-and-industry-4-related/, 2018.
[15] B.C. Ervural and B. Ervural, Overview of Cyber Security in the Industry 4.0 Era. In: Industry 4.0: Managing The Digital Transformation. Springer Series in Advanced Manufacturing. Springer, pp. 267–284, 2018.
[16] ENISA, Industry 4.0 – Cybersecurity Challenges and Recommendations, pp. 1–13, 2019.
[17] P. Haug, Cloud Computing One of the Success Factors for Industry 4.0, https://www.grin.com/document/385753, 2016.
[18] F. Rennung, C.T. Luminosu, A. Draghici, Service Provision in the Framework of Industry 4.0, SIM 2015 / 13th International Symposium in Management, Procedia - Social and Behavioural Sciences 221, pp. 372–377, 2016.
[19] M. Landherr, U. Schneider, T. Bauernhansl, The Application Centre Industrie 4.0 – Industry-driven Manufacturing, Research and Development, 49th CIRP Conference on Manufacturing Systems (CIRP-CMS 2016), Procedia CIRP 57, pp. 26–31, 2016.
[20] U.M. Dilberoglu, B. Gharehpapagh, U. Yaman, M. Dolen, The Role of Additive Manufacturing in the Era of Industry 4.0, 27th International Conference on Flexible Automation and Intelligent Manufacturing, Procedia Manufacturing 11, pp. 545–554, 2017.
[21] F.D. Pace, F. Manuri, A. Sanna, Augmented Reality in Industry 4.0, 2018.
[22] J. Wang, W. Zhang, Y. Shi, S. Duan, J. Liu, Industrial Big Data Analytics: Challenges, Methodologies, and Applications, Submitted to IEEE Transactions on Automation Science and Engineering, pp. 1–12, 2018.
[23] B. Bagheri, S. Yang, H.A. Kao, J. Lee, Cyber-Physical Systems Architecture for Self-Aware Machines in Industry 4.0 Environment, IFAC Conference, 48(3), pp.1622–1627, 2015.

Chapter 2

Industry 4.0: the next frontier and its technological impacts, the role of global standardisation and sustainable development

Alex Looi Tink Huey[1]

The Fourth Industrial Revolution or Industry 4.0 encompasses production/manufacturing-based industries, marrying advanced manufacturing techniques with digital transformation, driven by connected technologies to create intelligent manufacturing systems that not only are interconnected but also have the ability to communicate, analyse, forecast and use this information to drive further intelligent actions.

New business models and technologies such as the internet of things (IoT), big data, artificial intelligence (AI) (accurate diagnosis in healthcare systems, market and financial data analysis, self-driving cars, etc.) and additive manufacturing (aerospace and automotive components manufacturing, biocompatible materials for customised implants, life-saving devices in the medical sector, efficient and environmentally friendly energy and power sector components manufacturing, etc.) are driving the change of current business models and shifting the global economics and market structures.

According to McKinsey, Industry 4.0: Reinvigorating ASEAN Manufacturing for the Future, Industry 4.0 is expected to deliver between US$1.2 trillion and US$3.7 trillion in gains globally [1] (Figure 2.1). Of this, the Association of Southeast Asian Nations (ASEAN), whose member economies among the 10 countries have significant manufacturing-based economies, has the potential to capture large opportunities and productivity gains worth US$216 billion to US$627 billion.

2.1 Global standardisation in Industry 4.0

Industry 4.0 is opening the door to connectivity, innovativeness and economic strength, allowing businesses to be more flexible, efficient and resource-saving [2]. The key ingredient to this digital revolution is data. The efficient utilisation of transforming raw data into meaningful information is an essential enabler for future businesses.

[1]Malim Consulting Engineers, Kuala Lumpur, Malaysia

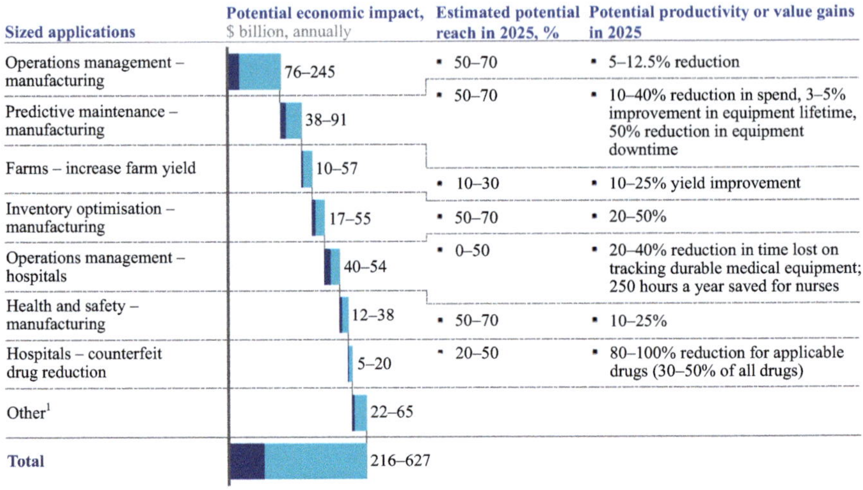

1 Other: livestock-location monitoring, livestock-health monitoring, smart pills for livestock, climate control for greenhouse gases, pay-as-you-go insurance, hospital-building security, hospital energy management and improved medical devices.

Note: ASEAN = the Association of Southeast Asian Nations. Estimates of potential economic impact are for some applications only and are not comprehensive estimates of total potential impact. Estimates include consumer surplus and cannot be related to potential company revenue, market size or GDP impact. We do not size possible surplus shifts among companies or industries, or between companies and consumer. These estimates are risk-adjusted or probability-adjusted. Numbers may not sum, because of rounding.

Figure 2.1 ASEAN potential productivity gains in 2025 – Industry 4.0 (McKinsey Digital – Industry 4.0: Reinvigorating ASEAN Manufacturing for the Future) [1]

To keep up with the advancement of Industry 4.0 and the transition to this new era, the manufacturing sector, for example, needs to undergo digital transformation with embedded sensors in manufacturing equipment and product components, automation, ubiquitous cyber-physical systems and analysis of all relevant data and information. These transformations are driven by four main clusters of disruptive technologies and enablers, namely data, computational power and connectivity; analytics and AI; human–machine interaction governed by touch interfaces and augmented reality (AR) and digital-to-physical conversion where advanced robotics and 3D printing come into place (Figure 2.2).

Meanwhile, the power utility sector must focus on upcoming power energy technologies, standardisation of high-level and high-speed technologies with information and communications technology (ICT) fusion and convergence, smart and intelligent power grids, energy storage systems, big data, AI and cyber security. The power utility sector is a critical infrastructure. Hence, ensuring efficient and continuous supply of power is essential.

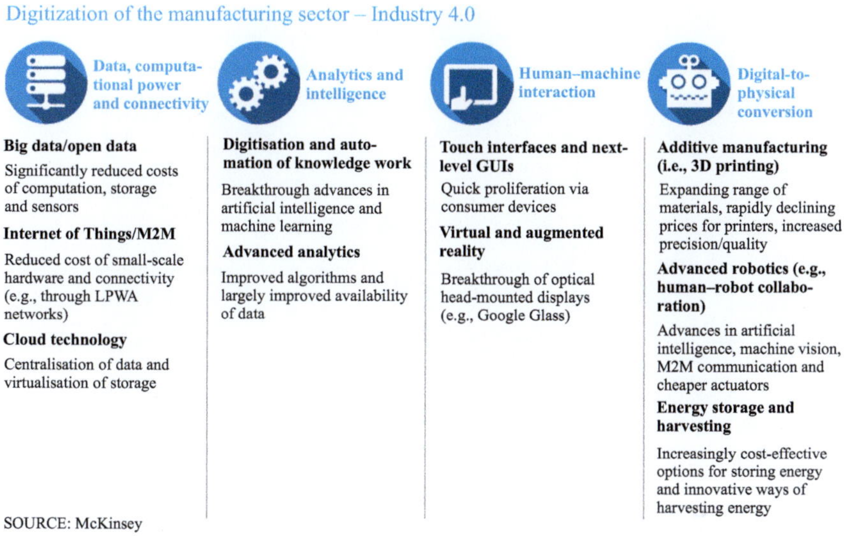

Figure 2.2 Digitisation of the manufacturing sector – Industry 4.0 (McKinsey Digital – Industry 4.0: How to Navigate Digitisation of the Manufacturing Sector) [3]

The wide use of new and advanced technologies requires an intelligent integration system, which can only be achieved if the relevant technologies, interfaces, frameworks and formats conform to globally accepted standards. The prerequisites for interoperability in an Industry 4.0 environment are internationally applicable and accepted norms and standards. Therefore, Industry 4.0 and standardisation must go hand in hand.

Industry 4.0 involves global economic transformation (Figure 2.3). Hence, national standardisation activities need to be harmonised with the international level to focus on stipulating the international collaboration and cooperation mechanisms and exchange of information. The focus on Industry 4.0 requires a coordinated and concerted action from all relevant stakeholders and government ministries and agencies to spearhead the national transition to Industry 4.0 through funding and incentives, talent and human capital, technology and standards and digital-infrastructure and ecosystem.

2.1.1 Technological impacts of Industry 4.0

The imminent arrival of 5G or fifth-generation wireless technology offers lightning-fast speed, extremely reliable connections, low latency and enables massive connectivity simultaneously. The 5G networks are expected to supercharge the IoT ecosystem by

Figure 2.3 Industry 4.0 involves global economic transformation

providing the network infrastructure needed to carry massive amount of data and serving the communication needs of billions of IoT-connected devices. This wireless technology opens a wide range of new possibilities such as robotics and autonomous systems, cloud technologies, remote surgery and remote medical applications, AR and self-driving vehicles. The International Electrotechnical Commission (IEC) technical committee (TC) 106 has been tasked to develop safety-testing standards for mobile devices, base stations and wireless communication systems to support the implementation and development of 5G networks.

The IoT is a system of interrelated computing devices, mechanical and digital machines, objects embedded with sensors, software and technology for the purpose of connecting and exchanging data with other devices and systems over the Internet. These IoT-connected devices range from household objects such as kitchen appliances and baby monitors to sophisticated industrial tools such as smart power grids, smart manufacturing, connected and smart logistics and preventive and predictive maintenance. Companies will deploy thousands or millions of wired and wireless sensors that produce large amounts of raw data, which is being turned into useful information providing value for the user. According to the IEC, the IEC's role in the IoT, an estimated 50 billion objects will be connected by 2020 [4]. The ISO/IEC JTC 1/SC 41 provides the standardisation in the area of IoT and related technologies, including sensor networks and wearable technologies to ensure connected systems are seamless, safe and resilient.

Robotics and autonomous systems are the areas that are constantly growing in the market industry globally. Due to the boundless capabilities of storing information, together with the implementation of AI, and improved human–computer interactions, robotics and autonomous systems are making their strong presence felt in all fields and with countless applications.

Cloud technologies are part of an enabler of disruptive innovation, which integrate services and democratise access to information, learning and communication.

Big data and analytics have been noted globally as commodities more valuable than oil, as data reigns in today's digital economy. Today's capability of collecting and storing tremendous amount of data, analysing data in faster and smarter ways, big data and analytics allow the transformation of historical and real-time data into valuable information of understanding, producing, selling, predicting and so on.

The AR is the integration of digital information with the user's environment in real time providing an interactive, reality-based display environment, which creates a bridge between virtual reality and data. The applications of AR are limitless. Industries are adopting AR technologies to enable virtual manufacturing, where engineers can design and evaluate new parts for industrial applications in a 3D environment; specialised and enhanced training and simulation; intelligent control and maintenance functionalities and many more.

Blockchain is a decentralised database that is encoded, unchangeable, consensual, verifiable, transparent and permanent. There are myriad potential uses of blockchain technology and one of them is 'Bitcoin'. In the logistics industry, blockchain is being used to establish trusted information such as where a product was made, when the product was made, when the product was shipped, where the product is located and when the product would arrive. Blockchain technology has also been implemented in the energy sector such as peer-to-peer (P2P) renewable energy (RE) trading and electric vehicle charging infrastructure. ISO/IEC JTC 1/ SC 27 working group (WG) oversees the standards of the development of blockchain and distributed ledger technologies.

The main pillars or technology drives of Industry 4.0 include autonomous and AI where production or manufacturing processes will become increasingly digitised and interconnected cyber-physical systems. The industry's recognition and the broad adoption of AI and automation will redefine how businesses and industries work. Hence, AI is expected to be one of the most crucial enablers and digital frontier in the evolution of information technology (IT).

AI is not only a single technology but also a variety of software and hardware technologies employed in applications. AI is generally understood to refer to a machine, which has the ability to replicate human cognitive functions to learn and solve problems. The components of AI such as neural networks, natural language processing, biometrics, facial recognition, gesture recognition, minibots and video analytics enable digital transformation.

A recent article by the Forbes, 'Artificial Intelligence Beats the Hype with Stunning Growth', suggests that investment in AI is growing very fast. The Stamford, Conn., firm found that AI implementations grew 37% during 2018 and 270% over the last 4 years [5]. IDC Corp., an international investment service, forecasts spending for cognitive and AI systems will reach US$77.6 billion by 2022, and notes that 60% of global GDP should be digitised by 2022, driving almost US$7 trillion in IT spending [5].

Today, all sectors rely heavily on AI, from finance, manufacturing and robotics to healthcare, transportation, household appliances and even our smartphones, which we carry everywhere we go! AI is evolving and expanding far more quickly than we have ever imagined. So, it is important that AI technology standardisation is needed to achieve and accelerate global adoption and digital transformation.

Figure 2.4 How do we implement human ethics in AI?

How do we implement human ethics in AI (Figure 2.4)? What moral decisions should be made by AI in self-driving cars on life-and-death situations? For example, if the brakes suddenly fail, should a car steer to the left and probably cause the death of one man or should it steer to the right and cause injuries to 20 people? The AI systems must be transparent to allow users to understand how AI decision-making is made.

Key barriers to the adoption of AI are trustworthiness and ethical considerations of the system as humans put their trust in machines. The AI ecosystem can be divided into three key areas involving technical, societal and ethical considerations. The technical consideration involves foundational standards and computational methods, techniques, architectures and characteristics of AI systems. The societal and ethical considerations are the concern of how adoption of AI affects and influences human lives on labour force, privacy, eavesdropping, algorithmic bias and safety.

The IEC has embarked on the development of smart manufacturing standards with the International Organisation for Standardisation (ISO) such as the participation of joint technical committee, ISO/IEC JTC 1 – the standards development environment for ICT to form WGs to look into harmonising existing reference models and oversee the development of the underlying architecture for smart manufacturing [6]. The IEC and ISO joint technical committee on artificial intelligence, ISO/IEC JTC 1/ SC 42, was established to develop international standards to cover areas which include AI, big data, use cases, governance implications, computational approaches of AI and ethical and societal concerns.

The IEC standardisation evaluation groups (SEGs) are established to identify new technical areas and anticipate emerging markets or technologies such as smart manufacturing, communication technologies and smart home/office building systems [7]. The IEC SEG 10 – Ethics in Autonomous and Artificial Intelligence Applications – is set up to evaluate the work that covers a broad area of new technologies to identify ethical issues and societal concerns throughout its work and to collaborate with broader technical committees within the IEC as well as to develop broadly applicable guidelines for IEC committees on ethical aspects related to autonomous and AI applications.

The Institute of Electrical and Electronics Engineers (IEEE) standards association (SA) has worked on projects that focus on the human well-being in the creation of autonomous and intelligent technologies [8]. IEEE-SA saw the release of ethics guideline for automation and intelligent systems, titled 'Ethically Aligned Design (EAD)'. This guideline focuses on high-level principles and recommendations for ethical implementation of autonomous and intelligent systems, which are covered in eight general principles: human rights, well-being, data agency, effectiveness, transparency, accountability, awareness of misuse and competence.

AI helps to streamline efficiency for businesses in smart manufacturing as it can provide insights into where improvements can be made and, more importantly, where businesses can go further in terms of production planning. Standardisation is of central importance for Industry 4.0, which requires an unprecedented degree of system integration across domain borders, hierarchic boundaries and lifecycle phases. The international standards development organisations are playing a key role in the transition to Industry 4.0.

Data centres are the backbone of Industry 4.0 as they pave the way for Industry 4.0 applications (Figure 2.5). As cloud computing and ecosystems continue to grow to be one of the key strategies for digital transformation, hyperscale data centres are expanding in capacity to meet the exponential growth of data volumes. Data centres today are now home to AI engines, vast cloud ecosystems, blockchain technologies, advanced connectivity architectures, high-performance computing platforms and many more. ISO/IEC JTC 1/ SC 38 serves as the focus, proponent and systems integration entity on cloud computing and distributed platforms.

To prepare for the next frontier in digital transformation, data centres must implement advanced network infrastructures to make their networks faster, more

Figure 2.5 Data centres are the backbone of Industry 4.0

Figure 2.6 United Nations sustainable development goals [9]

flexible, more robust, and more reliable and more cyber resilience and secure. Hyperscale data centres are big consumers of electrical energy. It is essential to make data centres more energy-efficient and less power-hungry to combat increasing global carbon emissions. ISO/IEC JTC 1/ SC 39 was set up to develop a new metric that measures and compares the energy efficiency and the sustainability of data centres.

To capture this, global standardisation provides an important contribution to ensure the smooth expansion of digital transformation and Industry 4.0 and to allow systems of different manufacturers to interconnect and interoperate without the need for special integration efforts, as well as mitigating climate change and reducing global carbon emissions, which are in line with United Nations (UN) sustainable development goals (SDGs) (Figure 2.6).

2.2 Industry 4.0 and sustainable development

The UN member states adopted the 2030 agenda for sustainable development, which provides a shared blueprint for peace and prosperity incorporating 17 SDGs, where all developed and developing countries shall work hand in hand with strategies to improve health and education, spur economic growth, reduce inequality, tackling climate change and preserving our environment.

A study conducted by the UN Industrial Development Organisation (UNIDO) and partners identified that Industry 4.0 can address the global challenges with the enhanced use of information and communication technologies but cautioned that its limits and risks to sustainable development must be better understood [10].

Industry 4.0 guiding principle focuses on boosting productivity, revenue growth and competitiveness and has to overcome the challenges of standardisation of systems, platforms, protocols, changes in work organisation, cybersecurity,

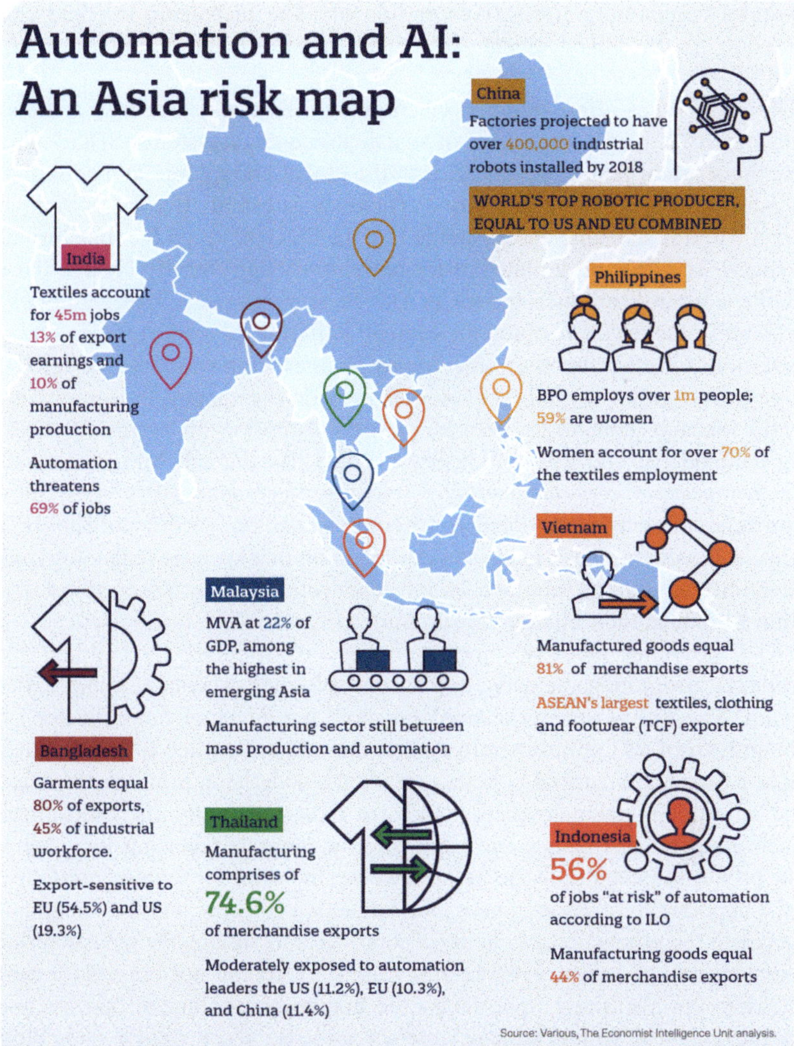

Figure 2.7 Automation and AI: An Asia risk map (United Nations Development Programme (UNDP) – Development 4.0: Opportunities and challenges for accelerating progress towards the sustainable development goals in Asia and the Pacific) [11]

availability and quality of skilled workers, research and investment, and the adoption of relevant legal frameworks to be successful.

Industry 4.0 has to cope with producing within the environmental constraints in order to drive sustainable development. The rate of exploiting natural resources must not exceed the rate of regeneration; the depletion of non-renewable resources must require comparable alternatives and so on. The increasing trend in energy demand due to the advancement of digital transformation requires an urgent adoption of low-carbon energy systems.

SDG 7 promotes affordable and clean energy. Renewable energy and energy efficiency are two main components of sustainable energy systems. SDG 7 aims to increase the share of renewable energies in the global energy mix and doubling the global rate of improvement in energy efficiency substantially. Combining both Industry 4.0 and sustainable energy guided by the SDGs through the transformation of the energy sector with digital transformation will significantly change the way people live, consume, produce and trade. The development in ICT technologies, 5G networks and blockchain technologies opens windows of opportunities and provides solutions to integrate renewable energy sources from wind to solar into small and large power grids. Digital technologies provide the capability to monitor and efficiently manage generation, delivery and consumption of energy to meet the varying demands of end users. Industry 4.0 also has the capability to assist the manufacturing sector in saving energy by transforming business processes. Some of the approaches such as energy storage offer advantages of security of supply, grid flexibility and reducing peak loads and optimisation of a specific technology where the behaviour of a large number of interconnected robots is controlled by an algorithm to reduce their energy consumption (Figure 2.7).

SDG 9 promotes industry, innovation and infrastructure by building strong infrastructure, promoting inclusive and sustainable industrialisation, and fostering innovation. The roll out of 5G technologies and the development of the IoT will spur the growth of ICT infrastructure investment and emergence of new innovative approaches. Strategic industrial policies provide the underlying infrastructures needed to foster innovation. Unemployment is also the most commonly discussed threat, as increasing adoption of robotics and automation systems is replacing jobs that have been the production engines in the past. However, the same technology portfolio also offers the opportunity of creating new jobs for skilled workers, and the integration of intelligence in the production machineries contributes to sustainable industrialisation.

The adoption of Industry 4.0 requires us to overcome not only the technological barriers but also more importantly the human-psychological barriers, which defines the ways in which humans interact and use these digital technologies. Today, digital technologies have been increasingly humanised with the maturation of elements such as virtual reality and augmented reality. Hence, businesses have to increase their digital transformation efforts through a human-centric approach to see through to their successful implementation.

References

[1] McKinsey & Company, 'Industry 4.0: Reinvigorating ASEAN Manufacturing for the Future' (McKinsey Digital, 2018).

[2] Winterhalter, C., 'A New Revolution in the Making', ISO Focus: The New Industrial Revolution, 2018, pp. 2–3.

[3] McKinsey & Company, 'Industry 4.0: How to Navigate Digitisation of the Manufacturing Sector' (McKinsey Digital, 2015).
[4] 'Architecting a Connected Future', https://www.iso.org/news/ref2361.html, accessed January 2020.
[5] 'Artificial Intelligence Beats the Hype with Stunning Growth', https://www.forbes.com/sites/jonmarkman/2019/02/26/artificial-intelligence-beats-the-hype-with-stunning-growth/#137fc4531f15, accessed January 2020.
[6] 'ISO/IEC JTC 1 Information Technology', https://www.iso.org/isoiec-jtc-1.html, accessed January 2020.
[7] 'IEC Standardisation Management Board', https://www.iec.ch/dyn/www/f?p=103:48:14443879686867::::FSP_ORG_ID:3228#3, accessed January 2020.
[8] 'IEEE Global Initiative for Ethical Considerations in Artificial Intelligence (AI) and Autonomous Systems (AS) Drives, together with IEEE Societies, New Standards Projects; Releases New Report on Prioritizing Human Well-Being', https://standards.ieee.org/news/2017/ieee_p7004.html, accessed January 2020.
[9] 'Sustainable Development Goals', https://sustainabledevelopment.un.org/?menu=1300, accessed February 2020.
[10] United Nations Industrial Development Organisation (UNIDO), 'Accelerating Clean Energy Through Industry 4.0: Manufacturing the Next Revolution' (UNIDO, 2017).
[11] United Nations Development Programme (UNDP), 'Development 4.0: Opportunity and Challenges for Accelerating Progress Towards the Sustainable Development Goals in Asia and the Pacific' (UNDP, 2018).

Chapter 3

Industrial revolution 4.0 – big data and big data analytics for smart manufacturing

Yu Seng Cheng[1], *Joon Huang Chuah*[1] *and Yizhou Wang*[2]

3.1 Smart manufacturing and cyber-physical system

Manufacturers worldwide are competing in a global and dynamic marketplace demanding products of highest quality and lowest price. Further impacted by time to market and innovativeness, manufacturing processes were progressively pushed towards their limit boundaries, and inducing substantial manufacturing uncertainty, if not controlled, may result in catastrophic failure. This scenario triggers the importance of adapting smartness in manufacturing for effective and efficient resource management, to improve manufacturing uncertainties to stay competitive and maintain profitable margins both locally and globally.

Industry 4.0 (I4.0) encompasses various technologies but centres on highly automated, digitalised manufacturing processes and advanced information communication technology. I4.0 is entrenched in the smart manufacturing (SM) concept, an intelligent manufacturing notion where the production lines can self-optimise and adapt in response to changing production processes and conditions. SM facilitates high manufacturing flexibility and capability to produce products with higher complexity and customisation in large scale, increase product quality and achieve better resource consumption.

SM can be defined as a new level of production model with a combination of various technologies to integrate cyber-physical capabilities enabling huge data collection from multiple sources in a network-connected, resource sharing environment, utilising advanced analytic techniques to devise strategic information for system improvement and decision-making for increased manufacturing flexibility and adaptability [1,2]. It is the reinforced and ubiquitous application of networked information-based technologies throughout the manufacturing domain to achieve optimum production capabilities. SM requires several underpinning technologies to establish communication of various entities within the manufacturing area. The communication and information exchange allow the behaviour of each object to change

[1]Department of Electrical Engineering, University of Malaya, Kuala Lumpur, Malaysia
[2]Center on Frontiers of Computing Studies, Peking University, Beijing, P.R. China

and adapt to different situations based on signals produced through past experiences and learning capacities empowered by these technologies, which were feedback to them. This impact leads to the set-up of communicatory systems between human-to-human and human-to-machines known as the cyber-physical system (CPS).

CPS is a concept derived from the functionality of the systems that allow human–machine accessibility and interoperability. Shafiq *et al.* [3] state that CPS is an establishment of a global network on the all manufacturing entities from production machinery to processes and to supply chain management (SCM) with the convergence of physical and digital worlds. Monostori *et al.* [4] state that CPSs are entities with the ability of having computational collaboration with the encompassing physical world through extensive connection, in the continuing form of providing and utilising and also concurrently accessing and processing data from the services available in the network. Lee *et al.* [5] presented an accurate description of CPS in a 5C architecture summarising the primary design levels of CPS: (1) *Connection*, (2) *Conversion*, (3) *Cyber*, (4) *Cognition* and (5) *Configuration*. Figure 3.1 illustrates this architecture.

The connection level serves as the interface between the physical environment and computational cyber world. Different communication mediums were used to collect the information such as sensors, actuators, embedded system terminals and wireless network for data exchange. Adequate protocol and device design are needed to accommodate the different types of data acquisition. The conversion level functions as a data-processing agent to convert data acquired into useful information specific to each process for suitable actions to be taken to improve productivity. As data from each of the entities represent its condition, the information devised can serve as the monitor for machine self-awareness. The cyber level functions as the central role in this architecture. It gathers all data from all components in the systems and creates the cyber symbols of the physical properties. These symbols replication allows the building of a growing knowledge base for each machine or sub-systems.

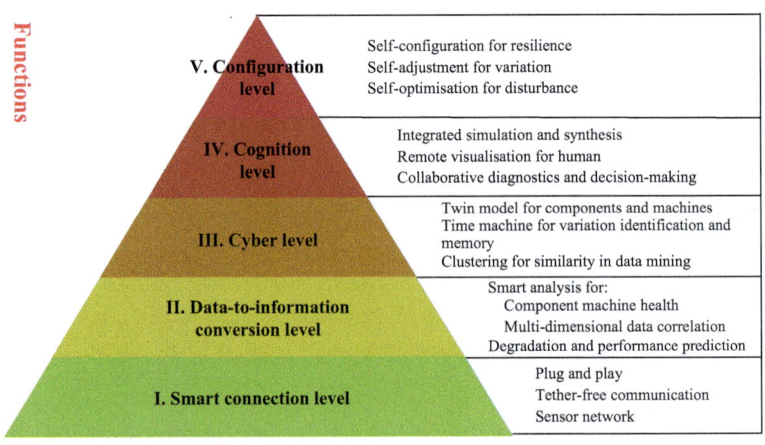

Figure 3.1 Architecture of CPSs [5]

Proper utilisation of the knowledge base established for system improvement actions can produce a more reliable and effective result. The cognition level acts as the brain of the architecture. Simulation and synthesis of all the information available provide in-depth knowledge of the systems. The learnings and outcome from it can then be used by experts to support decision-making and reasoning methods. The final level is the configuration level, a flexible control system bridging the commands, suggestions or decisions developed for execution in the physical systems.

The real-time and dynamic information interchange among the integrated CPS allows vertical and horizontal communication for an accurate business decision, multiparty collaboration, efficient and effective system improvements and maintenance within the complete SM domain [6]. Generally, in the context of SM, CPS established an innovative approach in integrating industrial automation system functionalities of all entities within the manufacturing system domain through networking to all surrounding operation physical reality allowing computational collaboration and communication infrastructure for interoperability. The systems are connected in a distributed modular structure in contrast with a unidirectional connection in conventional manufacturing facilities, allowing multisource data interaction and knowledge integration across the networks. This provides total visibility and control to the manufacturing processes, enabling timely problem solving and decision-making. The information is available and flows wherever and whenever they are needed across all areas of manufacturing processes.

Though CPS established the model of communication between the physical system and cyber world, it is the combination of several base technologies that bring life to the SM system. For instance, the Internet of Things (IoT) allows all the physical devices or equipment to be connected to a centralised platform for data collection and exchange. Big data (BD) produced by each of the entities contains invaluable information about the domain, which then is analysed through advanced techniques in BD analytics (BDA). The analytics outcome forms the signals to be feedback to the systems for adaptability, as explained earlier. Cloud computing (CC) allows for the information and signal accessibility by every party in the SM system in real-time, in any part of the world, as long it is connected to the network. Figure 3.2 illustrates the critical base technology for SM in I4.0. Ultimately, each of

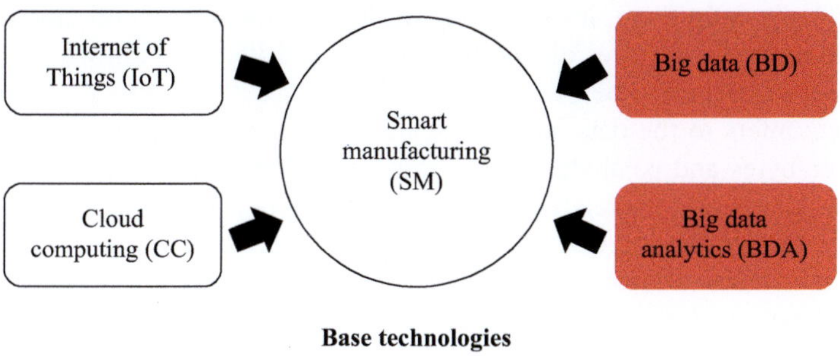

Figure 3.2 Key base technology for smart manufacturing in Industry 4.0

the base technologies jointly forms the core architecture of SM. The discussion herein will focus on explaining BD and BDA, followed by a case study on the proposal of BDA for optical measurement in solar cell manufacturing.

3.2 Overview of big data

The digitalisation of manufacturing processes enabled different devices and sub-processes within manufacturing facilities to be interconnected into a large digital platform. Prevalent integration of information and communication technology generates an enormous capacity of heterogeneous data from different objects within the domain. This massive amount of structured, semi-structured and unstructured data is known as BD. In contemplation to acquiring the correspondent data, due to its enormous capacity, it is a substantial challenge to attain, store and analyse these data in terms of both time and money [7]. As such, utilising novel technologies from today's internet era is crucial to bring values opportunities to the industries.

Multiple publications available in addressing the components and definition of BD and the description evolved rapidly and to an extent raised some confusion [8]. BD often understand by many, according to the volume and size of the data. While the amount and size of the data are inevitably essential, however, it is not the only crucial aspect. Both Cemernek *et al.* [1] and Gandomi and Haider [8] stated two versions BD definition, which most accurately describe BD, gathered first from definition by TechAmerica Foundation:

> Big data is a term that describes large volumes of high velocity, complex and variable data that require advanced techniques and technologies to enable the capture, storage, distribution, management, and analysis of the information. [9]

Another more common definition, aligned to I4.0 by Gartner IT Group, is as follows:

> Big data is high-volume, high-velocity and high-variety information assets that demand cost-effective, innovative forms of information processing for enhanced insight and decision making. [10]

From both definitions, it clearly shows BD comprises several characteristics with standard shared features of *Volume*, *Variety* and *Velocity*, known as the Three Vs. The Vs are described below.

Volume refers to the data size and scale. BD sizes usually are in the range of multiple terabytes and petabytes. There was no definite capacity threshold for BD volume classification. It is relative and varies by different factors, such as data type, data compression technology, data management and storage capacities. For instance, the increase in storage capacity will increase the threshold on BD volume characteristic, allowing more data to be stored. Besides data compression technology to slim the data size without losing the information, different data types, such as text data against video data which has different data density, and the data management technology to manipulate the data, are all possible factors to change the volume definition as 'big'.

Variety refers to data structure diversity and multidimensionality of the data contents. BD data exist in three categories: (1) Structured data – This type of data resides in the fixed fields within a pre-defined input framework or data model such as a record in a relational database, and an entry in the spreadsheets. It does not require data pre-processing or pre-conditioning to be used for explicit interpretation or analysis to obtain meaningful information. (2) Unstructured data – A type of unorganised, inconsistent format that does not follow any pre-defined data model or manner. Data is typically text and graphic-heavy, often not suitable for analysis without pre-processing or pre-conditioning to conform to the structural organisation requirements of the analytic techniques. Image, audio and video data are an example of unstructured data. (3) Semi-structured data – A type of data that does not conform strictly to any formal data standard but contains specific identifiers to distinguish between the semantic elements and instructional element within the data. Example for semi-structured data is Hypertext Markup Language, where the processing information was enclosed within the user-defined tags making it interpretable by the browser to render informative output.

Velocity is the rate of data generation and speed of processing the data to draw actionable outcomes measured by its frequency. The speed of data generation is closely related design and arrangements of the productions systems. When production cycle time is short, data generation will be faster and vice-versa. In both cases, CPS controls data generation and traffic. Therefore, the design of the CPS should have the adequate capability of handling the scenario. The proliferation of digital devices had contributed to the data velocity in both generation and processing. Single or group connected devices further processed the production information made available on the machines; this allows more relevant and in-depth information about a particular task or scenario, enabled subsequent actions on real-time analytics and evidence-based planning. Multilevel of new data generation and transmission across devices accelerate an unprecedented rate of data creation. The traditional data management system was unable to handle the enormous data feed to it instantly through this new form of data velocity. This pushes the new BD technologies into the scene to manipulate these high volumes of data in a 'perishable' way to create real-time intelligence. Figure 3.3 illustrates the combination of the 3 Vs to visualise the concept.

Besides the three general dimensions of BD, several key corporations had extended the dimensionality of BD into *Veracity*, *Variability* and *Value*. Each of the extended Vs is described below:

Veracity represents the unreliability enclosed within some data sources, also known as data quality. IBM concocted it. Veracity recognised as the most important factor hindering the broad adoption of BDA in the industry. Some identified areas affecting data reliability, such as data accuracy, integrity and uncertainty, need to be improved before a reliable implementation can be achieved [2]. For example, market demand for products such as luxury cars is very much driven by customer sentiments, e.g., preference of specific brands, past experiences and social status. All these involve human judgement, which causes significant uncertainties. However, these data contain

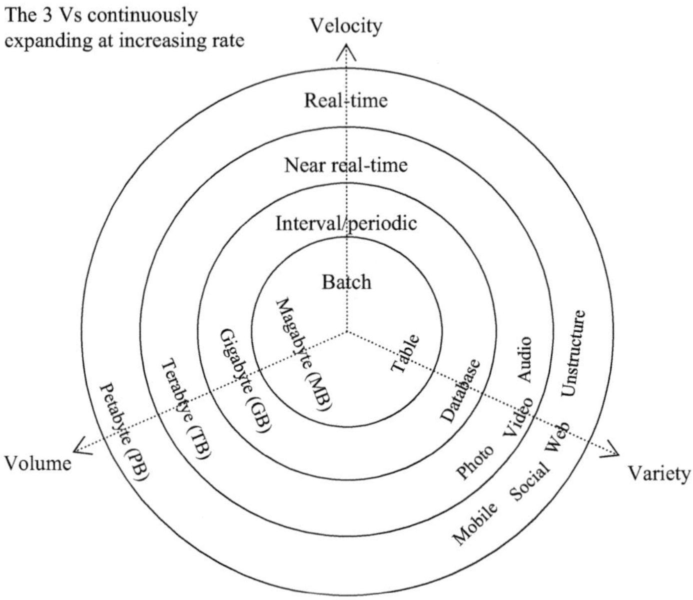

Figure 3.3 Graphical representation of the 3 Vs of big data

valuable information about the market. This other facet of BD needs to be addressed using advanced analytic techniques to draw specific and relevant information.

Variability and *Complexity* were the two BD aspects first introduced by SAS. In high volume and velocity data transmission, the data flow often exhibits the characteristic of periodic peaks and trenches. This inconsistency and variation is termed *Variability*. Data availability in today's digitalisation era come from a myriad of sources, producing data of different types and formats. This multisourced data generation is termed as *Complexity* and impose a critical challenge in BDA application. The data need to be pre-processed to match and connect to transform into meaning information.

Value was introduced by Oracle, defining value in BD collected at the beginning exhibit relatively low cost concerning the volume acquired, characterised as 'low-value density'. However, the data value increases when it is analysed in a large amount further down the pipeline, revealing the initially hidden economically sound insights and high-value information. In other words, *Value* represents the in-depth information obtained by applying BDA based on a continuous large volume of data analysed, where this information was unseen at the initial stage due to relatively low data volume. Data value is the most critical dimension of BD.

There was no universal benchmark of the 3 Vs to characterise BD. Volume as the fundamental of BD relates to all other dimensions; variety is related to its value and so on. Each of the BD dimensions was, in one way or the other, dependent on each other. The BD characterisation threshold depends upon the size, area and location of implementation, and these limits evolve through time. The baseline does exist to deal with BD, depending on the nature and capacity of each manufacturing domain. It is where traditional data management system and technologies can no longer produce satisfactory intelligence to maintain manufacturing efficiency, cost control and competitiveness. When this situation occurs, a trade-off has to be made between future values expected by the implementation of BD technologies over the cost to implement it.

Potential gains achievable through BD implementation will be more beneficial compared to traditional system and technologies over the long run. The use of BD will provide business edge through the value-added opportunities identified. With thorough analyses, it can give systematic guidance for relevant manufacturing activities to be taken in accordance to the system behaviour. From the actions taken within the whole product lifecycle, it allows achieving cost-efficient and fault-free operation processes. Also, it helps related personnel in making the correct decision and to solve problems related to the operations. One of the main focuses of SM is to set up an intelligent manufacturing system with accurate and timely decision-making through the support of real-time data acquired [11]. BD conceivably will be the next major contributor for future transformation and enhancement of SM.

3.2.1 Data-driven smart manufacturing

SM is a new manufacturing paradigm where all entities within the manufacturing domain are fully connected in a form or another to a network. The network connections comprise sensors, actuators and embedded systems. By leveraging base technologies such as CPS, IoT and CC, data can be collected across all different manufacturing stages. With advanced computational algorithms, these data can be transformed into meaningful information and insights on the manufacturing processes, creating data-driven manufacturing, providing intelligence in quality control, flexibility and adaptability enhancement, cost-saving and more. The data-driven approach allows the derivation of complex multivariate nonlinear data characteristic model across various entities within the domain without the need to thoroughly understand system behaviour. Direct modelling prevents a time-consuming and error-prone manual approach. Advancement in artificial intelligence (AI) technologies further complement the adoption of BD for data-driven SM. Figure 3.4 illustrates an overview of data-driven SM with BD.

The vital enabler in establishing intelligence in manufacturing is the data. Current advanced manufacturing systems in the digitalised era generates multiple types of data in enormous capacity. In the BD age, the availability and ability to collect a large pool of data had become straightforward and simpler. Together with new ICT technologies, the collection, processing and storage of data had also significantly enhanced. Manufacturing facilities can now take advantage of these improvements to obtain benefits from the value of data. The following categories classify the heterogeneous spatial data from manufacturing processes and equipment, and Table 3.1 summarises historical data acquisition in different manufacturing age together with its storage, analysis, transfer and management mediums:

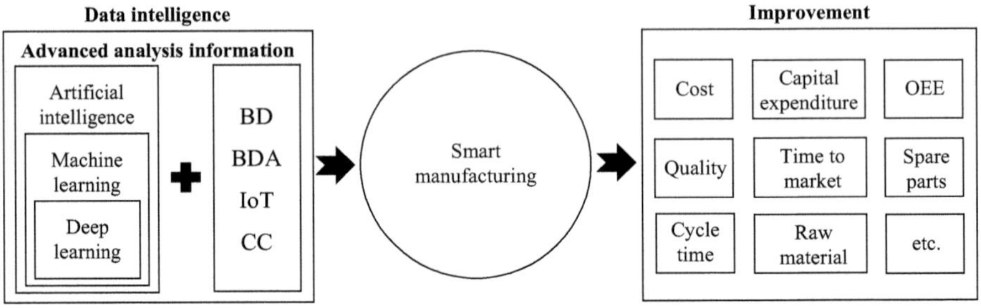

Figure 3.4 Smart manufacturing with big data

1. Management data with primary function related to management decision-making on pre-manufacturing control such as financial planning, material management, logistics, production planning; in manufacturing control such as material flow, product flow, maintenance; and post-manufacturing control such as sales and marketing, logistics and distribution, order dispatch and customer service. The data are collected from manufacturing information that possesses a variety of data such as manufacturing execution system (MES), enterprise resource planning (ERP), customer relationship management, SCM and product data management.
2. Equipment data is collected through CPS, which includes data related to real-time performance, operating conditions and maintenance history of production equipment. Data originated from the machine were compiled into a centralised information system that is designed and developed according to the manufacturing facility set up through the backbone communication channel such as MES. These collected data can then be further utilised to develop into sub-information systems to display task- or condition-specific information.
3. User data is collected from internet sources and serves as the input to the special manufacturing requirements such as customer locations, demographics, profiles, order history, product preferences and more. These data are generated mainly out from main manufacturing facilities in external servers that contain customers' information, primarily used for marketing and feedback purposes. With the advancement of CPS, these data can be further utilised after pre-processing and filtration, to integrate into the central production planning system. Production planning can then adapt to the latest market requirements and feed back on the newest production requirements to the manufacturing processes. This application enables high manufacturing flexibility and adaptability.
4. Product data are collected from the product and product-service systems through internet client database, IoT technologies that produce and feed-back in real-time. The data reflects on product performance, the context of use such as area and time and any other product/service-related indicators such as humidity, temperature and efficiency. These data serve as the source of information to obtain product-related information down in the actual

Table 3.1 Comparison of manufacturing data in different manufacturing era [14]

	Data source	Data collection	Data storage	Data analysis	Data transfer	Data management
Handicraft age	Human experience	Manual collection	Human memory	Arbitrary	Verbal communication	N/A
Machine age	Human and machines	Manual collection	Written documents	Systematic	Written documents	Human operators
Information age	Human, machines, information and computer systems	Semi-automated collection	Databases	Conventional algorithms	Digital files	Information systems
Big data age	Machines, product, user, information systems, public data	Automated collection	Cloud services	Big data algorithms	Digital files	Cloud and AI

field of application. These data provide invaluable information on the actual utilisation of the product for future improvement.

5. Public data is collected from government systems or open databases. Examples of information sources are government protocols, rules and regulations, intellectual property, scientific development, demographics, environmental protection, healthcare, industry standards and more. These related governmental data are vital to guarantee strict compliance of manufacturing processes and manufactured products to all regulations.

3.2.2 Data lifecycle

In a full-scale manufacturing environment, various sources will generate different data. The raw data extracted in its original condition is incapable of providing any useful information to users. Classification of these data must be carried out with the aim of knowledge discovery to transform into user interpretable content for meaningful information extraction (IE) [12]. In considering the BD from the perspective of the complete manufacturing product lifecycle, the transformation of raw data into useful information involves multiple steps. The entire journey is ranging from data generation, collection, storage, processing, visualisation to final application, known as Data Lifecycle [13]. Manufacturing data produced along the data lifecycle were utilised at various points for data criteria sorting. This sorting can highlight its value to the users' interest, such as what data are needed and where is it relevant in the different lifecycle phases. Figure 3.5 illustrates the lifecycle.

1. *Data sources*: Section 3.2.1 explained the typical manufacturing data type and their origin. Management and equipment are the two-primary manufacturing-related data types. A core information system framework such as MES connects all the management data produced. These systems were customised and developed according to manufacturing facility requirements. Additional sub-systems to provide lower level information targeted on each specific process group can also be generated by integrating to the core framework. Management data are typically structured data, although the form of data during transmission between systems to the scheme was semi-structured. Equipment sensors, actuators and embedded systems produce equipment data. Data type differs according to system design and application, and is unstructured. Most modern equipment with I4.0 compatibility contains an internal system interface to pre-process most equipment data into semi-structured form before sending to the core framework.
2. *Data collection*: Several ways were used to collect the manufacturing data due to its widespread availability across different data entities. CPS allows useful physical data into the cyber world. Manufacturing or management data collection is through core communication and information management systems such as MES. Individual entity data generation is supplied through a standardised industrial protocol interface such as SECS/GEM and sent to the MES centralised server. These data can then be used for further application in either semi or real-time through database technologies. The data are mainly semi-structured and unstructured machine data. The data retention period is depending on the server

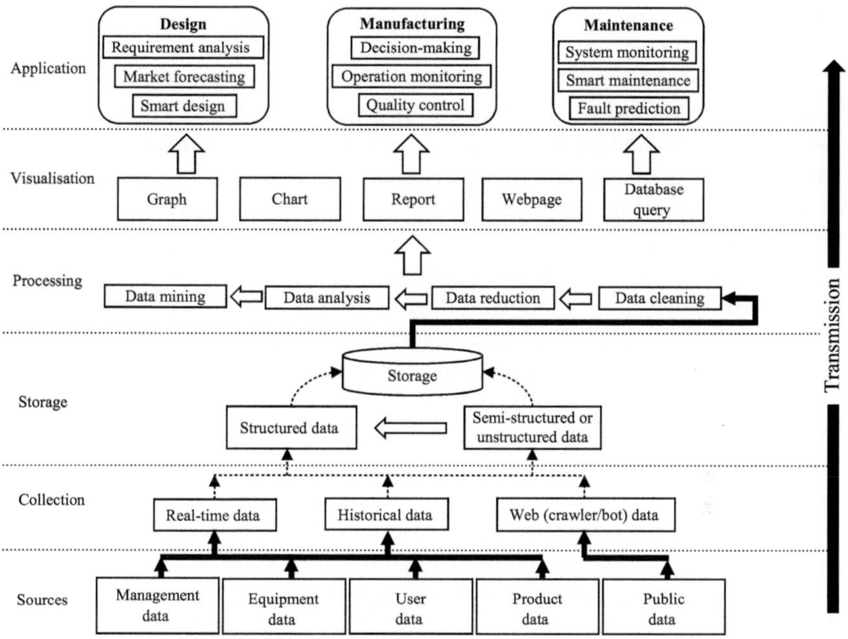

Figure 3.5 Manufacturing data lifecycle

the storage capacity. IoT is another form of standard data collection technology in SM. Equipment and product data were collected instantly for real-time status monitoring through sensing devices, smart sensors, and other embedded systems connected to the network. From the emerging mobile technology, data collection was possible through terminals like PCs, mobile phones, tablets and laptops. IoT-based data collection is generally a direct application of the data on an individual entity scale. These data can then be utilised for specific application development, mainly for system performance monitoring using software development kits or application programming interfaces. User and public data used web scraping method for data collection. Web scraping refers to the data extraction from websites to obtain desired information using bot or web crawlers. It is typically an automated process in the form of search and copy through the websites crawled and collected into a database for subsequent application. Web scraping usually is applied together with AI, enabling efficient data collection.

3. *Data storage*: Digitalisation of manufacturing systems generates a massive amount of data. These data, from raw equipment I/O to management system data, need to be stored and integrated securely to allow access and utilisation for subsequent actions. Storage capacity and speed are two of the critical factors in determining the data lifespan, retention period and functionality of the data. Efficient and secure data storage will facilitate speed and accessibility for

utilisation. There are three standard data type classifications: (1) Structured, such as numbers, managed strings, tables, which is direct database-centric manageable. (2) Semi-structured, such as graphs and XML. These data type are not direct extractable for end-user information. It mainly function as the intermediate interfacing medium within systems and software. (3) Unstructured, such as log files, audios and images. These data require a specialised and advanced approach to process and obtain required task-specific information. Due to the nature of direct IE for the structured data type, most companies focused and invested in the storage medium and accessibility for this data type. New object-based storage technology opens the window for efficient storage of semi-structured and unstructured data types. The data are stored and managed as an object, enabling flexibility in storage integration and pull for usage. Another essential storage technology is cloud storage. Applicable for all form of data structure, cloud storage allows energy-efficient, cost-effective, secure and flexible storage solution. Data are stored in a virtual storage location and with multiple accessibilities anywhere, anytime with an internet connection by different entities. A cloud storage service provider often includes value-added functionality such as automatic backup, guaranteed up-time and multiple device access features. It can also engage with a service provider system for data analytic functions such as deep learning training, data processing and more. All these features made cloud storage a highly scalable and shareable mode of data storage.

4. *Data processing*: Can be referred to as a series of actions or operations to unveil useful and hidden information in large data volumes for knowledge discovery. Raw data in their natural form are typically scattered, uninterpretable and meaningless. Various data-processing procedures are involved to convert the raw data into meaningful information. The extracted information is useful to drive improvement actions such as quality control, predictive maintenance and yield improvement in the manufacturing domain. First, data cleaning removes redundant, duplicated missing values and inconsistent information. Next, the data reduction procedure is used to transform the large data sets into a simplified, arranged and compatible format according to the task requirements for processing. The pre-processed data is then ready for exploitation through data analysis and mining to develop new information. The performance of new information generation depends on the analysis and a wide range of available mining techniques. Due to the large data volume, the conventional statistical techniques no longer the best option. Replacing it are modern machine learning (ML), large-scale computing forecasting models and advanced data-mining techniques such as clustering, regression and many more. Optimum usage of BD with compelling data analysis efforts, in-depth knowledge and insights on the manufacturing entities can be derived. This drives the achievement of high competency and intelligent manufacturing.
5. *Data visualisation*: The presentation of the data-processing outcome in a more explicit form to facilitate user comprehension and interpretation. The delivery of information is more effective in visualised form, such as graphs, diagrams, figures and virtual reality. Real-time data can be envisioned online through users'

dedicated channels or platforms, enabling instant information sharing. In addition, static data processed into visual form and shared across each entity promotes more straightforward perception, accessibility and user-friendliness. Advanced analytics software includes data visualisation creation function, easing the tasks of users.

6. *Data transmission*: Modern digitalised SM generates a large volume of data, and the generated data are flowing continuously across different platforms such as equipment, CPS and humans over different mediums in a distributed mode. The transmission of these data is critical to ensure effective communication, information sharing and interaction across all entities within the domain. The current advancements in communication technologies extensively centralised the infrastructure of real-time, secure and reliable transmission of all kinds of data. Correspondingly, distributed manufacturing resources in SM can be unified at virtually anytime and anywhere to attain high production flexibility and competency, keeping the competitiveness in the global market.
7. *Data application*: Data generated and existed in almost every facet of modern manufacturing. It had become the primary catalyst for controlling the three most crucial manufacturing criteria, namely *yield*, *quality* and *cost*. In maintaining the competency, data is applicable from the beginning of the product design stage until the final delivery stage. During the design stage, data analytics can be utilised to obtain information and understanding about the product to be manufactured, such as market demand, competitor offering and technology limitations. These insights allow the development of a unique and customer-oriented product. Next, during the production stage, a different form of manufacturing entities data can be utilised to optimise production. Real-time equipment data are tracked for overall equipment effectiveness monitoring. It can be further analysed over long-term data for predictive maintenance to eliminate unscheduled downtime of the machine, which will affect yield and cost. Process data are used to control each manufacturing step, ensuring an optimised process for product quality and manufacturing cost control. Analysis of process data can provide early warnings on the tendency of process shift affecting product quality, abnormally high utilisation of raw materials and low product efficiency. It also allows for diagnosis of the root causes of problems for rectification and current process limitation for further improvement. Operation data can be used for online yield monitoring to trigger respective equipment and process personnel for immediate attention. Production output data can be analysed to cope with market demand and production planning. In the final delivery stage, production, management and market data were analysed to determine optimum logistics planning, marketing time, production output and other SCM items. Collectively, in-depth data analytics enable execution of early stage and precautionary actions, preventive and predictive maintenance, fault prediction and operational improvements. In a nutshell, data drives the new modern manufacturing.

The value of BD for SM does not rely solely on its large data volume. The ability for effective data utilisation to devise meaningful information is the key to realising optimum outcome in SM. Standard conventional technologies provided by software and hardware developers mainly focused on structured data for predictive analysis, which form only the minor component of BD. The venture into unstructured data areas such as audio, video and unstructured texts was generally ignored. However, the trend is undergoing a change with the recent rise in AI. This rising trend observes more research efforts and practical industrial approaches utilising advanced AI techniques such as ML, to gather new dimensionality and acumen of unstructured data. This way is a critical approach to harvest the true potential of BD. Substantial evidence has shown the power and improvements achieved through AI approaches on unstructured data. The next section further discusses the BD analytical concepts, implementation, data-mining techniques and several other aspects.

3.3 Big data analytics

Data analytics is deemed as a branch of the science and engineering approach to discover hidden patterns, anonymous information and correlation within the systems. BD further deepens the demand for analytic capability with its defining characteristics, which result in the insufficiency of conventional data management and analytics to harvest BD full potentials for SM applications. The conventional approach restrictions, together with computational technologies advancements, led to the establishment of a new branch of analytic techniques known as BDA. BDA had gained much popularity and traction recently due to the advancement and generalisation of AI.

The capability to analyse large datasets will be one of the main factors to keep business competitiveness, growth and innovation [15]. BD, in its raw format, has not much value; it is meaningless without relevant analysis to gather its value to turn the information into usable knowledge [16]. Good skill on data analytics can determine meaningful insight of the processes, enable several essential benefits such as real-time tracking, predictive maintenance and many other actions to improve the systems and decision-making by the management. It promotes autonomous interoperability, flexibility, system improvement and cost reductions [17]. This phenomenon triggers the shift in conventional manufacturing paradigm of centralised production applications into distributed cluster SM concept, the key in SM. BDA that delivers intelligence to manufacturing processes is predicted to contribute a significant improvement to SM.

The analytics process to extract information from the various sources and types of data within the BD can be categorised into five classes, namely *text analytics, audio analytics*, *video analytics*, *social analytics* and *predictive analytics*. Two main sub-processes involved in the analytics process, which are data management and data analytics, are illustrated in Figure 3.6.

3.3.1 Text analytics

Text analytics is also known as text mining. It refers to the techniques to obtain information from character and numerical texture-based data. Examples of textual data are emails, newspaper, documents, books and more. Textual data available in every level of any organisations and also human lives is the primary source of new knowledge creation.

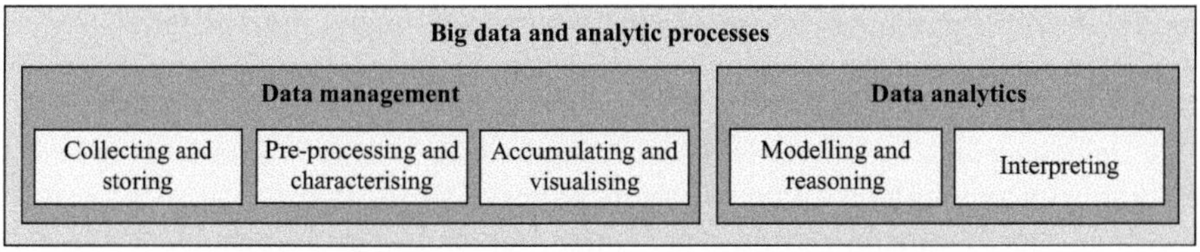

Figure 3.6 Data analytic types and processes

Efficient and effective analysis and interpretation of these data will reveal much-hidden info, providing enormous opportunities for improvements. Statistical analysis, computational linguistics and ML are three main fields involved in textual data analyses. The result from these analyses will provide clues, indications and confirmations to support decision-making with evidence. An example of useful text analytics application is in the field of business intelligence (BI), extracting information from financial news to predict the stock market performance [18]. Text analytics consists of four main techniques briefly described next.

IE is the technique used to obtain structured data from unstructured text. The IE algorithm was designed to recognise text entity and relation from the textual source, producing structured info. A most typical IE example is the algorithm for a calendar application to create a new entry, including essential information such as date and venue, automatically extracted from the email. Two sub-tasks performed in IE are entity recognition (ER) and relation extraction (RE) [19]. The ER locates and organises specific and pre-specific named entity that exist in the unstructured text into designated categories such as location, values, names and more. RE examines and extracts semantic relationships among entities in the text. The extracted relationship typically occurred among the entities of certain types and grouped into defined semantic categories. For example, in 'Steven resides in Malaysia', the extracted relationship can be *Person* [Steven], *Location* [Malaysia].

Text summarisation is the technique that automatically shortens long pieces of texts into a clear and concise summary, outlining only the main points. Application examples are news headlines, book synopsis, meeting minutes, and more – two conventional approaches used in this technique, namely the extractive approach and abstractive approach. Extractive summarisation creates a summary from the original text units, resulting in a subset of the original document in a shortened format. Extractive summarisation technique does not 'understand' the text. The summary was formulated by determining the prominent text units and connecting them together. The prominence of the sentence units is evaluated through its occurrence frequency and location in the text. Trained ML models are often used for this technique, to filter useful information in the texts and output it in summary, conveying the critical information. In contrast, abstractive summarisation extracts linguistic information from the text and generate the summary using new sentences produced by new text units or rephrasing, instead of distilling the critical sentence. Advanced Natural Language Processing (NLP) techniques are used to process the texts and generate the summary. Correspondingly, more coherent summaries can be produced. However, due to its complexity and harder adoption, extractive summarisation is more commonly used for BDA.

Question answering (QA) is a technique to answer questions posed in natural language in the digitalised system. Apple's Siri, Amazon's Alexa, Google's Assistant and Microsoft's Cortana are some examples of prominent QA systems. A variety of areas, such as healthcare, banking and education, have used QA systems. The intelligence of these systems lies in its application of sophisticated NLP techniques. QA system methods can be further divided into three approaches, namely, information retrieval (IR)-based, knowledge-based and hybrid. IR-based QA system typically comprises three sub-components. (1) *Question processing* component for identifying the details used to construct a query such as a case and focus of the questions and its answer type. (2) *Document-processing* component for retrieving relevant pre-written texts from a set of large volume of existing documents, using search information devised through the query made in question processing. (3) *Answer processing* component for filtering candidate answers input from the document-processing component. The answers are then ranked according to their relevancy, and subsequently, return the highest-ranked answer as the output. IR-based QA systems are commonly used in areas with a massive collection of pre-written texts such as law and education. Knowledge-based QA systems query the information contained within an organised source using the semantic description developed from the input questions. This system consists of two major components: (1) *Knowledge base*, which contains a vast collection of information about the designed field, commonly generated through human expert input, used as a reference source during inferencing process. (2) *Inference engine*, which functioned to infer judgement from information in the knowledge base based on the question input given to acquire correct answers. Knowledge-based QA systems are useful within a controlled domain due to no large volume of pre-written text input available; its information within the knowledge base is specifically designed according to the given field. Hybrid QA systems are the combination of both halves of IR-based and knowledge-based QA systems. The questions are semantically evaluated as in the knowledge-based system, whereas the answers are generated using IR-based approach.

Sentiment analysis (opinion mining) is the technique to analyse computationally on textual data, which contains subjective information such as opinion and thoughts. The analysis is to identify, filter, obtain and quantify analytically, using NLP, text analytics and computational linguistics techniques on the data for the voices of the customers. Businesses are progressively involved in acquiring more data to understand their customers' behaviour, specifically to their products or services. Only by understanding the customers can the business be profitable by selling products or providing services that fulfil customers' demands. This business behaviour led to the propagation of sentiment analysis. Due to its analytic nature on subjective data, sentiment analysis is more applicable to marketing, political and social science. But it is also applicable in manufacturing environments for workers' performance and feedback analysis. Sentiment analysis technique is further divided into three subgroups: (1) Document-level analysis is based on the whole document for overall sentiment classification of positive, neutral or negative. It is assumed that the document represents or describes about a single entity. (2) Sentence-level analysis is more complex compared to document analysis. It needs to first identify the single sentence of their subjective and objective, then determine the sentiment schism over the elements described in the sentence. (3) Aspect-based analysis distinguishes all views within a complete document and classifies the entity features according to the views, enables broader and overall information about the topic being analysed, in contrary to single entity or polarisation. This allows the analyser to acquire in-depth information about the topic in multiple perspectives.

3.3.2 Audio analytics

Audio analytics is the type of analysis performed on unstructured audio data to extract meaningful information for business improvement. This type of analysis is particularly important in business service sector. Human spoken language analysis is considered as part of audio analytics, which is termed *speech analytics*. With the rise of customer-centric business orientation, audio analytics can be applied in manufacturing environment in deriving constructive inputs from audio-based customer feedback for improvement. There are two general types of audio analytic techniques, namely *phonetic-based analysis* and *large-vocabulary continuous speech recognition* (LVCSR).

Phonetic-based analysis uses the sequence of sound in the audio or sentence, known as a phoneme, as the basic unit for analysis. Phonemes are the single unit to differentiate words in the analysis medium by the pronunciation. An indexing and searching phase is involved in this technique. In the indexing phase, the medium is translated into a sequence of phonemes. Then, in the searching phase, the algorithm search for the phonetic representation of the input search terms. This approach encodes the phonemes into a matrix of possibilities, which is then referred to by matching the search terms input into the analysis process. The advantages of this technique are the ability to search for words which is not predefined, if the phonemes are identifiable. The processing is also faster compared to LVCSR. The drawbacks of this technique are slower search process due to phonemes unable to be indexed like how a word can and occupy larger storage footprint.

LVCSR technique checks the probability of different word sequences using a statistical approach, enabling higher accuracy compared to a phonetic search approach. Similar to the phonetic-based approach, two phases of indexing and searching are involved. In the first indexing phase, special algorithm of automatic speech recognition was implemented to transliterate audio or speeches into matching words. The words were predefined in a list of dictionaries. Every word is recognised, and the nearest matching word will be used if the actual word is not found to produce full transcript of the speech, with a searchable index file. The second searching phase, standard text search algorithm, was used to process the search term in the transcript. Although this technique requires more processing resources than phonetic-based techniques, the generated speech transcript enables more effective and efficient use of the gathered information.

3.3.3 Video analytics

The advancement of computing and digital imaging technologies drives the implementation of video content analytics (VCA) to obtain important information from the video content. Together, with increased computing power, VCA can be implemented to produce meaningful information and insights almost instantly. Modern manufacturing facilities equipped with high numbers of CCTV and camera systems to monitor manufacturing processes, equipment conditions and material flows within the production floor. With these imaging facilities, production controllers will have all-important process monitoring within a centralised control room. The BD generated from the video feeds can be further analysed for improved BI and actionable insights. Examples of insights generatable from VCA are the forms of patterns, characteristics, attributes and specific events, allowing effective decisions to be made, and efficient implementation by the right people at the right time.

Apart from monitoring insights produced, automatic video indexing and retrieval is also another important area of VCA application. With the rise of multiple forms of video content and sources, indexing the mediums on searchable contents for IE is critical. The indexing process is performed based on different levels of data contained inside the video, such as the texts, audios, metadata and visual contents. Different analytics techniques can be combined to analyse the contents for indexing; for example, text analytics can be used for indexing the video on the transcripts and audio analytics on the soundtracks contained in the video.

In the system architecture for VCA, there were two main techniques: *server-based* and *edge-based*. For *server-based* configuration, a centralised and dedicated server was set up to receive all videos captured by each camera. VCA were performed directly on this server using all the videos received. Main consideration of this set-up is the video data transmission bandwidth limitation. Video data from the source are often compressed to cater for this limitation. This compression result in information losses within the video, which will affect the analysis accuracy. However, server-based configuration enables easier maintenance and most importantly the economies of scale, overcoming the bandwidth limitation. In contrast, *edge-based* configuration provide direct VCA applied locally at the imaging system using the raw data captured by the camera. This set-up does not induce any information losses as raw video data were used. The complete video content is accessible for analysis, contributing to improved analysis accuracy. However, set-up and maintenance costs are higher for this configuration, and the system processing performance is also lower, compared to *server-based* systems.

3.3.4 Social analytics

In understanding customers' requirements for business strategy development and process improvement, corporations utilise multiple approaches to get close to the existing clients and at the same time to attract potential clients. From this front-end approach, social analytics develop the first-hand information, which then transform into business strategy and subsequently proliferate to manufacturing to produce the required products.

Social analytics involved the analyses of different societal media, which contain both structured and unstructured data. Content exchange over the social media platform encompasses multiple level of communities and portals, containing huge amounts of data of different types. Due to the huge user bases and variety of information shared, social media is one of the biggest data generators. Social media platforms can be categorised into social networks such as Facebook and LinkedIn, content writing blogs such as Wordpress and Blogger, media sharing platforms such as YouTube and Instagram, knowledge sharing wikis such as Wikipedia and Wikihow, content review sites such as Trustpilot and Testfreaks, question-and-answer sites such as Ask.com, social bookmarking such as Pinterest and StumbleUpon, and micro-blogging such as Twitter and Tumblr [20]. The main source of information from social media analytics is the user-generated contents such as images, videos and sentiments, linked with its relationships and interactions between the network objects such as products, companies, peoples and places. With reference to this grouping, two analytical approaches are classified and further elaborated below.

Structure-based analytics. This is an analytical approach in synthesising and evaluating social network organisational characteristics to extract meaningful insights and intelligence from the relationship established among the involved entities. The analytical process uses networks and graph theory, representing individuals or entities within the network as nodes, and connecting them with ties based on the relationship or interaction established [21]. The nodes are represented as points, and ties are represented as lines in these networks and usually visualised using sociograms. Visualisation of the nodes and lines reflect the interest features among the entities, providing qualitative assessments of the networks [22]. This technique is also known as *social network analytics*.

Content-based analytics. This analytical approach focuses on the content posted by the users in different social media platform and analysing them to gather the information required and made useful for business purposes. Because these data were posted from different sources and people, they are often produced in large bulk, unstructured, stochastic and dynamic. Different analytical techniques as discussed previously are used to process these data to derive the information. The analytical process can be summed up to data identification, data analysis and information interpretation. Data identification is to divide and identify the available data into subsets, to set focus to facilitate analyses. Data analysis is the activity to transmute the raw data into information that can lead to new knowledge and business value. Information interpretation is the final step to draw conclusions and extract important perceptivity on the questions to be answered based on the data analysed. The data or information are often represented in easy-to-understand visualisation, such as charts or graphs, allowing final receiver of the information, who are generally non-technical people, to be able to comprehend and derive business decisions.

The techniques used to extract information from the social networks are generally defined into three main methods: (1) *community detection*; (2) *social influence analysis;* and (3) *link predictions.*

Community detection is one of the most popular topics in modern network science. It is a way to identify implicit association on the nodes within the network, which are more densely connected than to the others, that form specific community. Community in social networks is the group of users that interact more extensively with the pattern confining within a sub-network, compared to the rest of the network. These connection irregularities imply that natural division exists in a network. Social network size is enormous, containing millions of nodes and lines. Identifying these underlaying community structure helps to compile the network, enabling the discovery of present behavioural patterns and prediction of network properties. This can provide insight into how the communities behave and also how the network topology affects each other. Example of the benefit for community detection is enabling company to provide more suitable products according to the requirements and preferences of each different communities.

Social influence analysis is a technique to measure the influence of each individual within the social network and to identify the most influential individual. It models and evaluates the behavioural change caused by the key influencer and also each connected individual within the network. This technique uncovers the influential patterns and also the influence diffusion in the network, facilitating the application examples of proliferation of brand awareness and adoption over the targeted audience.

Link prediction is a technique to predict future possible linkages between the nodes in the network. It is also used to predict missing links between incomplete data. Social networks are a dynamic structure and evolve over time. It grows continuously with formation of new nodes and vertices. By understanding the association between the nodes – for example, how does the association characteristic change over time, and what are the factors affecting the associations and how the other nodes influencing the association – it will provide useful prediction on entities behaviour. This technique provides useful application in forecasting future product requirements.

3.3.5 Predictive analytics

Predictive analytics utilised diversified statistical techniques, from data mining to ML, to analyse current and historical data to predict possible future events and outcomes. Predictive analytics plays a main role in SM. With effective implementation, it will bring a major change to the current manufacturing horizon with unprecedented manufacturing efficiency improvement from all the aspects of cost, yield and quality.

Fundamentally, predictive analytics techniques are used to discover pattern and relationship in the data to provide analysts to understand the performance and generate correlated predictions, such as material requirements, equipment performance and product quality impact. Predictive analytics approach can be divided into two groups:

(1) Pattern discovery, where statistical techniques such as moving averages were used to reveal historical patterns in the data and subsequently uses the outcome to extrapolate for forecasting future behaviour; (2) capture interdependency by utilising statistical techniques such as linear regression to acquire interrelationship between analytical inputs over the outcome variables and exploit them to produce predictions on the events.

Predictive analytics techniques rely heavily on statistical methods to analyse and extract information from the data. Massive data volume, size and high dimensionality of BD leads to unique computational and statistical challenges on its analytics. Several factors show the conventional statistical method is not suitable for BDA, and a new analytical approach is required. First, the main difference between BD over conventional manufacturing process data is that BD contain all data of the population in the processes. Conventional statistical methods were based on the significance studies through data sampled from a population, and the devised information was generalised to conclude the behaviour of the population. This approach was used due to the inability to obtain full population data prior to digitalisation of manufacturing processes. In contrast, modern digitalised manufacturing processes are able to produce full population data along each process. Conventional statistical approaches based on sampled data significance will result in underutilised data. Small sample size data analysis in the conventional method will also limit the thoroughness of analytics in multidimensionality of the data, causing limited or loss of information. As a result, statistical significance interpreted in conventional statistical techniques were not appropriate to BD. Secondly, conventional method computational efficiency was not suitable for large data analytics. The emphasis of conventional statistical techniques on the underpinning probability models, rooted in statistical significance, determined through random sampled data over the represented uncertainties of the population. Applying full population data for inferential statistical analysis in the conventional approach will not yield meaningful outcome due to the small standard error derived from the almost known uncertainties of the population. The third factor links to the BD salient features of heterogeneity, spurious correlations, incidental endogeneity, and noise accumulation.

Heterogeneity in BD were caused by the data acquisition from different sources, representing different information on different object populations. The data were often unclear, containing missing values, redundancy and mendacity, resulting in difficulty for direct application. The inconsistency of the data will also cause the low-frequency sub-population data deemed as outliers within the large high-frequency population data. However, the steep ratio of the large frequency data to the outliers will overshadow the outliers influence with advanced analytical (AA) techniques.

Spurious correlation. The high dimensionality in BD produced by its huge data volume can cause spurious correlation. It is a phenomenon where high data dimensionality inducing false links of actual uncorrelated random variables, which then become correlated. This error will result in wrong statistical inferences leading to false prediction or forecast derived from the model.

Incidental endogeneity. This is another phenomenon induced by the high dimensionality of BD. In contrast to spurious correlation, incidental endogeneity is the genuine existence of unintended correlation between random variables due to the high dimensionality of the data, particularly explanatory variables with the errors. This symptom occurs as a result of omitted variables, measurement errors and selection biases. It frequently arises in BD due to (1) the digitalisation of measurement methodology that results in ease of data collection, and analysts tend to collect as much data as possible. The larger data capacity increases the probability of incidental correlation to residual noise. (2) Multiple different data-generating sources for BD increases the probability of measurement errors and data selection bias which also leads to incidental endogeneity.

Noise accumulation. Analysing BD to extract information for model prediction and forecasting involves concurrent estimates or tests with many different parameters. These large numbers of parameters result in calculation estimation errors, and it accumulates when a prediction or decision model uses these parameters. These noise accumulations will induce significant impacts on the models generated, especially with high dimensional data. The noise will reach an extent where it dominates over the true signals in the model, resulting in false or inaccurate decision or prediction from the model. In other words, the true significant variables with descriptive signals may be left unnoticed due to the noise accumulation.

The intrinsic characteristics of BD result in unique features that limit the conventional statistical techniques in analysing the data. BD also posed critical challenges to computational efficiency. From the factors discussed and highlighted, it is certain that new analytical techniques are needed to obtain meaningful insights from the predictive models generated using BD.

3.3.6 Big data advanced analytics for smart manufacturing

The modern SM challenge for complex decision-making corresponds to four main aspects: (1) The ability to realistically model the actual manufacturing systems for information and data gathering; (2) the ability to integrate consistent, reliable and valid manufacturing plant data; (3) the ability to process the gathered data to obtain required information within reasonable computational efforts; and (4) the ability to incorporate feedback systems into the manufacturing process for continuous improvement and to facilitate continuous decision-making procedures over time. BD and advanced analytics for prediction, prescription and detection is the underpinning groundwork needed to achieve the above four aspects [23].

The expansion of BD availability in its unique characteristics of *volume*, *variety*, *velocity*, *veracity*, *variability* and *value*, together with advancement of increasing computing and communication technologies, has been the catalyst to push manufacturing businesses to implement the combination of diverse and AA techniques to drive process and performance improvements into an SM context. These AA techniques can be described as:

- *Descriptive analytics*: The preliminary stage to create summary of historical data for extracting useful information of past events as well as preparing the data for further analysis for future events. To be used in back-casting past practices and forecasting future business demands.

- *Predictive analytics*: The use of an analytical approach with available data, usually heterogenous and large volume data, to predict possible future events such as trends, movements, behaviours and inclinations.
- *Prescriptive analytics*: Relating both descriptive analytics that provide information on past events and predictive analytics that forecast future events, prescriptive analytics determines the best solution or outcome from the available options based on the known parameters. It finds the best process output options based on the input information.
- *Detective analytics*: The approach to investigate the causes and sources of incidents or variations that restrict the processes' or systems' ability to achieve optimal results through the collected data to eliminate and rectify the inappropriate values used in predictive and prescriptive analytics.
- *Cognitive analytics*: The analytics of generating interpretations from existing data or patterns and to draw conclusions based on existing available knowledge and input back to the knowledge base as a new learned feature. It automates all analytics techniques for smarter decisions over time through the self-learning feedback-loop.

The implementation of AA for SM requires commitments from top management down to the production floor to ensure efficiency and effectiveness of the system. AA for BD will be able to provide answers to the questions raised for the company about past events (descriptive analytics), potential predicted future events (predictive analytics) and evidence-based decisions on how the company shall capitalise the short-, medium- and long-term plans (prescriptive analytics). SM efficiently interconnect manufacturing as the central process to horizontal SCM on both supplier for optimising raw material input and timely product output delivery to customers, in addition to vertical management communication for efficiency business decision inputs. AA will provide best controlled output in the SM model.

There are many applications of BDA in manufacturing industries, both conventional and SM. Areas of implementation include quality control, process optimisation, fault detection, equipment diagnosis and predictive maintenance. With the flooding of BD in the modern digitalised world, an efficient and effective use of these data can provide in-depth insights and information about the topic of interest. With BDA, all entity associations, product requirement forecast, marketing focus and event possible outcomes can be made, and related actions can be planned ahead of time. This approach enables manufacturing and maintenance cost reductions as well as business competitiveness.

References

[1] D. Cemernek, H. Gursch, and R. Kern, "Big data as a promoter of industry 4.0: lessons of the semiconductor industry," in *Proceedings – 2017 IEEE*

15th International Conference on Industrial Informatics, INDIN 2017, IEEE, 2017, pp. 239–244.
[2] J. Moyne and J. Iskandar, "Big data analytics for smart manufacturing: Case studies in semiconductor manufacturing," *Processes*, 2017.
[3] S. I. Shafiq, C. Sanin, C. Toro, and E. Szczerbicki, "Virtual engineering object (VEO): Toward experience-based design and manufacturing for industry 4.0," *Cybern. Syst.*, 2015.
[4] L. Monostori, B. Kádár, T. Bauernhansl, *et al.*, "Cyber-physical systems in manufacturing," *CIRP Ann.*, vol. 65, no. 2, pp. 621–641, 2016.
[5] J. Lee, C. Jin, and B. Bagheri, "Cyber physical systems for predictive production systems," *Prod. Eng.*, vol. 11, no. 2, pp. 155–165, 2017.
[6] H. Lasi, P. Fettke, H. G. Kemper, T. Feld, and M. Hoffmann, "Industry 4.0," *Bus. Inf. Syst. Eng.*, 2014.
[7] Q. Qi and F. Tao, "Digital twin and big data towards smart manufacturing and industry 4.0: 360 degree comparison," *IEEE Access*, 2018.
[8] A. Gandomi and M. Haider, "Beyond the hype: big data concepts, methods, and analytics," *Int. J. Inf. Manage.*, 2015.
[9] TechAmerica Foundation, "Demystifying big data: a practical guide to transforming the business of government," *WashingtonTech Am. Found.*, 2012.
[10] Gartner IT Glossary, "Definition of Big Data." [Online], 2012. Available: https://www.gartner.com/it-glossary/big-data/.
[11] Q. P. He and J. Wang, "Statistical process monitoring as a big data analytics tool for smart manufacturing," *J. Process Control*, vol. 67, pp. 35–43, 2018.
[12] J. Lee, E. Lapira, B. Bagheri, and H. Kao, "Recent advances and trends in predictive manufacturing systems in big data environment," *Manuf. Lett.*, 2013.
[13] A. Siddiqa, I. A. T. Hashem, I. Yaqoob, *et al.*, "A survey of big data management: Taxonomy and state-of-the-art," *J. Netw. Comput. Appl.*, vol. 71, pp. 151–166, 2016.
[14] F. Tao, Q. Qi, A. Liu, and A. Kusiak, "Data-driven smart manufacturing," *J. Manuf. Syst.*, vol. 48, pp. 157–169, 2018.
[15] J. Manyika M. Chui, B. Brown, *et al.*, "Big data: The next frontier for innovation, competition, and productivity," 2011. Available: http://www.mckinsey. com/Insights/MGI/Research/Technology_and_Innovation/Big_data_The_next_frontier_for_innovation.
[16] R. F. Babiceanu and R. Seker, "Big Data and virtualization for manufacturing cyber-physical systems: a survey of the current status and future outlook," *Comput. Ind.*, 2016.
[17] V. Alcácer and V. Cruz-Machado, "Scanning the industry 4.0: a literature review on technologies for manufacturing systems," *Eng. Sci. Technol.*, 2019.
[18] W. Chung, "BizPro: extracting and categorizing business intelligence factors from textual news articles," *Int. J. Inf. Manage.*, 2014.
[19] J. Jiang, "Information extraction from text," in *Mining Text Data*, Springer, 2013.

[20] G. Barbier and H. Liu, "Data mining in social media," in *Social Network Data Analytics*, Springer, 2011.
[21] E. Otte and R. Rousseau, "Social network analysis: a powerful strategy, also for the information sciences," *J. Inf. Sci.*, 2002.
[22] D. Z. Grunspan, B. L. Wiggins, and S. M. Goodreau, "Understanding classrooms through social network analysis: a primer for social network analysis in education research," *CBE Life Sci. Educ.*, 2014.
[23] B. C. Menezes, J. D. Kelly, A. G. Leal, and G. C. Le, "Predictive, prescriptive and detective analytics for smart manufacturing in the information age," *IFAC-PapersOnLine*, vol. 52, no. 1, pp. 568–573, 2019.

Chapter 4

Virtual and augmented reality in Industry 4.0

Mohankumar Palaniswamy[1], Leong Wai Yie[2] and Bhakti Yudho Suprapto[3]

There was a period in our history, where industrial works used to happen manually by labors at their houses. Later, it started to happen in industries with the help of machines. The process of transferring a labor-based agrarian economy to a machine-based is called as Industrial Revolution. This transformation initially started in Great Britain during the eighteenth century, followed by European countries such as Belgium, Germany, and France and slowly started to spread across the globe. Industrial revolution was the result of development in knowledge, equipment, machineries, technologies, and tools, especially social and cultural acceptance from people and benefits arising from it. Earlier, iron, steel, and wood were the primary raw materials used widely. Steam engine, combustion engine, and electricity were available at that time. Later, due to industrial revolution, steamship, locomotive, telegraph, and radio were developed. The idea of using science for development of humankind in the name of industry or factory started taking roots.

The economic transformation that took place in Western countries from 1780 to 1850 was termed as the First Industrial Revolution [1]. Realizing the importance of science and technology, with more research, second and third industrial revolutions took place. The duration of every industrial revolution was around five decades. This was the period of mass production.

4.1 Industry 4.0

There is a general belief that first industrial revolution was about steam and water, second industrial revolution was about steel, and third industrial revolution was about electricity, internet, and automobiles. In contrast, fourth industrial revolution has just begun [2] (Figure 4.1). The term Industry 4.0 was put forward by the German government in 2011 at Hanover fair [1]. The term Industry 4.0 is also called as "Industrial Internet," "Digital Factory," "Intelligent Factory," "Factory of Future,"

[1]School of Engineering, Taylor's University, Subang Jaya, Malaysia
[2]Faculty of Engineering and Built Environment, MAHSA University, Jenjarom, Malaysia
[3]Department of Electrical Engineering, Universitas Sriwijaya, Palembang, Indonesia

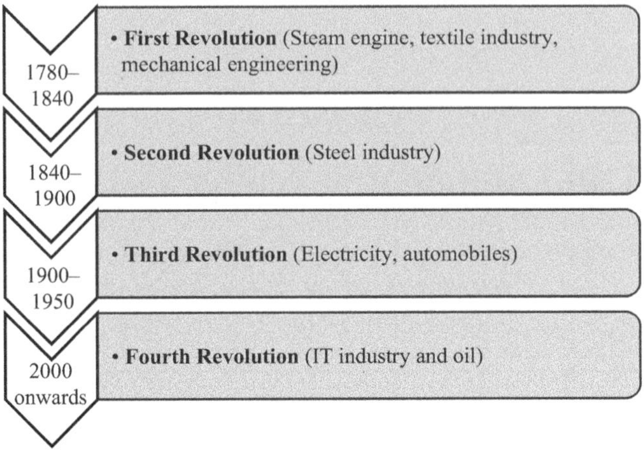

Figure 4.1 Industrial revolutions

and "Smart Manufacturing" by professionals [3]. Industry 4.0 or Fourth Industrial Revolution is based on internet of things (IoT), internet of services (IoS), cyber-physical systems (CPSs), and interaction and exchange of data between humans (C2C), machines (M2M), and between humans and machines (C2M) [4,5]. As the Industry 4.0 is still in its early stage, several research organizations, universities, and companies try to develop a fully automatic internet-based industry or factory, such as Siemens Amberg plant in Germany, where the 108,000 square foot factory can run and assemble components without any human input [6]. Industry 4.0 has several components such as cloud computing, CPS, IoT, system integration, additive manufacturing (AM), augmented reality (AR), virtual reality (VR), data analysis, AI robots, etc. (Figure 4.2).

Cloud computing – production-related data will be uploaded and shared in the cloud technology, where the receiving and reaction time will be few milliseconds. CPS – an advanced IT tool, which has a big storage capacity and high transmission speed. CPS provides solutions based on software-related applications while reducing the role of hardware or mechanical devices [8]. Smart products are components of CPS. It can connect between both the real and virtual world [9]. IoT – modern manufactures thought how to convert their product into a long-term revenue-providing service. It led them to IoT. IoT is the fundamental integration of smart devices, which are parts of the smart projects. For example, in a smart house, television, gaming console, and air conditioner that have an IP address can be connected to the IoT [10] and can be accessed from anywhere. Otis supplies elevators with sensors to the public, which send data to their cloud. This data is analyzed and Otis offers maintenance packages, which again generates extra revenue [11] (refer Figure 4.3). System integration – in Industry 4.0, several departments under a factory or several factories under same company become organized and interconnected with universal data integration. Additive manufacturing (AM) – with Industry 4.0, 3D printing will be widely used to manufacture customized products.

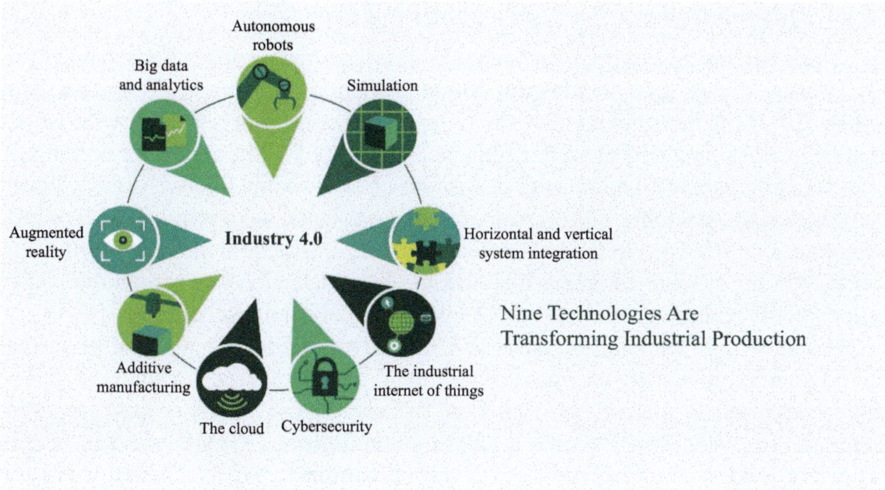

Figure 4.2 Components of Industry 4.0 [7]

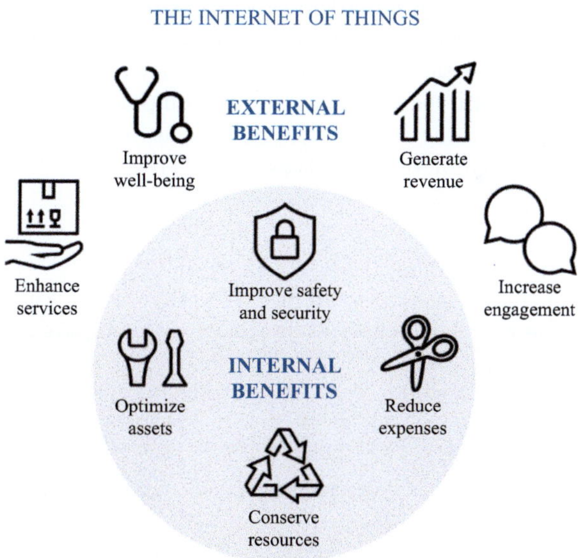

Figure 4.3 Internet of things [12]

AR, VR – AR- and VR-based services offer workers and operators to improve decision-making and limiting the work procedures. Data analysis – collecting and indexing data from different sources will become default in industries to support real-time decision-making. AI robots – robots will start to work with each other safely along with or without human intervention.

4.1.1 Augmented reality and virtual reality

The human and machine work interference in industry should be flexible and adaptive [13]. Because of this, several industries started to adopt using AR and VR to train their workers [14]. With this training, they can (i) speed up the work or reconfigure the work, (ii) support operators, (iii) execute augmented virtuality (AV) training for compiling or constructing parts, (iv) administer depository or stockroom effectively, (v) support diagnostics in the assembly, and (vi) minimize the risk in the work setting [15]. The key technologies on AV used in the industries are display interaction, tracking positioning and registration, human–computer interaction, object detection and recognition, calibration, model rendering, analysis on 3D space, and collision detection [16–19].

Attempts on 3D simulation dates to 1930, when Stanley G. Weinbaum used a pair of goggles to watch his movie “Pygmalion’s Spectacles” to undergo holographic experience. Latter AR was first introduced by computer pioneer Ivan Sutherland in 1968 [20]. He used a head mounted display (HMD) device, which was connected to a stereoscopic display from a computer. AR is a technology that overlays virtual image or information in real world. For example, adding an image or text to user display virtually in the real environment (RE). Here, RE represents surrounding of the user. The recent Pokemon GO game is a good example of AR usage [21]. In contrast, virtual environment (VE) is the surrounding completely generated by computer, where the user will not have any contact with the RE [22] (Figure 4.4). VR is a synthetic world, which may or may not look like a real world [23]. Initially, VR took over the world in 1990s by equipment such as Saga VR and Nintendo Virtual Boy. In the twenty-first century, using VR, people started to create a virtual representation of themselves called “Avatars.”

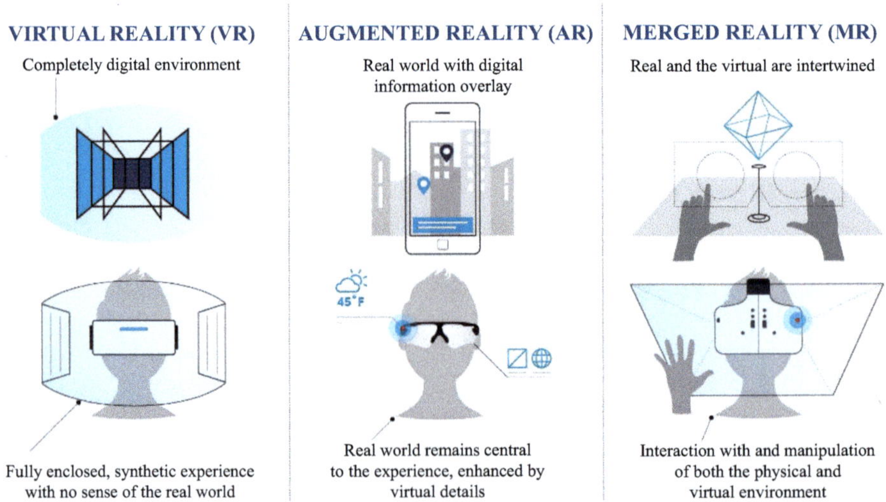

Figure 4.4 Difference between VR, AR, and MR [24]

4.1.1.1 Programming libraries

There are several AV programming libraries used in the industries. Among them ARToolKit, ARTag, osgART, and Vuforia are commonly used. ARToolKit is an opensource tracking library. It is the most common tool used to develop AR. Hirokazu Kato first developed it in 1999. ARToolKit was further developed with ARTag. ARTag is used to simplify the virtual objects within real world. OsgART is a C++ cross-platform library. It compiles the tracking library with 3D library. Vuforia is an AR software-development kit for mobiles. It compiles computer vision, planar changes, and 3D objects in real time [25]. With the recent development of mixed reality (MR), AR and VR are more commonly used in industries to reduce their processing, training, and decision-making time. With this in mind, this chapter reviews the achievements of AR and VR, tools used for constructing reality, and their field of use in the recent times.

4.2 AR and VR in Industry 4.0

AR and VR are very commonly used in the manufacturing industries for the purpose of quality control in the product, training or teaching the new recruits, designing the workspace, customization according to client need, maintenance of factory, manufacturing items, and marketing.

Quality assessment, which is done in traditional method, consumes a lot of time. A conceptual idea proposed and validated by Federica Ferraguti *et al.*, found that AR can be used in quality control [26]. With their approach, an operator can directly see the product if it has reached the quality specifications or any further refinements required. Factories are run not only by humans but also with machines such as robots. New theory puts forward that VR can be used for human–machine collaboration for adaptive product engineering [27].

Not only in designing, production, and quality control of a product but also in marketing the product AR and VR can be used. A 2013 study on footwear marketing used AR and VR technologies for marketing. Morenilla *et al.* [28] developed a stereoscopic vision system based on 3D+ software and footwear design. A desktop computer operating Windows 7 with i5 processor, 4GB RAM, nVidia Quadro 600 graphic card, 3dVision pack, BENQ double-frequency LED 3D monitor, and CodeGear C++ RAD Studio 2007 were used to view the footwear. To interact with the footwear, a VR glove from 5DT was used. It had five sensors for flexion and two for rotation. This single glove replaces the job done by keyboard and mouse. Different positioning of finger gestures corresponds to different view (Figure 4.5). To capture the user or consumer and view him/her in AR, a scene-capturing camera, LG Webpro 2, was used. To locate the position of foot, IR camera from Nintendo console and emitting device comprised of four LED diodes were used (Figure 4.6). After calibration and processing, final output is viewed in the monitor in real time, where the consumer is real, and the shoe is virtually developed (Figure 4.7). Using this method, consumer can see himself or herself wearing the shoes virtually.

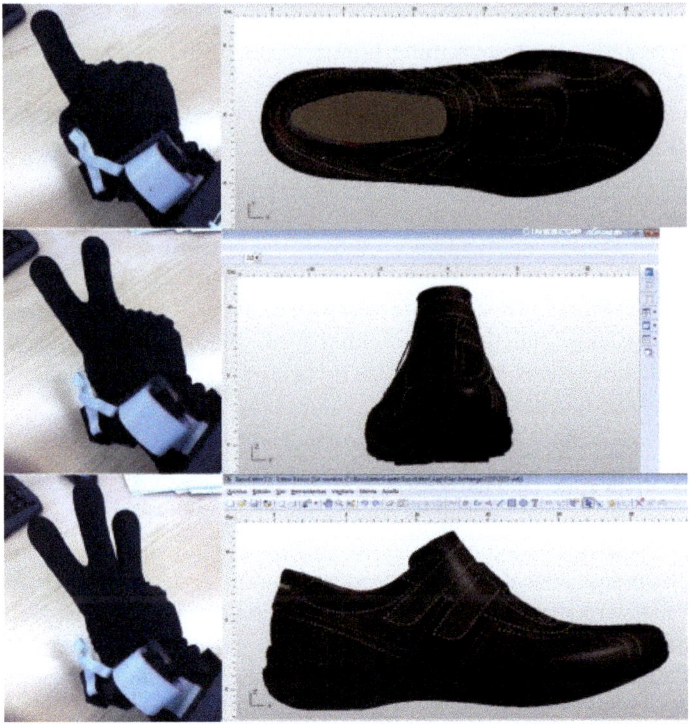

Figure 4.5 Glove gestures and the corresponding view [28]

Using AR and VR, an ergonomic workplace can be set up or ergonomics of the existing workplace can be found. Ceit ergonomic analysis application (CERAA) is used in mobile or with cave automatic virtual environment (CAVE), markers, Wii remotes, and assessment forms such as rapid entire body assessment (REBA), and rapid upper limb assessment (RULA) to find the office ergonomics [29,30].

AR is used in customization of an industry or factory, based upon the need. Using product manuals or descriptions in a PDF format is old fashioned. Recent 2019 studies incorporated AR by utilizing Unity 3D, Vuforia, HMD, and HoloLens to convert paper manuals into digital texts, signs, and graphical representations [31–33]. This method helps the operator to understand and use the manual easily [34]. AR and VR are used in designing a virtual workspace environment, where the operator or mechanic can understand how the parts are assembled [35,36]. Using AR and VR, remote maintenance is also possible [37,38]. VR is particularly used for industrial application training. Instead of onsite visit and consuming working hour, a virtual representation of the industry, warehouse, and its components is beneficial and valuable in training a fresher [39,40].

A 2019 study by Perez *et al.* [41] developed a VR of the industry, mimicking the same environment along with the robots to train the operators. Generally, robot is a mechanical structure with electronics, motors, controllers, and human–

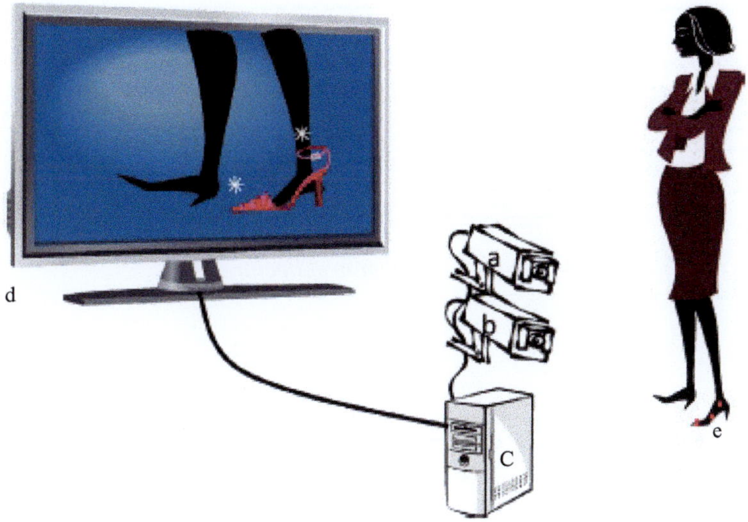

Figure 4.6 AR prototype. (a) Scene-capturing camera, (b) IR camera, (c) computer with footwear design, (d) double-frequency LED monitor, and (e) IR LED diodes [28]

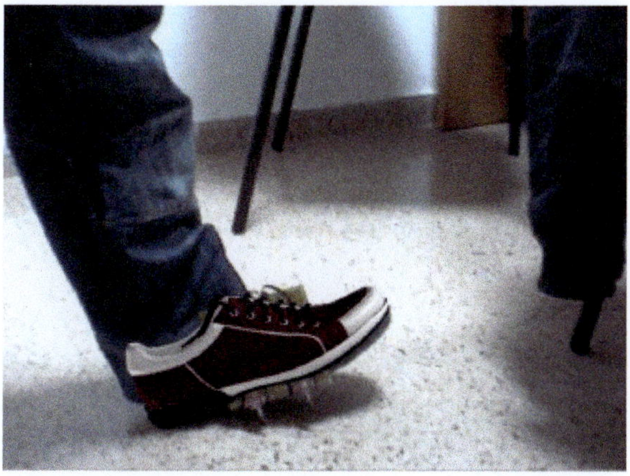

Figure 4.7 Final outcome [28]

machine interface (HMI). Perez *et al.* replaced this HMI with VR, which is connected to the robot. Their proposed system consists of two robots: a real and virtual, both controlled by same controller which was paired with VR (Figure 4.8). Sensors were connected to the real robot to obtain the pose and accuracy. To develop an identical workplace environment virtually, the workplace environment was initially

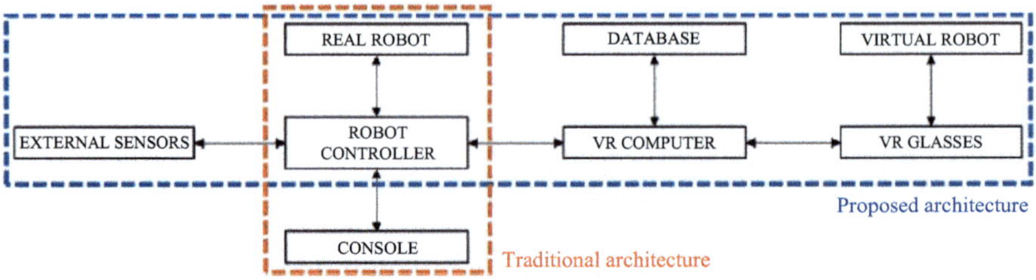

Figure 4.8 System architecture [41]

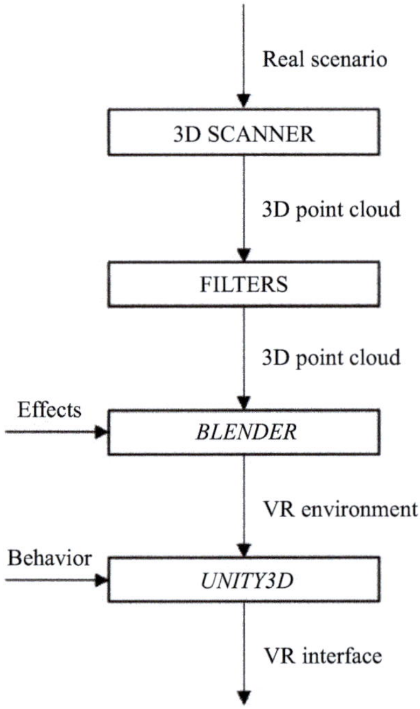

Figure 4.9 Process to develop VR interface [41]

scanned with FARO Focus3D HDR. The obtained 3D point cloud was processed with CloudCompare, modeled with Blender, and then processed with Unity3D to implement HMI (Figure 4.9). Using Oculus Rift and HTC Vive, the developed virtual was screened to operators for VR-simulated and VR-operated robot training (Figure 4.10). Twelve people underwent training and their feedback was recorded using questionnaire. They felt that VR training would increase efficiency of working environment.

AR, VR, and additive manufacturing are the most valuable aspects of aviation industry right now, as they play a major role in maintenance, detection of errors and safety in aircraft, and manufacturing small parts for small aircraft [42–44]. AR and

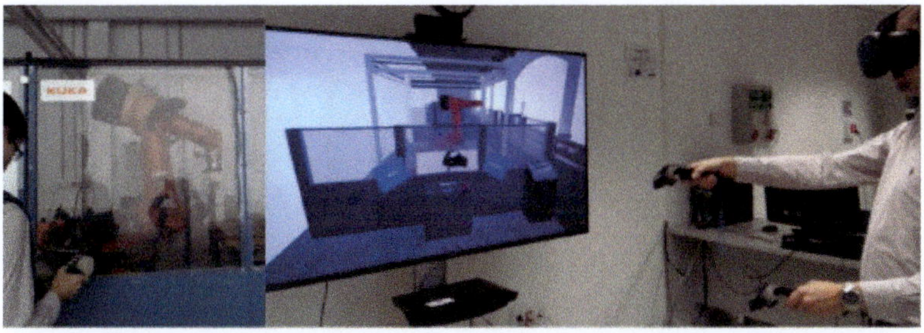

(a) (b)

Figure 4.10 Training. (a) With real robot and (b) with virtual robot [41]

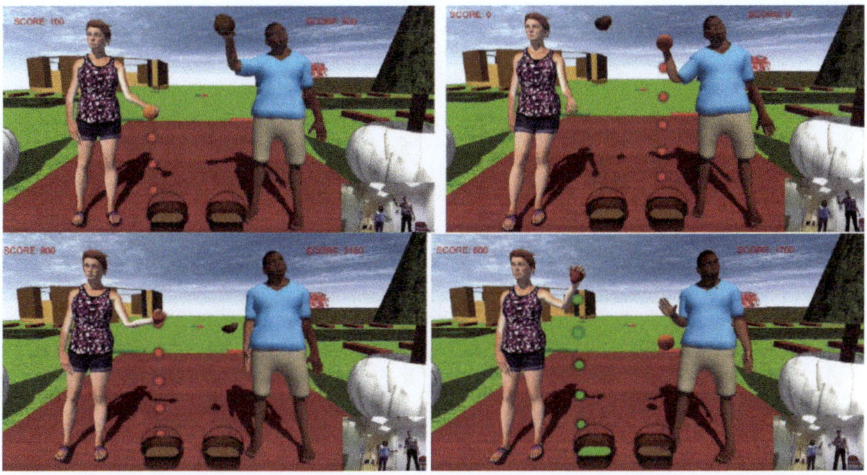

Figure 4.11 ReHabGame session, therapist, and patient [51]

VR technologies are used in construction and designing of buildings, machines, and robots. Particularly, AR is used in building information modeling (BIM) [45]. BIM information's provided in the software can be used during onsite visit. It can also be used latter for maintenance purpose. This in turn contributes to optimization and defect prevention [46–48].

AR and VR have revolutionized the medical field. They are utilized in analysis, rehabilitation, surgery, and treating phobias. Using leap motion sensor, IR camera, LED, and Unity 3D, hand motion analyses were performed [49]. Using Microsoft Kinect, Myo Armband, and Unity 3D, rehabilitation for stroke patients is provided. A game such as virtual environment is developed using Unity 3D, where the patients are supposed to play, such as throwing a ball or a fruit in a basket [50,51].

A 2018 study conducted by Esfahlani *et al.* [51] developed a VR-based game called "ReHabGame" to treat stroke patients. The game was developed using

Table 4.1 Augmented and virtual reality in different fields

Author	Year	Field	Technology	Tool	Sample	Objective
Lee *et al.* [48]	2011	Construction – Design	VR	ARToolKit, Microsoft Vision SDK, ROBOOP, Boost, V-Collide	–	Construct a mixed reality-based digital manufacturing environment
Morenilla *et al.* [28]	2013	Marketing	AR, VR	3D+, nVidia Quadro 600, 3D monitor, IR emitter, 5DT Virtual Glove	–	Implement AR, VR during footwear purchase
Fillatreau *et al.* [66]	2013	Industry – Maintenance	VR	Virtuose 6D Haptic Arm, CyberGlove II, ARTrack, VSL, C#	–	Framework for immersive checklist-based project reviews
Wang *et al.* [45]	2014	Construction – Maintenance	AR	Mobile, Tablet, HMD	–	Need for a structured methodology of fully integrated AR technology in BIM
Asgari *et al.* [46]	2017	Construction – Maintenance	VR	Leap Motion Controller, Myo, Nimble VR, Prio VR	–	Construction process optimization and defect prevention
Hui [40]	2017	Industry – Training	VR	Leap motion, Unity 3D, HMD, Gyroscope, Accelerometer, Nexus 6P mobile, Trinus	10	Compare experience and training effects in HMD-based and screen-based training
Giorgio *et al.* [27]	2017	Industry	VR	HTC Vive, Unity 3D	–	Human robot collaboration in product engineering by VR
Bun *et al.* [58]	2017	Medical – Therapy	VR	3DS Max, EON Studio, nVisor MH60, Oculus Rift	20	To treat acrophobia using VR
Gasova *et al.* [30]	2017	Industry – Design	AR, VR	CERAA, Mobile	–	Ergonomic assessment in industrial environment
Masoni *et al.* [37]	2017	Industry – Maintenance	AR	AR Goggles, Tablet, Unity 3D, Vuforia	–	Provide a remote maintenance in industry 4.0 using AR
Eschen *et al.* [42]	2018	Aviation – Maintenance	AR, VR	Oculus Rift, Oculus SDK, OpenGL, Game controller, Microsoft HoloLens, KUKA Robot	–	Detection of cracks in the maintenance of aircraft engines
Fazeli *et al.* [49]	2018	Medical – Analysis	VR	Leap Motion Sensor, IR Camera, IR LED, Unity 3D	Case Study	Hand motion analysis using VR

(Continues)

Karvouniari *et al.* [29]	2018	Industry – Customization, Ergonomics	VR	CAVE, VICON Bonita 3, Markers, Wii Remotes, Unity 3D, RULA, REBA	–	VR-based decision tool for exoskeleton integration in industry
Scurati *et al.* [32]	2018	Industry – Maintenance	AR	Questionnaire, Unity 3D, Vuforia, Mobile	–	Convert manuals into graphical symbols and use it in maintenance with AR
Shi *et al.* [50]	2018	Medical – Rehabilitation	VR	Kinect Sensor, Unity 3D, C#	–	VR-based user interface for the upper limb rehabilitation
Esfahlani *et al.* [51]	2018	Medical – Rehabilitation	VR	ReHabGame, Kinect, Myo Armband, Questionnaire	20	Rehab stroke patients using VR
Mourtzis *et al.* [36]	2018	Industry – Design, Training	AR	HoloLens, Unity 3D, CAD	100	Applying advanced visualization techniques in product design using AR
Wolfartsberger *et al.* [38]	2018	Industry – Maintenance	VR	Desktop PC (VR Ready), HTC Vive, Sensors, Controller. Unity 3D, 3DS Max	–	Implementation of a lightweight VR system in industrial engineering applications
Masood *et al.* [34]	2019	General – Manufacture	AR	Questionnaire, 5-point Likert scale	365	Factors influencing the success of implementing AR in industry
Gattullo *et al.* [33]	2019	General – Maintenance	AR	iFixit, PDF, Unity 3D, Vuforia, Questionnaire	22	Convert traditional text manuals to digital manuals in AR with compliance to Industry 4.0
Alarcon *et al.* [44]	2019	Aviation – Maintenance	AR	Questionnaire, 5-point Likert scale	56	AR application in space product assurance and safety activities
Ceruti *et al.* [43]	2019	Aviation – Maintenance, Manufacture	AR, Additive Manufacture	Game controller, Microsoft HoloLens, HoloLens clicker	Case Study	Demonstrate that AR and AM are viable tools in aviation maintenance
Kwiatek *et al.* [47]	2019	Construction – Maintenance, Design	AR	Tablet, AR application	61	Impact of AR on assembly in construction
Wolfartsberger [35]	2019	Industry – Design	VR	HTC Vive, Unity 3D, 3DS Max	–	VR-based tool to support engineering design

(Continues)

Table 4.1 (Continued)

Author	Year	Field	Technology	Tool	Sample	Objective
Perez *et al.* [41]	2019	Industry – Training	VR	KUKA KR500, Oculus Rift, HTC Vive, FARO Focus 3D, Blender	12	Replace the HMI directly with VR system which is connected with robot controller
Kopec *et al.* [53]	2019	Education – Training	VR	Proto.io, Epic Game's Unreal Engine, Unity Game Engine, HTC Vive, Mattel View	4	Virtual ATM training for older adults
Garcia *et al.* [54]	2019	Education – Training	VR	Unity Pro, Raspberry Pi, Oculus Rift, FESTO lab	–	Achieve a virtual training environment to teach pneumatic systems in university students
Roldan *et al.* [55]	2019	Education – Training	VR	Unity Game Engine, SteamVR, HTC Vive, Questionnaire	20	Training for industry operators based on VR
Urbas *et al.* [31]	2019	Industry – Customization	AR	Bluetooth-enabled Vernier Caliper, HMD, HoloLens, Vuforia, Unity Game Engine	–	Transfer of product manufacturing information from a 3D model and display the graphical presentation in AR
Ferraguti *et al.* [26]	2019	Industry – Quality Control	AR	HoloLens, Mixed Reality Toolkit, Vuforia, Xbox Controller	–	AR in quality control
Mourtzis *et al.* [39]	2019	Industry – Design, Operation	AR	Enterprise Dynamics 10, 4D Script, QR Codes, Mobile GPS	–	Operate a warehouse using AR
Nguyen *et al.* [56]	2019	Medical – Surgery	AR	Tap to Place, HoloLens Cursor	27	AR-assisted surgery
Ibanez [52]	2020	Education – Training	AR	IMMS survey, Questionnaire, ARGeo	93	Impact of AR in students' academic achievement
Masood *et al.* [67]	2020	General – Manufacture	AR	Microsoft HoloLens, Unity, Questionnaire, TCT, TLX	11	AR implementation in industry
Nguyen *et al.* [57]	2020	Medical – Surgery	AR	AR Roadmap	240	CAN for pedicle screw insertion during spine surgery

Abbreviation: AM, additive manufacture; AR, augmented reality; BIM, building information modeling; CAD, computer aided design; CAN, computer-assisted navigation; CAVE, cave automatic virtual environment; CERAA, Ceit ergonomic analysis application; HMD, head mounted display; IMMS, instructional materials motivation survey; REBA, rapid entire body assessment; RULA, rapid upper limb assessment; SDK, software development kit; TCT, task completion time; TLX, task

Unity game engine, which integrates motion / gesture capture, image processing, skeleton tracking, and IR emitter. Hardware components include only Microsoft Kinect and Thalmic Labs Mayo armband. Postural control, gestures, range of motion were tracked. ReHabGame include three main game designs: (i) reach grasp release, (ii) reach press hold, and (iii) reach press. Players reach, grasp, hold, and release virtual fruits in the virtual basket (Figure 4.11). This game involves flexion, extension, abduction, adduction, internal/external rotation, and horizontal abduction/adduction. Twenty stroke-affected patients participated in the study. Majority of the patients enjoyed the rehabilitation-based game and provided positive feedback.

Education is most important for all [68–74]. AR and VR outsmart the traditional black board and power point presentations. Irrespective of the age, AR and VR can be used to teach and train the young and old. AR is helpful for students' academic achievement [52]. VR is used in ATM training for older adults [53]. Virtual trainings are also provided to industrial operators [54,55]. A recent study developed a VR of mining process. Workers were trained and evaluated using questionnaire. A 2020 study used computer-assisted navigation (CAN) technique using AR Roadmap, Tap to Place, and HoloLens Cursor to perform spine surgery [56,57]. A VR-based study used Oculus Rift, nVisor MH60, EON Studio to treat acrophobia – the fear of height. Therapy is provided by developing a virtual high altitude scenario such as hill, tall building, and asking the patients to walk and cross the high altitude from one point to another [58]. Apart from industry, education, and medicine, AR and VR are used in other fields as well. Tourism [59,60], gaming [61], choreography [62], and customer satisfaction [63–65] are to name a few. The summary of this review is provided in Table 4.1.

4.3 Conclusion

This chapter assessed and reviewed the implementation of AR and VR in Industry 4.0. Researchers came up using low-cost devices and own algorithms rather than traditional commercial software such as CAVE, which cost more. Some papers proposed a conceptual theory, whereas others came up with quantitative and qualitative assessments. Their results suggest that AR and VR play a major role in minimizing the real-time decision-making and processing time. In industry-based setting, the utilization of AR and VR is more prominent. However, other than industry, the roots of AR and VR are not deep. Other fields such as education, tourism, and customer marketing have started to make use of AR and VR gradually, and it can be expected that eventually, the roots may deepen.

References

[1] A. Lele, "Industry 4.0," in *Disruptive Technologies for the Militaries and Security*, vol. 132, Singapore, Springer, 2019, pp. 205–215.

[2] J. Morgan, "What is the fourth industrial revolution?," Forbes, 19 February 2016. [Online]. Available: https://www.forbes.com/sites/jacobmorgan/2016/

02/19/what-is-the-4th-industrial-revolution/#5738154ff392. [Accessed 24 January 2020].

[3] The Boeing Center, "Industry 4.0: A brief history of automation and tech advancement," Olin Business School, 3 July 2017. [Online]. Available: https://olinblog.wustl.edu/2017/07/industry-4-0-•-brief-history/. [Accessed 24 January 2020].

[4] H. S. Ning, and H. Liu, "Cyber-physical-social-thinking space based science and technology framework for the Internet of Things," *Science China Information Sciences*, vol. 58, pp. 1–19, 2015.

[5] J. Cooper, and A. James, "Challenges for database management in the Internet of things," *IETE Technical Review*, vol. 26, no. 5, pp. 320–329, 2009.

[6] T. M. Hessman, "The dawn of the smart factory," Industry Week, 14 February 2013. [Online]. Available: https://www.industryweek.com/technology-and-iiot/article/21959512/the-dawn-of-the-smart-factory. [Accessed 24 January 2020].

[7] D. Kupper, "Embracing Industry 4.0 and rediscovering growth," [Online]. Available: https://www.bcg.com/en-in/capabilities/operations/embracing-industry-4.0-rediscovering-growth.aspx. [Accessed 24 January 2020].

[8] M. Mikusz, "Towards an understanding of cyber-physical systems as industrial software-product-service system," *Procedia CIRP*, vol. 16, pp. 385–389, 2014.

[9] S. Greengard, The Internet of Things, Boston: MIT Press, 2015.

[10] M. Martin, Building the Impact Economy, Berlin: Springer, 2016.

[11] M. James, "The internet of services in Industrie 4.0," Concept Systems, [Online]. Available: https://conceptsystemsinc.com/the-internet-of-services-in-industrie-4-0/. [Accessed 25 January 2020].

[12] A. Javed, "IoT 2018: The three most important security trends," Xorlogics, 19 January 2018. [Online]. Available: http://www.xorlogics.com/tag/internet-of-things/. [Accessed 24 January 2020].

[13] A. Yew, S. Ong, and A. Nee, "Towards a griddable distributed manufacturing system with augmented reality interfaces," *Robotics and Computer-Integrated Manufacturing*, vol. 39, pp. 43–55, 2016.

[14] C. Fischer, M. Lusic, F. Faltus, R. Hornfeck, and J. Franke, "Enabling live data controlled manual assembly processes by worker information system and nearfield localization system," *Procedia CIRP*, vol. 55, pp. 242–247, 2016.

[15] M. Demartini, F. Tonelli, L. Damiani, R. Revetria, and L. Cassettari, "Digitalization of manufacturing execution systems: The core technology for realizing future Smart Factories," in *XXII edition of the Summer School "Francesco Turco"*, 2017.

[16] J. Yu, L. Fang, and C. Lu, "Key technology and application research on mobile augmented reality," in *7th IEEE International Conference on Software Engineering and Service Science*, 2016.

[17] Z. Hong, and L. Wenhua, "Architecture and key techniques of augmented reality maintenance guiding system for civil aircrafts," *Journal of Physics: Conference Series*, vol. 787, no. 1, 2017.

[18] H. Zhang, and S. Z. Yi, "Research on key technology of intelligent terminal navigation based on augmented reality," *Advanced Materials Research*, vol. 629, pp. 848–852, 2012.

[19] Y. Wang, Y. Shen, D. Liu, S. Wei, and C. Zhu, "Key technique of assembly system in an augmented reality environment," in *2010 Second International Conference on Computer Modeling and Simulation*, 2010.

[20] I. E. Sutherland, "A head-mounted three dimensional display," in *AFIPS '68 (Fall, part I)*, 1968.

[21] A. Javornik, "'It's an illusion, but it looks real!' Consumer affective, cognitive and behavioural responses to augmented reality applications," *Journal of Marketing Management*, vol. 32, no. 9–10, pp. 987–1011, 2016.

[22] M. Parveau, and M. Adda, "3iVClass: A new classification method for virtual, augmented and mixed realities," *Procedia Computer Science*, vol. 141, pp. 263–270, 2018.

[23] D. A. Guttentag, "Virtual reality: Applications and implications for tourism," *Tourism Management*, vol. 31, no. 5, pp. 637–651, 2010.

[24] M. Philips, "Augmented reality vs virtual reality vs mixed reality — An introductory guide," 21 October 2018. [Online]. Available: https://uxplanet.org/augmented-reality-vs-virtual-reality-vs-mixed-reality-an-introductory-guide-5c3dce87bce0.

[25] L. Damiani, M. Demartini, G. Guizzi, R. Revetria, and F. Tonelli, "Augmented and virtual reality applications in industrial systems: A qualitative review towards the industry 4.0 era," *IFAC-PapersOnLine*, vol. 51, no. 11, pp. 624–630, 2018.

[26] F. Ferraguti, F. Pini, T. Gale, *et al.*, "Augmented reality based approach for on-line quality assessment of polished surfaces," *Robotics and Computer Integrated Manufacturing*, vol. 59, pp. 158–167, 2019.

[27] A. D. Giorgio, M. Romero, M. Onori, and L. Wang, "Human-machine collaboration in virtual reality for adaptive production engineering," in *27th International Conference on Flexible Automation and Intelligent Manufacturing*, Modena, 2017.

[28] A. J. Morenilla, J. L. S. Romero, and F. S. Perez, "Augmented and virtual reality techniques for footwear," *Computers in Industry*, vol. 64, pp. 1371–1382, 2013.

[29] A. Karvouniari, G. Michalos, N. Dimitropoulos, and S. Makris, "An approach for exoskeleton integration in manufacturing lines using virtual reality techniques," in *6th CIRP Global Web Conference*, 2018.

[30] M. Gasova, M. Gaso, and A. Stefanik, "Advanced industrial tools of ergonomics based on Industry 4.0 concept," in *TRANSCOM 2017: International scientific conference on sustainable, modern and safe transport*, Zilina, 2017.

[31] U. Urbas, R. Vrabic, and N. Vukasinovic, "Displaying product manufacturing information in augmented reality for inspection," in *52nd CIRP Conference on Manufacturing Systems*, 2019.

[32] G. W. Scurati, M. Gattullo, M. Fiorentino, F. Ferrise, M. Bordegoni, and A. E. Uva, "Converting maintenance actions into standard symbols for

augmented reality applications in Industry 4.0," *Computers in Industry*, vol. 98, pp. 68–79, 2018.

[33] M. Gattullo, G. W. Scurati, M. Fiorentino, A. E. Uva, F. Ferrise, and M. Bordegoni, "Towards augmented reality manuals for industry 4.0: A methodology," *Robotics and Computer-Integrated Manufacturing*, vol. 56, pp. 276–286, 2019.

[34] T. Masood and J. Egger, "Augmented reality in support of Industry 4.0 - Implementation challenges and success factors," *Robotics and Computer Integrated Manufacturing*, vol. 58, pp. 181–195, 2019.

[35] J. Wolfartsberger, "Analyzing the potential of virtual reality for engineering design review," *Automation in Construction*, vol. 104, pp. 27–37, 2019.

[36] D. Mourtzis, V. Zogopoulos, and E. Vlachou, "Augmented reality supported product design towards Industry 4.0: A teaching factory paradigm," in *8th Conference on Learning Factories 2018 – Advanced Engineering Education & Training for*, 2018.

[37] R. Masoni, F. Ferrise, M. Bordegoni, *et al.*, "Supporting remote maintenance in industry 4.0 through augmented reality," in *27th International Conference on Flexible Automation and Intelligent Manufacturing*, Modena, 2017.

[38] J. Wolfartsberger, J. Zenisek, and C. Sievi, "Chances and limitations of a virtual reality-supported tool for decision making in industrial engineering," *IFAC PapersOnLine*, vol. 51, no. 11, pp. 637–642, 2018.

[39] D. Mourtzis, V. Samothrakis, V. Zogopoulos, and E. Vlachou, "Warehouse design and operation using augmented reality technology: A papermaking industry case study," in *12th CIRP Conference on Intelligent Computation in Manufacturing Engineering*, Gulf of Naples, 2019.

[40] Z. Hui, "Head-mounted display-based intuitive virtual reality training system for the mining industry," *International Journal of Mining Science and Technology*, vol. 27, pp. 717–722, 2017.

[41] L. Perez, E. Diez, R. Usamentiaga, and D. F. Garcia, "Industrial robot control and operator training using virtual reality interfaces," *Computers in Industry*, vol. 109, pp. 114–120, 2019.

[42] H. Eschen, T. Kotter, R. Rodeck, M. Harnisch, and T. Schuppstuhl, "Augmented and virtual reality for inspection and maintenance processes in the aviation industry," in *6th International Conference on Through-life Engineering Services 2017*, Bremen, 2018.

[43] A. Ceruti, P. Marzocca, A. Liverani, and C. Bil, "Maintenance in aeronautics in an Industry 4.0 context: The role of augmented reality and additive manufacturing," *Journal of Computational Design and Engineering*, vol. 6, pp. 516–526, 2019.

[44] R. Alarcon, F. Wild, C. Perey, *et al.*, "Augmented reality for the enhancement of space product assurance and safety," *Acta Astronautica*, vol. 168, pp. 191–199, 2019.

[45] X. Wang, M. Truijens, L. Hou, Y. Wang, and Y. Zhou, "Integrating augmented reality with building information modeling: Onsite construction process

controlling for liquefied natural gas industry," *Automation in Construction*, vol. 40, pp. 96–105, 2014.

[46] Z. Asgari, and F. P. Rahimian, "Advanced virtual reality applications and intelligent agents for construction process optimisation and defect prevention," in *Creative Construction Conference 2017*, Primosten, 2017.

[47] C. Kwiatek, M. Sharif, S. Li, C. Haas, and S. Walbridge, "Impact of augmented reality and spatial cognition on assembly in construction," *Automation in Construction*, vol. 105, 2019.

[48] J. Lee, S. Han, and J. Yang, "Construction of a computer-simulated mixed reality environment for virtual factory layout planning," *Computers in Industry*, vol. 62, pp. 86–98, 2011.

[49] H. R. Fazeli, S. K. Venkatesh, and Q. Peng, "A virtual environment for hand motion analysis," in *6th CIRP Global Web Conference*, 2018.

[50] Y. Shi, and Q. Peng, "A VR-based user interface for the upper limb rehabilitation," in *6th CIRP Global Web Conference*, 2018.

[51] S. S. Esfahlani, T. Thompson, A. D. Parsa, I. Brown, and S. Cirstea, "ReHabgame: A non-immersive virtual reality rehabilitation system with applications in neuroscience," *Heliyon*, vol. 4, pp. 1–29, 2018.

[52] M. B. Ibanez, A. U. Portillo, R. Z. Cabada, and M. L. Barron, "Impact of augmented reality technology on academic achievement and motivation of students from public and private Mexican schools. A case study in a middle-school geometry course," *Computers & Education*, vol. 145, 2020.

[53] W. Kopec, M. Wichrowski, K. Kalinowski, *et al.*, "VR with older adults: Participatory design of a virtual ATM training simulation," *IFAC-PapersOnLine*, vol. 52, no. 19, pp. 277–281, 2019.

[54] C. A. Garcia, G. Caiza, J. E. Naranjo, A. Ortiz, and M. V. Garcia, "An approach of training virtual environment for teaching electro-pneumatic systems," *IFAC-PapersOnLine*, vol. 52, no. 9, pp. 278–284, 2019.

[55] J. J. Roldan, E. Crespo, A. M. Barrio, E. P. Tapia, and A. Barrientos, "A training system for Industry 4.0 operators in complex assemblies based on virtual reality and process mining," *Robotics and Computer-Integrated Manufacturing*, vol. 59, pp. 305–316, 2019.

[56] N. Q. Nguyen, J. Cardinell, J. M. Ramjist, *et al.*, "An augmented reality system characterization of placement accuracy in neurosurgery," *Journal of Clinical Neuroscience*, 2019.

[57] N. Q. Nguyen, S. M. Priola, J. M. Ramjist, *et al.*, "Machine vision augmented reality for pedicle screw insertion during spine surgery," *Journal of Clinical Neuroscience*, 2020.

[58] P. Bun, F. Gorski, D. Grajewski, R. Wichniarek, and P. Zawadzki, "Low-cost devices used in virtual reality exposure therapy," in *ICTE 2016*, Riga, 2017.

[59] S. M. Correia Loureiro, J. Guerreiro, and F. Ali, "20 years of research on virtual reality and augmented reality in tourism context: A text-mining approach," *Tourism Management*, vol. 77, 2020.

[60] F. Serravalle, A. Ferraris, D. Vrontis, and A. Thrassou, "Augmented reality in the tourism industry: A multi-stakeholder analysis of museums," *Tourism Management Perspectives*, vol. 32, 2019.

[61] L. L. Faican, and J. Jaen, "EmoFindAR: Evaluation of a mobile multiplayer augmented reality game for primary school children," *Computers & Education*, vol. 149, 2020.

[62] T. J. Dube, and G. Ince, "A novel interface for generating choreography based on augmented reality," *International Journal of Human-Computer Studies*, vol. 132, pp. 12–24, 2019.

[63] A. Rese, D. Baier, A. G. Schulz, and S. Schreiber, "How augmented reality apps are accepted by consumers: A comparative analysis using scales and opinions," *Technological Forecasting and Social Change*, vol. 124, pp. 306–319, 2017.

[64] S. G. Dacko, "Enabling smart retail settings via mobile augmented reality shopping apps," *Technological Forecasting & Social Change*, vol. 124, pp. 243–256, 2017.

[65] C. Flavian, I. S. Sanchez, and C. Orus, "The impact of virtual, augmented and mixed reality technologies on the customer experience," *Journal of Business Research*, vol. 100, pp. 574–560, 2019.

[66] P. Fillatreau, J. Fourquet, R. Le Bolloch, S. Cailhol, A. Datas, and B. Puel, "Using virtual reality and 3D industrial numerical models for immersive interactive checklists," *Computers in Industry*, vol. 64, pp. 1253–1262, 2013.

[67] T. Masood, and J. Egger, "Adopting augmented reality in the age of industrial digitalisation," *Computers in Industry*, vol. 115, 2020.

[68] W. Y. Leong, and D. P. Mandic, "Towards adaptive blind extraction of post-nonlinearly mixed signals," *2006 16th IEEE Signal Processing Society Workshop on Machine Learning for Signal Processing*, IEEE, pp. 91–96, 2006.

[69] S. Huang, D. H. Zhang, W. Y. Leong, *et al.*, "Detecting tool breakage using accelerometer in ball-nose end milling," *2008 10th International Conference on Control, Automation, Robotics and Vision*, IEEE, pp. 927–933, 2008.

[70] W. Y. Leong, J. Homer, D. P. Mandic, "An implementation of nonlinear multiuser detection in Rayleigh fading channel," *EURASIP Journal on Wireless Communications and Networking (EURASIP JWCN)*, USA, Vol. 2006, pp. 1–9, 45647.

[71] W. Y. Leong, and J. Homer, "Blind multiuser receiver in Rayleigh fading channel," *Australian Communications Theory Workshop (AusCTW'05)*, Brisbane, pp. 145–150, 2005.

[72] E. B. T. Dennis, R. S. Tung, W. Y. Leong, and C. M. T. Joel, "Sleep disorder detection and identification," *Procedia engineering*, Elsevier, no. 41, pp. 289–295, 2012.

[73] W. Y. Leong, "Implementing blind source separation in signal processing and telecommunications, *Thesis*, University of Queensland, Australia, 2005.

[74] P. Mohankumar, and W. Y. Leong, "3D modelling with CT and MRI images of a scoliotic vertebrae," *Journal of Engineering Science and Technology EURECA*, pp. 188–198, 2015.

Chapter 5

Cyber security: trends and challenges toward Industry 4.0

Boon Tuan Tee[1] and Lim Soon Chong Johnson[2]

In the present day, the development industry is facing sophisticated demand amidst increasing competition. Fast-paced technological advancement has led to the emergence of the Industry Revolution 4.0. Integrating information technology and business operations presents several challenges, of which cyber security is of primary importance. Cyber security refers to aspects of technology that address the confidentiality, availability, and integrity of data in cyberspace and is considered to be associated with other security aspects, as depicted in Figure 5.1 [1]. Several industries regard cyber security as crucial since this aspect helps protect confidential information specific to people or systems against abuse, attacks, theft, and misuse in the digital space. Network connectivity is growing steadily; therefore, there is a chance of data being more prone to cyber-attacks, where the data may be abused for financial or strategic gain [2].

A majority of the organisations consider cyber security as a part of the technology domain. Even though organisations know the potential risks and how those risks might affect the business, the typical propensity is not to pinpoint security vulnerabilities [2]. Recently, many large businesses have fortified their cyber security infrastructure. Substantial capital is required to formulate new strategies to facilitate security-specific technological advancement in information technology (IT) to contain the risk and impact of cyber-attacks.

Cyber security is primarily considered by most organizations as a technology issue. Although they are alert about the risks and effects toward their business/operation, the organizations tend not to reveal any security vulnerabilities [2]. Most large companies have considerably strengthened their cyber security capabilities in recent years. Huge amount of investment is spent on developing new strategies with technological development in information technology (IT) security to reduce the risk of cyberattack [3]. According to *financesonline.com* [4], government sector has the highest spending on cyber security with 11.9% in 2019 as shown in Figure 5.2. This is followed by telecommunication company with 11.8% and industries with 11.0%.

[1] Fakulti Kejuruteraan Mekanikal, Universiti Teknikal Malaysia Melaka, Melaka, Malaysia
[2] Fakulti Pendidikan Teknikal & Vokasional, Universiti Tun Hussein Onn Malaysia, Johore, Malaysia

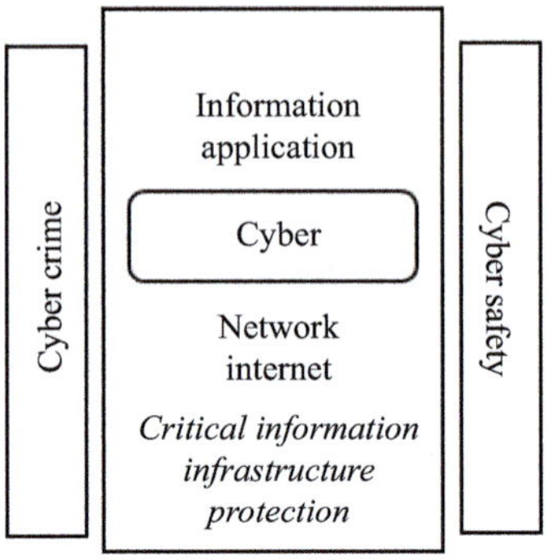

Figure 5.1 Cyber security domains

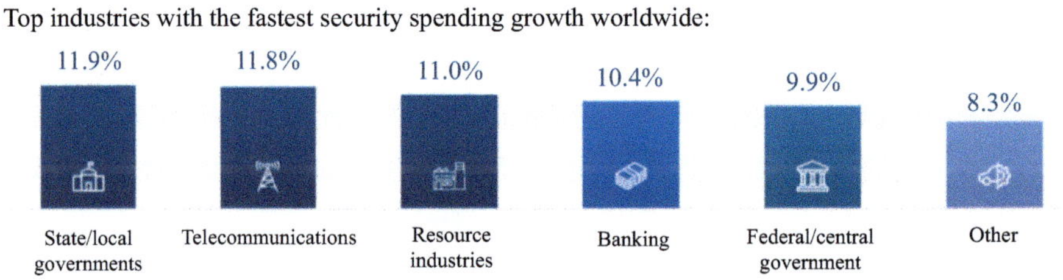

Figure 5.2 Worldwide spending on cyber security (Source: financesonline.com)

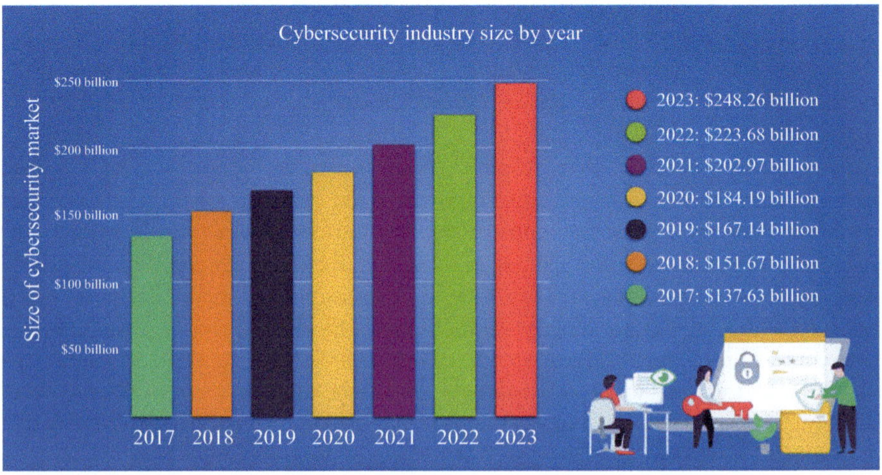

Figure 5.3 Cyber security market projection (Source: broadbandsearch.net)

Based on the analysis by *broadbandsearch.net* [4], the worldwide cyber security market is expected to reach USD167 billion in 2019. With the advanced digital era and the impact of cyberattacks, cyber security market is projected to be one of the lucrative industries for the coming years as shown in Figure 5.3.

5.1 Introduction

Industry 4.0 is a broad umbrella of technologies where automation, robotics, artificial intelligence, Internet-of-Things (IoT), data collection systems, sensors, sophisticated engineering, and precision computer control are extensively used [5]. Using one or more systems, industrial production can be transformed using digital technology for better efficiency and sustainability [5]. Nevertheless, digital transformation brings along significant exposure to cyber security risks. In the next section, there is a brief discussion about the prevalent challenges affecting security. Moreover, future challenges and technological development are discussed.

5.2 Recent trends

5.2.1 Web servers

A large fraction of the cyber-attacks target organisational web servers since these machines store and host sensitive data. Cyber criminals very frequently attack servers since data may be stolen or malicious code may be injected in the network. Typically, malicious code enters an organisation's server using legitimate web servers that the criminals have compromised. Data theft is a very significant issue since the media gets involved swiftly on cases specific to data theft. For these reasons, added attention must be paid to develop more secure tools and attack detection systems that can protect the servers and pre-empt potential attacks.

5.2.2 Cloud computing

The adoption of cloud computing has increased massively. More and more businesses are migrating to the cloud to provide services or for data storage. There is a clear trend towards the emergence of cloud computing. This latest trend poses another huge challenge for cyber security, as traffic can go around traditional points of inspection. Additionally, as the number of applications available in the cloud grows, policy controls for web applications and cloud services will also need to evolve in order to prevent the loss of valuable information. As cloud technology and services may provide immense opportunities, it is vital to successfully address the cyber threats that increasingly target the cloud environment.

5.2.3 Advanced persistent threat

Among the several types of attacks, an advanced persistent threat (APT) refers to an attack where one or more individuals form a long-term but illegitimate presence to extract sensitive information from that network [6]. The attackers must remain inconspicuous or undetectable for extended periods to attack systems using APT. While instant monetary gain is one aspect of the attack, APTs extend beyond the initial objectives and continue to compromise the systems for long [7]. Cybercriminals are employing increasingly vague and sophisticated techniques to compromise the networks; hence, security services must keep up or stay ahead of the attackers to detect present and potential future threats.

5.2.4 Smart mobile phones

Mobile technology has advanced immensely, and the smartphone is now a platform that provides a way to access information and services conveniently. Nevertheless, increased information availability in the smartphone ecosystem has led to this network becoming a target for cyber-attacks [8]. The risk of exposure to cyber threats is higher because the smartphone is used for personal tasks like social media browsing, email communication, and phone-based financial transactions. The increasing popularity of the smartphone has led to numerous attackers targeting this system. Due to the highly dynamic nature of technology, security in the mobile domain should also be enforced as a culture [8].

5.2.5 New internet protocol

Internet Protocol (IP) is the set of rules that permits devices to identify and connect to other devices. IPv6 is the latest version of the protocol and is the backbone for commercial networks and a large part of the internet. IPv6 succeeds the previous version IPv4 and will eventually replace IPv4 to become the default protocol at the network layer. IPv6 facilitates the end-to-end transfer of traffic across network interfaces. IPv6 adoption comes in the wake of an increased need for address space to connect the massive influx of devices. Furthermore, IPv6 offers advanced security, decreases the burden for routers, and offers provisions for enforcing Quality of Service (QoS) [9].

5.2.6 Code encryption

Encryption is a process employed for data security where information is transformed using a secret key to an unreadable format. Encryption is among the best ways to ensure data security [10]. Encryption requires an algorithm for data transformation, which is typically fed an encryption key that defines how the data would be transformed. Encryption essentially protects data integrity and privacy; however, highly complex encryption may create problems concerning cyber security. Data is also encrypted during transit from one end of the network to the other (e.g. internet, a wireless mic, cellular phones, intercom facilities). Encryption can help identify leakage [11].

5.2.7 Social engineering

Cybercriminals may employ social engineering to deceptively manipulate individuals and extract sensitive information. Social engineering is based more on hacking human psychology than technology. Technology-based attacks use malware to illegitimately gain access to data, systems, and entire buildings [12]. One form of social engineering attack could be an individual masquerading as a legitimate employee of a business (e.g. banks, clubs, e-commerce) and manipulate a user into providing sensitive details like password or credit card number over the phone. Similarly, attackers could use impersonation to facilitate entry into buildings secured by electronic entry systems. Such entry may be done through stolen passwords, access codes, etc.

5.2.8 Social media exploits

With regards to the earlier section, the increasing usage of social media would lead to personal cyber hazards. Social media espousal among organisations is increasing rapidly along with the risk of attack. With the radical progress in different platforms, business could expect to see a rise in social media profiles utilised as an avenue for social engineering strategies. For example, attackers can visit websites such as LinkedIn to detect all users which work at a firm and collect lot of personal information to conceive an attack. Of late, some social media portals like the Russia-based Yandex have been allowing normal users to scout for an individual's information, like social media account, just by uploading a photo of a person's face. To deal with such risks, firms would need to see beyond the fundamentals of policy and process development to more innovative technologies like data leakage avoidance, improved network monitoring and log file analysis. Security awareness training in workers is also an effectual means towards deterrence of social engineering.

5.2.9 Bad universal serial bus (USB) attack

Bad USB is also regarded as a malicious USB attack, in which a simple USB device (i.e. USB pen drives) can be converted into a keyboard by attackers to use these devices to type malicious commands and take control of victim's computer [13]. This is a new type of attack method that was first mentioned by security researcher Karsten Nohl at a Black Hat conference in 2014, and the attack coding was released in favour of public interest. The attack's basic principle was to apply reverse engineering for the firmware of USB microcontrollers, and thereby the firmware was reprogrammed to allow a USB device to imitate the functions of a keyboard. As this attack was conducted at the firmware level, storing of the malicious program can also be done in rewritable code at a basic level of USB input/output (I/O), in which even if the USB flash memory's whole storage content is deleted, it would not have cast any impact. This suggests that the malware is basically unpatchable. Thus, it is challenging to avoid such an attack due to the involvement of social engineering. For example, a USB device could be easily obtained at a throwaway price.

5.2.10 Air-gapped system attack

In the context of cyber security, an air gap could signify a network security measure that enables segregating a network or computer from any external Internet connection. Basically, since an air-gapped computer lacks network interfaces, only direct physical control is allowed. Computer systems are employed to safeguard different types of mission critical systems, e.g. military defence, stock market as well as systems used to control critical infrastructures like nuclear power plants [14]. If we ensure data backups are air-gapped, it makes it easy to perform system restoration when there is a system failure. Usually, a specified amount of space is set between the air-gapped computer system and outside walls, as well as between its wires and other technical equipment's wires in order to avoid unauthorised computer access or any potential intrusion via electromagnetic or electronic exploits.

Although air-gapped systems may seem like completely secured, but these systems also have flaws. As per a recent cyber security research, exploits can be implemented by employing specialised equipment in order to capture telecommunication signals that are emitted by air-gapped computer, like keystrokes or screen images, from that of demodulated electromagnetic radiation (EMR) waves [15]. As per a different research study, exploitation of isolated computers can also be done via infected sound card that emits EMR waves [15]. Information can also be gathered from isolated keyboards, screens and other computer components by employing continuous EMR wave irradiation. Social engineering can also be used to attack such a system. The Stuxnet worm, a malware targeting a specific industrial control system, is an infamous example, which is believed to have been brought via infected USB thumb drives from employees or distributed as free giveaways.

5.3 Cyber security solution technologies

Establishing cyber security could help avoid data breaches, cyber-attacks and identity theft as well as help in risk management. Over the years, development of different states of art technologies has been carried out in an effort to minimise the risk from cyber-attacks, while few of these have been described in the following sections.

5.3.1 Vulnerability scanners

A vulnerability scanner can be defined as a security tool that is employed to evaluate the security of an application, server or network. Basically, a vulnerability scanner can be perceived as a tool that can also think as a hacker would do. It evaluates the target in a similar manner as a hacker, i.e. identifying weaknesses or vulnerabilities [16]. Web application scanners can be defined as automated tools that are the first one to crawl a web application, and then scan its web pages to seek vulnerabilities that could exist in the application by employing a passive technique. Here, probe inputs are generated by scanners, after which the response is checked against such input to identify security vulnerabilities [17].

5.3.2 Intrusion prevention system

An intrusion prevention system (IPS) can be a software or hardware device that possesses all the features of an intrusion detection system (IDS). In some cases, IPS also allows preventing possible incidents. An IPS can react to a detected threat in many ways [18]:

1. to block future attacks, it allows reconfiguring other security controls in systems like firewall or router;
2. in network traffic, malicious content can be removed from an attack in order to filter out the threatening packets; or
3. other security and privacy controls in browser settings can also be reconfigured to avoid future attacks.

5.3.3 Intrusion detection system

A specialised tool called Intrusion Detection System (IDS) possesses the ability to read and interpret the contents pertaining to log files from servers, firewalls, routers as well as other network devices. Also, the IDS has the ability to maintain a database containing the information about known attack signatures. This allows comparison of different patterns of traffic, activity or behaviour and also quickly checking the logs for monitoring against such signatures in order to identify a close match between a signature and the recent or current behaviour. When a match is established, the system can issue alerts or alarms, triggering different types of automatic action like shutdown of Internet links or specific servers as well as launch back traces. It can also make other active attempts to recognise attackers as well as actively gather evidence pertaining to their nefarious activities [19].

5.4 Challenges

Cyber security problems can bring about many challenges to stakeholders like business firms, companies and end-users. According to us, challenges related to Industry 4.0 can be described as three main aspects: cyber security risk management, user privacy and digital forensics. The following sections will elucidate every challenge exhaustively.

5.4.1 User privacy

In the context of Industry 4.0 application, one of the tenets is offering customer-focused, highly customised quality services that address a customer's specific needs. This normally suggests that usage data collection is prevalent, in which it could act as a prerequisite before attempting customisation. A typical common example is installing an app in mobile phones, in which users need to agree, up to some extent, with permissions in order to use phone's certain features for data collection - for example, camera data access, microphone access, facial features data, fingerprint data access or even system file access, all of which allow the app to perform its best.

While data collected via such means are chiefly aimed to bring about better, personalised experiences, users may not always know about what the app could actually do. Normally, uploading of these personalised data is done to a central server, in which the company providing the app would also carry out certain analytics, by taking into account user profile information, crash reports, usage statistics, etc. In this context, users may have granted the access of personal data unknowingly. Even if the user is knowledgeable and understand the potential risk, they would have given the access unwillingly in order to make full use of the app.

Such a practice certainly poses risk to user privacy, in terms of multiple types of user data theft as well as digital surveillance. For the former, data theft is possible because of web server attacks. Usually, cyber criminals may steal valuable personal information, i.e. personal identification credentials, social security numbers, credit card or payment information and addresses, from web servers that have been compromised. Post execution of a successful data theft operation, such gathered information could either be leaked in the public domain or sold to interested parties for different reasons such as web marketing or scams.

The said context also presents a trade-off between preserving the user privacy and providing highly customised, quality customer services. Businesses need data to perform customer analytics and customise themselves to provide better quality services. However, customers also increasingly face user privacy concerns, which has resulted in a constant issue for business implementation. After numerous data leaks reported by famous web services like Yahoo, Facebook and Marriot International [20], customers these days have become more aware regarding the potential pitfalls whenever they share personal data in the public domain. This has posed constraints for business and firms in deciding how excessive personal data collection can be avoided, while also maintaining their service quality. From the point of view of users, they are also worried about how to detect such a risk and how they should respond when faced by such a risk.

There are some means of addressing this issue. First, education needs to be imparted for users regarding the risk it carries when sharing personal information. These days, service companies can easily access customers' digital record, e.g. likes and shares, posting, purchase record and location, through applications like online shopping websites and social media. Since users usually have to give consent to use their personal information in order to run these application, they need to be vigilant regarding what they share online. Since these digital records can be stored in data silos permanently, they need to realise that what they are about to share can also be viewed by anyone and the information would stay permanent on the Internet. They also need to be educated on how they need to stay vigilant before installing any application on their digital

Second, service providers need to comply with certain standards pertaining to privacy policy implementation. For example, web services use cookies, which allow them to store, track and share user behaviour and is also a potential privacy risk. Companies need to establish a clear policy regarding cookie usage in order to avoid excessive leak of user-related information. For example, they need to comply with the latest standards being followed, like European's General Data Protection Regulation (GDPR) [21], which has made an attempt to set the best practice to be followed to improve privacy rights as well as enforce consumer protection. Even though regulations such as these are regarded to be troublesome for implementation and could even alter the way companies conduct business, it could also be viewed as a form of reassurance pertaining to customers in order to ensure that their privacy is prioritised as well as protected during service delivery. This also helps to boost the confidence of customers and keep them loyal, as companies keep information transparent regarding how their personal data are being used, as well as provide explanation regarding the design and practice that allow managing customer data that are revealed in the entire service life cycle.

5.4.2 Cyber security risk management

Cyber security is generally regarded as a highly specialised field, and thus smaller companies and individuals tend to think that it is challenging to attain the necessary knowledge and skills to protect them from cyber-attack. Often, smaller organisations have limited resources, either financial or human, to handle cyber security. The company's senior management often does not fully understand the need/priority and tends to overlook the risks concerning cyber security. As a matter of fact, these companies often outsource their IT services to third party companies as they think they are more capable and this has now become a trend. With this regard, cyber security issues concerns are overlooked, as they tend to think that this has mitigated their risks.

The hard truth is that companies, generally small or medium enterprise (SME) that tend to outsource their IT services, make deal with IT services companies that are in turn SME or start-up organisations with limited experience in good cyber security practices [1]. For example, as there are resource constraints, the onus of handling cyber security related responsibilities falls into the IT personnel as part of duties and tasks. In the view of best cyber security practice, such a delegation needs to be discouraged, while making an attempt to segregate duties. If such a requirement is not complied with, it could result in serious implications.

For example, in 2012, an infamous case occurred in Malaysia, i.e. hacking of cloud hosting services exabytes.my, wherein numerous virtual machines (for cloud services) got deleted [22]. An employee of the organisation performed the attack, who had already been terminated but the system login credentials had not been changed and still remained valid. Because of this incident, the company suffered irreparable damages, wherein just ~80% of customer data could be recovered from old backups, while the remaining of their client's data was lost permanently. Such

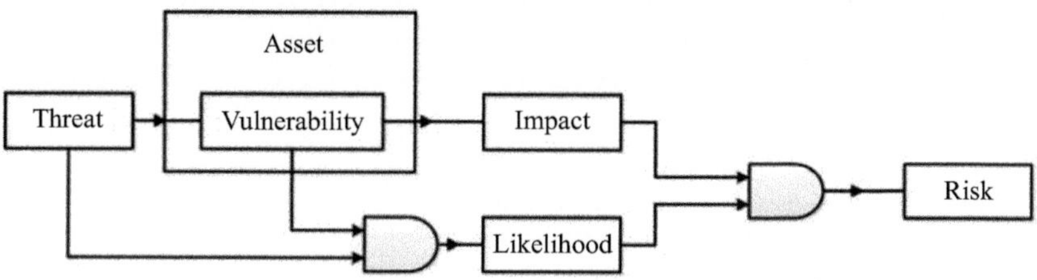

Figure 5.4 A general view of the risk environment [1]

an incident could have been avoided easily if the company had complied more stringently with the cyber security standards as well as best practices.

In an age wherein business companies are seeking to transform their services digitally, risk management and cyber security awareness have become more important issue to be handled. A general view pertaining to risk environment by Sutton is shown in Figure 5.4, which also provides a nice summary regarding the basic elements in a risky environment. Based on the figure, the threat acts on an asset's vulnerability, which results in an impact. When a motivation of attack exists, along with vulnerability, it results in the chances of the threat to be carried out. After this, impacts and likelihood get combined to yield risk. The risk assessment procedure usually encompasses five steps, as shown in [1]: context establishment, risk identification, risk scrutiny, risk assessment and risk treatment. It includes practical steps pertaining to the entire process flow of risk management in the cyber security arena, which also helps organisations to protect their data assets.

5.4.3 Digital forensics

Digital forensics can be defined as a branch of forensic science that deals with the process of identification, preservation, extraction as well as documentation of computer evidence, which the court of law can use for cases [23]. In terms of cyber security, it is deemed as an evidence retrieval process that extracts from digital media like mobile phone, computer, network or a web server, as well as the verification process that would collect evidence post a cyber-attack. With growing number of Industry 4.0 applications anticipated in the near future, this has become a significant area that has gained much attention in recent years. The cyber security practice is complemented by digital forensics in order to safeguard safety and integrity of computing systems and operations, as it allows processing, extracting and interpreting factual evidence through compromised computing systems in order to ascertain cybercrime actions in the court of law.

In general, digital forensics entails five essential steps of (1) identification, (2) preservation, (3) analysis, (4) documentation and (5) presentation [24]. In the first step, the aim of the investigation and the resource needed for forensics task are identified, after which the retrieved data from digital media are isolated, preserved and secured. In the third step, the retrieved data are analysed, which involves employing appropriate tools and techniques as well as interpretation of data analysis results. Next, based on analysed results, crime scene is documented as well as sketching, photographing and crime-scene mapping are performed. In the final step, explanation of summarised findings as well as conclusion is presented.

These days, the digital forensics field has become more complicated and difficult because of the advent of cheap computing devices with Internet access (e.g. IoT devices), rapidly changing computing technology applications each year and easy availability of hacking tools [24]. All of these contribute to impacting the field's best practice as forensics professionals need to frequently update their knowledge and skills to keep up with the more sophisticated cyber exploits. Since companies have recently started to implement cyber security risk management, digital forensics is considered as a new aspect that companies can think of to become sustainable with the advent of the era of Industry 4.0, as well as to make sure that business operations always remain uninterrupted.

5.5 Conclusions

Industry 4.0 has brought about an innovative era that is well connected, has responsive supply networks and allows smart manufacturing, all of which improve production process and services. Even when digital capabilities have been enhanced significantly throughout the supply chain and manufacturing processes, it has also increased the risk of cyber threats and the industry may not have been fully prepared. As the expansion of threat vectors has rather been radical with the advent of Industry 4.0, cyber security should be made a significant part of the strategy, operation and design, and taken into account from the start of any new connected device/application with the Industry 4.0 driven initiative.

References

[1] D. Sutton, "Cyber Security: A Practitioner's Guide", BCS Learning & Development Ltd, 2017.

[2] 101 Impressive Cybersecurity Statistics: 2020 "Data & Market Analysis", [Online] URL: https://financesonline.com/cybersecurity-statistics/#cyberattacks

[3] "50 Essential Cyber Security Facts for Business Owners", [Online] URL: https://www.broadbandsearch.net/blog/business-cyber-security-facts-statistics

[4] B.C. Ervural & B. Ervural, "Overview of Cyber Security in the Industry 4.0 Era", Industry 4.0: Managing The Digital Transformation, Springer Series in Advanced Manufacturing, Springer International Publishing, 2018.

[5] N. Benias & A. P. Markopoulos, "A Review on the Readiness Level and Cyber-Security Challenges in Industry 4.0", 2017 South Eastern European Design Automation, Computer Engineering, Computer Networks and Social Media Conference (SEEDA-CECNSM), Kastoria, pp. 1–5, 2017.

[6] Advanced Persistent Threat (APT), [Online] URL: https://www.imperva.com/learn/application-security/apt-advanced-persistent-threat/

[7] I. Ghafir & V. Prenosil, "Advanced Persistent Threat Attack Detection: An Overview", International Journal Of Advances In Computer Networks And Its Security, Vol. 4, Issue. 4, pp. 50–54, 2014.

[8] J. Wright, M. Dawson, Maurice & Omar, Marwan. (2012). "Cyber Security and Mobile Threats: The Need for Antivirus Applications for Smart Phones", Journal of Information Systems Technology and Planning. Vol. 5., pp. 40–60.

[9] S. Hermann & B. Fabian, Benjamin. "A Comparison of Internet Protocol (IPv6) Security Guidelines", Future Internet. Vol. 6., pp. 1–60. 10.3390/fi6010001, 2014.

[10] Lantronix, "Encryption and Device Networking", [Online] URL: https://www.lantronix.com/wp-content/uploads/pdf/Encryption-and-Device-Networking_WP.pdf
[11] G.K. Reddy & U. Reddy, "A Study of Cyber Security Challenges and Its Emerging Trends on Latest Technologies", Published in ArXiv, Computer Science.
[12] J. Fruhlinger, "Social Engineering Explained: How Criminals Exploit Human Behavior", [Online] Date: Sept 25th, 2019. URL: https://www.csoonline.com/article/2124681/what-is-social-engineering.html.
[13] S. Shroeder, " BadUSB Can Turn Thumb Drives Into Cyberweapons", [Online] Date: Oct 03rd, 2014. URL: https://mashable.com/2014/10/03/bad-usb/
[14] R. Pompon, "Attacking Air-Gap-Segregated Computers", [Online] Date: Sept 05th, 2018. URL: https://www.f5.com/labs/articles/cisotociso/attacking-air-gap-segregated-computers
[15] M. Rouse, "Air Gapping (Air Gap Attack)", [Online] URL: https://whatis.techtarget.com/definition/air-gapping
[16] K.C. Bourne, Application Administrators Handbook, Elsevier, 2014.
[17] A. Razzaq, A. Hur, H. F. Ahmad, & M. Masood, "Cyber Security: Threats, Reasons, Challenges, Methodologies and State of the Art Solutions for Industrial Applications", 2013 IEEE Eleventh International Symposium on Autonomous Decentralized Systems (ISADS), Mexico City, Mexico, pp. 1–6, 2013.
[18] A. Patel, Q. Qassim, & C. Wills, "A Survey of Intrusion Detection and Prevention Systems", Information Management & Computer Security, Vol. 18, No. 4, pp. 277–290, 2010.
[19] A. Carasik-Henmi, T.W. Shinder, C. Amon, R.J. Shimonski, & D.L. Shinder, The Best Damn Firewall Book Period, Syngress, 2003.
[20] D. Swinhoe, "The 14 Biggest Data Breaches of the 21st Century", [Online] Date: March 20th, 2020. URL: https://www.csoonline.com/article/2130877/the-biggest-data-breaches-of-the-21st-century.html
[21] European Commission, "Data Protection: Rules for the Protection of Personal Data Inside and Outside the EU", [Online] URL: https://ec.europa.eu/info/law/law-topic/data-protection_en
[22] C.F. Heoh, "Cloud Hosting Hacked – Customer Data Lost", [Online]. Date: Oct 8th, 2012. URL: http://storagegaga.com/cloud-hosting-hacked-customer-data-lost/
[23] D. Tobok, "What Is Digital Forensics?", [Online] Date: Feb 15th, 2018. URL: https://cytelligence.com/resource/what-is-digital-forensics/
[24] Guru99, "What Is Digital Forensics? History, Process, Types, Challenges", [Online] URL: https://www.guru99.com/digital-forensics.html

Chapter 6

The role of IIoT in smart Industries 4.0

Mohsen Marjani[1], *Noor Zaman Jhanjhi*[1], *Ibrahim Abaker Targio Hashem*[2] *and Mohammad T. Hajibeigy*[1]

Economic growth is the backbone of any country, which is mainly linked with industrial power, production and efficiency. Industries are changing from old fashioned to new technological perspectives with new era requirements. Industries equipped fully with technology are known as smart industries. The Industrial Internet of Things (IIoT) is the source that makes regular industries into smart industries by providing them cost cutting, remote access, production management, supply chain and monitoring, as well as reducing energy consumption cost, etc.

IIoT brings the revolution in the industry to enhance its production cheaply and allow the industry to access and share the data remotely. Smart industry, or Industry 4.0 (I4.0), is an emerging paradigm that has brought a revolution in the overall manufacturing and production industry. Currently, a great number of industries are converted into smart industries, while the rest is in the phase of converging. This book chapter will elaborate in detail the existing state of the art of IIoT, different IIoT applications, use cases, open research issues and challenges as well as future directions and aspects of the smart industry. In addition, this chapter will elaborate on the strong link between the Internet of Things (IoT), IIoT and smart industries.

6.1 Introduction

The IoT plays a vital role in several application domains. It attracts very high interest due to its ease in configuration and deployment as well as in providing greater access to the different applications of life. IoT is further expanded to provide several solutions to the industry, which are known as Industrial IoT (IIoT). This section of the chapter will briefly introduce IoT in different aspects such as what IoT exactly means, the journey of IoT towards IIoT to I4.0, which is also known as industrial evolution, and finally an explanation of the differences between IoT and IIoT.

[1]School of Computer Science, SCE Taylor's University Subang Jaya, Selangor, Malaysia
[2]College of Computing and Informatics, Department of Computer Science, University of Sharjah, Sharjah, UAE

6.1.1 Internet of Things means what?

The concept of living the smart and easy life, where most of the routine work can be completed or assisted smartly in a timely manner, which were not possible before. For example, it used to be a dream to know about your home or office and control the different appliances of your home or office remotely. This all became possible with the concept of the smart home, where one can control and monitor all desired items remotely from anywhere in the world. IoT is what powers the smart home to make it smart. At the same time, the dream of monitoring and controlling the environment out of the home and office, such as news about route information, accident information on the route, connections to the car's sound system, showing the calculated road maps, etc. This all became possible based on IoT. In the same way, IoT is used in most of our daily life applications [1]. One can get a very clear idea of the influence of IoT applications used daily [2] in life by referring to the applications [1] in Figure 6.1.

6.1.2 Journey of IoT to IIoT to Industry 4.0

The term IoT was introduced by Kevin Ashton in 1999. His initial idea was how to connect the physical objects to the network or popularly known as the internet. That idea brought the revolution, and the world of computing was entirely changed with the integration of the different sensors with different objects. The objects started communicating with the internet by sensing and passing the information to the networks. This revolution does not require more programming skills and other formal computing, which is normally required for programming. The communicating objects are known later as the smart object and then the concept moved towards the smart home concept, where most of the objects of the home can easily

Figure 6.1 Internet of Things applications [2]

communicate and connect with the internet. Later, the expansion of this idea boomed with GPS, smart cars, fitness monitoring and e-health applications; thus IoT has evolved to lead the concept of the next industrial revolution, which is known as IIoT. IIoT can further be understood in a better way by referring to Figure 6.2.

6.1.3 Difference between IoT and IIoT?

IoT is the source of the smart environment as briefed in earlier sections, which enables the objects to communicate with other objects with the help of different types of sensors attached to it. Those attached sensors enable the object to pass the information to other components as well as to share the data on the internet. The real difference among IoT and IIoT is the operational area, or the scope of the application; once IoT is enabled for the industry on the large scale and address the industry issues, such as manufacturing, supply-chain management, production, etc., then it will be referred to as IIoT. IIoT brings wonders to the industry, where the industry can be monitored and controlled remotely on a wider scale. This increased the progress of the industry on a large scale in all sectors. However, currently, IIoT is not fully implemented but has increased in scope. Most countries have already implemented IIoT and make their industries as smart I4.0, while the others are speedily following to match smart industry goals. The emergence of IoT can be imagined by scientists, who are expecting that by 2020; IoT devices will be more than 50 billion around the globe, which will be almost seven times higher than the population of the world (Figure 6.3).

6.2 Overview of IoT in smart Industrial 4.0

IoT plays a vital role in several application domains. It attracts very high interest due to its ease in configuration and deployment as well as vast in providing greater access to the different applications of life. IoT further expanded to provide several solutions to the industry, which are known as Industrial IoT (IIoT). This section of

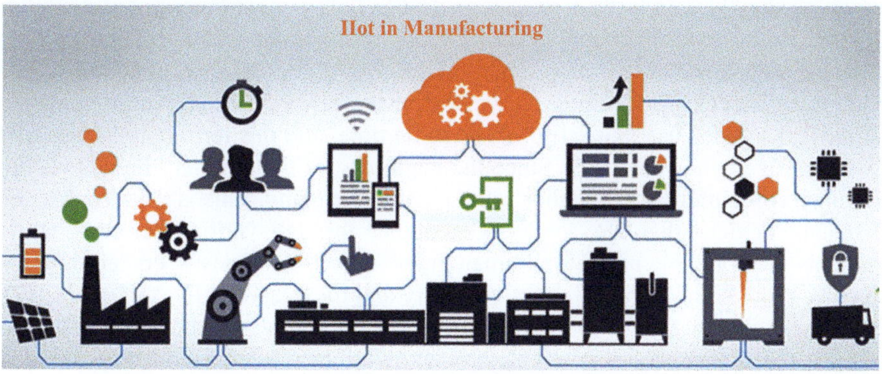

Figure 6.2 Industrial Internet of Things [3]

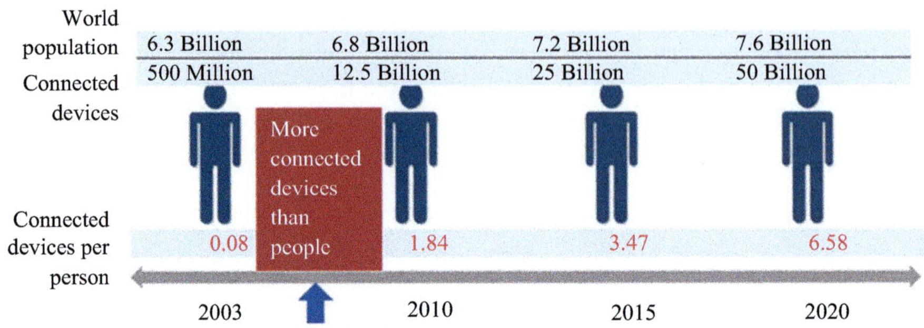

Figure 6.3 Internet of Things growth rate [1]

the chapter will briefly introduce IoT in its different aspects such as what IoT exactly means, the journey of IoT towards IIoT to I4.0, which is also known as industrial evolutions, and finally the difference among IoT and IIoT for the better understanding of the readers.

In the early nineteenth century, English workers, called luddites, were opposed to technological change and attacked factories and demolished machinery as a measure of protest. This act not only did not stop technological advancement but also accelerated the trend.

Since the eighteenth century, which was the beginning of the invention of the mechanical tools that enabled thermal and kinetic energy, the advancement of the industrial segment has reshaped our lives. The ensuing inventions were part of the first industrial revolution. Later, inventions through the mid-nineteenth century, notably the development of the electrical technological production system, comprised the second industrial revolution, which reshaped the entire industry by enabling mass production [4].

The utilisation of the Programmable Logical Controller (PLC) in the year 1969 enabled humans to work together between information technology and electronics, accelerating a rise in industrial automation that continues today. This evolution is known as the third industrial revolution [5,6]. Today, manufacturing companies confront innovative challenges such as cropped innovation and technology life-cycle and a demand for custom products at the cost of large-scale production. These requirements were good recipes to give birth to the new industrial revolution named IoT. IoT as an idea has existed for quite some time, and this terminology was first used as early as the year 1999. Since the beginning of the internet, thoughts about machines communicating together were a vision for many. This is the new process of having devices that are connected together through various communication protocols without the need for human intervention. In today's era automation becomes more commonplace in both industry and the everyday lives of people [7]. A large portion of this is the ability of cloud computing to allow for vast data storage, the miniaturisation of electronics with less power consumption and Internet Protocol (IP) version 6, allowing for enough addresses to register every conceivable device one would want to connect to the internet [8]. IoT has expanded into the manufacturing world, which results in a manufacturing revolution that is called I4.0. This has spurred a lot of industrial automation, which is generally referred to as IIoT. The industrial revolution started in Britain within the late 1700s when manufacturing was generally done in people's homes using basic machines and hand tools, and continued on to an existing time in the platform of IIoT. A short history of the industrial revolution helps us better realise how the industry has unfolded in the past three centuries, which continues to advance into the generations to come. The first industrial revolution occurred in the late eighteenth century, and mechanical production was the main focus. At the beginning of the twentieth century, the industrial revolution 2.0 emerged and the era of electronics has been introduced with the ability to have mass production. Less than 50 years ago, the third industrial revolution began with the introduction of the PLC that enabled IT-based manufacturing automation. Recently, I4.0 was merged with a concentration on IoT and cyber-physical system (CPS) paradigms to analyse, monitor and automate business processes at large [9]. Figure 6.4 shows a brief overview of the Industrial Revolution from 1.0 to 4.0. These technological revolutions will remodel the assembly and supply processes into sensible factory environments, which will increase productivity and potency.

In this realm, businesses are able to automate processes for safety, speed or decrease the cost of connecting machines and processes that before were not capable of communicating. IIoT has a market forecast approaching $100 billion by the year 2020, so it has everyone's attention right now [7].

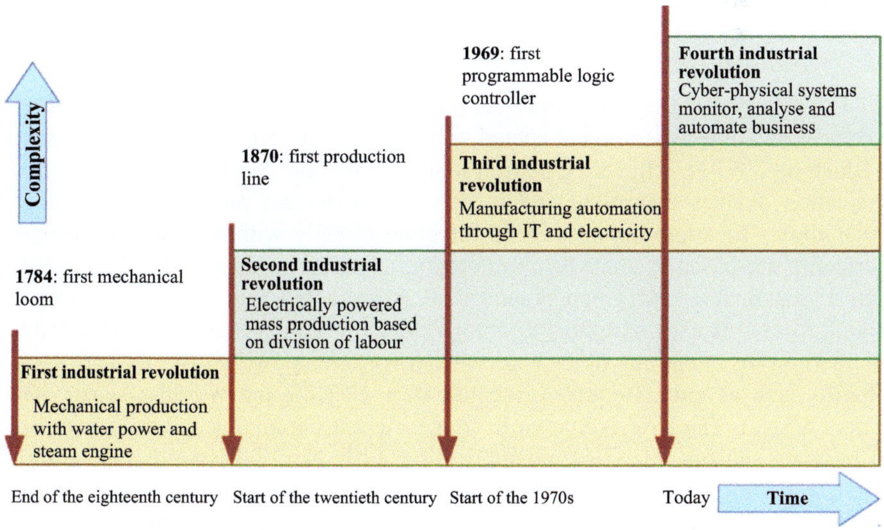

Figure 6.4 The four industrial revolutions result of the smart factory of the future and cyber-physical production systems [9]

The growth of low-cost and high-speed sensors has aided businesses to automate processes more efficiently and with minimal cost. In the past, several workers were needed to visually track an object and perform time-consuming inspections. Now, machine vision allows for real-time inspections with the ability to separate materials that are deemed defective. All of this is happening without human intervention. Some places have even gone as far as having the movement of material by autonomous guided vehicles without concerns for workers' safety. The future vision of a manufacturing plant with a minimal human workforce is growing as robots and other automation techniques are developed and deployed into such workplaces. The ability to gather all of this data in real time and have machines automatically make decisions is the true spirit behind what is driving the growth of IIoT. In this scope, the production system should be reliable, available and optimised for application safety. The IIoT is an industrial machine which is associated with the enterprise cloud storage area for both data retrieval and data storage. IIoT has the potential for a big change in the management of the global supply chain. The Lean Six Sigma approach in the global supply chain employing IIoT and I4.0 creates an ideal process flow that is perfect and highly optimised, which minimises defects and wastage.

In summary, the intelligent production technology, human, machine and product itself are joining forces in an independent and intelligent network focusing on the development of a component, system and solution which addresses industrial base infrastructure, automation solution, intelligent controller and functional safety against cyber-attack. This manufacturing revolution shifts economics, increases productivity, fosters industrial growth and modifies the profile of the workforce, which ultimately changes the competitiveness of companies and industries in any corner of the world.

6.2.1 State-of-the-art industrial Internet of Things

IIoT has emerged as the new trend over the past few years in which the continuum of devices and objects is interconnected with various communication solutions such as Bluetooth, Wi-Fi, ZigBee, global systems for mobile (GSM), etc. These communication devices controlled and sensed all the devices and objects remotely, which allows for more opportunity to integrate directly with the external devices over computer-based systems to improve the industrial production such as quality control, sustainable and green practices. It is predicted that billions of devices ranging from sensors, smartphones, laptops and game consoles will be connected to the internet via numerous diverse access networks empowered by network technologies such as radio-frequency identification (RFID) and wireless sensor networks (WSNs). The IIoT is known in four different paradigms: Internet-oriented, smart assets, sensors, analytics and applications [10]. Normally the implementation of IoT technology can be understood very close to modern society, where intelligent assets and things are integrated virtually to form systems that could offer monitoring, gather, exchange and analyse data, delivering valuable insights that enable industrial companies to make smarter business decisions faster [11]. Various adaptations of wireless network technologies have placed IIoT as the next revolutionary technology by utilising the benefits of internet technology [12].

Meanwhile, WSNs include a large number of devices with sensing capabilities, wireless connectivity and limited processing capability, which allow nodes to be deployed close to the environment [13]. WSN supports miniature low-cost and low-power devices that connect from hundreds to thousands to one or more sensors. The sensor has a radio transceiver to send and receive signals, a microcontroller, an electronic circuit for sensor interfacing and an energy source, usually a battery or an embedded form of energy harvesting. The significant mechanism about connecting devices and objects is to be a backbone of omnipresent computing, enabling intelligent environments to recognise and identify objects and retrieve information. In addition, technologies such as sensor networks and RFID will address these emerging challenges that focus on the integration of information and communication systems [14].

A new framework named Health IIoT monitoring is introduced [15], in which electrocardiogram (ECG) and other healthcare data are collected through mobile devices and sensors and transmitted securely to the cloud by healthcare professionals for seamless access. Healthcare professionals will use signal enhancement, watermarking and other related analytics to prevent identity theft or clinical error. Both experimental evaluation and simulation validated the suitability of this approach by deploying an IoT-driven ECG-based health monitoring service in the cloud.

A new clock synchronisation architecture of a network for the IoT is offered [16], which is based on organisation, adaptation and region levels. The adaptation level architecture is to provide a solution for the internet adaptability of things; the organisational architecture is to organise and manage the clock synchronisation system and the region-level architecture is to ensure clock synchronisation. The goal is to provide reliable clock synchronisation security and accuracy.

A novel meta-model-based approach is proposed to integrate architecture objects on the IoT [17]. The idea is to feed into a holistic digital enterprise architecture environment semi-automatically. The main objective is to provide adequate decision support with development assessment systems and the IT environment for complex business and architecture management. Thus, architectural decisions for IoT are closely linked to code implementation to enable users to understand how enterprise architecture management is integrated with the IoT. However, there is a need for more work to ensure conceptual work to federate IoT enterprise architecture by extending federation model architecture and approaches to data transformation.

A peer-to-peer scalable and self-configuring architecture for a large IoT network is suggested by Cirani *et al.* [18]. The goal is to provide mechanisms for automated service and resource discovery that do not require human intervention to configure them. The solution is based on the discovery of local and global services that enable successful interaction and mutual independence. The important thing about this solution is that through the experiment conducted in the real-world deployment, the configuration shows. However, the possibility that the error could occur is the main factor in terms of IoT architecture being a more reliable solution.

In [19], the IoT study enabled the smart home scenario to analyse the requirements of the IoT application. As such, the CloudThings architecture is proposed to accommodate CloudThings PaaS, SaaS and IaaS to accelerate IoT applications based on the cloud-based IoT platform. The cloud integration in IoT provides a viable approach to facilitate the development of Things applications. The fundamental developments in approaching the architecture of CloudThings are based on pervasive IoT applications running and composing. However, the system requires wireless network communications that are embedded in the IoT to deal with heterogeneity.

In [20], the IoT based on video streaming is examined, which builds on top of content-centric networking architecture. The experiment has been conducted based on the dynamic adaptive streaming over HTTP mechanism to split video pieces and multiple bit-rates. The divided videos are then streamed by the dynamic adaptive streaming client from a typical HTTP server via centric networking. The result shows that the centric networking can fit into the IoT environment and that its integrated cashing mechanisms are reduced. Nonetheless, additional experiments are required to evaluate different networks related and compare with the existing one.

Modular and scalable virtualisation-based architecture is proposed in [21]. The modularity provided by the proposed architecture, in conjunction with Docker's lightweight virtualisation orchestration, simplifies management and allows distributed deployments. The features of availability and fault tolerance are ensured by distributing the application logic across different devices where there is no effect on system performance from a single microservice or even device failure.

6.2.2 Applications of IIoT for smart Industries 4.0

I4.0 is a new digital industrial technology that enables faster, more flexible and efficient process to produce higher quality goods at reduced costs and makes it possible to gather and analyse data across machines.

Applications of IoT involve connecting people to the devices they have had for the past years but providing them with information that was not easily accessible before. In the new industry developments taking place that affect manufacturing, the way industries were organised and anything else around us, such as smart house, smart grid, smart building, etc. with connectivity of all kinds of things driven by the presence of wireless communicating by computational power, which embed it into the new smart and flexible products. These directly impact how the products are actually manufactured, which manufacturing process can leverage these technologies and concepts that are driven by IoT.

Fully embedded within the IoT and CPSs facilitated manufacturing, unrestrained paradigms just like the industrial plant of the longer term and I4.0 envision data-intensive business intelligence environments which sensible customised merchandise is created through smarter processes and procedures [22]. I4.0 is established on the logic of CPS in which there will be arbitrary, self-optimised and self-sufficiently operating systems that communicate with one another and ultimately optimise production as a whole. Technological innovations in miniaturisation along with wireless communication have empowered mobile devices, sensors and applications to communicate with each other and continuously be in ourselves and each other's interaction range. Hi-tech context-aware applications and services have been built on top of these actuators and wireless sensors [23].

Nowadays manufacturing companies are confronted with adaptive burdens created by swift technological changes, i.e., volatile demand, highly customised products and short product lifecycles [24]. The manufacturing companies are able to stand the future market challenges, they must be able to produce small group sizes of a product or a single item in a timely and cost-effective manner, adequate functionality, connectivity and flexibility with customers and suppliers to meet these requirements [25]. To meet these existing challenges, the present systems are required to be more complicated, which will be more difficult to control and monitor. Therefore, traditional manufacturing systems with devoted production lines going to be impotent to compete in the future high technological era. Therefore, the industry needs a radical change to adapt and addresses existing challenges. The answer to these challenges is I4.0. The basic principle of I4.0 is to use the recent emerging information technologies such as IoT, Internet of Services (IoS), CPSs, big data, Semantic Web, cloud computing and virtualisation so that engineering and business process are deeply integrated making production operate in an efficient, modifiable, environmentally friendly, low cost and yet high quality [26]. The vision of I4.0 is to take an internet to the lowest point which every single actuator and sensor is a contributor in the IoT. In this environment, every device has its own unique IP address so that the automated system and products could communicate with each other efficiently. The IIoT is an industrial machine which is associated with the enterprise cloud storage area for both data retrieval and data storage with the potential for big changes in the management of the global supply chain. The Lean Six Sigma approach in the global supply chain employing IIoT and I4.0 create an ideal process flow that is perfect and highly enhanced which minimises from defects and wastage. The application of the IIoT could be in many aspects of the infrastructure of the society and help to improve human life as shown next.

6.2.2.1 Smart factory

The enactment of the CPS technologies and IoT in manufacturing systems has tallied new capabilities, permitting the management of complex and flexible systems to fulfil rapid changes. With the advancement of I4.0 and also the emergence of the smart factory model, the standard philosophy of producing systems has been modified. The smart factory introduces changes to the factors and parts of conventional producing systems and incorporates these necessities of smart systems so it will contend within the future. It also carries modules which are intelligent and standardised that may connect and mix quite simply with every other module and carries a specific functionality to form an online network within the factory. An increasing quantity of analysis in each academe and industry is devoted to transitioning the conception of the good manufactory from theory to tangible product

[24]. One of the most important challenges is that the smart factory system should reach a high level of sharing and exchanging information among their merchandise, the infrastructure of their merchandise and processes, their control system and time period application [27]. Totally machine-controlled production in future, which can save cost, time, tools and processes, are outlined digitally throughout.

6.2.2.2 Requirements of a smart factory

Factories are needed to extend their smartness and automation on what is doable these days with existing technology and sensors.

6.2.2.3 Modularity

This refers to the planning of system elements. Modularity may be outlined because of the capability of system elements to be separated and combined simply and quickly [7].

6.2.2.4 Interoperability

This implies to each the flexibility to share technical data among system elements together with product and the flexibility to share business data between producing enterprises and customers. CPS enables affiliation over the IoT and also the IoS [28].

6.2.2.5 Decentralisation

System parts (material handling merchandise, modules, etc.) can create selections on their own, unsubordinated to an impact unit whose choices are going to be created autonomously [28].

6.2.2.6 Driverless cars

The autonomous automobile is one of the foremost futurist applications of IoT. These cars that will appear as such a product in the near future truly exist these days and are primarily under prototype or development stage. The cars do not have operators and are smart enough to take you to your desired location on their own. These cars are equipped with smart devices and high technologies such as gyroscopes, sensors, Internet, and cloud services. They use these facilities to sense immense chunks of data on traffic, pedestrians, conditions of the road such as potholes, speed breakers, corners, and sharp turns, and instantaneously process them. The gathered information is transferred to the controller which takes an interrelated driving decision. Machine learning and artificial intelligence are essential aspects of driverless cars.

6.2.2.7 Smart hotel rooms

Hotel owners deliver higher safety and security by replacing card-based door keys with smartphone apps which the occupier can enter the room by pointing out his/her smartphone to the room locks.

In addition, the lighting and heating systems are connected to the internet to enable a personalised experience to guests. With this feature, you will be able to set lighting and control the temperatures of the room based on your preferences. In addition, you will be able to sign up simply and obtain travel suggestions and a lot of information on offered hotel deals [29]. For hotels and energy management, this results in consistent solutions for efficient operation through to integration of production systems. This increases energy efficiency while reducing operating costs.

6.2.2.8 Automation and adaptive with maximum efficiency

Adaptive behaviour provides flexibility, handles any kind of disturbance for maximum efficiency and ultimately strengthens production. Instead of having a central controller as a conventional method, there is intelligent collaboration and distributed controllers which adapt independently to process equipment and devices. In this realm, complex processes and unexpected events are easier to be optimised and manage, which means producing batch sizes of one at the equal low cost of mass production. The intelligent and autonomous production system is economical over the long term.

6.2.2.9 Smarter manufacturing and production with machine learning and data mining

More programme and intelligence are being integrated into industrial production to produce systems to cut back the cost and improve the quality, potency and suppleness of production [22]. In the intelligent control system, the RFID tag-equipped workpiece passes through all production stations. The intelligent control system is able to identify at which production station the workpiece will be processed.

6.2.2.10 Optimising industry processes with cloud computing and big data

Although the IoT model and modern advances in machine to machine (M2M) communication enable concurrent monitoring of smart factories, resource consumption and the effective optimisation of business processes generally rely on fairly data-intensive processes for which the computational resources available on-site are not sufficient. In this perception, big data and cloud computing are preparing technologies for the I4.0 paradigm. It is not solely the place where the majority of industrial device data decisions and important information is ingested and analysed but it also offers the adaptability to scale on demand for diverse workloads that can automate and optimise business processes. It allows information analytics with inevitable performance, even with growing industrial wireless networks of interconnected things, leading to a cost-effective offer chain.

The overall application of the smart industry will be expanded to be communicative and secure, easy to install, prototyped virtually, easy to operate and resource efficient.

6.3 Use cases of IIoT for smart Industries 4.0

This section will describe several use cases related to IIoT for smart industries, with different aspects of applications.

6.3.1 Automotive navigation system

In order for GPS navigation to provide a direction to the driver, it requires an automotive navigation system which gives a position data to locate the car on a road in the map database. However, this action demands an internet connection to order to fully operate. Using personal IoT architecture allows the navigation system to connect to the internet. The idea is the remote-control procedure of car navigation uses a car navigation device, smartphone and remote-control server and client to easily have connectivity to the internet. This navigation device can be remotely controlled via smartphone, which does not require any wireless communication technology [30].

6.3.2 Supply-chain management and optimisation

NetObjex's IoT platform can provide easy and simplified access to real-time supply-chain information. This is done by tracking equipment, materials and products accurately as they move through the supply chain. NetObjex's I4.0 solutions give your organisation greater control, insight and visibility of data throughout the entire supply chain. Effective and accurate reporting helps you collect and feed critical product lifecycle management information, business resource planning and other systems. This helps track the flow of material, interdependencies and cycle times of production. Access to this data will help you to reduce inventory, predict problems and minimise capital needs.

6.3.3 Asset tracking and optimisation

I4.0 solutions from NetObjex via PiQube and the NetObjex Matrix Digital Twin Platform help manufacturers at every stage of the supply chain to become more efficient and profitable. This helps them to keep a close check of the opportunities associated with logistics for quality, inventory and optimisation. In addition, your organisation's employees with IoT in place can reap the benefits of better global visibility in their assets. Also, tasks related to standard asset management such as disposal, asset transfers, adjustments and reclassification can be centrally managed and streamlined in real-time.

6.3.4 Driving enterprise transformation

IBM is offering implementation assistance to customers with a huge impact in terms of increasing operating efficiency, revolutionising business models (BMs), improving operations in the industry and reflecting the experience of their customers. Daimler was challenged to handle some of the internal operations using IoT technologies. The company uses IBM services to launch car2go, an on-demand fleet of environmentally friendly smart cars that users can book from mobile apps. By using IoT architecture that includes sensors and wireless communication, the company is able to monitor vehicle performance, provide an accessible vehicle network and analyse data to increase the car's efficiency [17,31].

6.3.5 Connecting the form to the Cloud

John Deere Company, which produces distinct green and yellow farm equipment, has implemented IoT technologies, which change the way traditional agriculture was operated. The company uses IoT to connect its vehicles to a mobile application called JDLink. This new technology enables farmers and dealers to access, use and diagnose data for each machine remotely. The solution is based on the IoT technologies that include wireless data streaming of production data, monitoring and weather reporting in real-time. As a result, the value of their operation is enhanced through ensuring the equipment is operating in a reliable manner [32].

6.4 The current uses and limitations of IIoT in smart Industry 4.0

This section presents some of the existing uses and limitations of IIoT in smart Industry 4.0. These limitations can be related to infrastructure, networking and internet, energy concerns, human resources, economic environment, costs and software requirements. The following are some of the key limitations in the current uses of IIoT in smart Industry 4.0.

6.4.1 Connectivity

Most of the modern IIOT use a centralised model to offer connectivity to the various sensors and systems such as wearables, smart cars, smart homes and in the big scheme smart cities. The model might be efficient given the amount of power u using IIoT where M2M communications dominating the field to connect all these different objects [4]. However, as the number of devices increases to billions using network simultaneously may cause a significant bottleneck in IoT connectivity, efficiency and the overall performance [5].

6.4.2 Autonomous power

The main limitations related to IIoT are concerned with autonomous action powered by context. Thus, a certain level of automation with IoT is used as an object to form a connected and manage the systems. The technology use evolves from WSN to RFID and narrow-band IoT that provide the capabilities to sense and communicate over the internet. Over the next decade, the big limitation IoT has to overcome is getting the context and performing actions independently. A system that learns and adapts without having to be configured is known as smart and automated system [6].

6.4.3 IoT hardware

Many industries are adopting IoT to make systems smarter; however, cloud computing and data integrity still remain a challenge. Given the enormous amount of data collected from different sources, it is essential to properly underscore the importance of separating useful and actionable information from irrelevant data. It is critical to calibrate your IoT sensors on a regular basis, just as you would any other kind of electrical sensor. Many sensors are embedded in a variety of devices, including meters, current clamps, chart recorders, power monitors and more, and the data flow between all this hardware is difficult to synchronise without the assistance of a professional team [7].

6.4.4 Security

IoT devices are becoming pervasive and extending from the cyberworld to the physical world, creating new types of security issues and concerns that are more complex. The IoT was initially promoted as a hyper-secure network suitable for the storage and transmission of confidential data sets [8]. While it is true that the IoT is safer than the average internet or LAN connection, it was not exactly the bullet-proof shell that some users expected. Some of the most important security concerns involve the IoT and the cloud. A recent analysis predicts the takedown of just one cloud data centre a loss of up to $120 billion in economic impact [9].

6.5 Open research issues and challenges of smart Industries 4.0

According to a general belief, using IoT devices and systems in the industry increases productivity by changing normal equipment to smart and interconnected ones with higher efficiency and lower maintenance costs. The production process will be more agile and the generated IoT data offer much deeper insights. However, pros and cons of embedding IoT systems into the industry are still under investigation and the implementation challenges such as cultural resistance to technological change, cyber-attacks and other infrastructural issues that may occur are being analysed [33].

According to research by Thoben *et al.*, these research issues can be categorised into three groups: (1) methodological, (2) technological and (3) business-based. However, some of the identified issues can be categorised in more than one group. Hence, that research categorises them based on the most significant factors according to the author's knowledge [34]. Table 6.1 lists and categorises most of the research issues and challenges of smart Industries 4.0. Some of the challenges can be grouped in multiple categories, and the majority of the issues are technological as IoT devices, technologies and data are key components of the technological aspect of the smart Industries 4.0.

The IIoT sensors enable a manufacturing firm to collect all possible sorts of data from all physical and virtual objects involved in the production process. For instance, the temperatures of the devices and speed of the machines can be easily recorded using IIoT sensors. These data are collected to be used in data analytics in real-time [35]. This possibility opens an interesting window of industrial and business opportunities to improve the operational efficiency and to enhance decision makers' power by extracting meaningful information from IIoT data.

The remaining of this section presents some of the existing research issues and challenges of smart Industries 4.0.

Table 6.1 Categorical open research issues and challenges of smart Industries 4.0

Research challenges	**Categories**		
	Technological	**Business-based**	**Methodological**
Interoperability	√		√
Big IIoT data analytics	√	√	√
Data availability	√		
Data quality	√	√	
Data complexity	√		
Dynamic data	√		
Data validation and verification	√	√	√
Data security	√	√	√
Data heterogeneity	√		√
Data privacy	√	√	√
Reference models			√
Visualisation	√		√
Service/app marketplaces		√	√
Sensors/actuators	√		
Requirements engineering			√
Investment issues		√	
Servitised BMs		√	

6.5.1 Interoperability

This challenge refers to the interoperability between different systems of the companies where some of the firms use proprietary solutions available on the market and some other corporations use open-access or self-developed solutions. These solutions are selected according to different dimensions such as nature of the business, types of generated data and operational process. Therefore, when these different systems are supposed to work together, interoperability becomes a major challenge which must be addressed properly to enable smart manufacturing. For instance, integrating transportation systems and human interface devices used by different merchants require interoperability issues to be answered effectively. In a more dynamic and complex manufacturing environment, various integrations may be needed to resolve the issue to simplify the development and the process of smart manufacturing supply networks successfully [34].

6.5.2 Big IIoT data analytics

Big IIoT data analytics involves the processes of collecting data, transferring them into the centralised cloud-based data centres and deploying preprocessing, data mining, analysis, visualisation and reporting [36,37]. Big data analytics implicates the processes of searching a database, mining and analysing data dedicated to improving manufacturing firms' performance [38]. Big IoT data analytics offers elegant analytical tools to process big IoT data in real-time. This opportunity produces sensible timely results for decision makers [39]. Big IIoT data analytics is one of the main components of smart data-based manufacturing and I4.0 initiatives [40]. Although big IoT data analytics is categorised under technological research issues, it can be branded under methodological research issues due to the requirement of designing relevant analytical methodologies. In some cases, having better analytical methods is more important than the type of analytics' tools and algorithms. In fact, sometimes selecting methods to connect old and new developed algorithms according to smart manufacturing certainties are critical. It is very important to make humans involved in the cycle of the data analytics process of captured data via manufacturing IoT sensors and devices. This secures a strong data visualisation and in-depth insights into the outcomes gate.

Automation of the process, availability of real-time data and automated monitoring using high-level algorithms that support human decisions are key characteristics of smart Industries 4.0. Having a safe interaction between humans and smart systems, which is involved with process automation such as robotics process automation, requires reliable and enhanced analytics algorithms [41]. Several studies have reported and described issues of data analytics when it comes to big data and smart industries [36,37,39,42–46]. But the most important issue of big data analytics in smart manufacturing is integrating different dependent and independent data from different sources such as energy usage, material flow, quality inspection data and other added information in real-time [34,47,48]. The increasing availability of raw data and associated issues such as data quality, data complexity, dynamic data, data validation and verification is another important challenge.

6.5.3 Data security

Smart industries are involved with real-time valuable data generated by smart IoT devices and sensors using internet communication protocols via wireless and ad-hoc networks. These data are being stored more and more onto the data centres using cloud-based services. Using data analytics on these valuable data in smart industries provides more meaningful information and at the same time, it motivates criminal parties and individuals to try to use this opportunity via illegal ways. Therefore, smart industries systems must be continually enhanced to avoid unauthorised access to valuable data and to detect any other security threats [34]. In fact, smart manufacturers prefer limiting access to their data, supplying them on the isolated data centres and servers and giving access to only eligible customers and users to see only a very specific and limited analyses results [49]. The other data security challenge is related to the mechanisms of preventing any unauthorised access to the control systems and connected devices and machines which are other targets of criminals to sabotage the manufacturing machines and processes.

6.5.4 Data quality

Developing IoT systems with the ability to generate reliable data is not easy, and collected data from IoT devices mostly suffer from lack of quality. [50,51]. Therefore, the quality of the data produced by IIoT systems in smart industries is as important as other issues and challenges such as big IIoT data analytics' [34,52]. Using low-quality data in smart industries endanger monitoring and control systems as well as any data-centric optimisation. Hence, it is required to develop data quality algorithms to monitor the quality of the generated data in the smart manufacturing processes and machines automatically. This improves the level of trust and accuracy of the decisions made based on the data.

Also, data heterogeneity in smart industries is another aspect of data quality in the entire product lifecycle. Integrating various data sources with different formats and semantics is still a significant challenge for advanced data analytics. Therefore, semantic mediator systems must be developed and embedded in the preprocessing stage prior to starting data analysis phase. These mediators can be updated according to the requirements of the data analytics.

6.5.5 Visualisation

There is a strong need for data visualisation to provide illustrated information to the users in a smart, sustainable and robust way [53]. Visualisation challenges can be considered as both technological and methodological research issues. Visualisation is a vital vehicle to communicate complex manufacturing information [54]. For example, complex results of data analytics need to be shared with various shareholders of the company [55]. This is a challenge as different shareholders may have different permission to have access to the visualisation and the existing results [56]. This is the same for other users such as administrators who manage the visualisations systems or engineers who are responsible for production planning. Different users require access to different information, visualisation results and even user interface compared to sales or factory managers.

Visualisation is a critical component of smart industry and requires both industries and academia work closely together to drive proper visualisation research to understand what types of information need to be presented to different shareholders. They also need to design and implement a rigorous visualisation solutions which can break down abstract sensor-based data, deliver additional value and appropriate information. It is required to highlight here that visualisation processes and outcomes are extremely reliant on individual shareholders' needs. Therefore, it is not possible to use a single visualisation solution for all possible scenarios when it is expected to have a positive impression on the operations [34,42,57].

6.5.6 Privacy issues

It is clear that data privacy issues are related to data security. However, data security issues are more about the technological aspect of the data process in smart industries in which the focus is on protecting and preserving sensitive manufacturing data. But data privacy issues describe some challenges related to exchanging data, information and knowledge internally and externally. Revealing valuable data or information to noneligible parties or individuals can create a lot of problems for the data owners. For example, having access to the valuable data by competitors can give them the opportunity of using reverse engineering to produce the same products or extracting essential information which can be even more problematic. Therefore, a privacy mechanism must be developed to ensure anybody having access to any portion of the data can use them only for the predefined purposes that the data is shared for. For instance, video data of the aircraft surveillance that belong to an airline are deposited in storage by a service provider. These data are regulated according to the passengers' privacy, with laws differing between countries. Sharing some of the information for some reasons such as quality improvement can be beneficial [58]. This requires assurance policies that must be designed via interdisciplinary research involving law, business, engineers and computer science experts.

6.5.7 Investment issues

This is a general business-based issue in smart industries. Complexity and interdisciplinary nature of smart industries instalments have barriers for any small- and medium-sized enterprise (SME). Deploying smart industrial frameworks in an SME, such as the Cyber-Physical Logistics System (CPLS), needs an important investment. The purpose of having the CPLS is increasing flexibility through independent decisions and enabling a reduction of inventories by independent real-time resolving errors. Measuring the trust in collaboration and the possibilities of organisational improvements are not easy. Robust testbeds as put together by the Smart Manufacturing Leadership Coalition or lighthouse projects can be a proper start to conduct benchmarks and successful instances to highlight the potential of such an investment. Besides that, theoretical research about the quantification and the return on investment on smart industrial applications is needed especially regarding the effects of collaboration in complex and dynamic supply networks in the SMEs [34].

6.5.8 Servitised business models

Traditionally, BMs of manufacturing industries mostly focus on producing or assembling customised physical products and revenues can be generated by selling those products. But the costs of materials, devices and skilful staffs is very high. Thus, the efficiency of the processes and the supply-chain association have effects on the competitiveness of the company significantly [34,59]. At the same time, traditional BMs are pushed by the global harmonisation of technological standards and the reduction of trade barriers to moving towards modern BMs. It is suggested by several researchers manufacturing companies in developed economies to have a vital role in the value chain by adding on more services to their products. By these extensions, they no longer need to compete merely on manufacturing cost [60,61].

According to a study, a proposed framework using the so-called product-service system (PSS) concept describes the realisation, integrated development and offers specific product-service bundles as a solution to the customer. This proposed framework is based on five fundamental developments: (1) the shift from a world of products to a world including solutions, (2) outputs to outcomes, (3) transactions to relationships, (4) suppliers to network partners and (5) elements to ecosystems

Figure 6.5 Future smart factory [65]

[62]. The concept used in this framework is so similar to the idea of smart industries where the so-called CPS offers a solution for a particular problem via their application's outcome. The relationships between other systems and their environment can be built by CPS to be replaced with one-off sales transactions. For instance, having access to lifecycle usage may lead to improving manufactural processes. It also may offer add-on services for the core product. Therefore, a network ecosystem can be created around the CPS that connects customers, suppliers and other partners.

In another research, the term 'Cyber-Physical Product-Service System' (CPSS) has been invented to integrate the PSS concept and smart industries [34,63]. Now, if a manufacturing firm changes from manufacturing of products to proposing CPSS solutions and converts its supplier base into a network ecosystem of partners, it has to analyse and adjust elements of its BM to continue having profit and also to stay competitive. These elements include the new value proposition, key resources, different customer segments and relationships, distribution channels, key resources, costs structure and also revenue streams [64].

6.6 Future directions of IIoT in smart Industries 4.0

The IoT is the concept where the network/internet extends into the real world, embracing everyday objects [65]. With the help of IoT now, physical items are able to keep connected as part of the virtual world, and one can control them remotely as a physical access point to internet service. IoT truly brings a revolution for the industry where they reduce the cost, make monitoring easy, increase the production, fully monitor supply-chain management, as well as manufacturing processes of different industries, which has a very direct and high impact on the economy of the world.

The future concept of the smart industry can be expected as shown in Figure 6.5, where all processes will be managed in a smart manner, being controlled by smart robotics.

In the current era, the importance of the data, remote access and timely delivery is an essential part for the growth of any organisation, which will have a direct impact on the growth of the industry as well as indirect impact on the economy of the country. IIoT leads the smart industries to its high growth rate by providing all necessary support to the industry data besides of their automation and remote monitoring. IIoT is the source of transformation for the industries, which will improve multiple aspects of the industry including automated and transportation, healthcare, manufacturing, retail, supply chain, infrastructure, insurance,

Table 6.2 Internet of Things: a source of transformation

Industry	**Key changes**	**Potential benefits**
Automotive and transportations	Real-time driving behaviour, traffic and vehicle diagnostics	Reduced pollution, improved customer experience, increased safety and additional revenue streams
Healthcare	Remote monitoring of staff and patient's ability, to locate and identify the status of the equipment	Improved employee productivity, resources usage and outcomes that result in efficiency gains and cost-savings
Manufacturing	Quick response to fluctuations in demand; maximised operational efficiency, safety and reliability, using smart sensors and digital control systems	Enhanced agility and flexibility, reduced energy-consumption and carbon footprint
Retail	Stock-out prevention through connected and intelligent supply chains	Ability to predict consumer behaviour and trends using data from video surveillance cameras, social media, internet and mobile devices
Supply chain	Real-time tracking of parts and raw material, which helps organisation preempt problems, address demand fluctuations and efficiently manage all stages of manufacturing	Reduced working capital requirement, improved efficiencies and avoidance of disruptions in manufacturing
Infrastructure	Smart lighting, water, power, fire, cooling, alarms and structural health system	Environmental benefits and significant cost savings with better utilisation of resources and preventive maintenance of the critical system
Oil and gas	Smart components for oil and gas	Reduced operating costs and fuel consumption
Insurance	Innovative services such as pay-as-you-go insurance	Significant cost savings for both insurers and consumers
Utilities	Smart grids and smart meters	More responsive and reliable services; significant cost savings for both utilities and consumers resulting from demand-based and dynamic pricing features

utilities, etc. For further details, please refer to Table 6.2; most of the findings in it are with the courtesy of ZDNet through Ericsson and M2M magazine.

Furthermore, the IoT applications have very vast scopes including the IIoT applications. A recent study explores [66–73] the top 20 major applications of the IIoT for current and future which will change the way of working style and will have high rising growth and contribution towards the economy. The top 20 suggested industrial applications include

1. **ABB**: Smart robotics
2. **Airbus**: Factory of the future
3. **Amazon**: Reinventing warehousing
4. **Boeing**: Using IoT to drive manufacturing efficiency
5. **Bosch**: Track and trace innovator
6. **Caterpillar**: An IIoT pioneer
7. **Fanuc**: Helping to minimise downtime in factories
8. **Gehring**: A pioneer in connected manufacturing
9. **Hitachi**: An integrated IIoT approach
10. **John Deere**: Self-driving tractors and more
11. **Kaeser Kompressoren**: Air as a service
12. **Komatsu**: Innovation in mining and heavy equipment
13. **KUKA**: Connected robotics
14. **Maersk**: Intelligent logistics
15. **Magna Steyr**: Smart automotive manufacturing
16. **North Star BlueScope Steel**: Keeping workers safe
17. **Real-Time Innovations**: Micro grid innovation
18. **Rio Tinto**: Mine of the future
19. **Shell**: Smart oil field innovator
20. **Stanley Black & Decker**: Connected technology for construction and beyond

6.7 Conclusion

The economic growth of a country is directly linked to its industrial growth, while the use of new technology is essential for industrial growth. Industries could produce well efficiently and smartly, and known as smart industry by adopting new era technologies such as the IIoT, etc. Hence, industries are changing with the new era requirements from older fashion to the technological perspective. Industries equipped fully with technology are known as smart industries, and IIoT is the one which makes ordinary industries smart by providing them cost cutting, remote access, production management, supply chain, monitoring, reducing energy consumption cost, etc. IIoT brings the revolution in the industry to enhance its production sharply and allow the industry to access and share the data remotely. This book chapter contributed by enlightening the core aspects of smart industry or I4.0 besides of IIoT based state-of-the-art industrial revolution. Further, this book chapter explores more in-depth concepts by providing different use cases of IIoT-based I4.0. In addition, the chapter also provides current uses, limitations, open issues, challenges and future directions of IIoT on smart industries in great detail.

Acknowledgement

We are thankful to Taylor's University, Malaysia, for providing the platform and resources to complete our book chapter.

References

[1] Z. A. Almusaylim and N. Zaman, "A review on Smart home present state and challenges: Linked to context awareness internet of things (IoT)," Wireless Networks, vol. 25, pp. 3193–3204, 2019.

[2] A. Irniger, Core system part of SAP, [Online]. [Accessed March 2019] Available: (2019) https://bit.ly/2TMciNg.

[3] DreamzTtech Solutions, IIOT in manufacturing [Online]. [Accessed March 2019] Available: (2019) https://bit.ly/2lZWVRY.

[4] A. Banafa, "What is next for IoT and IIoT," in Enterprise Mobility Summit, Australia, 2015.

[5] S. Mumtaz, A. Alsohaily, Z. Pang, A. Rayes, K. F. Tsang, and J. Rodriguez, "Massive Internet of Things for industrial applications: Addressing wireless IIoT connectivity challenges and ecosystem fragmentation," IEEE Ind. Electron. Mag., vol. 11, no. 1, pp. 28–33, 2017.

[6] L. Hou, S. Tan, Z. Zhang, and N. W. Bergmann, "Thermal energy harvesting WSNs node for temperature monitoring in IIoT," IEEE Access, vol. 6, pp. 35243–35249, 2018.

[7] A. S. Petrenko, S. A. Petrenko, K. A. Makoveichuk, and P. V. Chetyrbok, "The IIoT/IoT device control model based on narrow-band IoT (NB-IoT)," in 2018 IEEE Conference of Russian Young Researchers in Electrical and Electronic Engineering (EIConRus). Russia: IEEE, 2018, pp. 950–953.

[8] A. Hassanzadeh, S. Modi, and S. Mulchandani, "Towards effective security control assignment in the Industrial Internet of Things," in 2015 IEEE 2nd World Forum on the Internet of Things (WF-IoT). Milan, Italy: IEEE, 2015, pp. 795–800.

[9] H. Boyes, B. Hallaq, J. Cunningham, and T. Watson, "The industrial internet of things (IIoT): An analysis framework," Comput. Ind., vol. 101, pp. 1–12, 2018.

[10] L. Atzori, A. Iera, and G. Morabito, "The internet of things: A survey," Comput. Networks, vol. 54, no. 15, pp. 2787–2805, 2010.

[11] M. Munir, S. Baumbach, Y. Gu, *et al.*, Series in machine perception, and artificial intelligence, 'Data analytics: Industrial perspective & solutions for streaming data'. W.S.P.C. Singapore, 2018, 83, (1).

[12] H.-C. Hsieh and C.-H. Lai, Internet of things architecture based on integrated PLC and 3G communication networks. Tainan: IEEE, 2011, pp. 853–856.

[13] C. Zhan, Y. Zeng, and R. Zhang, "Energy-efficient data collection in UAV enabled wireless sensor network," IEEE Wireless Commun. Lett., vol. 7, no. 3, pp. 328–331, 2018.

[14] P. Nikitin, S. J. Kelly, and R. Martinez, Systems and methods for controlling radio-frequency identification (RFID) tag communication. Google Patents, 2018.

[15] M. S. Hossain and G. Muhammad, "Cloud-assisted industrial internet of things (IIoT) – Enabled framework for health monitoring," Comput. Networks, vol. 101, pp. 192–202, 2016.

[16] S. K. Mani, R. Durairajan, P. Barford, and J. Sommers, An architecture for IoT clock synchronization. USA: ACM, 2018, pp. 17.

[17] A. Zimmermann, R. Schmidt, K. Sandkuhl, M. Wißotzki, D. Jugel, and M. Möhring, Digital enterprise architecture-transformation for the internet of things. Adelaide, SA: IEEE, 2015, pp. 130–138.

[18] S. Cirani, L. Davoli, G. Ferrari, *et al.*, "A scalable and self-configuring architecture for service discovery in the internet of things," IEEE IoT J., vol. 1, no. 5, pp. 508–521, 2014.

[19] J. Zhou, T. Leppanen, E. Harjula, *et al.*, CloudThings: A common architecture for integrating the internet of things with cloud computing. IEEE, 2013, pp. 651–657.

[20] P. Martinez-Julia, E. T. Garcia, J. O. Murillo, and A. F. Skarmeta, Evaluating video streaming in network architectures for the internet of things. Taichung: IEEE, 2013, pp. 411–415.

[21] J. Rufino, M. Alam, J. Ferreira, A. Rehman, and K. F. Tsang, Orchestration of containerized microservices for IIoT using Docker. Canada: IEEE, 2017, pp. 1532–1536.

[22] D. Preuveneers and E. Ilie-Zudor, "The intelligent industry of the future: A survey on emerging trends, research challenges and opportunities in Industry 4.0," J. Ambient Intell. Smart Environ., vol. 9, no. 3, pp. 287–298, 2017.

[23] D. Merico, "Tracking with high-density large-scale wireless," J. Ambient Intell. Smart Environ., vol. 2, no. 4, pp. 441–442, 2010.

[24] H. Gaub, "Customization of mass-produced parts by combining injection molding and additive," Reinf. Plast., vol. 60, no. 6, pp. 401–404, 2016.

[25] A. Schumacher, S. Erol, and W. Sihn, "A maturity model for assessing industry 4.0 readiness and maturity of manufacturing enterprises," Procedia CIRP, vol. 52, pp. 161–166, 2016.

[26] G. Adamson, L. Wang, and P. Moore, "Feature-based control and information framework for adaptive and distributed manufacturing in cyber-physical systems," J. Manuf. Syst., vol. 43, pp. 305–315, 2017.

[27] M. Alexandre, Industries 4.0 framework, challenges and perspectives. Wiesbaden: RheinMain University, 2014.

[28] M. Mabkhot, A. Al-Ahmari, B. Salah, and H. Alkhalefah, "Requirements of the smart factory system: A survey and perspective," Machines, vol. 6, no. 2, pp. 1–22, 2018.

[29] S. Aravind, 5 Best Examples of IoT Applications in the Real World, [Online]. [Accessed 7 March 2019] Available: (2018) https://analyticstraining.com/5-best-examples-iot-applications-real-world/.

[30] S. Seol, Y. Shin, and W. Kim, "Design and realization of personal IoT architecture based on mobile gateway," Int. J. Smart Home, vol. 9, no. 11, pp. 133–144, 2015.

[31] I. Lee and K. Lee, "The Internet of Things (IoT): Applications, investments, and challenges for enterprises," Bus. Horiz., vol. 58, no. 4, pp. 431–440, 2015.

[32] A. Weersink, E. Fraser, D. Pannell, E. Duncan, and S. Rotz,, "Opportunities and challenges for Big Data in agricultural and environmental analysis," Annu. Rev. Resour. Econ., vol. 10, pp. 19–37, 2018.

[33] D. Tomic, The benefits and challenges with implementation of Internet of Things (IoT) in manufacturing industry. USA: Annual Review, 2017.

[34] K.-D. Thoben, S. Wiesner, and T. Wuest, "'Industrie 4.0' and smart manufacturing-a review of research issues and application examples," Int. J. Autom. Technol., vol. 11, no. 1, pp. 4–16, 2017.

[35] J. Haight and H. Park, "IoT analytics in practice," Blue Hill Res. Bost., 2015.

[36] E. Ahmed, I. Yaqoob, I. A. T. Hashem, *et al.*, "The role of big data analytics in Internet of Things," Comput. Networks, vol. 129, pp. 459–471, 2017.

[37] M. H. Ur Rehman, E. Ahmed, I. Yaqoob, I. A. T. Hashem, M. Imran, and S. Ahmad, "Big data analytics in industrial IoT using a concentric computing model," IEEE Commun. Mag., vol. 56, no. 2, pp. 37–43, 2018.

[38] O. Kwon, N. Lee, and B. Shin, "Data quality management, data usage experience and acquisition intention of big data analytics," Int. J. Inf. Manage., vol. 34, no. 3, pp. 387–394, 2014.

[39] M. Marjani, F. Nasaruddin, A. Gani, *et al.*, "Big IoT data analytics: Architecture, opportunities, and open research challenges," IEEE Access, vol. 5, 2017.

[40] Y. Zhang, S. Ren, Y. Liu, and S. Si, "A big data analytics architecture for cleaner manufacturing and maintenance processes of complex products," J. Clean. Prod., vol. 142, pp. 626–641, 2017.

[41] D. Surdilovic, G. Schreck, and U. Schmidt, "Development of collaborative robots (COBOTS) for flexible human-integrated assembly automation," in ISR 2010 (41st International Symposium on Robotics) and ROBOTIK 2010 (6th German Conference on Robotics). Germany, 2010, pp. 1–8.

[42] I. A. T. Hashem, I. Yaqoob, N. B. Anuar, S. Mokhtar, A. Gani, and S. U. Khan, "The rise of 'big data' on cloud computing: Review and open research issues," Inf. Syst., vol. 47, pp. 98–115, 2015.

[43] G. Wang, A. Gunasekaran, E. W. T. Ngai, and T. Papadopoulos, "Big data analytics in logistics and supply chain management: Certain investigations for research and applications," Int. J. Prod. Econ., vol. 176, pp. 98–110, 2016.

[44] S. Fosso Wamba, A. Gunasekaran, T. Papadopoulos, and E. Ngai, "Big data analytics in logistics and supply chain management," Int. J. Logist. Manage., vol. 29, no. 2, pp. 478–484, 2018.

[45] T. Wuest, D. Weimer, C. Irgens, and K.-D. Thoben, “Machine learning in manufacturing: Advantages, challenges, and applications,” Prod. Manuf. Res., vol. 4, no. 1, pp. 23–45, 2016.

[46] C. Loebbecke and A. Picot, “Reflections on societal and business model transformation arising from digitization and big data analytics: A research agenda,” J. Strateg. Inf. Syst., vol. 24, no. 3, pp. 149–157, 2015.

[47] F. J. Riggins and S. F. Wamba, “Research directions on the adoption, usage, and impact of the internet of things through the use of big data analytics,” in 2015 48th Hawaii International Conference on System Sciences. USA: Hawaii, 2015, pp. 1531–1540.

[48] M. E. Porter and J. E. Heppelmann, “How smart, connected products are transforming companies,” Harv. Bus. Rev., vol. 93, no. 10, pp. 96–114, 2015.

[49] T. Handel, M. Schreiber, K. Rothmaler, and G. Ivanova, “Data security and raw data access of contemporary mobile sensor devices,” in World Congress on Medical Physics and Biomedical Engineering 2018. Czech Republic: Prague, 2019, pp. 397–400.

[50] S. Lin, J. Gao, and A. Koronios, “The need for a data quality framework in asset management,” in Australian Workshop on Information Quality. Australia, 2006, vol. 1.

[51] P. Brous, M. Janssen, and P. Herder, “Next generation data infrastructures: Towards an extendable model of the asset management data infrastructure as complex adaptive system,” Complexity, vol. 2019, 2019.

[52] T. Wuest, S. Wellsandt, and K.-D. Thoben, “Information quality in PLM: A production process perspective,” in IFIP International Conference on Product Lifecycle Management. Doha, Qatar, 2015, pp. 826–834.

[53] M. Bouloukakis, N. Partarakis, I. Drossis, M. Kalaitzakis, and C. Stephanidis, “Virtual reality for smart city visualization and monitoring,” in Mediterranean Cities and Island Communities. Greece: Springer, 2019, pp. 1–18.

[54] J. Davis, T. Edgar, J. Porter, J. Bernaden, and M. Sarli, “Smart manufacturing, manufacturing intelligence and demand-dynamic performance,” Comput. Chem. Eng., vol. 47, pp. 145–156, 2012.

[55] R. Y. Zhong, S. Lan, C. Xu, Q. Dai, and G. Q. Huang, “Visualization of RFID-enabled shopfloor logistics Big Data in cloud manufacturing,” Int. J. Adv. Manuf. Technol., vol. 84, no. 1–4, pp. 5–16, 2016.

[56] A. Brodsky, G. Shao, M. Krishnamoorthy, A. Narayanan, D. Menascé, and R. Ak, “Analysis and optimization in smart manufacturing based on a reusable knowledge base for process performance models,” in 2015 IEEE International Conference on Big Data (Big Data). Santa Clara, CA, 2015, pp. 1418–1427.

[57] T. Wuest, K. Hribernik, and K.-D. Thoben, “Accessing servitisation potential of PLM data by applying the product avatar concept,” Prod. Plann. Control, vol. 26, no. 14–15, pp. 1198–1218, 2015.

[58] T. Wuest, Identifying product and process state drivers in manufacturing systems using supervised machine learning. Springer, 2015.

[59] R. Boyer and M. Freyssenet, “The emergence of new industrial models, Actes du Gerpisa, no. 15,” Evry Univ. d’Evry–Val d’Essonne, pp. 75–142, 1995.

[60] K.-D. Thoben, J. Eschenbächer, and H. Jagdev, “Extended products: Evolving traditional product concepts,” in Proceedings of the 7th International Conference on Concurrent Enterprising: Engineering the Knowledge Economy through Co-operation, Bremen, Germany, 2001, pp. 429–439.

[61] S. Vandermerwe and J. Rada, “Servitization of business: Adding value by adding services,” Eur. Manage. J., vol. 6, no. 4, pp. 314–324, 1988.

[62] T. S. Baines, H. W. Lightfoot, S. Evans, *et al.*, “State-of-the-art in product-service systems,” Proc. Inst. Mech. Eng., B: J. Eng. Manuf., vol. 221, no. 10, pp. 1543–1552, 2007.

[63] C. Gorldt, S. Wiesner, and I. Westphal, “Product-Service Systems im Kontext von Industrie 4.0: Auf dem Weg zu CPSS,” Industrie 4.0 Management 1/2016: Product-Service Design, 2016, pp. 15–18.

[64] A. Osterwalder and Y. Pigneur, Business model generation: A handbook for visionaries, game changers, and challengers. John Wiley & Sons, 2010.

[65] Rexroth a Bosch Company, Factory of the Future, [Online]. [Accessed March 2019] Available: (2019) https://bit.ly/2HwjpmA.

[66] IIoT World Today, The top 20 industrial IoT Applications [Online]. [Accessed March 2019] Available: (2019) https://bit.ly/2HGVS5n.

[67] M. Almulhim and N. Zaman, “Proposing secure and lightweight authentication scheme for IoT based E-health applications,” in 20th International Conference on Advanced Communication Technology (ICACT), Chuncheon-si Gangwon-do, Korea (South), 2018, pp. 481–487.

[68] M. O. Alshammari, A. A. Almulhem, and N. Zaman, “Internet of Things (IoT): Charity automation,” Int. J. Adv. Comput. Sci. Appl. (IJACSA), vol. 8, no. 2, 2017. http://dx.doi.org/10.14569/IJACSA.2017.080222.

[69] Y. Q. Lu and X. Xu, “A semantic web-based framework for service composition in a cloud manufacturing environment,” J. Manuf. Syst., vol. 42, pp. 69–81, 2017.

[70] Y. K. Liu, X. Xu, L. Zhang, L. Wang, and R. Y. Zhong, “Workload-based multi-task scheduling in cloud manufacturing,” Rob. Comput. Integr. Manuf., vol. 45, pp. 3–20, 2017.

[71] R. Y. Zhong, Y. Peng, F. Xue, *et al.* “Prefabricated construction enabled by the Internet-of-Things,” Autom. Constr., vol. 76, pp. 59–70, 2017.

[72] Y. K. Liu and X. Xu, “Industry 4.0 and cloud manufacturing: A comparative analysis,”. J. Manuf. Sci. Eng., vol. 139, no. 3, pp. 034701-1–034701-8, 2017.

[73] Y. Koren, W. Wang, and X. Gu, “Value creation through design for scalability of reconfigurable manufacturing systems,” Int. J. Prod. Res., vol. 55, no. 5, pp. 1227–1242, 2017.

Chapter 7

Simulation in the 4th Industrial Revolution

Chee Pin Tan[1] *and Wen-Shyan Chua*[2]

7.1 Introduction

As defined by the Boston Consulting Group [1], and shown in Figure 7.1, one of the technological pillars of the 4th Industrial Revolution is simulation, where it is stated that

> Simulations will be used more extensively in plant operations to leverage real-time data and mirror the physical world in a virtual model, which can include machines, products, and humans. This will allow operators to test and optimize the machine settings for the next product in line in the virtual world before the physical changeover, thereby driving down machine setup times and increasing quality.

According to www.dictionary.com, one of the definitions of simulation is

> the representation of the behavior or characteristics of one system through the use of another system, especially a computer program designed for the purpose.

Simulation itself is not a new notion; it has been there since the invention of computers (before the 4th Industrial Revolution). However, with the arrival of the 4th Industrial Revolution, where things change at breakneck speed, the importance of simulation is even now more amplified. In this chapter, we will give examples of the various types of simulations available as well as some examples, and describe the benefits of simulation, especially in the context of the 4th Industrial Revolution.

7.2 Types of simulation

A simulation is typically a model, which mimics the characteristics and behavior of a system/process, without actually executing or running the system/process. There are various types of models, which will be described in detail in this section.

[1]School of Engineering and Advanced Engineering Platform, Monash University Malaysia, Bandar Sunway, Selangor, Malaysia
[2]Malaysian Smart Factory, Selangor Human Resources Development Corporation, Shah Alam, Selangor, Malaysia

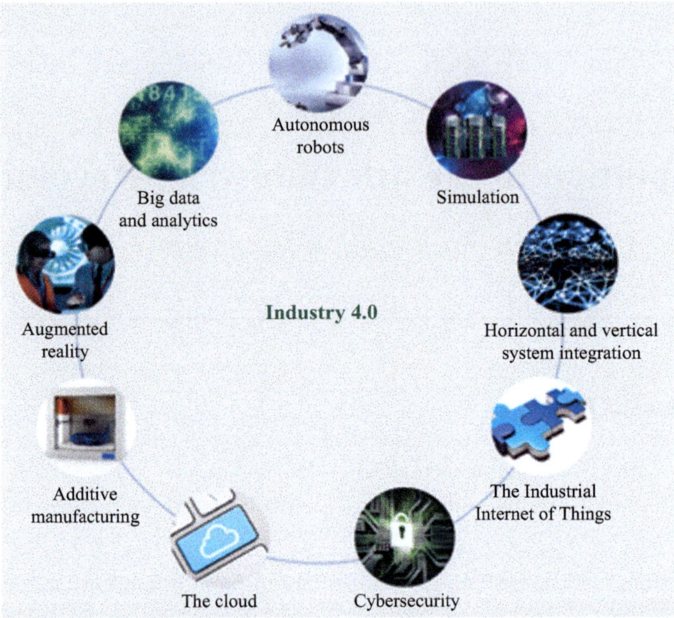

Figure 7.1 The nine technological pillars of the 4th Industrial Revolution

7.2.1 Simulation of a physical system

Mathematical models of a physical system can take many forms, for instance, dynamical systems, statistical models, or differential equations. In the simulation software and packages, the user simply needs to key in (or program) the mathematical information and the software will solve the information (or equations) and produce the result.

7.2.1.1 Controller design

An area where simulation is popular is the area of control theory, where the objective is to design controllers for dynamic systems so that they will behave in a way that is desired. Typically, the behavior of these dynamic systems is governed by differential equations. A popular software for this kind of simulations is MATLAB® [2], which has been widely used to solve differential equations. Also available in MATLAB is the Simulink® toolbox, which has drag-and-drop blocks for users who may need more visual representations of the model.

Consider a popularly used but simple example, which is the mass-spring-damper system, shown in Figure 7.2. The block has a mass m (kg) and its displacement is denoted by x (m). The spring has a constant k (N/m) and the damper has a constant b (Ns/m). The block is also subject to a force F (N). The spring generates a force that is proportional to the displacement of the block, namely, kx,

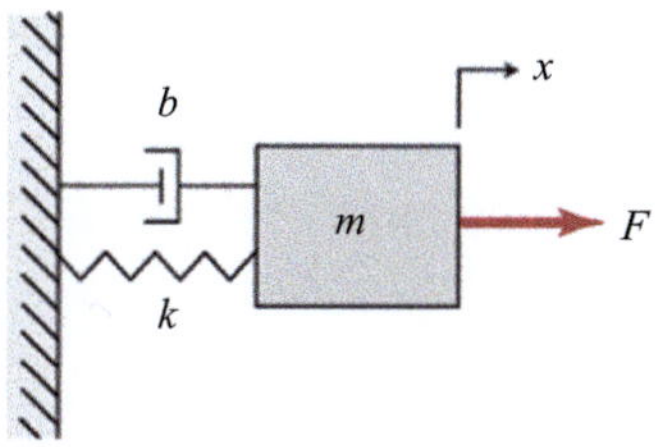

Figure 7.2 Mass spring damper (Source: Google Images)

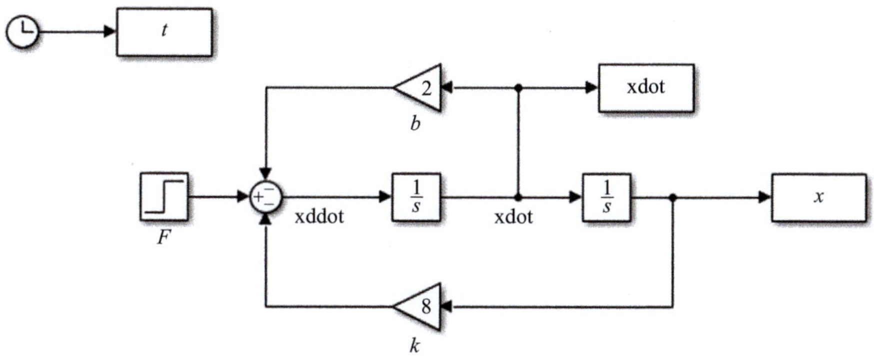

Figure 7.3 Simulink representation of mass-spring-damper system

and the damper generates a force that is proportional to the velocity of the block, namely, $b\dot{x}$. It is clear that the summation of forces (on the block) is $F - b\dot{x} - kx$.

By using Newton's second law, the sum of forces is equal to the mass times acceleration and, hence, Newton's second law being applied to the block will yield a differential equation that determines the motion of the block, namely

$$F - b\dot{x} - kx = m\ddot{x} \tag{7.1}$$

Equation (7.1) can be implemented in the MATLAB Simulink environment in graphical form, using integrators and mathematical gain blocks, which is shown in Figure 7.3.

Now, it is desired to simulate (and predict) the response of the system. Supposing that the constants take the values $m = 1, b = 2, k = 8$ and subject to a unit (magnitude) step input at time $t = 5$.

Figures 7.4 and 7.5, respectively, show the displacement and velocity of the mass in response to the force applied at time $t = 5$. From Figure 7.3, other signals can also be inspected, for example, the acceleration. In addition, one can also change the applied force (to a sine wave or to a noisy signal) and Simulink will show the response of the system (displacement, velocity, acceleration). The parameter values (e.g., mass, spring constant, damper constant) can also be changed and

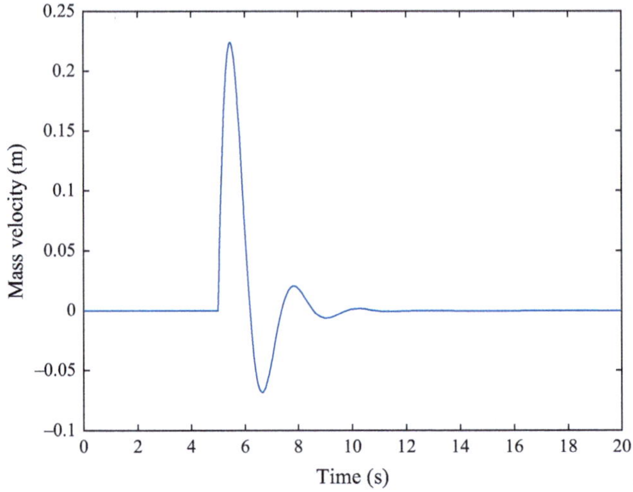

Figure 7.4 Displacement of mass in response to step input at $t = 5$ s

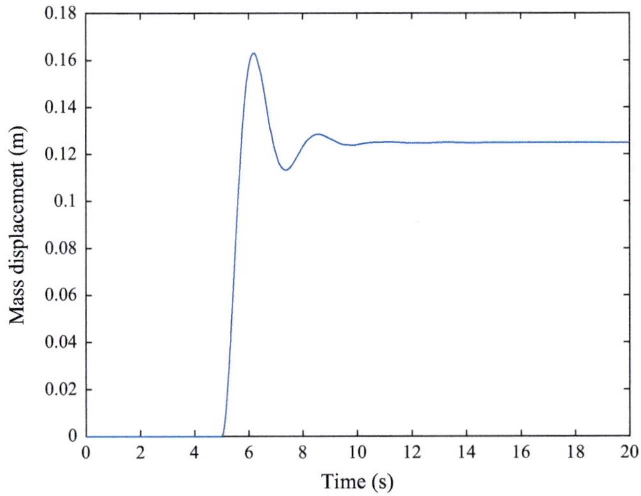

Figure 7.5 Velocity of mass in response to step input at $t = 5$ s

the response can be plotted based on it. Hence, from this example, it can be concluded that simulation enables one to

- gain a deep insight into the system, such as the interaction of internal signals and
- predict the response of a system for different inputs and parameter values.

From the model that has been developed, controllers can now be designed and tested extensively until satisfactory results are obtained. Here, we show an example

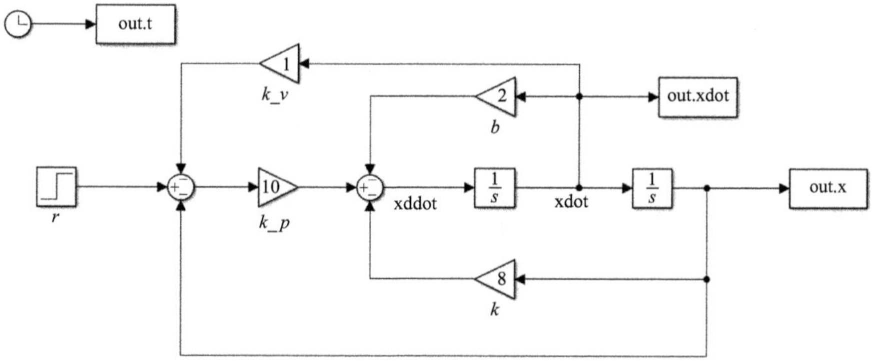

Figure 7.6 Mass-spring-damper, with a proportional controller

of controller design, and it is desired to get a lower overshoot; as shown in Figure 7.4, the overshoot is 20%, which is very high.

Suppose that the control input (which is the force F) is now a function of feedback of the mass position, i.e., $F = k_p(r - x)$ where r is the reference (target) value of x; this configuration for F is known as *proportional control*, where the control input (F in this case) is proportional (multiplied by k_p) to the tracking error (which is r in this case). The simulation diagram is now modified to incorporate this, as shown in Figure 7.6.

By setting the following varying values; $k_p = 0.5, 1.0, 10.0$, the simulation can be easily repeated and the results for the response of x are shown in Figure 7.7.

It is evident from Figure 7.7 that the various values of proportional gain k_p do not reduce the percentage overshoot and, in fact, increase it. Therefore, a different controller configuration is used, namely $F = k_p(r - x) - k_v\dot{x}$, where a velocity feedback term has been added (and, hence, we get a proportional-velocity feedback), as shown in Figure 7.8.

Several simulations were then carried out, with $k_p = 10$ being set, and k_v taking values of $k_v = 0.2, 1.0$, and the results are shown in Figure 7.9, where it is now evident that the velocity feedback has significantly reduced the overshoot.

From this, it can be seen that in addition to enabling the prediction of output for different inputs and parameters, simulation has enabled the testing of different controller structures (from proportional controller to proportional-velocity controller) seamlessly and almost effortlessly.

7.2.1.2 Mechanical systems

Extending from mathematical simulation, graphical simulation is a physics-based animation of real-world environments, which is frequently used in areas such as dynamical systems and fluid dynamics. In computational fluid dynamics (CFD) [3], engineers perform simulations to visualize the flow of fluids, which can then be used to speed up (and optimize) the design of efficient aircraft or race cars, for example,

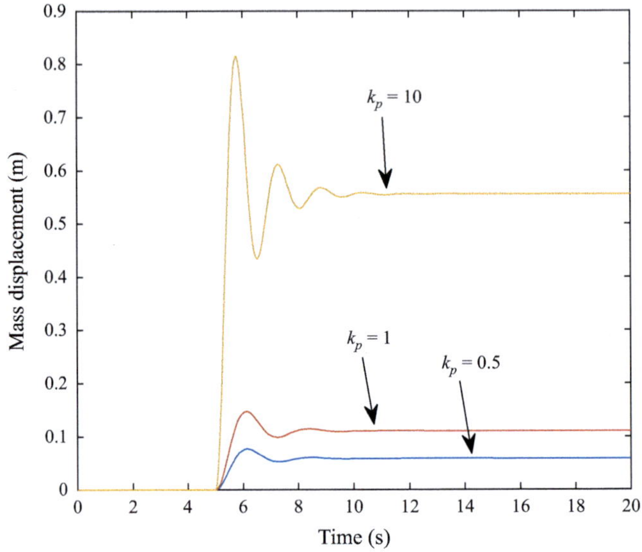

Figure 7.7 Response of mass displacement x for various values of proportional gain

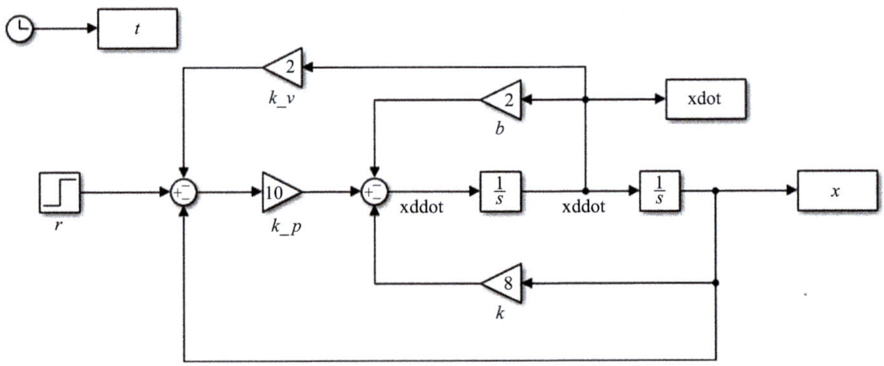

Figure 7.8 Mass-spring-damper, with a proportional-velocity control

as shown in Figure 7.10, which depicts the analysis of airflow around a Formula-1 race car.

In multibody dynamics simulation, an engineer is able to visualize the motion of several bodies (either rigid or elastic), considering mechanical-related factors such as weight, mass, forces, friction, and inertia. Multibody simulations have been widely used in many sectors such as transport and automative, being used as an integral part of aircraft design (see Figure 7.11), designing automotive suspension design [4] and studying train-track interaction dynamics [5]. It is also used extensively in

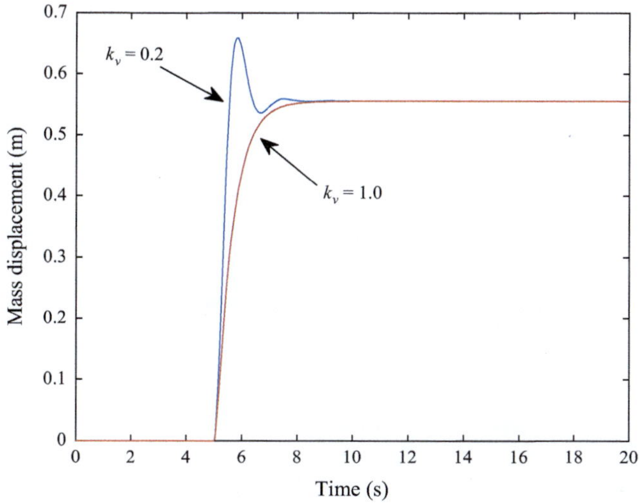

Figure 7.9 Response of mass displacement x for various values of velocity feedback

Figure 7.10 CFD analysis of airflow around a F1 car (Source: Google Images)

biomechanics, for example, to noninvasively estimate bone strain [6], study knee deterioration/degeneration [7] (Figure 7.12), and control walking motion [8].

There are many software capable of performing multibody dynamics simulation; COMSOL [9] and ADAMS [10] are among them. These software also solve equations of motion in order to generate the simulation results. Users can enter these equations, as well as any associate constraints, and the software will solve them and the results will be reflected in the simulation.

Figure 7.11 Multibody simulation of an aircraft

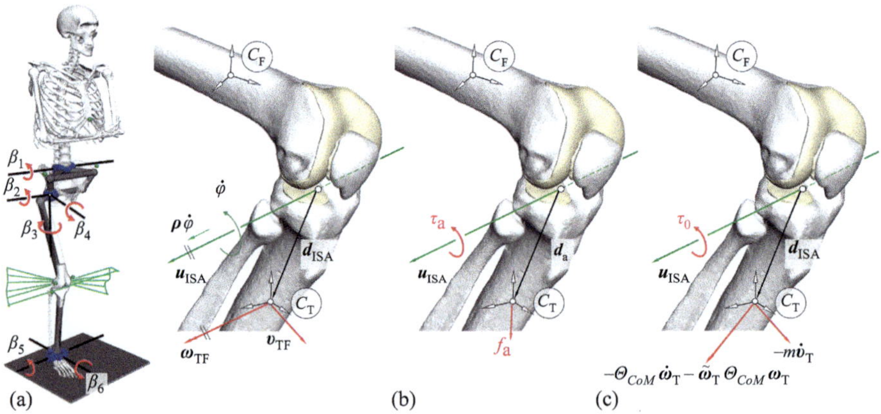

Figure 7.12 Multibody simulation of a knee movement (Source: Google Images)

7.2.1.3 Manufacturing systems

Simulation is also used extensively in the manufacturing industry, where simulation is the computer-based modeling of a real production system. In the manufacturing industry, the need for efficiency has never been greater, and will continue increasing. Material, transportation, and labor costs continue to rise each year, and the competition also intensifies – there is an overwhelming pressure need to produce more, produce better, and produce cheaper. In order to maintain the competitive edge, companies need to ensure that the costs associated with time, equipment and investments are being considered and optimized. Manufacturers can use simulation to analyze and experiment with their processes in a virtual setting, always with the purpose of meeting production goals at the lowest possible cost. As this is done in software, it eliminates (or significantly reduces) the need to run the real physical process – thereby saving time and cost. Simulation also provides a means to test and implement principles of lean manufacturing and Six Sigma. There are also many commercially available software packages for manufacturing simulation, such as FlexSim® [11] and Arena® [12] just to name a few.

In a case study [13], a manufacturer of household appliances wanted to redesign a significant portion of its refrigerator liner final assembly process, and create and implement an effective and appropriate production schedule for that process. The objectives are to (a) determine the optimal amount buffer space for liners and (b) locate additional floor space for new equipment purchases.

The production system produces refrigerator liners of various sizes, transfers those whole liners to an area where they are cut, taped, and pressed, and then transfers the liners to an insertion area (Figure 7.13).

In Figure 7.13, the challenge was in the second subprocess, to determine an appropriate mix of various liner sizes to be moved into the "press" area, in order to maximize the utilization of system equipment, as changeovers can take a lot of time. There is a buffer area before the press area, which is to "bank" liners for later use during off-shift or slow production, due to upstream failures or bottlenecks. More buffer space was needed for overflow storage and additional floor space had to be located for new equipment purchases. The company was willing to invest a significant amount of equipment and workforce staffing to redesign the production plant; however, the amount of equipment and workforce was not known.

Arena simulation software was used to develop a user-friendly simulation model. The model was highly detailed and could evaluate the dynamic flows of products through the system, evaluating material handling as well as production

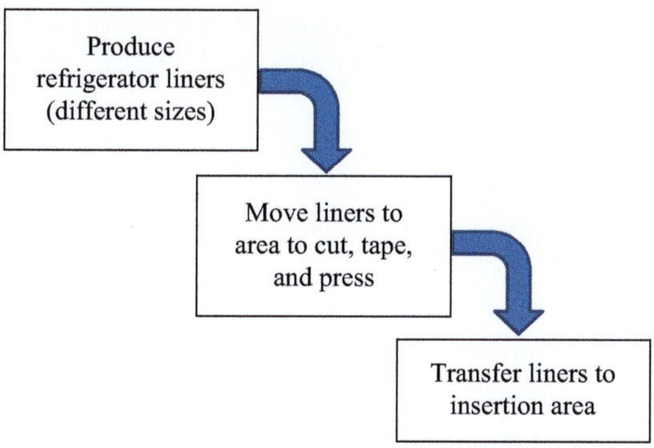

Figure 7.13 Production process of refrigerator liners

operations. The fine level of detail was needed to capture the system sensitivities for the production operations in the system. The analysis and calculations showed the required buffer space for various production scenarios and for multiple equipment layouts. To validate the model, a detailed animation displayed each liner as it moved through the system (and exposed bottlenecks in the system) and showed the dynamic status of the buffers, thus verifying the model. Finally, by running an anticipated production schedule, it was possible to produce a design with the minimum system resources necessary to meet production goals. Various cost trade-offs were calculated with the model, balancing equipment and conveyor costs versus production throughput and volume.

In another case study [14], a cosmetics distribution company used FlexSim to create a simulation model of their distribution center, which was then used as a decision-support tool. By using the model, decision-makers in the company could quickly make changes in input variables (such as product slotting, order volumes, order mix, staff/scheduling plans, operator productivity, and many others) and accurately quantify the effects. The model is flexible and can be changed and analyzed in a matter of minutes. A dashboard was created to instantly inform the operator/engineer about the current state of the system, what happened during the workday, operator utilization rates, and how balanced are the pick areas and pick zones.

7.2.1.4 Transport systems

Simulation is also used in the traffic and transport discipline, where urban planners and transport engineers use simulations to plan, design, and operate transportation systems. In transportation, simulation is very important, because its models are very complicated, unpredictable, and sometimes chaotic and, therefore, it is near impossible to obtain a mathematical (or analytical) solution. In these simulations, the designer can vary conditions (such as, number of lanes, traffic light timings, location of intersections, etc.) and view its effect. Traffic simulation are often done in two levels – the macroscopic level and microscopic level. Microscopic simulation is done at a low level, which involves individual vehicles, where the behavior of vehicles is investigated in response to changes in variables. On the other hand, macroscopic simulation is at a higher level, where the simulation is performed for a traffic network. Various software packages are available for traffic simulation, such as SUMO [15] (Simulation of Urban Mobility) and Vissim [16] just to name a few (see Figures 7.14).

7.2.2 Interactive simulation

Interactive simulation usually involves a human, where the simulator responds to the actions of the human. This is somewhat similar (or related) to virtual reality where the virtual environment is the simulation platform. Such simulations enable one to visualize the environment without having to step into it. For example, there are driving simulators, motorcycle simulators, and flight simulators. Although many of these are used for entertainment and games, they are also very useful for training. For example, a flight simulator can be used as a significant part of a pilot's

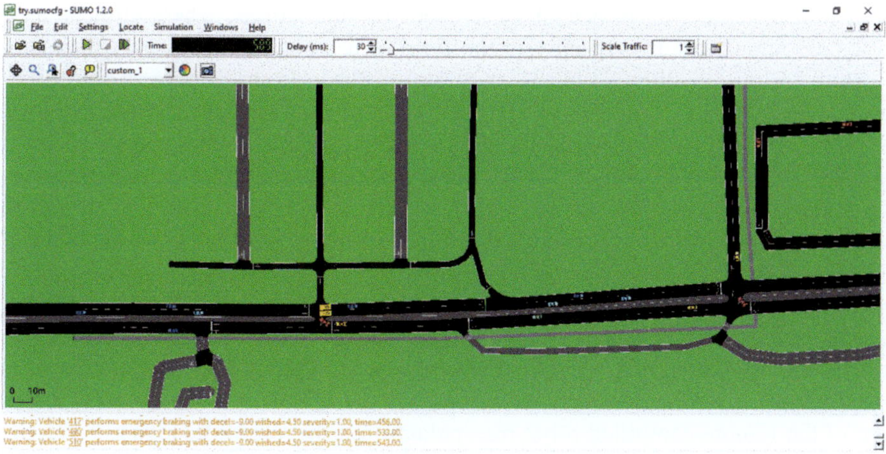

Figure 7.14 Microscopic traffic simulation using SUMO (Source: Google Images)

Figure 7.15 Flight simulator of a Boeing 747 aircraft (Source: Google Images)

training; even though real field training is required, a simulator (Figure 7.15) can provide a major part of the training, thus saving cost and time.

7.2.2.1 Physically responding simulations

Physically responding simulations are those that produce mechanical motions, or vibrations, or sounds, in response to a person's actions.

Clinical simulation is one of these types of simulations, where clinical scenarios are simulated. It is usually in the form of a human mannequin, which can simulate various conditions such as (multiple) cardiac arrests, or being a subject of surgery (Figure 7.16), in order to train medical and nursing students to respond to different conditions that they may face in the future professional careers.

Another example of physically responding simulation is the one that generates physical reactions when the user performs a certain action. For instance, haptic-based systems simulates touch, by producing forces, vibrations, or motions to the user, in response to the user's actions. An important application of haptics is teleoperation, where the user remotely controls a system (or various agents) in environments that are hazardous, such as controlling robotic manipulators to handle radioactive material or controlling a swarm of robots in rough environments. In these tasks, it would be very beneficial if the forces or motions are fed back to be felt by the user/operator. In the case of the remote-controlled robot manipulator (see, for example, Figure 7.17), by providing a feedback of the reaction force, it would enable the user/operator to determine the appropriate level of gripping force to be applied.

Another application of haptic teleoperation is in surgical robotics, where the surgeon performs surgery by remotely controlling a robot, which is able to achieve much higher precision. One of the key enabling technologies is the force feedback provided to the surgeon. The Da Vinci (Figure 7.18), by Intuitive Surgical, is one such existing surgical robot in clinical use.

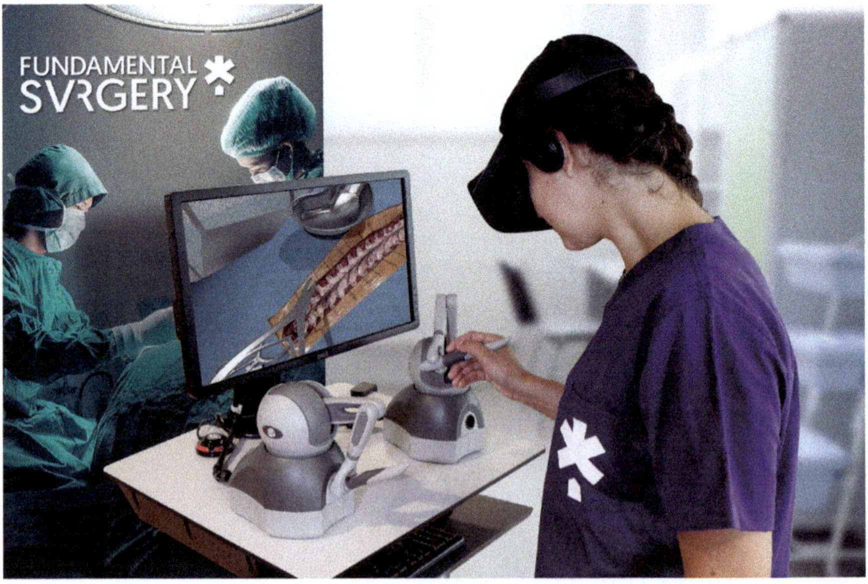

Figure 7.16 Medical students practicing on a computerized mannequin (Source: Google Images)

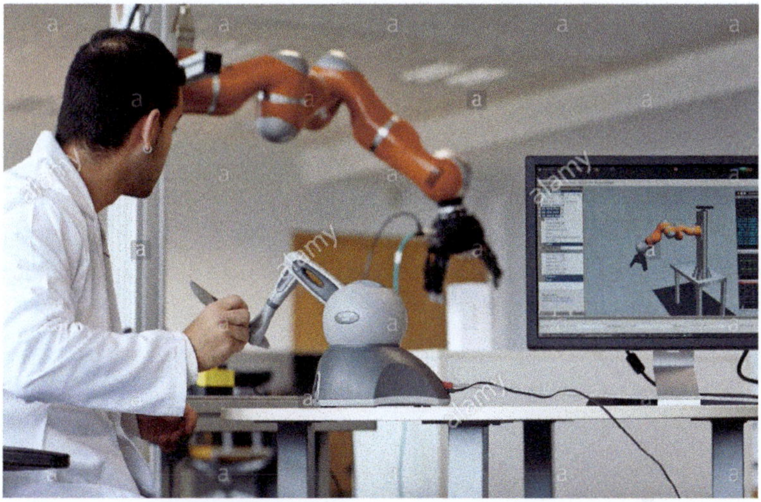

Figure 7.17 Teleoperation of an industrial robot (Source: Google Images)

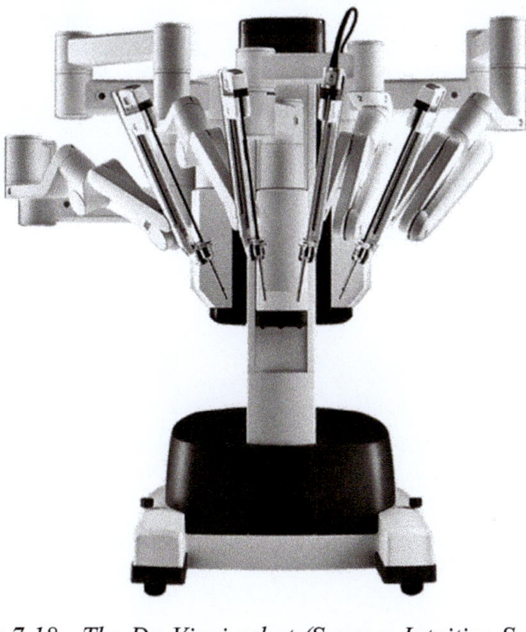

Figure 7.18 The Da Vinci robot (Source: Intuitive Surgical)

7.2.2.2 Virtual reality simulation

Virtual reality is a platform that provides an immersive experience for the user. It has been used in many fields. For instance, Balfour Beatty Rail uses virtual reality to simulate the structure that is to be built, even before a shovel is lifted. See

Figure 7.19 for a concept example of virtual reality (VR) being used for construction planning.

Urban planning is another area that benefits from virtual reality. The American Society of Landscape Architects calls virtual reality a "powerful tool for landscape architects, architects, planners and developers — really anyone involved in designing our built and natural environments." The immersive nature of VR, which allows viewers to encounter a simulated 3D landscape from multiple points of view, can be a very useful tool to city planners (Figure 7.20). They can use it to redraw streets and neighborhoods, offering real and imagined views of existing and proposed developments [17].

Figure 7.19 Virtual reality in construction (Source: Google Images)

Figure 7.20 A virtual reality view of an urban center (Source: Google Images)

Figure 7.21 Virtual reality at the Franklin Institute (Source: Franklin Institute)

Virtual reality can also be used to enhance the experience of visitors of historical sites, where visitors can now actively engage with the exhibits, for example, to "experience" historical sites or events; this can be used to attract a much larger audience (especially younger children), as opposed to traditional exhibition methods which usually only attract the older generation. One such place that has added this feature is the Franklin Institute (Figure 7.21).

7.3 Benefits of simulation

From the previous description, it is obvious that simulation has immense benefits. In this section, we will share some of them.

7.3.1 Predictive maintenance

In industrial processes, unplanned downtime is very costly, both in terms of production/service disruption (loss of revenue) and costs to restore service of assets. A non-functioning production process/equipment may still mean revenue losses of thousands of dollars per minute. Regular scheduled maintenance can help avoid (or reduce) unplanned downtime, but does not guarantee that the process/equipment will not fail. Hence, it would be very useful and valuable to be able to predict when a process or equipment will experience a breakdown, so that necessary and appropriate corrective maintenance actions can be taken; this also reduces the frequency of maintenance works. All these benefits can be achieved by predictive maintenance.

To perform predictive maintenance, operating data in normal conditions is obtained from the process/equipment. Following that, the data is processed, to create a simulation model of process/equipment (which is also commonly known as a Digital Twin). The model can also be developed using first principles or physical equations that govern the behavior of the process/equipment (in addition to the obtained operating data). The model will then be used to simulate faulty conditions, in order to obtain data during faulty operation (which we will term as faulty data).

The faulty data will then be used to develop a predictive model or a prediction algorithm. The algorithm is then deployed to run in parallel with the process/equipment, receiving real-time data of the process/equipment, and processing and analyzing the data to predict a fault is occurring, and the time before the process/equipment experiences a breakdown.

Figure 7.22 shows a workflow for predictive maintenance (adapted from [18]).

In [18], the predictive maintenance workflow is demonstrated on a triplex pump with three plungers. A 3D multibody simulation is developed (using a CAD model from the manufacturer) and optimized. Following the development of the multibody simulation (digital twin), fault behaviors are added to the model, to simulate the faulty conditions, such as seal leakage and blocked inlet. From that, (faulty) data is generated for all possible combinations of faults (individual faults, as well as those faults that occur simultaneously). Numerous simulations were carried out to account for the quantization effects in the sensor – this would have been almost impossible (and very expensive) to do without the simulation model (or Digital Twin). The faulty data is then used to train a machine-learning algorithm that will recognize patterns of faulty data. After the machine-learning algorithm is developed, it is then tested against more simulation data to verify its effectiveness, and once that is done, the algorithm can be deployed to the system.

7.3.2 Prediction

Simulation enables users to predict the output of a system for given inputs and parameters. One such example is in New Zealand [19], where uninterrupted supply

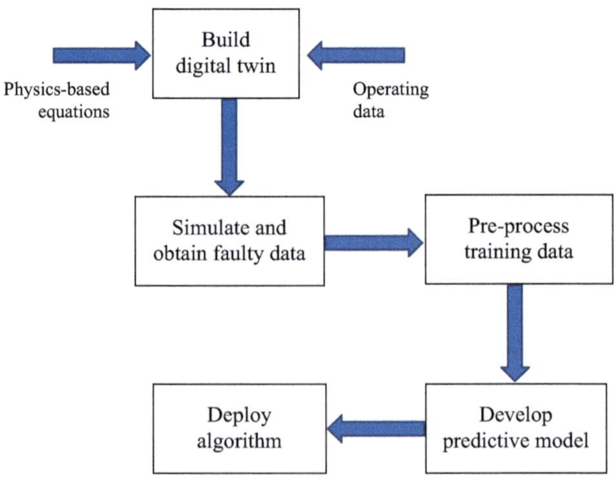

Figure 7.22 Workflow of predictive maintenance (adapted from [18])

by the national power grid is ensured. The generation power and load consumption must be kept in balance, where there must be sufficient reserve power to cater for any (increase in) demand, and yet the reserve power must not be too high as it is costly to hold generated power in reserve. Hence, there is a need to precisely calculate the minimum reserve required to ensure that the grid is kept reliable. Traditionally, to achieve this, spreadsheets were used to calculate the required minimum reserve, which required a staff member to perform many manual steps and run some scripts. This caused rigidity, where it was difficult to incorporate new generators, and to comply with ever-increasingly complex rules.

To overcome this difficulty, a reserve management tool (RMT) was developed in MATLAB and Simulink comprising a detailed model of the entire grid, encapsulating generators, load and links between New Zealand's two main islands. Every 30 min, simulations are run in the RMT, based on minute-by-minute information, to determine the currently required reserve. In those simulations, measured data (measured every minute) is injected into the RMT and verified against actual generator performance. This approach enabled the grid model and individual generator models to be improved. By doing this, reliable power supply is ensured (sufficient reserve) in a cost-effective way.

7.3.3 Design, development, and training

As explained in earlier sections, simulation enables the repetitive running of a system, quickly and at no cost. This therefore enables a user (or designer) to modify parameters and visualize and evaluate the performance of the system. From this, the user/designer can test the simulation (of a product or system) for various parameters and settings, quickly, without cost, and safely. This is particularly invaluable for sensitive systems such as aircraft or spacecraft that are expensive, sensitive, and potentially dangerous. The repetitive simulations from which the user/designer can visualize/evaluate the performance can speed up the design and development process, thus shortening the time to bring a product (or system) to market.

The ability of simulation to repetitively run a system quickly, safely, and at no cost also has benefits in human training, by significantly speeding up and reducing the cost of human training. By having an established simulator, it enables a person to be trained much faster. Using the example of a flight simulator, the following are the advantages of using simulation for human training, especially in safety-critical missions.

1. Easy replication of scenarios: Trainee pilots must be familiar with many scenarios in order for them to qualify and they need to experience these scenarios. A simulator can easily produce these scenarios and consistently replicate them. In addition, through the simulator, a customized training program can be tailored to the needs of the trainee, thus further contributing to the speeding up of training.
2. Faster training: Due to the ease and consistency of the replicated scenarios, the pilot can be trained much faster. Simulation also allows the trainee to make mistakes without any major consequences and, hence, this further speeds up the training.

3. Less expensive training: Some scenarios can be very costly to replicate, and the simulator eliminates this problem.
4. Safer training: From the previous point, some scenarios might even be dangerous to replicate (especially aircraft-related emergencies), and once again, a simulator eliminates this problem.

However, it is acknowledged that real field experience is always required in training and there are limits to what simulation can achieve. Having said that, simulation does reduce the time and cost and risk of training and, therefore, is still an integral component of training, especially in mission-critical applications.

References

[1] Embracing Industry 4.0 and Rediscovering Growth https://www.bcg.com/capabilities/operations/embracing-industry-4.0-rediscovering-growth.aspx

[2] MATLAB – MathWorks – MATLAB & Simulink https://www.mathworks.com/products/matlab.html

[3] What is CFD | Computational Fluid Dynamics? https://www.simscale.com/docs/content/simwiki/cfd/whatiscfd.html

[4] M. Blundell, and D. Harty (2004). *The Multibody Systems Approach to Vehicle Dynamics*, Elsevier..

[5] A. Lau, and I. Hoff (2018). *Simulation of train-turnout coupled dynamics using a multibody simulation software*, Modelling and Simulation in Engineering, vol. 2018, Article ID 8578272.

[6] R.A. Nazer, T. Rantalainen, A. Heinonen, H. Sievanen, and A. Mikkola (2008). *Flexible multibody simulation approach in the analysis of tibial strain during walking*, Journal of Biomechanics, vol. 41, pp. 1036–1043.

[7] M.E. Mononen, P. Tanska, H. Isaksson, and R.K. Korhonen (2016). *A novel method to simulate the progression of collagen degeneration of cartilage in the knee: Data from the osteoarthritis initiative*, Scientific Reports, vol. 6, article no. 21415.

[8] M. Wojtyra (2007). *Multibody simulation model of human walking*, Mechanics Based Design of Structures and Machines, vol. 31, pp. 357–379.

[9] COMSOL Multiphysics Modeling Software. https://www.comsol.com/

[10] Adams – The Multibody Dynamics Simulation Solution https://www.mscsoftware.com/product/adams

[11] Manufacturing Simulation – FlexSim Simulation Software https://www.flexsim.com/manufacturing-simulation/

[12] Arena Simulation https://www.arenasimulation.com/industry-solutions/industry/manufacturing-simulation-software

[13] Arena Simulation https://www.arenasimulation.com/industry-solutions/resource/manufacturing-process-optimization

[14] Case Study: Critical Warehouse Decision Making | FlexSim https://www.flexsim.com/case-studies/critical-warehouse-decisions/

[15] SUMO – Simulation of Urban Mobility sumo.sourceforge.net/

[16] PTV Vissim. http://vision-traffic.ptvgroup.com/en-us/products/ptv-vissim/
[17] How Immersive Virtual Reality Can Be a Boon to City Planners https://www.govtech.com/dc/articles/How-Immersive-Virtual-Reality-Can-Be-a-Boon-to-City-Planners.html
[18] Predictive Maintenance Using a Digital Twin – MATLAB & Simulink https://au.mathworks.com/company/newsletters/articles/predictive-maintenance-using-a-digital-twin.html
[19] Transpower Ensures Reliability of New Zealand National Grid with Reserve Management Tool – MATLAB & Simulink. https://au.mathworks.com/company/user_stories/transpower-ensures-reliability-of-new-zealand-national-grid-with-reserve-management-tool.html

Chapter 8
The role of artificial intelligence in development of smart cities

Rahulraj Singh Kler[1] *and Joon Huang Chuah*[1]

8.1 Industry 4.0 and smart cities

Industry 4.0 represents the fourth revolution that has occurred in the manufacturing industry. This revolution has happened due to the digitization of the entire manufacturing process which is causing a significant transformation in the way products are being produced today. From the first industrial revolution that involved mechanization through steam power to the second industrial revolution that involved mass production and assembly lines using motors and electricity, a huge leap in manufacturing occurred in the third industrial revolution where computers and automation were grandly introduced into the industrial scene. Programmable logic controllers (PLCs) and robots were used to significantly improve the rate, quantity and quality of manufacturing, which consequently brought a big change in the standards of living in cities today. As products enabled by automation can be more quickly manufactured and brought to the market at a lower cost, the price of products in the market have significantly dropped allowing a large fraction of the population to be involved in the advent of technology, such as computers and smartphones. Through the course of time, this has brought us to the fourth industrial revolution which takes what started in the third revolution and enhances it with smart and autonomous systems fuelled by data and artificial intelligence (AI).

Today, computers have the ability to connect and communicate with each other at a very large scale, due to the rapid development of the Internet, mobile communication technologies, computing power and storage capacities. The development has progressed to an extent that computers today are capable of making fast human-like decisions without human involvement, a feat that would have been impossible in this scale an industrial revolution ago. Through the expansion of cyber-physical systems, the Internet of Things (IoT) and AI, Industry 4.0 has become a reality, leading to the existence of the smart factory. The smart factory is a factory that is more efficient and productive, and less wasteful. It operates through smart and

[1]Department of Electrical Engineering, Faculty of Engineering, University of Malaya, Kuala Lumpur, Malaysia

interconnected machines that are capable of making decisions, communicating and logging data into a unified database autonomously.

The establishment of Industry 4.0 brings along a new concept called Smart Cities, which refers to cities that utilize large amounts of data and communication technologies available to improve the performance and quality of life and urban services to ultimately reduce resource consumption, wastage and overall costs. According to the summary of the Belt and Road initiative session on urban industrial solutions under the United Nations (UN) Industrial Development Organization, the smart city concept is an urban development paradigm integrating new innovative ideas, concepts and emerging technologies such as IoT, the Internet of Services (IoS), the Internet of People (IoP) and the Internet of Energy (IoE), which ultimately form the concept of the Internet of Everything (IoE), also known as Industry 4.0 or the Fourth Industrial Revolution. This concept aims to provide effective and high-quality public services and infrastructure in real time and, thereby, a better quality of life for citizens and a move towards sustainable cities [1]. With growing innovation and with the introduction of newer technologies, smart cities are already a reality today, and they continue to expand and become more redefined. According to CB Insights market sizing data, the global smart cities market size is projected to be worth US$1.4 trillion in the next 6 years [2].

Cities can be classified as smart cities when they are able to collect and analyse large volumes of data from many different industries to optimize usage of resources and time with the goal to maximize sustainability and environmental efficiency, and minimize wastage of time and resources. The large volumes of data are obtained from a big number of sensors, devices and software that are interconnected to share the data to a single source, allowing for all the data to be analysed quickly and systematically. For a city to truly become a smart city, it needs to fulfil these criteria for a large number of spaces, including traffic management, parking management, energy management, waste management and infrastructure planning. According to a research paper from Bialystok University of Technology, the ISO37120 standard is the most practical method to measure a city's performance [3]. Figure 8.1 shows the themes and number of indicators in ISO37120:2014 Sustainable Development of Communities: Indicators for City Services and Quality of Life.

One key aspect that defines the 'smartness' of a smart city is the ability to analyse large volumes of data in short time spans to provide real-time optimizations to services. Although the power of computing available today makes it possible for very fast data analysis, most forms of data appear in unstructured forms which require human-like abilities to understand natural language, recognize patterns and make decisions. Using AI and machine learning techniques, this is made possible. Machine learning algorithms (MLAs) are transforming the way we capture, process, inspect and analyse data to provide accurate predictions, which make such data analysis possible in a large number of industries. In fact, AI and machine learning are already being used today from the water and energy industry, to traffic management, law enforcement and the healthcare industries with proven results. With large and rapid inrush of data each second in cities, AI plays a key role in the development of smart cities.

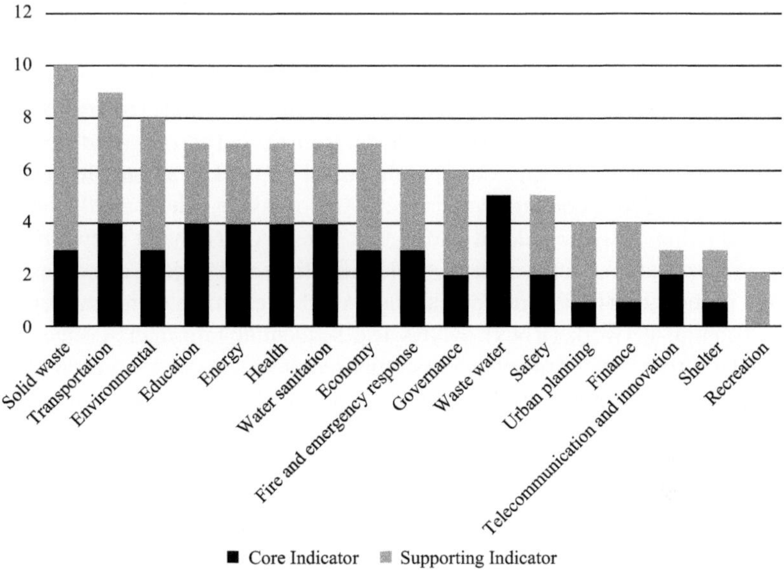

Figure 8.1 Themes and the number of indicators in ISO37120:2014 Sustainable Development of Communities [3]

8.2 Artificial intelligence

The term 'Artificial Intelligence' was first coined by John McCarthy in the Dartmouth Artificial Intelligence Project Proposal in 1956, when he invited a group of researchers to attend a summer workshop called the Dartmouth Summer Research Project on Artificial Intelligence to discuss the details of what would become the field of AI [4]. The researchers consisted of experts from various disciplines, such as language simulation, neuron nets and complexity theory. The term 'AI' was picked for its neutrality, that is, to avoid a name that focused too much on either of the tracks being pursued at that time. The proposal for the conference stated: 'The study is to proceed on the basis of the conjecture that every aspect of learning or any other feature of intelligence can in principle be so precisely described that a machine can be made to simulate it. An attempt will be made to find how to make machines use language, form abstractions and concepts, solve kinds of problems now reserved for humans, and improve themselves' [5].

However, AI has only become more popular recently due to increasing volumes of data, development of more advanced algorithms, and refinement of computing power and storage. In addition, as the necessity for automating the more specific tasks in fields such as manufacturing increases, AI is gaining more attention in today's world. In the past, the development of AI has evolved in stages. From the 1950s to the 1970s, the growth of AI revolved around neural networks. From the

1980s to the 2010s, machine learning became popular and, today, the boom of AI is driven by deep learning [6].

A neural network consists of interconnected layers of nodes, designed to very simply replicate the connections of neurons in a brain. Each node represents a neuron and the connections in the neural network begin from the output of one artificial neuron to the input of another [7]. An example of a neural network is a convolutional neural network (CNN). CNNs contain five distinct types of layers, which are the input, convolution, pooling, fully connected and output layers. CNNs have shown to be promising in applications such as image classification and object detection, natural language processing (NLP) and forecasting. Another example of neural networks is recurrent neural network (RNN). RNNs use sequential information such as time-stamped data from sensors or sentences composed of sequence of words. RNNs work differently from traditional neural networks, in which not all inputs to RNNs are independent of each other, and the output for each element depends on computations of the preceding elements. They are widely used in forecasting, time series applications, sentiment analysis and other text applications.

Machine learning is a technique of data analytics that teaches computers to learn from experience, that is, from recurring patterns. MLAs utilize computational methods to learn directly from data without initially requiring a model in the form of a predetermined equation. As the sample size increases, the algorithm adapts and the performance of the algorithm improves. Machine learning uses two types of techniques, that is, supervised learning and unsupervised learning. Supervised learning methods such as classification and regression develop a predictive model based on both input and output data, so it gains the ability to predict output data based on input data, whereas unsupervised learning methods such as clustering group interpret data based only on input data so that they can find hidden complex patterns that exist in data. Common algorithms for classification are support vector machines (SVMs) and k-nearest neighbour. Common algorithms for regression techniques are linear regression and nonlinear regression. Common algorithms for clustering are hierarchical clustering and subtractive clustering.

AI is gaining importance for a number of reasons. Through the use of years of collected data, AI automates repetitive learning and the process of discovery. However, AI is more than just hardware-driven robotic automation that simply automates manual tasks. AI has the power to perform high-volume and frequent tasks reliably and without fatigue. More importantly, AI can even automate tasks that require cognitive abilities that were only possible previously by humans. Moreover, AI incorporates intelligence into existing products. In most cases, AI will not be packaged as an individual product. Instead, products that already exist in the market will be embedded with AI capabilities. Examples of such incorporation today are the addition of Siri to newer generations of Apple products and the addition of Watson to IBM's computing services. With the large amounts of data that are being created and stored today, along with tons of data that already exist, AI in the form of automation, conversational platforms, bots and smart machines will improve many technologies in homes and at the workplace in a plethora of fields from security intelligence to finance and banking.

Furthermore, AI can adapt through progressive learning algorithms so that instead of programmers coding the rules based on the data, the data can do the programming by generating rules based on past occurrences. AI looks for structures and patterns in data, which turns the algorithm into a predictor or a classifier, as it can then search for similar patterns in new data to make decisions. For example, an AI algorithm programmed to play chess can teach itself the next set of moves that it needs to make based on the previous moves and whether the outcome of the moves were positive or negative. Similarly, an algorithm for a movie streaming service such as Netflix programmed to recommend movies to users can learn the kind of movies certain users prefer to watch and recommend movies of similar genre to provide greater convenience to the user. As the models receive more and different kind of data, they adapt. As a user learns a new genre of movies and begins to have a preference for this genre; the algorithm learns of this preference and begins to recommend different types of movies from this newly preferred genre. An AI technique of adapting to data is backpropagation, which allows the model to adjust through training and added data when the initial results are incorrect.

As computing power and storage has skyrocketed in the previous decades, analysing greater volumes and depth of data using neural networks that have many hidden layers has become possible in recent years. Just a few years ago, building a fraud-detection neural network algorithm with just five hidden layers was almost impossible. The same task can easily be done today with the computing power and big data that exist. Also, large amounts of data are necessary to train deep learning models with many hidden layers since they learn directly from the data. The more data that are fed to the models, the more accurate the models become. For example, products such as Amazon's Alexa, Google Search and Google Photos that obtain large amounts of data daily through interactions with people, and are based on deep learning, are getting more accurate the more they are used. In the medical field, AI techniques from deep learning such as image classification and object recognition can now find cancer on magnetic resonance imaging (MRI) with similar accuracies as seasoned radiologists. Ultimately, with AI, the quality of predictions is almost completely dependent on the quality of the data. This makes data become intellectual property. Since data now have a more important role than ever before, these can create a competitive advantage. The data of greatest quality mean greatest quality of predictions, classifications and recommendations.

Common applications of AI is machine vision and object recognition, NLP and cognitive computing. The role of AI in smart cities that are elaborated in the Section 8.2.1 mostly utilizes these three applications that are elaborated hereafter.

8.2.1 Machine vision and object recognition

According to Tractica News, the computer vision hardware and software market is projected to be US$48.6 billion by 2022, at a compound annual growth rate of 32.9 per cent [8]. Computer vision is a field of research involving efforts to provide computers with the ability to visually sense the world through cameras, in the way humans use eyes to interact with the world. This is done by automatically extracting, analysing and understanding valuable information from images. One of the most popular applications of machine vision today is object detection. Object detection is an aspect of machine vision that involves detecting and classifying objects in an image or video such as humans, vehicles and different types of food. Object detection enables the automation of repetitive tasks such as crowd counting that would previously only be possible through the use of more intricate and expensive sensors and impossible though the use of ordinary cameras. Object detection has been around for a long time; however, it has only become very popular in recent years through the discovery of state-of-the-art AI-based object detection models that make the process of object detection extremely accurate and possible at real-time speeds. Today, tools such as the Tensorflow Object Detection application programming interface (API) make training an object detection model possible in an average computer equipped with a graphics processing unit (GPU) very quickly. As a result, object detection and recognition technologies have become widely available for anyone to experiment with and enable themselves to become a popular field of research with rapid growth. Different techniques of object detection that exist include the region-based convolutional neural network (RCNN), Fast RCNN, Faster RCNN, single shot multibox detector (SSD), you only look once (YOLO) and RetinaNet. Some of these techniques are explained hereafter.

A deep learning approach to perform object detection is using the RCNN object detection model (Figure 8.2). Before understanding the operation of a RCNN model, it is necessary to understand how a CNN model works. A CNN model takes an image as input, divides the image into various regions, considers each region as a separate image, passes the regions through a CNN and classifies them into various classes, and combines the regions with the classes to get the original image with the detected

Figure 8.2 Example of object detection from Tensorflow Object Detection API [9]

objects [9]. Using a CNN-based model has a disadvantage, that is, objects in the image may have varying spatial locations and aspect ratios. This problem can be solved by breaking the image down into a large number of regions; however, this requires a large amount of computational power and takes up a significant amount of time. RCNN solves this problem by only passing certain regions of interests (RoI) to the CNN. This is done using a region proposal method such as selective search. Selective search takes an input image, generates initial sub-segmentations to produce different regions of the image, then combines similar regions based on colour, texture, size and shape similarity to form larger segmentations, and these segmentations produce the final object location, that is, the RoI. A RCNN model is developed by taking a pre-trained CNN model and re-training it by training the last layer of the neural network based on the number of classes that are to be detected. When the model is deployed, it takes an image, obtains the RoIs using a region proposal network (RPN) such as selective search, reshapes the regions of interest so they match the CNN input size and uses a SVM to classify objects. A separate binary SVM is trained for each class to be detected. This produces approximate bounding boxes in the locations of the detected objects. A linear regression model is used to obtain tighter bounding boxes that more accurately bound the detected objects in the image. Because a large number of steps are taken in detecting the images, a RCNN model can take up to 50 s to make predictions for each new image, making the model infeasible for object detection and cumbersome for training the model with large datasets.

Because of the disadvantages of RCNN, the Fast RCNN model was developed. Instead of using three different models such as RCNN, Fast RCNN uses a single model that extracts features from the regions in an image, divides them into the different classes and returns the boundary boxes for the identified classes in a single go. Hence, instead of needing to pass thousands of regions of an image through a CNN which is computationally expensive, the entire image is passed through the CNN once. This is done by passing the image through the region proposal method, which produces the RoIs. Then, a RoI pooling later is applied to all these regions to reshape them according to the input of the region proposal method, and the regions are simultaneously passed to the CNN. At the top of the CNN, a softmax layer is used to output the classes. Also, a linear regression layer is used in parallel to output the bounding box coordinates for each detected class. This makes Faster RCNN significantly faster than RCNN. However, because selective search is still used in Fast RCNN as in RCNN, and since selective search is a time consuming process, the detection is slowed down and takes about 2 s to make predictions for each new image.

To solve this problem, Faster RCNN was developed, which runs even faster than Fast RCNN. As compared to Fast RCNN which uses selective search to generate the RoIs, Faster RCNN uses a method called RPN. First, the image is passed through the CNN, which produces the feature maps of the image. The feature maps are then passed to the RPN, which uses a sliding window over the feature maps, and generates k-Anchor boxes of different shapes and size at each window. Two predictions are made for each anchor – (a) the probability that the anchor is an object and (b) the bounding box regressor for adjusting the anchors to they better fit the object. This creates bounding boxes of different shapes and sizes that are then passed on to the RoI pooling layer.

The pooling layer extracts fixed size feature maps for each anchor. The feature maps are passed to a final layer that contains a softmax layer and a linear regression layer. These layers produce the classes of the detected objects and the bounding box coordinates for each object. With the improvements made from Fast RCNN to Faster RCNN, predictions for each new image take approximately 0.2 s using the Faster RCNN model.

The SSD object detection model is designed for real-time object detection. The term 'Single Shot' means that tasks of object localization and classification are done in a single forward pass of the neural network, 'MultiBox' is the name of a bounding box regression technique and 'Detector' means that the network is an object detector that classifies the detected objects. Unlike the Faster RCNN model which uses a RPN to create boundary boxes and then uses these boxes to classify objects which creates predictions of great accuracies, the SSD eliminates the RPN from the model. This is because the RCNN model runs at about 7 frames per second, which is extremely slow for real-time object detection. However, removing the RPN comes with a cost of accuracy. The SSD compensates this loss by including multiscale features and default boxes. This makes the SSD able to achieve real-time processing speed, and an accuracy even better than that of the Faster R-CNN model. On standard datasets such as PascalVOC and COCO, the SSD scores over 70 per cent mean average precision (mAP) at 19 frames per second [10].

One of the most modern approaches of object detection, yielding speeds and accuracies greater than other approaches, is YOLO. As its name suggests, contrary to the previous models, YOLO takes the entire image in a single instance and predicts the class probabilities and bounding box coordinates for the detected objects in the image. YOLO works by taking an input image, dividing the input image into grids, for example 3 $\times$ 3 grids, and then performing image classification and localization on each grid. Then YOLO predicts the class probabilities and bounding box coordinates for the detected objects. On standard datasets such as VOC2007, YOLO has shown to produce a mAP of 63.4 per cent and detection at 45 frames per second.

Another aspect of computer vision that is growing in popularity is human pose estimation. Human pose estimation is the prediction of body parts and joint positions of a person from an image or a video. It is a growing problem with a growing number of applications in safety and surveillance, healthcare, transportation and sports. Examples of these applications include real-time video surveillance, detection of stroke and heart attacks, assisted living, advanced driver assistance systems (ADAS) and sports analysis. Human pose estimation is generally classified into two categories: (a) 2D-pose estimation and (b) 3D-pose estimation. 2D-pose estimation involves the estimation of a human pose in x- and y-coordinates, whereas 3D-pose estimation involves the estimation of a human pose in x-, y- and z-coordinates for joints in a red/green/blue (RGB) image. Human pose estimation is considered a challenging problem to solve due to variations between humans that exist in images, and other factors such as occlusions, clothing, lighting conditions, small joints and strong articulations.

Previously, classical approaches were used for articulated pose estimation, such as the pictorial structures framework [11]. This is a structured prediction task involving the representation of an object through a collection of 'parts' that are arranged in a deformable configuration. 'Parts' are templates of human joints that are matched to those in an image and springs are used to show the spatial connections between parts. This method can model articulation, producing an approximation of a human pose. Another classical approach for human pose estimation is the deformable part models [12]. This approach used a mixture model of parts that express complex joint relationships. Similar to the pictorial structures framework, this model is a collection of templates in a deformable configuration; however, this model is more complex as it consists of more 'parts' with global and part templates.

However, the classical approaches have their limitations such as lack of expressiveness and today, research in pose estimation greatly involves the use of CNNs. After the introduction of the DeepPose human pose estimation method based on deep neural networks, most research on human pose estimation started to rely on deep learning. Today, ConvNets have replaced most programmed features and graphical models as the main building block of most pose estimation systems, producing large enhancements of pose estimation performance on standard benchmarks. Some of the most popular deep-learning-based human pose estimation approaches are DeepPose, Convolutional Pose Machines, Efficient Object Localization using Convolutional Networks, Human Pose Estimation with Iterative Error Feedback and Stacked Hourglass Networks for Human Pose Estimation.

Although many research papers applying deep learning and other AI approaches for human pose estimation were published, DeepPose was the first major paper that applied deep learning in human pose estimation, yielding state-of-the-art performance and beating existing models. DeepPose formulates the problem of pose estimation as a CNN-based regression problem towards body joints. A cascade of regressors are used to refine the pose estimations. The research paper on DeepPose (Figure 8.3) states that CNNs provide a holistic reasoning of human poses, so that even if joints are unclear, small or hidden, the CNNs are able to reason and estimate the position of the joints [13].

Typically, pose estimation involves the inference of locations of body parts or landmarks and the quality of the prediction is based on metrics involving comparisons between the predicted and ground truth location in the image plane. Another active field of research is the estimation of human joint angles, which involves the estimation of angles made by segments of the human body at the joint landmarks in world coordinates. This can be done in two different ways: (a) directly regressing the joint angles from an image and (b) estimating the joint angles using geometry. The first method requires training a joint angle estimation model by feeding it with images of human poses and the joint angles, which can be recorded using sensors such as the *Qualisys* motion capture (*MoCap*) system, through which the deployment of the model would yield the joint angles from the input images. The second method involves performing human pose estimation on an image, which produces a 2D or 3D spatial arrangement of all the body joints as the output. This can easily be achieved through the use of trigonometry, producing the necessary joint angles in three dimensions, if the human pose estimation is done in three dimensions.

Figure 8.3 Example of human pose estimation from DeepPose [13]

An application of human pose estimation is in making a robot follow the trajectories of a human pose, so that it replicates motions of a person. Robots that are useful in this area are dual-arm robots with multiple axis per arm, such as ABB's YuMi, with two arms and seven axes per arm. This application can be achieved by taking frames of a person and passing it to a human pose estimation model, which yields a 3D spatial arrangement of the body joints, computing the joint angles of the arms using geometry, and sending these angles to each appropriate axis of the robot arm. The axes of the robot arm are then programmed to move to these angles. Performing this in real-time would yield real-time replication of human pose on a dual-arm robot. This application may gain large importance in training a robot to produce more human-like motion, which becomes useful in the healthcare field when programming robots for assisted living for the elderly or in performing robotic surgeries.

8.2.2 Natural language processing

NLP is a field that enables machines to understand, analyse and manipulate human language. It is a large field that encompasses many disciplines, from computer science to computational linguistics. The ultimate goal of NLP is to address the gap that exists between human communication and machine language. While humans speak a large number of diverse languages, ranging from English to Chinese to Spanish, computers inherently work in and understand machine language, which at the lowest levels of machines, are complex combinations of ones and zeros. Over 70 years ago, programmers punched cards indicating ones and zeros to communicate with machines. NLP is aimed at enabling computers to understand human communication and, at more complex levels, human interaction, and at the same time enabling humans to harness the large amount of computing power available today to accomplish tasks ranging from shuffling to the next song, to translating a paragraph of an email into a different language while maintaining the tone of the text.

NLP is considered a difficult problem to solve due to the nature of human language. The diversity of the rules of human languages makes it difficult for NLP tools to be programmed to understand and analyse a wide range of texts and conversations. When dealing with the basic rules of grammar, such as the addition of the letter 's' at the end of a noun to indicate plurality, simple if-else conditions can be written when developing the NLP tools. However, dealing with sarcasm in the passing of information is a complex problem to solve and would require for the NLP tool to understand the patterns of a conversation and the topics recently conversed about in order to detect the sarcasm. When developing an NLP tool to solve complex problems in processing text or conversations, AI is a key ingredient because of its ability to recognize patterns. Deep neural networks are widely being used today to develop more powerful NLP systems that are capable of having a human-like conversation with people. Examples of these applications are automated AI-powered chatbots used by many service providers today to provide instant responses to queries of their customers.

Although NLP has been around since the 1950s when Alan Turing published an article titled 'Computing Machinery and Intelligence' which proposed what is now called the Turing test as a criterion of intelligence, it has recently been gaining a lot of attention due to increasing interest in human-to-machine (H2M) communications, in addition to the large volumes of data being stored each day, increasingly powerful computing power available, and enhanced algorithms. Also, with the exponential growth of AI and deep neural networks that have proven to be a promising when used in NLP tools, the NLP market is growing at a rapid rate. According to Gartner's 2018 World Artificial Intelligence Industry Development Blue Book, the global NLP market will be worth US$16 billion by 2021. NLP is very quickly becoming a vital tool for businesses, for applications such as sentiment analysis, chatbots, customer service, managing advertising funnels and market intelligence.

One of the simpler tasks of NLP with AI is text classification. Similar to how object detection models classify objects within an image, text classification through NLP involves the classification of text into different predefined categories. This becomes important in applications such as spam detection or sentiment analysis where lines or paragraphs of text are the input of the NLP model, and the output of the model is a 'true' or 'false' for spam detection systems or 'happy' or 'sad' for sentiment analysis systems for example.

Another task of NLP is topic discovery and modelling. Topic modelling is the task of using unsupervised learning to extract the main topics that occur in text or a collection of documents; hence, it can be easily compared with clustering. Most of the time, data appear in an unstructured form, such as a large range of topics that are posted in a blog website every day or a large range of products that are posted on a merchant website. By performing topic modelling on the data, we can get a set of topics that are more common and allow for the large amount of data to be classified within these topics. Several algorithms for topic modelling that exist are latent dirichlet allocation (LDA), which is based on probabilistic graphical models, and latent semantic analysis (LSA) and non-negative matrix factorization (NMF), which are based on linear algebra.

8.2.3 Cognitive computing

Cognitive computing is the simulation of human thought processes in a machine or a computerized model. It is defined as the set of industrial strength computational componentry required to deliver cognition as a service (including the three Ls – language, learning and levels) to customers. Cognitive computing enables cognition as a service and powers cognitive services that augment and scale human expertise [14]. The three key pillars of cognitive computing are machine learning, NLP and hypotheses generation with evidence-based explanation capabilities. Using cognitive characteristics of the human brain such as data mining, NLP, pattern recognition and other self-learning algorithms, cognitive computing allows computers to make decisions based on events or conditions, just as a human would. However, replicating human thought processes is not an easy task, as we still do not have a full understanding of it. Despite computers gaining unparalleled speeds at calculations and processing far beyond human capabilities, they have not been able to perform tasks that humans consider simple, such as understanding natural language or detecting objects in an image.

An example of a cognitive system is IBM's Watson, which relies on neural networks and deep learning algorithms. It processes data by comparing it to a teaching set of data. Hence, the more data IBM's Watson is fed, the more it learns and the closer its decisions are to the correct decision. Another company, VantagePoint AI, uses cognitive computing through MLAs to solve the challenge of providing accurate, fact-based investment recommendations in the financial sector. Picking a stock that performs well and is expected to have increasing value involves a lot of guessing and studying of stock patterns by investors. Edge Up Sports is another company in the sports industry using cognitive computing to help coaches and personal trainers develop their team. It leverages AI to help reduce sports injuries in the field using sensor-tracking cameras on athletes that translate their motion into performance insights to deliver health and activity insights to coaches. It is predicted that vast cognitive abilities of global CaaS (cognition as a service) providers will be cheap and available via APIs for every device from the nano-scale up to the giant global applications and services. In such a world, cognition as a service could be as ubiquitous as electricity is for us today [15].

In the education sector, cognitive computing can become a heavy driving force in the teaching and learning process to maximize a student's potential. Similar to a teacher that has to go through numerous training and years of experience before they become skilful in educating students, a cognitive computing program needs to be fed with large amounts of data regarding the process of development of students and the syllabus in an education system before it can effectively aid in the process of education. However, a teacher cannot simply instantly transfer all their knowledge and education to a new, unexperienced teacher does. A new teacher would need to go through years of training before they become as skilful as a seasoned teacher. However, a well-trained cognitive programme can quickly and easily be replicated without needing to perform the entire training process again. This makes it possible for cognitive computing to be used in classrooms as personalized assistants for each individual student, focusing on the strengths and weaknesses of each student and providing the necessary level of difficulties in exercises so that the student improves quickly [16]. This can relieve the stress that teachers face when teaching a class with a large number of students. A single teacher also cannot cater to the needs of each individual student without neglecting the progress of other students. Such a cognitive computing application in education can also develop many techniques, such as creating lesson plans for students. Today, AI in the education process already exists as mobile applications, such as Mika and Thinskter Math, virtual tutors that adapt to the needs of students and help them to improve in subjects.

Cognitive computation is expected to reshape the business sector in three ways. Firstly, cognitive computation will improve employee capabilities, contributions and performance. Deep learning algorithms will speed up and increase the quality of work by automating low-value tasks such as collecting statistical data or updating client records with financial or demographic data. Predictive coding, an AI mechanism, will help lawyers mine through large volumes of data to speed up the process of exploration to provide clients with increased value and better services. Secondly, cognitive computation will enable higher-quality data analysis. Organizing data into meaningful chunks is not an easy process, especially if the data come in many different structured and unstructured forms. Deep learning algorithms will enable accurate, timely and meaningful analysis of large volumes of data, saving big amounts of time. Thirdly, cognitive computing will enhance overall business performances. Cognitive computing will help businesses to capture relevant information and use it to make good decisions when delivering better produces or higher-quality services to the market using market statistics and the results of previous decisions made by businesses. In today's era, the most successful businesses are those that combine strategic planning and smart technologies embedded with cognition. Two companies that have both these features are Google and Amazon, two among the best performing companies today.

8.3 Role of AI in smart cities

8.3.1 Safety and surveillance

Safe city is a pivotal concept of a smart city. Safety and security are extremely crucial components of the quality of life in a city. Hence, a smart city needs to be a safe city to live in. The focus of the Campbell Institute Symposium of the National Safety Council revolved around the implications of IR4.0 on safety [17]. The label most ascribed to this era is 'Cyber-Physical System'. According to projections from the UN, 68 per cent of the world's population will live in urban areas by 2050. With the growth of more and smarter cities, governments are investing more in public infrastructure to improve public services, such as transportation and healthcare, operations and the overall urban experience. With more interconnected devices through the bloom of IoT, more data are being transported through the Internet than ever before. With the advancement of telecommunications technology and an increase in the number of interconnected devices, leaders and citizens of cities are being introduced to new security and privacy risks each day. According to a study conducted by McMaster University, 88 per cent of people in Canada are concerned about their privacy in the smart city context [18].

As the implementation of smart cities continues to accelerate, cities are finding it more difficult to identify, respond and prevent cyberattacks and privacy risks for the following reasons [19]:

1. Security is still not a priority to city planners. Unprotected devices are still very vulnerable as they can be accessed from Shodan.io.
2. Cities are not prepared for the huge volumes of data being collected.
3. Security teams are unprepared to deal with cybersecurity vulnerabilities.
4. Physical threats to connected systems are not thoroughly addressed.

With the rapid growth of AI and IoT in recent years, AI has shown many times to be capable of having real-world applications in aiding and even completely supervising security systems. As we approach human limitations in performing these applications, AI is showing to become a viable replacement of humans in the field of security. According to NBC news, the police department in many cities across the United States is already short-staffed, making it more difficult to ensure public safety all the time in all places [20]. Some ways smart cities can use AI along with IoT and sensors to improve overall safety and security in cities include using video analytics to monitor camera surveillance in real-time, using sensors and cameras to monitor large crowds to detect potential threats, and using patrol robots and drones for monitoring potentially unsafe areas. Three major areas in safety and security that are expected to be vastly improved through the application of AI are as follows:

1. Crime detection
2. Crime prevention
3. Cybersecurity.

8.3.1.1 Crime detection

One of the main goals of using AI is detecting the occurrences of crime and supporting authorities in arriving at the location of the crime in the least amount of time possible. For example, upon the firing of a gunshot, it is difficult for the police to pinpoint the exact location of the gunshot and evaluate the criticality of the situation. The *Shotspotter* advanced surveillance technology is capable of solving this problem by sensing gun violence through the use of acoustic sensors and AI [21]. It sends an alert to the police department and other authorities within 45 s of a gunshot incident. The alert contains important situational intelligence regarding the incident, such as the precise location, number of gunshots fired and single or multiple shooters [22].

Figure 8.4 Integration of real-time gunfire detection to GE's LED lighting and cloud-based platform [23]

It is also very flexible in installation, as it can be independently installed on rooftops, poles, streetlamps and other areas out of sight. *Shotspotter* also has exclusive arrangements with General Electric, featuring the integration of real-time gunfire detection to GE's LED lighting and cloud-based platform (Figure 8.4) [23].

Furthermore, vision systems using AI have the capability of detecting weapons and notifying the authorities before any occurrences of casualties. For example, Athena Security has developed an AI camera system that is capable of recognizing guns and alerting authorities through their cloud system. The camera system can be connected to third-party security systems, so that intervention such as stopping elevators or locking doors can be executed. Athena Security claims that their system's gun detection has an accuracy of 99 per cent, effectively minimizing occurrences of false positives [24].

8.3.1.2 Crime prevention

Besides detecting crime, AI has also proven to be useful in the overall prevention of crime in cities. Using the pattern-recognition ability of MLAs, crime patterns can be deconstructed, heatmaps of crime criticality can be constructed and future offences in cities can be predicted.

In Vancouver, the police department has deployed a machine learning solution which performs spatial analysis to predict where residential break-and-enters will occur so police patrols can be placed accordingly to minimize crime [25]. The system was developed by geospatial engineers and statisticians who developed an algorithm to properly pinpoint property crime patterns. The system predicts crime within a 100-m radius and 2-h time window based on the past occurrences of crime. According to Esri, the developer of the crime-forecasting system, the system has shown an accuracy greater than 80 per cent.

In California, Aveta Intelligence deploys AI in predicting terrorists' activities. This is done through descriptive, diagnostic and predictive analytics with cognitive intelligence and machine learning to dynamically learn and respond to different types of security threats [26]. The company utilizes many tools, such as graph analysis, neural networks, heuristics, complex event processing, simulation, recommendation engines and machine learning to assist in improving security in cities. Aveta Intelligence offers two core products: Trust and Sentinel, where Trust determines potential threats within an organization through social behaviour, predictive and competitive analytics, and Sentinel predicts physical security threats for law-enforcement professional and campus security by analysing physical data such as social media, incident reports, event schedules and closed circuit television (CCTV) footages to prevent crimes.

Another company that uses AI in crime prediction for prevention is Predpol, short for predictive policing. Predpol claims that its crime analytics algorithms can improve crime detection by 10–50 per cent in some cities. Using years of historic data such as the type, location and time of crime occurrences, and other socio-economic data, the algorithm, initially designed to forecast earthquake aftershocks, performs crime analysis and predicts the location, time and type of crime that will occur over the next 12 h, and as new data are input, the algorithm is updated daily [27]. The technique used is called 'real-time epidemic-type aftershock sequence crime forecasting' by the company. Crime predictions are displayed on a map through coloured boxes, each representing an area of 46 square meters. 'High-risk' locations are bounded in red boxes and police patrol are encouraged to spend at least 10 per cent of their time there.

8.3.1.3 Cybersecurity

As the number of connected devices rapidly increases, with the number of connected IoT devices expected to rise to almost 20 billion by 2020 [28], cybersecurity is becoming more critical in the development of smarter cities. Cyberattacks in one region can have an amplifying effect on other regions, leading to severe data loss, reputational damage risks and financial impacts for businesses, and can even disrupt critical city services and infrastructure in a plethora of domains, including healthcare, transportation, residential, power, utilities, law and enforcement services. Such vulnerabilities may cause mortal danger and breakdown economic and social systems.

Today, the most common method of network protection is by deploying a firewall, either installed as a software or physically connected to the network as a hardware. Firewalls track network connections and track which connections are allowed on network ports, blocking other requests. Generally, network administrators of a server control authorizations of incoming connections to certain ports in the system. If a hacker manages to break through the system's firewall and network security, the following line of defence is the antivirus system that is installed in the server. The antivirus system is, typically, designed to search for malicious code in software or suspicious connections, aimed to remove a malware or virus before it can attack the system, such as a ransomware, or spread to other interconnected machines within the organization. Antivirus systems also prevent suspicious connections from obtaining personal or proprietary information from a system without the owner's permission. Typically, organizations also store backups of their systems and have an organized backup management system to prevent data loss in the case of an attack, for example, a ransomware. With a disaster recovery plan, they can quickly recover from outages or data breaches. The techniques currently used in cybersecurity ultimately depend on the expertise of the individuals operating the network's servers and the skills of the individuals are built through vigorous training and experience. The most successful organizations constantly educate their new and existing employees on the threats that exist online and how they can protect themselves and the organization.

Most existing cybersecurity tools require interaction with a human for system and network configurations, firewall policies, backup schedules and ultimately ensuring that the system is running successfully. Using AI in cybersecurity systems, human interaction with the system can be minimized. Also, as employees in an organization need to be constantly trained to operate network security and as employees in an organization are constantly changing, an advanced AI-based cybersecurity network may be superior since it can undergo transfer learning and only becomes more accurate in time as it collects more information.

AI tools today can already handle a large amount of tasks that are done by network administrators such as event monitoring and incident responses. Artificially intelligent firewalls will have in-built MLAs that allow them to recognize patterns in network requests and automatically block those that may be a threat. Also, AI systems have natural language capabilities which are expected to play a big role in the future of cybersecurity. By scanning large pools of data across the Internet, cybersecurity systems embedded with NLP tools can learn how cyberattacks occur and provide alerts to the organization before a significant issue arises. A company that develops NLP-embedded cybersecurity tools is Armorblox. According to Dhananjay Sampath, the cofounder of Armorblox, email messages and files are the top vectors for stealing data [29]. Based on a Trend Micro survey, 91 per cent of cyberattacks begin with a phishing email with malicious file attachments or links to malware. According to Armorblox, US$12 billion has been lost in business email compromise in the past 2 years. Armorblox is an NLP engine that obtains insights from organizational communications and data. It learns what is mission-critical for an organization and offers policy recommendations. In the case of a potential breach, it sends alerts to relevant people in the organization so preventive measures can be taken before an attack. According to Maurice Stebila, Chief Information Security Officer of Harman, Armorblox uses natural language understanding to tackle a security layer that has been unreachable by other security systems, that is, the human layer – context of communications between humans. Hackers are aware of this weakness and natural language understanding is capable of solving this problem.

An intrusion detection and prevention system (IDPS) is an installed software or connected hardware system that detects occurrences of system breaches and attempts to prevent them. Typically, an IDPS should have these characteristics to provide efficient security against cyberattacks as shown in Table 8.1. Over the past

Table 8.1 Characteristics of an ideal intrusion detection and prevention system

Characteristics of IDPS	Description in terms of smart factory
Real time	Intrusion detection while attack is in progress Intrusion detection immediately after attack
Reliable	Minimum false positives
Independent	Minimum human supervision Continuous operation Self-monitoring ability to detect breach attempts
Recoverable	Ability to restore system after accidental crashes Ability to restore system after cyberattack
Compliant	Monitoring of users' behaviour with regard to security policies of the system
Adaptable	Accommodate to change in user behaviour over time Adjust to system changes

few years, IDPS systems have been integrated with MLAs, genetic algorithms, fuzzy logic and artificial neural networks to increase the rate of detection and minimize the detection time for the systems [30].

According to Capgemini's *Reinventing Cybersecurity with Artificial Intelligence Report* [31], the five highest AI use cases for improving cybersecurity are fraud detection, malware detection, intrusion detection, network scoring risk and user/machine behavioural analysis. Twenty use cases across IoT, IT and operational technology have been analysed and ranked according to the complexity of the implementation and benefits in terms of time reduction. According to Capgemini's analysis, Forbes recommended five of the highest potential use cases with high benefits and low complexity. Fifty-four per cent of enterprises in the United States have already implemented the use cases [32]. Figure 8.5 shows a comparison of the recommended use cases by the level of benefit and relative complexity.

8.3.2 Healthcare

For a long time, hospital and clinics have been collecting and archiving large volumes of data including patients' information and treatment processes information. With available computing power, the rapid development of AI techniques and vast datasets from the health industry, MLAs can be used to revolutionize the way hospitals and healthcare institutions utilize the data to make predictions that can identify potential health outcomes. Overall, AI is expected to transform the healthcare industry through the following aspects:

1. Wellbeing
2. Early detection
3. Diagnosis
4. Decision-making
5. Treatment

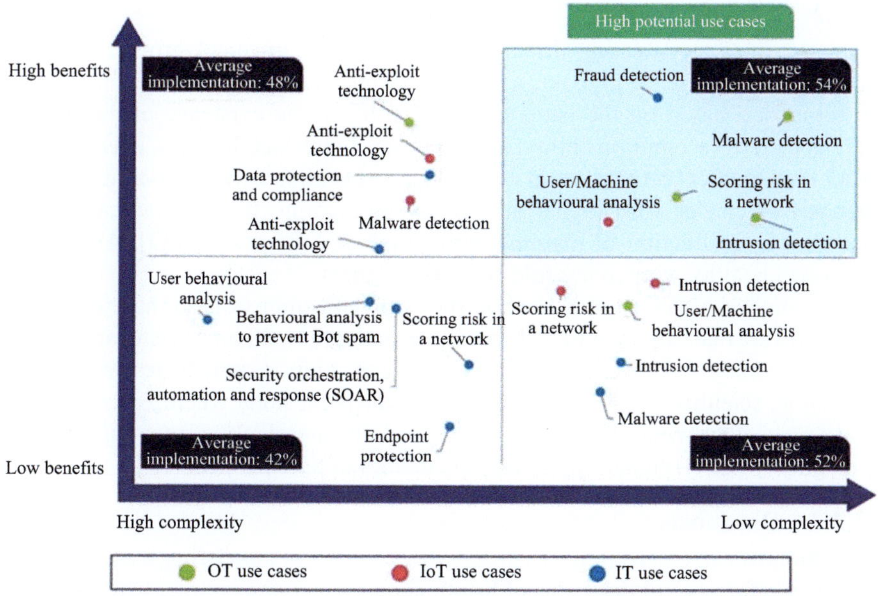

Figure 8.5 Recommended use cases for artificial intelligence in cybersecurity [32]

6. End of life care
7. Research
8. Training

8.3.2.1 Wellbeing

One of the greatest potential benefits of AI is to assist people in keeping their health in check so doctor appointments become less frequent. AI-powered applications in mobile devices can track many vital aspects of health and provide suggestions or changes that need to be made to improve the wellbeing of the user and improve the state of their health in the long term. The use of AI and the Internet of Medical Things (IoMT) in consumer health application is already in use and is helping people.

Mobile healthcare apps that utilize AI include TalkLife, which helps users with mental health issues, depression, self-harm and eating disorders by offering them a safe place to talk about their problems, and Flo Period Tracker, which makes accurate and reliable predictions of menstruation, ovulation and fertile days using machine learning, even if users have irregular cycles. In addition, AI in mobile applications enable healthcare professionals such as doctors to obtain information regarding the day-to-day patterns of a patient's diet, exercise routine and other health-related aspects and with this information, they are able to provide better feedback, support and guidance for staying healthy.

8.3.2.2 Early detection

When it comes to detecting diseases such as cancer, medical professionals use years of their experience in looking for patterns that show abnormalities in a scan or diagnosis, and based on the patterns found, confirm the existence of the disease. With its pattern recognition ability, AI can be used to detect diseases, even in stages earlier than medical professionals can. In fact, AI is already being used in the early diagnosis of cancer with higher accuracies. According to the American Cancer Society, a large fraction of mammograms produce false results [33]. As a result, one in two healthy women are told they have cancer.

Researchers at the Houston Methodist Research Institute in Texas have developed an AI software that can review and translate mammograms 30 times faster and with 99 per cent accuracy, reducing the need for unnecessary biopsies which are invasive [34]. In addition, scientists at Mayo Clinic have conducted a study where AI was used to detect early-stage heart diseases quickly and inexpensively using only electrocardiography (ECG) signals, producing accuracies of 85.7 per cent [35].

8.3.2.3 Diagnosis

In addition to early detection of chronic diseases, AI is also being used for large-scale, powerful general diagnosis of health conditions. For example, IBM's Watson for Health is aiding many healthcare organizations in applying cognitive technology for powerful diagnoses. Watson is capable of storing far more medical information, including every treatment case study and response, every medical journal and every symptom and reviewing them at speeds exponentially faster than humanly possible [36]. In addition, Google's DeepMind Health is partnering with clinicians and researchers to solve healthcare problems and bring patients from tests to treatment in lesser time using neural networks that mimic the human brain. Scientists are also using AI methods to vastly increase the accuracy of specific disease diagnoses. An AI system has been developed using artificial neural networks (ANN) algorithms to diagnose heart diseases from phonocardiogram signals, yielding accuracies of up to 98 per cent [37].

8.3.2.4 Decision-making

Artificially intelligent cognitive technologies are being introduced in the field of healthcare to reduce the human role of decision-making, ultimately aimed to minimize the factor of human error in decision-making in healthcare. According to a Johns Hopkins study, medical errors are the third leading cause of death in the United States, but they are not necessarily due to bad medical professionals; instead, they are often caused by cognitive errors, such as failed heuristics and biases, lack in safety and other protocols and an unjustified contrast in practice patterns of physicians. AI has shown to be promising in performing cognitive workload, producing better accuracies and overall patient experience. Studies have shown that computer-aided readings of radiological images are just as accurate as those performed by human radiologists [38]. This may appear, however, as a danger to the job security of medical professionals and physicians all over the world.

Table 8.2 Alternative roles for human clinicians in which variant forms of augmentation take place

Roles	Role description
Process design	Focus on how AI supports the process in question
Human capability	Employ unique human skills which AI systems may lack Skills include empathy and interpersonal intelligence
Colleague	Evaluate machines' immediate output Determine if data seem reasonable Use information to augment or inform their own judgement
Niche	A role which no technology has been developed or will be developed due to unfeasibility
Development	Development of AI techniques in healthcare Researchers or collaborators with AI vendors

However, instead of a large-scale job loss occurring due to the automation of human work, AI may provide an opportunity for more human-centric approach of augmentation [39]. With AI systems taking over the cognitive workloads, clinicians will become freer to perform the higher-level tasks, as shown in Table 8.2.

8.3.2.5 Treatment

Besides reviewing past health data to assist medical professionals in diagnosing diseases, AI can also help them take a more inclusive approach for disease management, improve coordination of healthcare plans and help patients in managing and complying with their long-term treatment programmes. Also, AI can be used to predict the effectiveness of medical treatments and their outcomes. Using modelling, Finnish researchers at the University of Eastern Finland, Kuopio University Hospital and Aalto University have developed a method to compare different treatment alternatives and to identify patients who will benefit from treatment. Relying on AI, the method is based on causal Bayesian networks [40].

From simple laboratory robots to highly complex surgical robots, robotic surgical treatment methods have been used in medicine for more than 30 years and they can either assist a human surgeon or perform operations independently. They are also used in repetitive tasks such as in rehabilitation and physical therapy. AI will improve the performance of surgical robots, providing greater optimization in robotic motion and in the future, they will be able to perform more complex surgeries independently.

8.3.2.6 End of life care

According to World Bank, the life expectancy in Malaysia was 75.3 years in 2016 and it has been showing a steady rise since the 1990s [41]. As life expectancy is increasing globally, people are dying more slowly and in a different way. Today, as people approach the end of life, they suffer from conditions such as osteoporosis, heart failure and dementia. This phase of life is often encompassed by loneliness and depression.

Although AI will not replace human physicians in end of life care, it will enable them in making better decisions. Medical AI will deploy MLAs to analyse structured data and cluster patients' characteristics to predict probabilities of disease outcomes. This will alert physicians about a deteriorating medical condition before it becomes severe and incurable. For example, MLAs that can detect the occurrence of deteriorating osteoporosis or heart failure can enable doctors to take early action before the conditions become worse and the patient grows old with incurable phases of these conditions.

In addition, AI along with robotics has great potential to revolutionize end of life care by aiding the elderly to be independent for longer and reducing the necessity for admissions in care homes or hospitals. The NLP tools of AI can enable conversations and other social interactions with aging people to keep their minds sharp and slow the growth of conditions such as dementia.

8.3.2.7 Research

Bringing a new drug from research labs to patients is a long and costly process. According to the California Biomedical Research Association, it takes an average of 12 years for a drug to travel from a research lab to the patient. Only five in 5,000 of the drugs that begin preclinical testing make it to human testing and just one of these five is approved for human usage. And, on average, it will cost a company US$359 million to develop a new drug from the research lab to the patient [42].

At the CASP13 meeting in Mexico in December 2018, Google's DeepMind beat experienced biologists at predicting the shapes of proteins, the basic building blocks of disease. In the process of looking for ways for medicines to attack diseases, sorting out structures of proteins is an extremely complex problem. There are more possible protein shapes than there are atoms in the universe, hence making prediction an enormously computation-intensive task. For the past 25 years, computational biologists have spent a large amount of their time developing software to more effectively performing this task. A tool that can accurately model protein structures could accelerate the development of new drugs. DeepMind, with its limited experience in protein folding but equipped with state-of-the-art neural network algorithms, did more than what 50 top labs from around the world could accomplish [43].

Drug research and discovery may be one of the more recent applications for AI in healthcare, but it shows promising results. This application of AI has the potential to crucially cut both the time for new drugs to enter the market and their costs.

8.3.2.8 Training

AI, through neural network algorithms, can be trained to understand natural speech and interact with another human almost as a human can, and this creates an environment for those in the healthcare sector that are undergoing training to undergo naturalistic simulations in ways that simple computer-driven algorithms are not able to. The emergence of natural speech understanding and synthesis and the potential of an artificially intelligent computer to draw almost instantly from a large database of scenarios can create challenges for trainees based on wide range of scenarios and data by asking and responding to questions, advices and decisions.

The training sessions are also adaptable and easily accessible as the neural networks can learn from the trainee's responses and adjust the challenges to maximize the trainee's learning rate and as the training can be done anywhere through the embedding of AI in smartphones and other mobile devices. This means that the trainee can undergo training in smaller and quicker sessions or even while walking or driving.

8.3.3 Big data

Industry analyst Doug Laney articulated the definition of big data as the three Vs [44]:

1. Volume – Data are collected from a large variety of sources, such as machine data, sensor data, social media and business transactions. Storing such large volumes of data were once a big problem. However, today with large storage capacities and technologies such as Hadoop, the process has become a lot easier.
2. Velocity – New data are being introduced at extraordinary speeds today; it is imperative to handle these data quickly. Continuous input of logs from applications, RFID tags, smart metering, social media data and sensor data make it very important to handle the inrush of data in near real-time.
3. Variety – The introduction of data is happening in both structured and unstructured formats. Structured data include organized data in traditional databases, and unstructured data include text documents, articles, emails and videos.

In addition to the three Vs above, SAS Institute, Inc. considers two additional dimensions in big data [45]:

1. Variability – Along with the surplus in velocities and varieties of data, the inflow of data can be very unpredictable. Although certain forms of data are predictable, such as the period between logs of sensor data, other forms of data may flow inconsistently, such as a trending topic in social media. Such forms of data flow may peak at any point of the day and this makes big data management difficult to perform. Handling unstructured data simply adds to the challenge.
2. Complexity – Data may come from a large number of sources, making it challenging to link, match, cleanse and transform data across systems. Nevertheless, it is important to associate relationships and hierarchies or an organization can quickly lose control of the data.

Oracle, on the other hand, states that two more Vs of big data have emerged over the past few years [46], which are value and veracity. It is well known that data are extremely valuable; however, until that value has been realized, the data are of no use. More importantly, it needs to be known if the data are truthful and reliable. Using unreliable data would provide organizations with inaccurate insights which may wind up doing more harm than good. Finding value in big data is more than just analysing the data. It involves a whole exploration process that requires insightful analysts, business users and executives who can recognize patterns, make informed assumptions and predict behaviour.

According to research from CB Insights, 32 companies that leverage AI for their core businesses exceeded the US$1 billion valuation mark in 2018. They leverage machine learning and other AI solutions in different fields including healthcare, autonomous vehicles, cybersecurity and robotics [47]. These AI-based companies have an assortment of business models and industries; however, they share a common advantage, which is also shared by the most innovative and valuable organizations in the world, such as Alphabet, Amazon, Microsoft, Netflix, Salesforce and Samsung. The advantage is that they leverage AI in big data analytics. These organizations have the led in combining large volumes of data and AI capabilities into differentiated solutions with large market values.

The MIT Sloan Management Review calls the union of big data and AI 'the single most important development that is shaping the future of how firms drive business value from their data and analytics capabilities' [48]. According to Forrester's Business Technographics survey of over 3,000 global technology and business decision makers from 2016, 41 per cent of global firms are already investing in AI and another 20 per cent are planning to invest in the next year [49]. Based on Forrester research analyst Brandon Purcell's reports on the current strong interest in AI and what can expected from it, 'The Top Emerging Technologies in Artificial Intelligence' and 'Artificial Intelligence Technologies and Solutions, Q1 2017', the most large organizations begin integrating AI into their businesses through chatbots for customer service, what is called 'conversational service solutions' [50]. Moving away from hard-coded rule-based chatbots that are pre-programmed to search for keywords and provide answers from databases, chatbots today are artificially intelligent with the use of NLP and deep learning. Many companies are even using the image, video and speech analytics capability of AI to obtain insights from unstructured data.

Large volumes of data are being introduced every day, and each day, the amount of added data increases. According to Domo, Inc., every person on earth will generate 1.7 MB of data every second by 2020 [51]. All the added data have the potential of solving challenges that are almost impossible to solve today, if properly harnessed. This is where AI comes in. Deep neural networks are capable of performing accurate pattern recognition in many situations; however, they first need to be trained using data. The more data used in the training process, the more accurate the model in making predictions. The abundance of data collected supplies AI with the training it needs to identify differences and see finer details in patterns. Normally, it is difficult to extract useful information from huge datasets along with unstructured data. However today, AI is aiding large companies in obtaining insights from data that were sitting in huge data banks for a long time. In fact, databases today are changing and are becoming more versatile and powerful. As compared to the conventional relational databases, today, powerful graph databases exist. These databases are

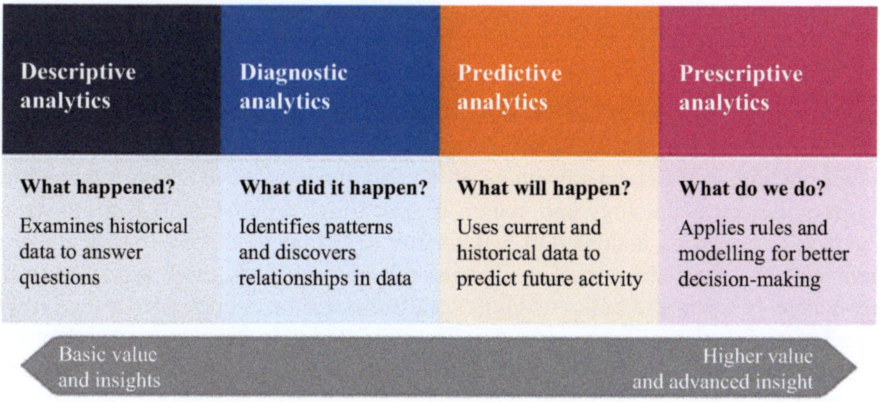

Figure 8.6 Categories of adoption of artificial intelligence in big data analytics

embedded with AI and are capable of associating data points and discovering relationships.

According to Singularity University, the process of adopting AI in big data analytics can be divided into four different categories, as shown in Figure 8.6 [52]. The most basic category that provides the most basic insights for an organization is descriptive analytics, where data from the past are used to describe the events of the organization. For example, a company *A* whose sales has been dropping for years can use a descriptive analytics system that shows exactly how much sales the company has been losing on a yearly, quarterly and monthly bases. The next category is diagnostic analytics, which searches for patterns in data to provide a story. For example, company *A* can use diagnostic analytics to study and analyse data regarding market conditions and past transactions to provide more detailed insight into the patterns in which the company has been losing sales, which tells the story on how company A has been losing sales and what has caused it. This way, company *A* can begin searching for solutions to the problem by eliminating the sources of the problem. Next, predictive analytics uses existing data to predict future occurrences. Company *A* can also use predictive analytics to predict the future of their businesses, so they can decide how they will need to maintain or change their business strategies to adapt with market changes. Finally, prescriptive analytics provides the greatest value and insights through cognitive computing to decide the next course of action for the maximum benefits. The use of prescriptive analytics can help company *A* in the process of managing the company, through rules, models and policies that have been decided by the company. At this stage, the AI system has the role of a consultant in the company. Many organizations use these strategies at different levels and in different combinations; however, the most successful organizations in big data analytics have a firm grasp in predictive and prescriptive analytics. With more in-depth applications of these strategies, organizations tend to gain greater understanding of customers' behaviours, greater insight of business performances currently and in the future, key performance indicators for decision-making and a greater overall actionable advantage over their less data-driven competitors.

In the future, the integration of AI and machine learning techniques into big data analytics can either mean the success or failure of an organization. With such large volumes of data flowing in today, many organizations are already making use of advanced analytics to take advantage of the data that they have. This includes forecasting relevant technology and market trends, which would enable them to proactively take necessary actions and make better decisions to adapt and stay ahead of competitors even before market changes take place. Also, they would have the advantage of using predictive analytics to minimize the occurrence of the HiPPO effect, which is the reliance on the Highest Paid Person's Opinion, rather than data that tell the actual story of current business conditions. In addition, with the computation power that exists today, AI systems would easily be able to track progress and velocities of every project taking place in an organization through every development phase and predict the value and outcome of each project so that organizations can maximize valuable projects and minimize white elephant projects.

8.3.4 Transportation and infrastructure

An application of AI in infrastructure industry is detecting structural defects. Using computer vision along with deep learning, structures or individual pieces of equipment can be continually scanned for defects so that problems can be identified before they become serious. Most of the time, structural defects build up with time and go unnoticed, and by the time they become severe enough to be noticed, fixing the defect becomes an expensive, difficult and time-consuming process. If the defect is discovered in its earlier stages, the problem can be solved more easily. Today, reality modelling is already being used to detect faults in concrete or cracks in a structure, stressing the gravity of the problem. When a defect has been discovered and fragmented, engineers can decide the scale of the defect, along with other information such as the shape and size of the defect. With these information, the engineers can more easily and quickly formulate a solution.

In addition to detecting structural defects, AI is being used in performing remote and autonomous inspection to reduce inspection durations and reduce occupational risks of human inspectors. Customer company AiviewGroup is improving in surveying, maintaining and inspecting more than 4,000 infrastructural buildings each year using the unmanned aerial vehicle (UAV) from Leica Geosystems to generate very precise data for bridge inspections [53]. They have also developed algorithm to evaluate the pictures taken by the UAVs and detect issues against a library of defect taxonomy, which checks for problems with cement, iron for the bridge, vegetation and the environment in general, according to Marietti, a bridge inspector with AiviewGroup.

Furthermore, the Housing and Development Board (HDB) blocks in Jurong East, Singapore, have embarked on utilizing camera-embedded drones to complement building inspection efforts, particularly for hard to reach places or those that may pose more risk to human inspectors [54]. The drone was operated by HUS

Unmanned Systems and image analytics is performed by an AI system from H3 Zoom.AI. The drone takes thousands of pictures which are uploaded to the AI system hosted on the cloud for analysis to be done. The system then identifies defects in a report generated automatically after each inspection. To protect residents' privacy, a masking software blurs out details if images of the residents are captured during inspection.

A popularly growing application of AI in the transportation industry is autonomous vehicles. Although fully autonomous cars and trucks are still in their infancy, the number of new autonomous vehicles being produced is increasing at a rapid rate. According to a study from Allied Market Research, the global market for autonomous vehicles will be worth US$54.23 billion in 2019 and increase to US$556.67 billion by 2026 with a compound annual growth rate of 39.47 per cent during that period [55]. This means that the value of the global autonomous vehicle market will grow more than tenfold from 2019 to 2026. Prototypes have also been built for autonomous larger vehicles such as rubbish collection vehicles, and Sweden and Germany have piloted driverless buses.

Autonomous vehicles are the vehicles that are capable of accelerating, braking and manoeuvring themselves around roads without human intervention. They rely on specialized sensors such as light detection and ranging (Lidar) sensors, radars, GPS antennas and cameras to understand their current location, the surroundings and the objects around them to decide the next action to be taken. When using lidar for autonomous vehicles, companies first equip ordinary vehicles with lidar sensors that collect information and generate a map of the surroundings. Once the map is complete, autonomous vehicles can use the map to track their surroundings and compare the map with the actual surroundings to understand where it is, and the objects that lie around it. This is accompanied with GPS to improve the estimation of the vehicle's position. Using live feed from cameras installed around the vehicle, autonomous vehicles can understand the objects that are around it, such as moving cars and motorcycles, and pedestrians through object detection and recognition. It can also get information such as how near the object is from the vehicle and how quickly it is approaching the vehicle. Because of the complexity of events that can occur in real life, programming the actions of an autonomous vehicle in every possible situation through a set of rules would be near impossible. Companies such as Waymo and Uber rely on machine learning to learn behaviours through large amounts of data regarding roads and behaviours of other objects on roads.

An increase in the usage of autonomous vehicles would bring a large number of benefits to not only road users, but ultimately the economy of a country and the environment. As the number of autonomous vehicles increases, overall traffic on roads will reduce. Researchers Dan Work of Vanderbilt University and Benjamin Seibold of Temple University have determined in their research paper that even a percentage of autonomous vehicles as small as 5 per cent could have a significant impact in eliminating waves and reducing the total fuel consumption by up to 40 per cent and the braking events by up to 99 per cent [56]. Naturally, human drivers tend to create stop-and-go traffic due to natural oscillations in human driving. This tends to occur in cases when someone makes a lane change or merge. The researchers found that by controlling the pace of the autonomous car in their field experiments, the autonomous car would dissipate stop-and-go waves so traffic does not oscillate as it does when all cars are driven by humans. In addition, a widespread use of autonomous vehicles could eliminate up to 90 per cent of accidents in the United States alone, preventing up to US$190 billion in damages and health costs annually and saving thousands of lives according to a new report from consulting firm McKinsey & Co [57]. With a reduce in traffic jams and more optimized driving algorithms, autonomous vehicles will have significantly reduced emissions. According to a paper, the use of electric autonomous taxis alone could reduce greenhouse gas emissions by 87–94 per cent per mile by the year 2030 [58]. These benefits ultimately come from the use of AI in optimizing the motion of the vehicle to minimize travel time, consumption and overall traffic congestion.

In addition to the above-mentioned benefits of AI in vehicles, AI would also lead to greater efficiencies of electric cars. Dr Michael Fairbank and Dr Daniel Karapetyan from the University of Essex have worked with Spark EV Technology to develop algorithms to help electric vehicles travel further between charges and eliminate the stress of drivers not knowing whether they have enough power to complete their journey [59]. The software collects live data on multiple variables, such as traffic conditions, weather conditions, tyre wear and driver behaviour to predict how much energy is required for each journey. Spark uses machine learning to compare the prediction against the actual energy requirements to increase the accuracy for each future journey.

AI is also used in public transportation systems to plan and optimize the timing of transportation systems, ultimately saving passenger time and operation costs. According to Retail Sensing, their automated passenger counter system is saving a UK bus company thousands of pounds per bus [60]. It automatically counts the number of people that get on and off a bus to crosscheck passenger numbers with ticket machine transactions. When the company realised that the number of passengers was greater than the ticket transactions, they took action to remedy the missing fare revenue. By counting the number of people that get on and off the bus, the transportation company can collect the data to aid in scheduling and forecasting. They can also easily generate reports for external funding agencies such as regional authorities and monitor ridership over time. It is also important in analysing performance, giving measures such as passengers per mile, cost per passenger and number of passengers per driver. The counting system uses object recognition to recognize people, and then count the number of people entering or exiting the bus based on the direction of their motion in the video feed. Retail Sensing claims that the video counting system has an accuracy of 98 per cent, which is greater than manual counting which is only about 85 per cent accurate.

AI is also used in the navigation space to provide users with the fastest route to their destination. Mobile navigation applications such as Waze and Google Maps utilize AI pathfinding algorithms to suggest the most convenient routes to reach the desired destination. The decision is made based on a number of factors, including the user's preference for paths without tolls and traffic conditions. With the large amount of data being collected by navigation applications with their large number of daily users, they can continually train their machine learning and deep learning algorithms to improve users' experience by analysing and testing the large amount of data. Also, users are able to set an alarm for the time they are required to leave their location to arrive at their destination at a certain time. These navigation platforms constantly collect traffic data and store them based on certain features, such as the time of the day, day of the week, whether it is a public holiday and weather conditions. By training AI algorithms based on the collected data, they are able to accurately predict how traffic conditions will vary in the future and provide accurate travel time to users. This also occurs in the e-hailing space. Companies such as Uber and Grab are able to perform accurate estimations of the price of the ride when booking a cab through machine learning. Not only that, they also use AI algorithms in selecting the best driver for each passenger to minimize waiting time and detours.

In contrast to navigation applications such as Waze and Google Maps which rely on large amounts of user geolocation data to compute traffic conditions, camera-based traffic predictions systems can be used, by performing traffic analysis on video feed captured by traffic cameras installed on roads and highways. Such camera-based traffic prediction systems operate by performing object recognition on the video frames and counting the number of different types of vehicles on different roads. This is necessary to get separate traffic data for each road and to understand how different types of vehicles affect traffic. For example, trucks have a greater effect on traffic congestion as compared to motorcycles since they cover a larger area and typically accelerate slower than motorcycles. The approximate speed of vehicles are also computed by calculating the distance travelled by the vehicle between frames and multiplying it with the frame rate. The average number of vehicles and average vehicle speed is mostly sufficient to understand traffic conditions in a particular road. These data can be used to compute a traffic congestion coefficient for each road in an intersection to understand how traffic flows on roads and intersections. The traffic data can then be used to train a deep learning model, which can then be used to predict traffic conditions based on different factors, such as the time of the day and weather conditions (Figure 8.7).

An added benefit to using traffic cameras in intersections with traffic lights is that the traffic data collected can directly be used to control traffic lights. Normally, the goal of controlling traffic lights is to ensure that traffic on all roads are balanced. This is why conventionally, different traffic lights in an intersection are programmed with the same duration of red and green light. However, this is usually not the case since traffic density in different roads varies throughout the day based on the direction of the roads. For example, in the mornings, traffic density may be higher on a road heading from the outskirts to the city as compared to the road heading from the city to the outskirts since many residents in the outskirts work in the city. In the evenings, the opposite would occur since the residents would return to their homes after work. A greater duration of green light should be provided to roads with greater traffic density to maximize traffic flow through an intersection. Since the system can calculate the number of vehicles in each road with a high accuracy, this data can be used to control the rotation timing of traffic lights (Figure 8.8).

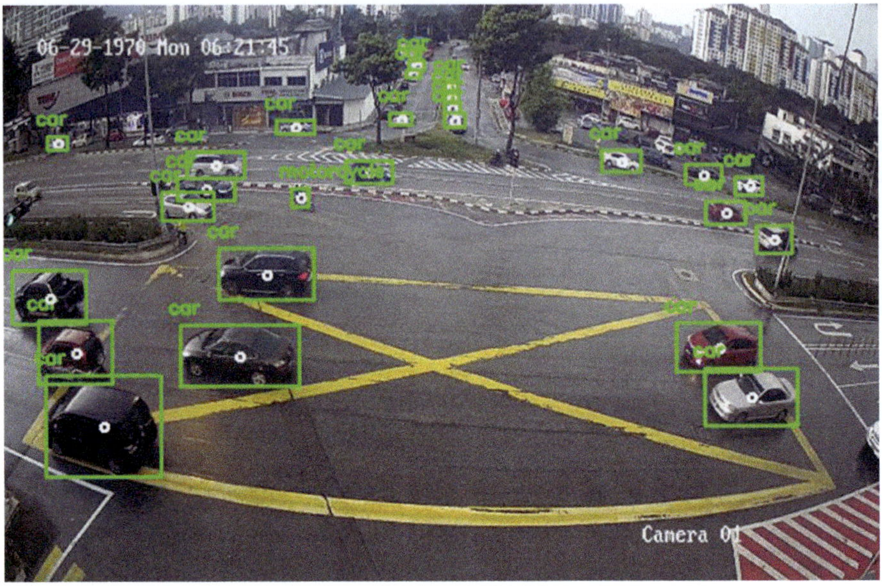

Figure 8.7 Object recognition on traffic feed to determine the number of different classes of vehicles

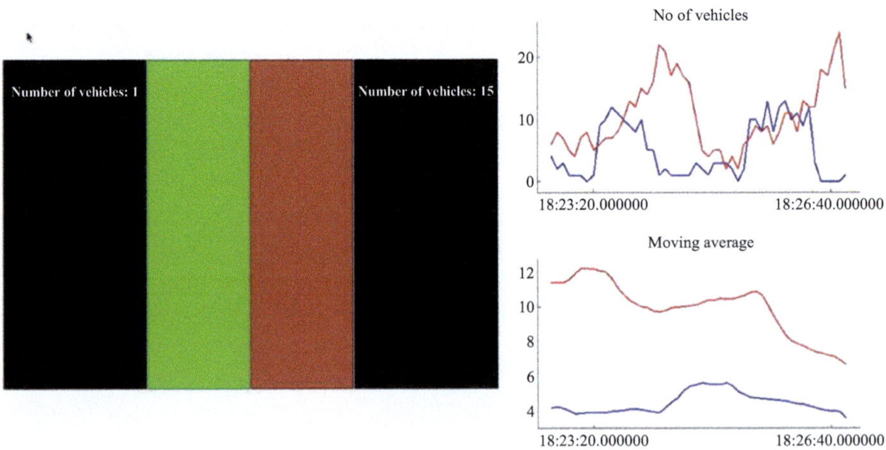

Figure 8.8 Visualization of live traffic data on roads based on a traffic analysing camera-based system

8.3.5 Energy planning and management

Because of the low efficiency and destructive effect of nonrenewable energy sources on the environment as debated at the UN climate talks in Poland, many nations have been actively making efforts in the increasing use of renewable energy sources such as solar, wind and hydroelectric energy, which are beneficial to the environment. However, using renewable energy sources poses a challenge to the energy industry due to the unpredictable nature of consumers' energy usage and renewable energy sources. AI is used to solve this problem and improve the overall process of energy generation, transmission, distribution and consumption, while lowering the overall cost of the processes and minimizing energy loss. It is used in a variety of applications in the energy industry, including forecasting energy supply and demand fluctuations so providers can ensure uninterrupted energy availability, and optimizing usage of electricity to reduce wastage. For example, Google's DeepMind has cut its data centres' energy consumption by 15 per cent using a MLA [61].

Research is also being conducted in using artificially intelligent systems to mitigate the urban heat island phenomenon, which is defined as a metropolitan area that is a lot warmer than the rural areas surrounding it [62]. Heat is generated from vehicles, such as cars, buses and trains, and buildings, in large cities such as New York, Beijing and Kuala Lumpur. Due to the heat generated from vehicles and the insulating properties of building materials, heat is built up and captured in large cities forming urban heat islands which worsens air and water quality and stresses native species such as aquatic life that are accustomed to cooler temperatures. To mitigate this problem, intelligent control systems integration and optimization for zero energy buildings are being studied. Intelligent control technologies such as learning-based methods including fuzzy systems and neural networks, model-based predictive control (MPC) technique and agent-based control systems are plausible techniques to satisfy several objectives, including increasing the energy and cost efficiency of building operations and the comfort level in buildings [63]. These technologies can be used in controlling actuators of shading systems, ventilation systems, auxiliary heating/cooling systems and household appliances to optimize the cost and usage of electricity.

While they consume electricity, electric vehicles will also become mobile energy storage units for the smart grid by exporting power to the grid. As the adoption of electric vehicles increase, the grid will become more flexible in responding to supply and demand of energy. The vehicles would be equipped with mature vehicle-to-grid (V2G) and grid-to-vehicle (G2V) technologies to solve the challenges posed using renewable energy techniques. For example, if there is a sudden unexpected surge in demand for electricity in a city, energy providers may not be instantly able to supply the large amount of electricity to the consumers because of insufficiency in available energy. They can instantly increase the generation of electricity from hydroelectric stations and even coal power stations; however, there may still be a delay in the transmission of the power from the stations to the consumers. In this case, the smart grid can temporarily extract stored electricity in electric vehicle batteries that are connected to the grid, and provide the electricity to fulfil the surge in demand. By the time the generated energy is ready to be used, the smart grid can continue to charge the electric vehicle to replace the extracted energy. Francisco Carranza, the director of energy services at Nissan Europe, has confirmed that electric vehicles could be utilized to benefit the grid. A study in Denmark conducted by Nissan and Enel S.p.A., Italy's biggest energy utility company, showed that utilities can use parked electric vehicles to pump power into the grid and even provide vehicle owners with a monetary return in the process [64].

While the use of electric vehicles to support smart grids in energy provision may solve the big challenge faced by smart grid systems, it poses a new challenge to electric vehicle owners. On the one hand, it is known that electric vehicles have more limited range as compared to gasoline-powered vehicles, which means that extracting stored energy from electric vehicles would further reduce available energy stored in electric vehicles, potentially causing users to be stranded if they were to travel longer distances. On the other hand, with a large number of electrical vehicles to be connected to the grid, local distribution grids will face peaks that may result in high electricity prices and electrical overload. Large volumes of research are being conducted today to develop strategies to solve these problems. Among the most popular techniques involve the use of AI to render electric vehicles and the systems that manage collectives of electric vehicles smarter. Based on a survey on managing electric vehicles in the smart grid using AI, several AI techniques with different complexities and effectiveness have been developed to solve a wide range of problems related to electric vehicles and smart grids, including [65]:

1. Energy-efficient electric vehicle routing and range maximization to route electric vehicles in order to minimize energy loss and maximize energy harvested during a trip. This also includes maximizing the efficiency of batteries to maximize battery life.
2. Congestion management to manage and control the charging of electric vehicles so as to minimize queues at charging stations and discomfort of drivers.
3. Integrating electric vehicles into the smart grid to schedule and control the charging of electric vehicles so that peaks and possible overloads of the electricity network may be avoided, while minimizing electricity cost.

8.4 Opportunities and risks

In Section 8.3 of the chapter, the application of AI in various aspects of cities in the formation of smart cities has been discussed. It is clear that AI and machine learning techniques are not only being rapidly developed, but are already being adopted by cities to save operational costs, increase sustainability and ultimately improve the quality of life. Today, the larger corporations mainly use smart analytics tools in examining their data since they have more data to deal with and the capacity to afford such expensive systems, while the smaller businesses and companies may still not have the necessity and capacity to utilize such systems. However, as the power of machine learning tools continue to advance and more companies and organizations begin to adopt these systems, operational costs of these smart analytics tools will decrease, enabling the smaller businesses to afford them. In addition, as businesses continue to become more data-dependent to maximize their revenues, they will eventually begin to lack the ability to analyse large volumes of data and will eventually require AI to support their operations.

As NLP and object recognition algorithms continue to improve, organizations and governments will have greater confidence in using them to improve services and take over tasks when they can perform the task with greater performance. For example, most traffic lights today are programmed using PLCs and they are programmed to have fixed timings for each lane in a junction. This may not necessarily be the best method of minimizing traffic, as the traffic light in an empty lane may be green while the traffic light in a congested lane is red. A traffic police officer controlling traffic would perform a better job than such a traffic light system as they can allow more congested roads more 'green light time'. However, an AI system that analyses traffic data to control traffic lights can outperform both systems. With numerous interconnected traffic cameras across the city, along with traffic geolocation data, AI systems can optimize traffic light timings for all traffic lights in all junctions of a city to minimize the effects of congestion in one junction on another junction of the city. However, governments would need to be very confident about such a system before they implement it in large scales. If the traffic-controlling AI system fails, it would cause major city-wide traffic congestion. Companies that develop these AI systems would begin by testing the system in simulations, followed by smaller-scale tests before they fully implement such a system.

The kind of AI mostly being developed today is known as narrow AI, which means that is designed to perform a specific task, and fails in performing other types of tasks. Most examples of AI use described in this chapter, such as facial recognition, NLP and traffic prediction are classified as narrow AI. However, many researchers are working on developing general AI, which can perform a multitude of tasks well. While narrow AI systems are able to outdo humans at specific tasks, general AI systems would be able to outdo humans at a large number of cognitive tasks.

Although AI is seen to be very beneficial in automating everyday tasks, many big names in science and technology such as Elon Musk, Bill Gates and Stephen Hawking, along with many leading AI researchers have expressed their concerns regarding the risks posed by AI. The concerns rose when developments of AI that experts predicted to be decades away were achieved merely 5 years ago, leading to the possibility of super-intelligence within our lifetimes. Most researchers agree that a super-intelligent AI would be unlikely to portray human emotions, so they do not expect AI to become intentionally kind or malicious. According to Future of Life, experts consider the following two scenarios when considering how AI may become a risk [66]:

1. The AI is programmed to do something destructive. An example of a technology aligned with this scenario is autonomous weapons, which are capable of causing mass casualties. As AI techniques become more advanced, certain parties will come to realize the potential of AI in causing large-scale geophysical damages, shortly leading to the invention of AI autonomous weapons. In addition, an AI arms race could lead to an AI war, which could cause casualties exceeding wars fought by people. Even with narrow AI, this risk is present. With the growth of

general AI, autonomous weapons would gain more autonomy, causing humans to conceivably lose control of the situation.

2. The AI's goal becomes misaligned from the purpose it was created. In this scenario, unlike the previous scenario, the AI is programmed to perform a beneficial task; however, it develops a destructive method for achieving its goal. This may occur when the AI is given a task to solve or perform and it focuses solely on solving the problem without considering the consequences of its actions in other aspects. If a super-intelligent system is given a task to perform a large geoengineering project, it may cause harm to the ecosystem as a side effect, as its sole purpose is to complete the given task, no matter the consequence.

AI safety research is an area of study that focuses on ensuring that the goal of AI is always aligned with ours, and at the same time, the AI needs to have a concern for humanity and the environment. From the scenarios above, it can be seen that the concern about advanced AI does not come from the fear that it may become evil, but that it achieves a level of competence where humans are not capable of stopping it. While tigers are at the top of the food chain in jungles, they are overpowered by humans because our intelligence significantly surpasses theirs. As AI far surpasses human intelligence, they may gain a similar form of control towards us. A key goal of AI safety research is to ensure that the motivation of AI in completing its tasks never surpasses the safety of humankind and the environment.

References – Industry 4.0 and Smart Cities

[1] (2017). *The belt and road initiative: Industry 4.0 in sustainable and smart cities*. United Nations Industrial Development Organization. Retrieved from https://www.unido.org/sites/default/files/files/2018-03/Unido_BRI_summary_web.pdf

[2] CB Insights (2019, January 9). What Are Smart Cities? *CB Insights*. Retrieved from https://www.cbinsights.com/research/what-are-smart-cities/

[3] Hajduk, S. (2016). "The concept of a smart city in urban management". Business, Management and Education. 14: 34–49. doi: 10.3846/bme.2016.319.

References – Artificial Intelligence

[4] Marr, B. (2018, February 14). The Key Definitions Of Artificial Intelligence (AI) That Explain Its Importance. Retrieved from https://www.forbes.com/sites/bernardmarr/2018/02/14/the-key-definitions-of-artificial-intelligence-ai-that-explain-its-importance/#56cf5c3b4f5d

[5] McCarthy J., Minsky M.L., Rochester N., Shannon C.E. (2006). "A proposal for the Dartmouth summer research project on artificial intelligence, August 31, 1955". AI Mag. 27(4):12.

[6] Artificial Intelligence – What it is and why it matters. *SAS*. Retrieved from https://www.sas.com/en_my/insights/analytics/what-is-artificial-intelligence.html

[7] Neural Networks – What they are and why they matter. *SAS*. Retrieved from https://www.sas.com/en_my/insights/analytics/neural-networks.html

[8] Tractica (2016, June 20). Computer Vision Hardware and Software Market to Reach $48.6 Billion by 2022. *Tractica*. Retrieved from https://www.tractica.com/newsroom/press-releases/computer-vision-hardware-and-software-market-to-reach-48-6-billion-by-2022/

[9] Tensorflow, Tensorflow Object Detection API, GitHub repository, https://github.com/tensorflow/models/tree/master/research/object_detection

[10] Liu W., Anguelov D., Erhan D., Szegedy C., and Reed S. (2015). "SSD: Single shot multibox detector".

[11] Belagiannis V., Amin S., Andriluka M., Schiele B., Navab N, and Ilic S. (2014, June 23–28). 3D Pictorial Structures for Multiple Human Pose Estimation, *Proceedings of the 2014 IEEE Conference on Computer Vision and Pattern Recognition.*

[12] Yang, Y., and Ramanan, D. (2013). "Articulated human detection with flexible mixtures of parts". IEEE Trans. Pattern Anal. Mach. Intell. 35(12): 2878–2890.

[13] Toshev, A., and Szegedy, C. (2014). DeepPose: Human pose estimation via deep neural networks. In Computer Vision and Pattern Recognition (CVPR), 2014 IEEE Conference on, 1653–1660. IEEE.

[14] Spohrer J., and Banavar, G. (2015). "Cognition as a service: An industry perspective". AI Magazine. 36: 71–86. doi: 10.1609/aimag.v36i4.2618.

[15] Spivak, N. 2013. Why Cognition as a Service Is the Next Operating System Battlefield. Gigaom. December 7, 2013 URL: http://gigaom.com/2013/12/07/why-cognition-as-a-service-is-the-next-operating-system-battlefield/

[16] Sears, A. (2018, April 14). "The Role of Artificial Intelligence in the Classroom". *ElearningIndustry*. Retrieved April 11, 2019.

References – Safety and Surveillance

[17] Mathis, T. L. (2018, April 3). Safety 4.0: Updating Safety for Industry 4.0. *EHS Today*. Retrieved from https://www.ehstoday.com/safety-technology/safety-40-updating-safety-industry-40

[18] CBC Radio (2019, April 26). Most Canadians skeptical about smart cities when it comes to their privacy. *CBC News*. Retrieved from https://www.cbc.ca/radio/spark/the-spark-guide-to-life-episode-eight-smart-cities-1.5107883/most-canadians-skeptical-about-smart-cities-when-it-comes-to-their-privacy-1.5107891

[19] Maxim, M. (2019, July 22). Making the smart city safe and secure. *Forrester Research*. Retrieved from https://www.go.forrester.com/blogs/making-the-smart-city-safe-and-secure/

[20] Ali, S. S. (2017, March 19). Police Shortage Hits Cities and Small Towns Across the Country. *NBC News*. Retrieved from https://www.nbcnews.

com/news/us-news/police-shortage-hits-cities-small-towns-across-country-n734721

[21] US application/patent 8134889, Showen, Robert L. (Los Altos, CA, US) Calhoun, Robert B. (Oberlin, OH, US) Dunham, Jason W. (San Francisco, CA, US), "Systems and methods for augmenting gunshot location using echo processing features", published 2012-03-13, issued 2012-03-13

[22] SmartCity (2017, December 15). The Use Of AI In Safety And Security. *SmartCity Press*. Retrieved from https://www.smartcity.press/safety-in-smart-cities-using-artificial-intelligence/

[23] Brandom, R. (2015, October 26). GE's Smart City project now includes streetlights with gunshot detection. *The Verge*. Retrieved from https://www.theverge.com/2015/10/26/9616298/general-electric-shotspotter-smart-city-gunshot-detection

[24] Locklear, M. (2018, September 27). AI camera can spot guns and alert law enforcement. *Engadget*. Retrieved from https://www.engadget.com/2018/09/27/ai-camera-detect-guns-alert-law-enforcement/?guccounter=1&guce_referrer=aHR0cHM6Ly93d3cuZ29vZ2xlLmNvbS8&guce_referrer_sig=AQAAANTLI8ENfj9WnXFHYNCmbEVWMG8Hzj85Y-vmeiob2lcUv-aCHyDIbH2STNHPTPDbGUWiwrjOXB1PEznO3KOJp2wQCtXXNXboQsf3NG0zsRp_pt2-MlJPNf-NK0hvFHZZc7Ba2tG5Zfso-0xUSDAUgjsp4C3m1CGKSTqlmE5rEsda

[25] Meuse, M. (2017, July 22). Vancouver police now using machine learning to prevent property crime. *CBC News*. Retrieved from https://www.cbc.ca/news/canada/british-columbia/vancouver-predictive-policing-1.4217111

[26] Forrest, C. (2016, June 7). Can AI predict potential security breaches? Armorway is betting on it. *Tech Republic*. Retrieved from https://www.techrepublic.com/article/armorway-grabs-2-5-million-to-expand-ai-security-platform/

[27] Smith, M. (2018, October 30). Can we predict when and where a crime will take place? *BBC*. Retrieved from https://www.bbc.com/news/business-46017239

[28] Pandey, P., Golden, D., Peasley, S., and Kelkar, M. (2019, April 11). Making smart cities cybersecure. *Deloitte Insights*. Retrieved from https://www2.deloitte.com/us/en/insights/focus/smart-city/making-smart-cities-cyber-secure.html#endnote-sup-2

[29] Wiggers, K. (2019, February 20). Armorblox raises $16.5 million for NLP that secures human communication. *Venture Beat*. Retrieved from https://venturebeat.com/2019/02/20/armorblox-raises-16-5-million-for-nlp-that-secures-human-communication/

[30] Dilek S., Çakir H., and Aydin M. (2015). "Applications of Artificial Intelligence Techniques to Combating Cyber Crimes: A Review", *International Journal of Artificial Intelligence & Applications (IJAIA)*, 6 (1): 21–39.

[31] Reinventing Cybersecurity with Artificial Intelligence. Reinventing Cybersecurity with Artificial Intelligence. *Capgemini*. Retrieved from https://

www.capgemini.com/research/reinventing-cybersecurity-with-artificial-intelligence/
[32] Columbus, L. (2019, July 14). Why AI Is the Future of Cybersecurity. *Forbes*. Retrieved from https://www.forbes.com/sites/louiscolumbus/2019/07/14/why-ai-is-the-future-of-cybersecurity/#2d0d3c95117e

References – Healthcare

[33] Limitations of Mammograms: How Often Are Mammograms Wrong? Retrieved from https://www.cancer.org/cancer/breast-cancer/screening-tests-and-early-detection/mammograms/limitations-of-mammograms.html#references
[34] Griffiths, S. (2016, August 26). This AI software can tell if you're at risk from cancer before symptoms appear. *Wired*. Retrieved from https://www.wired.co.uk/article/cancer-risk-ai-mammograms
[35] Withers, N. (2019, January 9). Artificial intelligence used to detect early-stage heart disease. *European Pharmaceutical Review*. Retrieved from https://www.europeanpharmaceuticalreview.com/news/82870/artificial-intelligence-ai-heart-disease/
[36] Watson Health: Get the facts. *IBM*. Retrieved from https://www.ibm.com/watson-health/about/get-the-facts
[37] Abdel-Motaleb, I., and Akula, R. (2012). "Artificial intelligence algorithm for heart disease diagnosis using Phonocardiogram signals," 2012 IEEE International Conference on Electro/Information Technology, Indianapolis, IN, pp. 1–6. doi: 10.1109/EIT.2012.6220714.
[38] Obermeyer, Z., and Emanuel, E. J. (2016). "Predicting the future — Big data, machine learning, and clinical medicine". The New England Journal of Medicine. doi: 10.1056/NEJMp1606181
[39] Davenport, T. H., and Glover, W. J. (2018). "Artificial intelligence and the augmentation of health care decision-making". The New England Journal of Medicine Catalyst. Retrieved from https://catalyst.nejm.org/ai-technologies-augmentation-healthcare-decisions/
[40] University of Eastern Finland. (2018, November 16). Artificial intelligence predicts treatment effectiveness. ScienceDaily. Retrieved August 26, 2019 from www.sciencedaily.com/releases/2018/11/181116110630.htm
[41] World Development Indicators. In Google Public Data. Retrieved from https://www.google.com/publicdata/explore?ds=d5bncppjof8f9_&met_y=sp_dyn_le00_in&idim=country:MYS:SGP:IDN&hl=en&dl=en
[42] *New Drug Development Process*. California Biomedical Research Association. Retrieved from https://ca-biomed.org/wp-content/uploads/2018/02/FS-DrugDevelop.pdf
[43] Langreth, R. (2019, July 15). AI Drug Hunters Could Give Big Pharma a Run for Its Money. *Bloomberg*. Retrieved from https://www.bloomberg.

com/news/features/2019-07-15/google-ai-could-challenge-big-pharma-in-drug-discovery

References – Big Data

[44] Laney, D. (2001). "3D data management: Controlling data volume, velocity and variety". 3D Data Management: Controlling Data Volume, Velocity and Variety. META Group. Retrieved from https://blogs.gartner.com/doug-laney/files/2012/01/ad949-3D-Data-Management-Controlling-Data-Volume-Velocity-and-Variety.pdf

[45] Big Data – What it is and why it matters. *SAS*. Retrieved from https://www.sas.com/en_my/insights/big-data/what-is-big-data.html

[46] The Definition of Big Data. *Oracle*. Retrieved from https://www.oracle.com/big-data/guide/what-is-big-data.html

[47] CB Insights (2019, February 14). "The increasingly crowded AI Unicorn Club". CB Insights. Retrieved from https://www.cbinsights.com/research/ai-unicorn-club/

[48] Bean, R. (2018, February 5). "How big data and AI are driving business innovation in 2018". MIT Sloan Management Review. Retrieved from https://sloanreview.mit.edu/article/how-big-data-and-ai-are-driving-business-innovation-in-2018/

[49] Forrester (2017, December). Global Business Technographics® Software Survey.

[50] Weldon, D. (2017, April 21). Artificial intelligence: Fulfilling the failed promise of big data. Retrieved from https://www.information-management.com/news/artificial-intelligence-fulfilling-the-failed-promise-of-big-data

[51] Domo Resource – Data Never Sleeps 6.0. Retrieved from https://www.domo.com/learn/data-never-sleeps-6

[52] (2019, March 21). AI and Big Data: A Powerful Combination for Future Growth. Retrieved from https://su.org/blog/artificial-intelligence-and-big-data-a-powerful-combination-for-future-growth/

References – Transportation and Infrastructure

[53] AiviewGroup using Leica Geosystems UAV solution for bridge inspections. Retrieved from https://leica-geosystems.com/case-studies/reality-capture/aiviewgroup-using-leica-geosystems-uav-solution-for-bridge-inspections

[54] Kwang, K. (2018, August 10). Heady days: Use of drones to detect defects on HDB blocks takes flight. *Channel News Asia*. Retrieved from https://www.channelnewsasia.com/news/technology/heady-days-use-of-drones-to-detect-defects-on-hdb-blocks-takes-10599070

[55] Jadhav, A. (2018). Autonomous Vehicle Market by Level of Automation (Level 3, Level 4, and Level 5) and Component (Hardware, Software, and Service) and Application (Civil, Robo Taxi, Self-driving Bus, Ride Share,

Self-driving Truck, and Ride Hail) – Global Opportunity Analysis and Industry Forecast, 2019–2026. Autonomous Vehicle Market by Level of Automation (Level 3, Level 4, and Level 5) and Component (Hardware, Software, and Service) and Application (Civil, Robo Taxi, Self-driving Bus, Ride Share, Self-driving Truck, and Ride Hail) – Global Opportunity Analysis and Industry Forecast, 2019–2026. Allied Market Research. Retrieved from https://www.alliedmarketresearch.com/autonomous-vehicle-market

[56] Stern R.E., Cui S., Monache M.L.D., et al. (2017). Dissipation of stop-and-go waves via control of autonomous vehicles: Field experiments. CoRR abs/1705.01693. URL http://arxiv.org/abs/1705.01693.

[57] Bertoncello M., and Wee D. (2015, June). Ten ways autonomous driving could redefine the automotive world. *McKinsey & Company*. Retrieved from https://www.mckinsey.com/industries/automotive-and-assembly/our-insights/ten-ways-autonomous-driving-could-redefine-the-automotive-world

[58] Greenblatt J.B., and Saxena S. (2015). Autonomous taxis could greatly reduce greenhouse-gas emissions of US light-duty vehicles.

[59] University of Essex (2017, October 12). Artificial intelligence leads to more efficient electric cars. Retrieved from https://www.essex.ac.uk/news/2017/10/12/artificial-intelligence-leads-to-more-efficient-electric-cars

[60] Retail Sensing (2019, January 11). Bus passenger counter saves transport company thou- sands of pounds per bus. Retrieved from https://www.retailsensing.com/people-counting/video-passenger-counting-saves-bus-company-thousands/

References – Energy Planning and Management

[61] Wakefield J. (2016, July 20). Google uses AI to save on electricity from data centres. *BBC*. Retrieved from https://www.bbc.com/news/technology-36845978

[62] Rutledge K., McDaniel M., Boudreau D., *et al.* (2011, January 21). National Geographic. In K. West & J. Evers (Eds.), *National Geographic*. Retrieved from https://www.nationalgeographic.org/encyclopedia/urban-heat-island/

[63] Zheng, Z., Wang L., and Nyuk, H.W. (2016). "Intelligent control system integration and optimization for zero energy buildings to mitigate urban heat island". Procedia Engineering. 169: 100–107. doi: 10.1016/j.proeng.2016.10.012.

[64] Douris, C. (2017, October 5). How Electric Cars Could Help the Power Grid Become More Efficient, Less Expensive. *Forbes*. Retrieved from https://www.forbes.com/sites/constancedouris/2017/10/05/how-electric-cars-could-help-the-power-grid-become-more-efficient-less-expensive/#6cb88de121e5

[65] Rigas E. S., Ramchurn S. D., and Bassiliades N. (2015). "Managing electric vehicles in the smart grid using artificial intelligence: A survey," in IEEE

Transactions on Intelligent Transportation Systems, 16(4): 1619–1635. doi: 10.1109/TITS.2014.2376873. Retrieved from http://ieeexplore.ieee.org/stamp/stamp.jsp?tp=&arnumber=7000557&isnumber=7174612

Reference – Opportunities and Risks

[66] Benefits & Risks of Artificial Intelligence. Retrieved from https://futureoflife.org/background/benefits-risks-of-artificial-intelligence/

Chapter 9

How industrial robots form smart factories

Yaser Sabzehmeidani[1]

For many years, industrial robots have been utilised to change mass manufacturing and register automation to carry out unique procedures more quickly and safely without any human lapse. The mass manufacturing produced from industrial robots lowers the cost of the product and increases the speed of delivery without much error. Production procedures, technology development, smart supply chain and trade aspects have been changed by Industry 4.0. Artificial intelligence (AI) and machine learning transformed the older ways of executing processes and operating industrial robots. To this end, novel research has been carried out and plans have been made for future robot generation, automation lines and smart factories. Although Industry 4.0 is not a widely used term and known idea, it is outstandingly capable of enhancing human life. All manufacturing procedures and supply levels are prognosticated to be impacted in the production. A relation of Industry 4.0 and smart factories in terms of industrial robots is described in this chapter and the benefits and applications are addressed. The influence of industrial robots on smart factories and the future expectations are later discussed in the text.

9.1 Industry 4.0 and industrial robots

What are the major impacts of Industry 4.0 on smart machinery and how smart factories are generated by the industrial robots? In the beginning, the phrase 'Fourth Industrial Revolution', or production digitalisation, especially 'Industry 4.0', was introduced through a German project by Germany Trade and Invest (GTAI) started in 2011. In this project, Industry 4.0 is defined as a paradigm change, which is actualised by technological advances that comprised a reversal of traditional manufacturing process logic. That is to say, it is not only expected for the machines to do the processing but the product also has to be communicated through the machinery to inform it on what exactly to do [4]. In the project known as vision of German 2020 Strategy, the final results of this commission will integrate high technology and international leadership. Programs such as 'Smart Nation Program' in Singapore, 'Industrial Value Chain Initiative' in Japan, 'Made in China

[1]Faculty of Engineering and Information Technology, Mahsa University, Jenjarom, Malaysia

2025' in China, and 'Smart Manufacturing' in the United States are amongst the similar steps taken by other nations [2,3].

Moreover, the GTAI project mentions that Industry 4.0 stands for the technological rise from embedded systems to cyber-physical systems which is a method that joins embedded manufacturing technologies and smart manufacturing procedures [6]. To put it in another way, Industry 4.0 is a state where production systems and the products produced by them are not merely joined, rather Industry 4.0 draws physical information into the digital field and conveys, analyses and uses this type of information to take more smart action back to the physical world to run a transformation from physical to digital and reverse [5].

Industry 4.0 is a mixture of numerous innovative technological developments such as information technology (IT) and communication technology, cyber-physical systems, network communications, massive data and cloud computation, modelling, virtualisation, simulation, and enhanced equipment for human-computer communication and collaboration. The notion of Industry 4.0 paves the way for various positive transformations to today's production such as mass customisation, elastic manufacturing, enhanced manufacturing speed, more high-quality products, lowered error rates, optimised efficiency, data-driven decision-making, more customer closeness, novel value production strategies and enhanced work life. Moreover, issues concerning business paradigm transformations, safety, security, legal concerns and standardisation, along with a number of human resource planning problems exist.

Through steam power and loom machine, Industry 1.0 dealt with mechanisation during the eighteenth and nineteenth centuries. Mechanisation superseded the major kind of energy, which was human power, and it facilitated faster and safer performance of processes. In the meantime, the manufacturing rate increased remarkably afterwards. The other revolution took place towards the end of nineteenth century through the industrialisation of electricity. Electricity was utilised in the factories; assembly lines were used for the purpose of mass manufacturing. Utilisation of microelectronics and Programmable Logic Controllers (PLCs) occurred in Industry 3.0 with the remarkable advance in the field of computers, robotics, PLCs, computer numerical controls (CNCs) and greater automation. Nowadays, advantages are taken from the sensor's explosion plus the network ability and remote access to information from these machinery. Figure 9.1 offers a general view of the timelines of industrial revolutions.

With the development of technology and the advent of machine-learning algorithms, the majority of routine jobs were carried out by automation, computers, robots, and machinery. It is predicted that these types of jobs are specialised with repeated processes and without any high-level skills. Registering industrial robots into routine jobs will be very efficient in Industry 4.0. Robotics, cyber-physical systems, automation, and Internet of Things (IoT) created smart factories. Following Industry 3.0 (computers and the internet), Industry 2.0 (mass production and electricity), Industry 1.0 (mechanisation and water/steam power), smart factories is the fourth industrial revolution (Figure 9.1). Industry 4.0 producers are attempting to join their machinery to the cloud and create Industrial IoT (IIoT) [3].

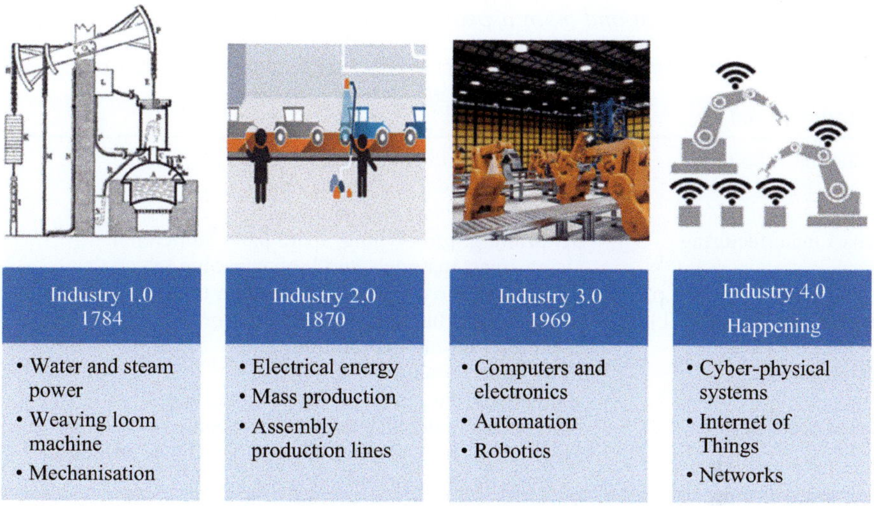

Figure 9.1 A schematic view of the timelines of industrial revolutions

The future generation of production and factories is expected to be smart in the industry scheme 4.0. All of the facilities, namely machines, devices, robots, processes and computers, have to be joined altogether in groups or one by one. Massive joining of systems brings about a rich community for huge data analysis, predictable loops and self-correcting processes amongst a number of other probabilities. Initialisation of Industry 4.0 is already acknowledged despite enormous transformations for development in the future [1].

It seems very challenging to convert raw material into a final product. Navigating a way of process design for high costs with the quickest delivering process at the time of the customisation of the production line on the basis of the consumer requests is very difficult. Despite remarkable advance and breakthrough of production and manufacture, more is predicted from Industry 4.0. A reorganisation of structure, methodology and performance is required in line with Industry 4.0 potentials in every stage starting from product progression and sales up to sourcing and delivering [7]. Table 9.1 illustrates descriptions of major aspects in terms of smart factory.

With the introduction of novel developed technologies, Industry 4.0 is expected to obtain better performance and productivity regarding data gathering, data analysis and process correction. In smart factories, production will be made digital and AI will be executed. Industry 4.0 impacts all the factory production stages such as design, production, engineering, marketing, operations, finance and so on. The primary concern regards separate development of each stage along with others with no idea on how to reach agreement. This issue can be catered for by a centralised smart networking system [8].

Table 9.1 Description and main aspects of Industry 4.0

Main aspects of Industry 4.0	Description in terms of smart factory
Industrial robotics	Cooperated and collaborative robots Being intelligent and autonomous Integrated sensors and detectors
Smart manufacturing	3D printing: prototyping, spare parts fabrication Using smart materials and biocompatible materials Quick design overview (faster design to product stage)
Internet of Things (IoT)	Everything is being connected (network) Flexibility of automation line
Data analysis	Mass data evaluation AI included for real-time decision-making
Cyber security	Operation in networks and open systems High level of networking between intelligent machines, products and systems
Simulation	Simulation of value networks Optimisation based on real-time data from intelligent systems

Zero Down Time Solution is a cloud-based software manufactured by General Motors (GM), the major manufacturer of US automotive industry for the maintenance criteria. Data gathered from robots across GM's factories for the prediction of the future is analysed using this platform. Since launching this platform in 2014, it is being developed on a daily basis and it goes on to yield unbelievable results. The programme analysed the data collected for the provision of return on investment (ROI) and avoidance of downtime. Any downtime may cost more than 20,000$ for every minute in the automotive industry. ROI was made useful by the software for the prediction of failure before it happens. Previous background was used by the AI brain for estimation and finding the maintenance while it is highly required for manufacturing. Although it is not possible yet upon collection of sufficient data from robot performance, the objective is to form a system for self-diagnosis according to slight changes in the procedures adjusted to enhance thorough efficiency [9].

9.2 Smart factories

Gathering and analysing data is one of the major criteria for Industry 4.0. Information is gathered from all stations by AI. Afterwards, the information is extracted, organised and analysed to yield outcomes. A highly advanced analytic technique is required for processing and deciding to analyse the huge data. Machine learning is made possible through massive data analytics. It means the ability to know data patterns and use past experiences to obtain precise predictions. Instead of writing millions of code lines for predictable decision trees, machine learning permits self-training of an algorithm to individually deal with the forthcoming issues or operate facilities well [10].

Radio frequency identity technology is amongst the widely used novel technologies in production industry with the aim of digitalisation. This technology has been utilised in various sensors, actuators, communicators, transmitters, tags and barcodes. The data gathered from such devices informs us of the inventory levels to realise the smooth execution of an operation or the repair time of equipment. Actual time supervision and optimisation can be conducted by connecting all devices and machinery through a central system in a smart factory. Everything could be joined in a smart factory while the machinery, resources and procedures could be automated and combined. This kind of production setting could be more efficient, could be foreseeable and would be able to manufacture higher quality goods with less equipment downtime than what we know of nowadays [11].

Implementation of the concept of one-by-one communication is not predicted in the Industry 4.0, whereas wiring-up equipment plus machinery is required. It is known to be a remarkable opportunity for producers. It should take into account the automation opportunities and carry out research on the technology required to combine automation into a fine operation. An independent mobile robot passing by the opening doors is a very simple instance. Otherwise, such doors would have been handled by human beings when transporting the components while even already made products have to be combined in order to permit robots to communicate with them along with automatic opening and closing when moving the products. Safety facilities such as fire alarm or emergency situation also could be integrated into the robots (Figure 9.2).

What are the expectations from smart factories and what are the major components in Industry 4.0? Industry 4.0 is capable of positively influencing supplying individual customer needs, manufacturing flexibility, decision-making optimisation, resource productivity and efficiency, opportunities to create value through novel services, demographic shifts in the workplace, human work force, work–life balance and a competitive economy with high salaries [6]. Four major elements predicted to bring about a great difference are highlighted here. They are

- efficiency and productivity,
- safety and security,
- flexibility and
- connectivity.

9.2.1 Efficiency and productivity

Making the efficiency higher in smart machinery is one of the main objectives in the industry. Various kinds of sensors are implemented by the smart machinery to supervise themselves to avoid any kind of failure. We can have more dependable and elastic automation/assembly lines for more efficient performance through data collection from smart machinery and networking. Smart machinery are used at the front line in the factory floor and embedded intelligent are highly required with the machine sensors and actuators. With this level of facilities, it would be possible to

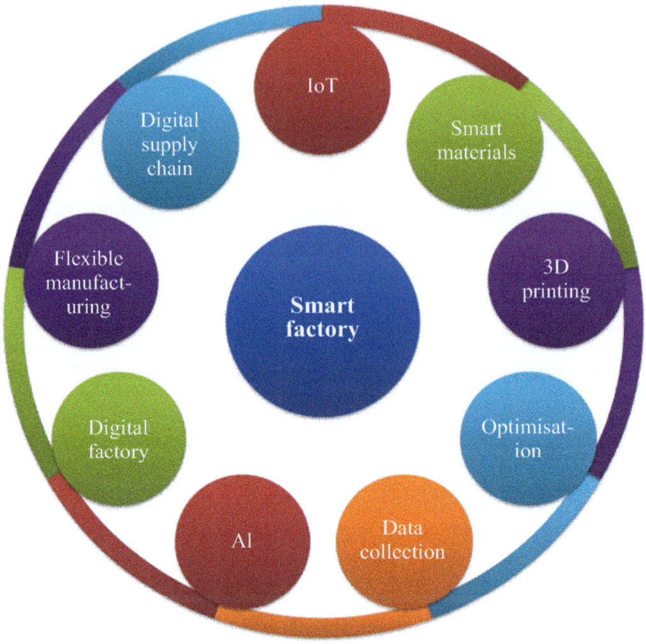

Figure 9.2 Principles of smart factory in the landscape of Industry 4.0

avoid or minimise any damages, preventive maintenance concerning the machinery and their downtime. Traditional supply chain and smart factory are compared in Figure 9.3. This adds value to the system.

Manufacturing will be faster given the advanced digitalisation and product simulation/virtualisation. Detection of the value of each design transformation and the impacts of manufacture speed is likely. Transparency will be yielded across the whole manufacturing through collecting, preprocessing, and analysing all accessible factory shop-floor data. Identification of the bottlenecks and potential betterment spots is allowed for. As an example, most of the time the design features are already specified by the clients. A suitable level of intelligence is required in smart machinery to rapidly analyse the collected data. Upon felling out of set criteria in the case of sensor-sharing data, it would be more efficient to manage the overall system and easier to handle the system.

9.2.2 Safety and security

Enhancing performance or effectiveness, safety and security should be taken into consideration. Safety sensors such as emergency, laser scanners hand detectors and cameras added to automation elements including PLC, drives, PCs and more have to be regarded to enhance safety of operators. Utilising these abilities is made possible by smart factories while data security is the major challenge in gathering data.

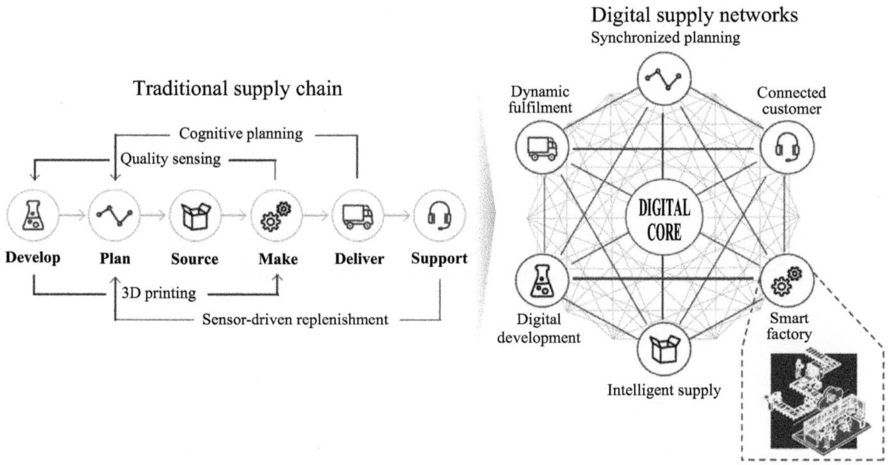

Figure 9.3 Comparison of traditional supply chain with the digital supply network

Nowadays, data security is the primary inhibitor of end-user adoption of novel networking technologies and work procedures. Achieving manufacturing advantages by networking components and machines brings about risks. Security needs to be taken into consideration at various levels especially with IoT (or IIoT) and increasing levels of connectivity. Security has to be provided at various levels while hardware, software, and services have to be incorporated. Machine manufacturers (and automation component vendors) have to make sure that the users are aware of the security vulnerabilities. Network infrastructure can be managed to lower a breach risk.

Cyber security is among the issues that worries the manufacturers: it should be guaranteed that they would not be infiltrated or their factories could not be taken control of. Furthermore, we have to make sure that the manufacturing equipment does not threaten human resources or the working setting. It should be ensured that employees are constantly trained on safety issues [6].

9.2.3 Flexibility

Smart machinery are expected to comply with the current installations or machines from multiple original equipment manufacturers (OEMs); product users request products that are capable of being installed within a short time period. Today, monolithic or single design is not allowed by the machinery lifecycle. Patterns of proven design from simple software functions up to completely functional units which provide a description of mechanics, electrical, motion and interfaces, traits and behaviour are taken advantage of.

Modularity is an enabler in which the model of reutilisation of software and hardware in another situation is in need of a novel level of thinking. The idea of precise and rigid interfaces with well-described behaviour to be examined is driven from the IT world. Its space is found in automation with some adaptation.

Through the foundation of smart factories, a more elastic manufacturing is made possible by the smart machinery which paves the way for the production of a wide range of products through a given production equipment, faster production procedures while reacting to shifts and temporary needs [6]. Thus, companies begin to manufacture for a greater number of customers and change in accordance with temporary market rise and fall. For instance, in the case of low request for a given product, the manufacturing line can be readily transformed to make another product with more capacity and faster delivery of the components to the clients than what is expected. This way, clients will trust the manufacturers more.

9.2.4 Connectivity

Broader (ethernet-based) networks will be directly integrated to the smart machinery. Hence, data sharing and manufacturing planning are made possible. This is more than the capacity of standalone machinery and automation. The IT and Operations Technology is integrated in the smart machinery which makes the access to manufacturing data possible. This set of data can be utilised in many management environments. The idea of utilisation of mobile devices at work is welcomed by machine operators and factory floor engineers. Monitoring and managing the performance of personnel do not depend on closeness to a machine anymore. Issues can be diagnosed by the machine engineers from a long distance and guidance can also be thus offered to quicken the conduction of a solution. Thus, downtime and losses coming from component failure are minimised.

A gourmet hamburger can be easily prepared by a multi-task robot (Momentum Machines) in a matter of 10 s very soon replacing a whole McDonald's team. Pioneer of AI technologies Google managed to win a patent to initialise manufacturing worker robots with their own identities. More and more, smart machinery has commenced their path on labour and become more complicated and developed than the generations before. Needless to say, that in smart factories human monitoring and customisation are required at some stage. Supervisors can manage robotic activities with the help of handheld technology such as smartphones and tablets and carry out changes with common inexpensive and specialised technology. The use of smart technology speeds up the speed of manufacturing managers in a fast-paced production setting in which they used to be tied to a computer or a stationery data system.

9.3 Internet of Things

Independent manufacturing strategies enabled by IoT is a crucial feature of Industry 4.0. Under this condition, every machine, object, device and computer need to be joined and allowed communication. We can still find the traces of the IoT in our world with the digitally connected devices increasing more and more. A very low number of such devices come into contact with us while they are expected to be more. The independent robots which are common in all industries are a simple instance for industrial production and smart factories.

Previously, the automation procedure and programming used to be managed by PLCs. PLCs behave as the head of the industrial automation, but their disadvantage is that they are likely to be wired up. Very high-skilled programmers are required to operate them and write the logic ladders. Given the intricateness and strictness, executing and transforming these solutions cost a lot. Robots need to be joined for communication in smart factories of Industry 4.0. Through connection to a central server, database or PLC, the coordination and automation of robots' actions would be probable to a larger extent than anytime. The tasks can be intelligently completed in an arranged way with the least human input. Autonomous mobile robots (AMRs) can transport the materials along the factory floor while simultaneously evading the barriers, collaborating with fleet mates, and trying to find out where pickups and drop-offs are required in any time.

Assembly and manufacturing workers can concentrate on actual assembly and manufacturing given AMRs' reception of work signals from real-time manufacturing systems and production implementation system. For instance, current PLC systems can be connected with robots through Aethon's Logic OS integration toolkit. The command and control communication between automation equipment, tools and sensors are facilitated by Logic OS while being supervised by PLCs and independent mobile robots (Figure 9.4).

A few layers and filters must be taken into account for the initial analysis of the raw data and report according to the requirements to prevent combining manufacturers with the huge received data. Rather, data can be understood and taken value of by the companies in a very strong manner. Robots are known to be the outstanding physical

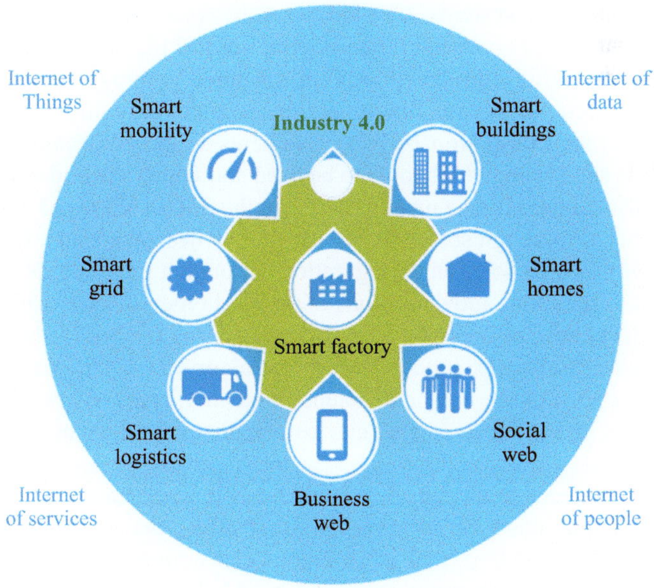

Figure 9.4 Distribution of internet platform in the scheme of Industry 4.0

and cognitive partner by the manufacturers in the way of forming the digital factory. Such machinery are essential, due to their ability to carry out tasks, gather and analyse data productivity, quality, creditability, plus their cost and other facilities in the procedure with the insights accessible for understanding and performance.

9.4 Artificial intelligence

AI has been implemented into production industry the examples of which are numerous indicating that this technology is currently being used across the sector. High-end technology companies such as Tesla, Intel and Microsoft internationally pioneer advancements by either investing or manufacturing or application. Neural networks have been utilised by Siemens for the supervision of the efficacy of their steel plants for years. This previous experience is being used to revolutionise producing AI sector with the use of AI in supervising variables (e.g., temperature) on their gas turbines coordinating machinery operation for high efficacy with no unexpected by-products.

System masters are also used to find possible issues and their solutions even before they are recognised by a human operator. Due to the capability of AI in detecting machinery wear before it gets out of control, the application of this technology has yielded positive outcomes for intelligent factories such as the reduction of maintenance costs. The indications are that the global turnover of the 'smart manufacturing' market is believed to have reached a projected $320 billion by the year 2020.

Machine learning is known to be among the widespread applications of AI for production while this technique is mainly relied on to make predictions on the part of predictive maintenance systems. This technique has many pros; one of which is the significant reduction of costs and removing the need of planned downtime in many cases. Upon preventing a failure through a machine-learning algorithm, no interruption might occur in system functions. Once repair is required, it ought to be focused providing information on what elements to be inspected, repaired and replaced while specifying which tools to be used and which methods to be followed.

AI can also be used by the manufacturers in the design phase. AI algorithm, also known as generative design software, can be used by designers and engineers for the exploration of probable solution configurations through a precisely described design brief as input. Limitations and definitions of material types, manufacturing strategies, time and budget constraints can be taken into account in the brief. Machine learning can be used to test the set of solutions given by the algorithm. Additional information is given in the testing phase on the efficiency and inefficiency of ideas and design decisions. Thus, more developments can be actualised until an optimal solution is reached.

Data access needs to be crucially considered in the manufacturing sector. No doubt the notorious costs of AI systems could affect the revenues of small and medium enterprises (SMEs), that is, in a financial competition with international corporations had they been more enhanced by AI. Besides terminator references, safety is another major issue in AI technology. Asimov's 'Three Laws of Robotics' is

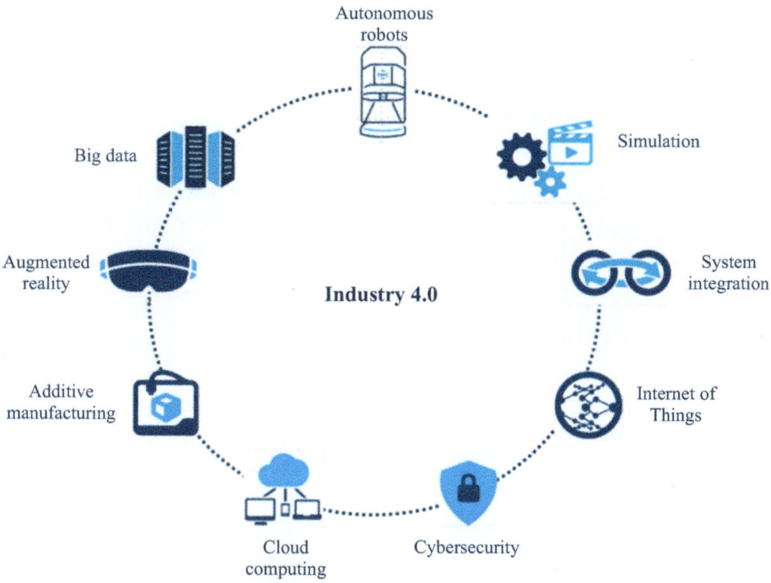

Figure 9.5 The main principles of Industry 4.0 affected by AI

currently neglected and liability issues are not addressed when a machine malfunction might result in a worker injury (Figure 9.5).

Business sectors must be prepared for probable personnel transformations besides direct regulation. It has been predicted that AI will 'cost jobs'; however, the fact is that it will create as many jobs as the ones lost. Namely, new facilities will be dealt with the current staff requiring the assessment of operation influence on each business.

As an example, AI algorithms have been made use of in the optimisation of the supply chain of production operations with the aim of assisting them in reacting to and predicting market changes. Demand patterns classified by date, location, socio-economic attributes, macroeconomic behaviour, political status, weather patterns and more need to be considered by an algorithm to estimate the market demands. This remarkable information can be used by manufacturers in the optimisation of inventory control, staffing, energy consumption and raw materials with better financial decisions concerning the company's methodology.

Given the complexity of AI use in industrial automation, manufacturers are required to cooperate with specialists for customised solutions. Construction of the required technology costs a lot and the majority of manufacturers do not have the essential skills and knowledge in-house.

An Industry 4.0 system is composed of a number of components/stages that must be reconfigured to match the manufacturers' requirements:

- Historical data collection
- Live data capturing via sensors

- Data aggregation
- Connectivity via communication protocols, routing and gateway devices
- Integration with PLCs
- Dashboards for monitoring and analysis
- AI applications: machine learning and other techniques

Manufacturers will attempt to cooperate with experts who realise their objectives and who can make a clearly described roadmap with a fast-paced advancement process connecting the AI integration into related key performance indicators (KPIs).

9.5 Smart materials and 3D printing

Through 3D printing, which is an additive manufacturing procedure, objects are produced layer by layer. Different kinds of 3D printing exist and some of which use thermoplastic materials while others use polymeric materials. 3D printing which uses newly founded smart materials creates new options and makes accessing different industries possible. Utilisation of intelligent materials that provide reshaping or redesigning next to 3D printing is very efficient and inexpensive. Given their productivity and performance, novel intelligent materials and developed fabrication strategies replace the traditional kinds of production. Scholars from Self-Assembly Lab at the Massachusetts Institute of Technology (MIT) hold that 'self-folding' water pipes are likely to be produced one day. This would be possible without using sensors, actuators, software or microprocessors while merely relying on the design of the materials' physical features.

Furthermore, Dr. Tibbits [9] supervised a group of researchers at MIT who started 4D printing that helped advancing the intelligent materials. 4D printing is a developed form of 3D in which the final dimension is 'time'. When facing a stimulus (heat, ultraviolet light or water), time is shown on the print and the print is made adjustable making it a dynamic structure with compatible features and functionality [10,11]. Through this development, digital manufacturing is more widely applied; however, multidisciplinary knowledge and skills are required for this purpose (e.g., mathematics, mechatronics, mechanical and chemical engineering).

Biomedical field has been profoundly affected by 3D printing. The adaptation of 3D printers to education and training programmes has been witnessed in numerous medical schools and centres. Inkjet 3D printing has been used for developing and creating tools for surgical training and planning [1] at the Centre for Image Guided Innovation and Therapeutic Intervention of the Hospital for Sick Children, Toronto, Canada. Numerous novel intelligent engineering materials have been shown and researched: smart valves to control hot and cold flow [12] or acidic and basic flow [13], adaptive pipes [14], sensors [15,16] and soft robots [17,18]. Open source tools in which hardware and software are connected have been made with the aim of platform-based product design [19].

This procedure has been lately used by scholars at Harvard University in Cambridge, Massachusetts, to print a flower which automatically unfolds while contacting with water. This development was followed by the use of a hydrogel in which we have a mixture of cellulose fibres, tiny particles of clay and plastic monomers. Besides water, there are other triggers that change the shape of 4D materials: materials with the corresponding design can also be activated by heat, motion, ultraviolet light and any other kind of energy.

What is intriguing at the same time is 'self-mending' materials – for example, plastics – that have the ability to mend themselves. Combining tiny nanotubes filled with unhardened polymer into the plastic is one of the ways of ensuring this impact. In case the plastic is damaged, and the tubes are wide open, the liquid polymers are freed to repair the material. In a unique process, traditional 3D printing has been used along with ultrasonic waves by the researchers at the University of Bristol for the purposes of bringing the nanotubes into the right position. In the liquid plastic, the patterned force fields inside the nanotubes have been created by the waves. This procedure impact not only the nanotubes but also the carbon fibres. Conventional composite materials, formerly manufactured by hand, can be produced by 3D printing process. This in turn is an actual success (Figure 9.6).

Because of their unique physical features, thermoset composites can be compared with metals or thermoplastics. Due to the stability of thermosets under high temperature condition, they can be beneficially applied for industrial purposes. The popularity of thermosetting resins is due to being processed at low pressures and viscosities. Convenient impregnation of reinforcing fibres such as fibreglass, carbon fibre or Kevlar is therefore allowed. During the curing process, we can find polymers in thermoset plastics that can be cross-linked to shape an irreversible chemical bond. Upon the application of heat, the problem of the product remelting is removed through the cross-linking. This feature lends thermosets for high-tech applications such as electronics and appliances.

The materials' physical features are remarkably enhanced by thermoset plastics, while chemical resistance, heat resistance and structural integrity are improved. Given the resistance of thermoset plastics to deformation, they are mainly used for

Figure 9.6 Simulation of assembly using virtual reality

sealed products. Thermosets could be utilised in thick and thin wall capabilities. They have perfect aesthetic appearance with the high levels of dimensional stability. It is worth mentioning that thermoset plastics cannot be recycled, remoulded or reshaped, which makes their surface finish more and more challenging.

References

[1] Germany Trade and Invest, *Smart factory*. Retrieved from https://www.gtai.de/GTAI/Navigation/EN/Invest/Industries/Industrie-4-0/Industrie-4-0/industrie-4-0-what-is-it.html?view=renderPdf, accessed December 20, 2018.

[2] Learn about Industry 4.0. Retrieved from https://dupress.deloitte.com/dup-us-en/focus/industry-4-0.html.

[3] J. Lawton, *The role of robots in the industry 4.0*. Retrieved from https://www.forbes.com/sites/jimlawton/2018/03/20/the-role-of-robots-in-industry-4-0/#7d7335aa706b. [Accessed 20 December 2018].

[4] Boston Consulting Group (2015), *Industry 4.0: The Future of Productivity and Growth in Manufacturing Industries*, 2015. Retrieved from https://www.bcgperspectives.com/content/articles/engineered_products_project_business_industry_40_future_productivity_growth_manufacturing_industries/

[5] C. Coleman *et al.*, *Making maintenance smarter: Predictive maintenance and the digital supply network*, Deloitte University Press, May 9, 2017. Retrieved from https://dupress.deloitte.com/dup-us-en/focus/industry-4-0/using-predictive-technologies-for-asset-maintenance.html.

[6] S. Wang *et al.*, "Implementing smart factory of Industrie 4.0: An outlook," *International Journal of Distributed Sensor Networks* (2016). Retrieved from http://journals.sagepub.com/doi/pdf/10.1155/2016/3159805.

[7] A. Radziwona *et al.*, "The smart factory: Exploring adaptive and flexible manufacturing solutions," *Procedia Engineering*, vol. 69, pp. 1184–90, 2014. Retrieved from http://www.sciencedirect.com/science/article/pii/S1877705814003543.

[8] German Industry 4.0 Working Group, *Securing the future of German manufacturing industry: Recommendations for implementing the strategic initiative INDUSTRIE 4.0, 2013*. Retrieved from http://www.acatech.de/fileadmin/user_upload/Baumstruktur_nach_Website/Acatech/root/de/Material_fuer_Sonderseiten/Industrie_4.0/Final_report__Industrie_4.0_accessible.pdf

[9] A. Mussomeli, S. Laaper, and D. Gish, *The rise of the digital supply network: Industry 4.0 enables the digital transformation of supply chains*, Deloitte University Press, December 1, 2016. Retrieved from https://dupress.deloitte.com/dup-us-en/focus/industry-4-0/digital-transformation-in-supply-chain.html.

[10] B. Sniderman, M. Mahto, and M. Cotteleer, *Industry 4.0 and manufacturing ecosystems: Exploring the world of connected enterprises*, Deloitte University Press, February 22, 2016. Retrieved from https://dupress.deloitte.

com/dup-us-en/focus/industry-4-0/manufacturing-ecosystems-exploring-world-connected-enterprises.html.

[11] "Adidas's high-tech factory brings production back to Germany: Making trainers with robots and 3D printers," Economist, January 14, 2017. Retrieved from http://www.economist.com/news/business/21714394-making-trainers-robots-and-3d-printers-adidass-high-tech-factory-brings-production-back.

[12] The New York Times, Consortium Wants Standards for 'Internet of Things', 2014, Retrieved from http://bits.blogs.nytimes.com/2014/03/27/consortium-wants-standards-for-internet-of-things/?_r=0

[13] G. Ramasubramanian, Machine learning is revolutionizing every industry, *Observer*, 2016. Retrieved from http://observer.com/2016/11/machine-learning-is-revolutionizing-every-industry/.

[14] R. Hadar, and A. Bilberg, *Glocalized manufacturing: Local supply chains on a global scale and changeable technologies, flexible automation, and intelligent manufacturing*, FAIM 2012, Helsinki, June 10–13, 2012.

[15] World Economic Forum, *Top 10 Emerging Technologies of 2015*, 2015. Retrieved from http://www3.weforum.org/docs/WEF_Top10_Emerging_Technologies_2015.pdf

[16] Available online: https://www.bosch.com/stories/industry-for-individualists/

Chapter 10
Integration revolution: building trust with technology

Ishaan Gera[1,2] *and Seema Singh*[3]

Artificial intelligence, machine learning, robotics and blockchain are all products of the Fourth Industrial Revolution. Although it is difficult to say which has a wider impact on the world of work and present status of the society, each is contributing in its own significant manner to changing the way we view and interact with technology. Blockchain, in this sense, can be characterised as a trust-building mechanism, which has eased the way we conduct transactions bringing in more reliability, transparency and trust to the system. The technology, which once formed the basis of bitcoin and digital currencies, has gone beyond its remit to create new fields. More important, more than developed economies, blockchain is to benefit developing countries by helping to reduce the instances of corruption and leakages with its efficient check and balances system. The chapter would focus on the impact of blockchain on different areas to build what we call trust economies and its limitations. We also focus on the nature of technology, and how marginalised sections may be left out if the governments do not take initiative to educate them about new technologies and systems.

10.1 Introduction

The onset of the Third Industrial Revolution was characterised by the introduction of computers, fourth, on the other hand, has relied on an array of technologies that together build on the computer revolution. Professor Klaus Schwab, Founder and Executive Chairman of the World Economic Forum, remarks that 'Previous industrial revolutions liberated humankind from animal power, made mass production possible and brought digital capabilities to billions of people. This Fourth Industrial Revolution is, however, fundamentally different. It is characterised by a range of new technologies that are fusing the physical, digital and biological worlds, impacting all disciplines, economies and industries, and even challenging ideas about what it means

[1]JRF scholar (Economics), Delhi Technological University, Delhi, India
[2]Assistant Editor, Financial Express, India
[3]Professor (Economics), Delhi Technological University, Delhi, India

to be human' [1]. Blockchain, a decentralised ledger system, is the technology that powers the bitcoin, but over the years has come to represent many more use cases. Being a ledger system, it creates an immutable record of transactions which are not only fast but also secure. Thus, blockchain can be characterised as a trust-building mechanism, which has eased the way we conduct transactions bringing in more reliability, transparency and trust to the system. In fact, blockchain forms the basis of opening new sectors in the field of production, organisation, services and finance. As more nations, companies and individuals rely on technology, it becomes that much more important to bridge the trust deficit. Blockchain is a mechanism that works on a decentralised system that establishes trust between counterparties. It has relevance not only for developed economies but also has use cases for developing countries, which are marred with instances of leakages from the distribution systems, inefficient platforms and corruption.

Although blockchain may have captured the imagination of many, its applications have been limited to scientists and engineers. Most research in blockchain has been on the issue of improving the technology or focusing on nitty-gritty of its functioning. This does suffice but the absence of information in the social science field inhibits further growth and knowledge. Further, the absence of papers that focus on the impact of blockchain on society and those that may form the end-consumer for the technology leads to greater ambiguity. This chapter intends to fill the gap in blockchain enterprises. In this background, the chapter will discuss the evolution of blockchain technology, the new initiatives that can transform systems using blockchain, and its impact on the pre-existing structures in the society. The chapter is divided into five sections, with the second detailing the methodology and objectives of the study. The third section highlights what is a blockchain, the origins of blockchain and its evolution beyond digital currencies. While the fourth explains how trust economies can be built with this technology, including its impact on developed and developing countries with a special focus on India. For purposes of clarity, this section entails the uses of technology along productions processes, organisation, services, finance governance and the issue of its permeation to the lowest section of the society. The fifth points hindrances to the efficient utilisation of blockchain and how different institutions can overcome these limitations. The sixth section concludes the discussion pointing to the selection bias that occurs with the overuse of technology and how the marginalised are excluded from its benefits due to the ill-informed decisions and unpreparedness of the government to familiarise them with new means.

10.2 Objectives and methodology

Blockchain starting as ledger system for bitcoin has come a long way in terms of its application. The distributed ledger system which creates blocks of data that are virtually impossible to hack into, has worked as a secure and efficient system to work through chunks of data and handle big data storage. Thus, blockchain follows a spate of innovations that have occurred across different fields. Although there is a paucity of data available on the blockchain, there are enough popular articles available to showcase the use of technology. Importantly, with the advent of quantum computing, many journals specifically detailing new technologies and use cases in blockchain have popped up. Being a crypto currency-based system, it is quite understandable that blockchain has followed most of its developments in the world of finance. We have followed a similar method to decode the impact of blockchain for sectors, starting with a detailed review of the development of blockchain, followed by its use cases in finance, business and governance. As sharing economies have become a buzzword for development, the entire focus of this exercise relies on how blockchain can improve these sharing economies, more specifically as consumers conduct business over the Internet, there is a need for new mechanisms of trust. Thus, detailing a review of popular articles, journals and research papers, this chapter tries to link new use cases of blockchain with the notion of trust economy, while also detailing on the shortcomings of the system.

10.3 Evolution of the blockchain

The blockchain story is linked to the story of digital currencies. Although bitcoins have gained a bit of international notoriety of being of disrepute, blockchain, on the other hand, has garnered the attention of many. The system developed to aid digital currency has not only helped the digital currency revolution but also led to the development of a new finance system. In order to understand why blockchain has been so important, it would be crucial to uncover the story of digital currency and blockchain (Figure 10.1) [2].

The whole system of Internet is based on backstops on ensuring your information is safe and secure. So, when it came to developing a digital currency that could be traded across the world, it would have been surprising if the creators would not have gone for safety and security. Ergo, creators had to come up with a way to grant security and privacy to a system, while also allowing transactions on a global scale. Importantly, they had to do it in such a manner that the system cannot be rigged, and any foul play does not lead to the creation of a bubble and siphoning off millions. Also, it was too much authority with one agency or person, so the system had to be decentralised. The bitcoin revolution also led to the blockchain revolution, with the latter supporting the infrastructure and architecture of this worldwide phenomenon. The whole system, which often involves unscrupulous ways, ironically worked on trust which was ensured with transparency. The blockchain ensured a decentralised ledger to be created that was visible to the counterparties for verification, with additional security of each block being separate from each other.

If we were to explain this easily, think of an Excel sheet, which is protected or locked so that not everyone can edit it, but few who have access to the sheet can make changes. Now, if we were to write some piece of information on the first tab of this

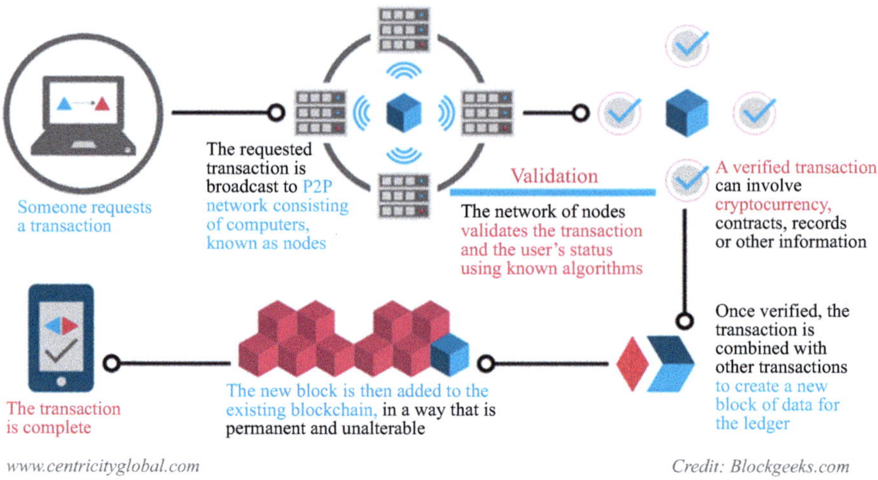

Figure 10.1 How a blockchain works (Source: Blockgeeks.com)

sheet, this information would have to be verified by the few who are part of this sheet. Once this is verified that information cannot get erased or deleted, any additions or editing would be reflected and would have to be approved by all. While the spreadsheet would be visible as a ledger to all, its contents shall remain between the parties involved unless they agree to share it with others. Similarly, the next tab of this sheet would be another block of transaction, with the same rules as the previous one [19].

Thus, by deploying this system not only the creators of bitcoin ensured transparency but also created the first hints of a decentralised system, where transactions could happen seamlessly and without many issues. In addition, with an easy flow of transactions and security, international payments became much more secure. As even if one were to hack the system, they could only hack one block of data, while others remain practically out of reach for hackers [3].

The blockchain revolution was not just restricted to bitcoin; it also led to the development of a digital currency system outside the realms of bitcoins. Over the last few years, many copycats and new ideas revolving bitcoin have emerged. Even banks have come to rely on the system to create a web of seamless transactions. One point to consider can be the efforts by almost all major global banks to create their own currency systems, on the lines of bitcoin. And, this revolution has been triggered by the ease of transaction and a cut-down in the transaction time for customers and banks [4].

If one was to review the system of international transactions and international payment settlements, one would find that it was riddled with too many backchecks to prevent the case of any fraudulent transactions [5]. Let us assume that a transaction is going to take place from a home country to a foreign country. Now, the bank in the home country would process the transaction here which would then go to the clearing house in the home country followed by a clearing house in the international country and finally get remitted to the bank in the foreign country. In the case of blockchain, if it was the same bank, a bank like Citi can bypass the whole process with transactions of its digital currency that lie within the remit of the rules of each of the countries and that too within minutes of settlement. On the other hand, the current system takes days to dispense off the amount. But where this closed group systems lack is the interoperability of coins. While Citi could easily transfer the amount from Citi Home to Citi International, it cannot do so for any other bank. Thus, bitcoin still has an advantage if banks do not come together on the same platform to provide seamless services.

This issue also brings to fore an important point of classification of blockchain systems. While we have been discussing blockchain as a single entity, in the hands of private players it has become a much wider network with more intricacies.

10.3.1 Closed and open systems

Ever since its evolution and adoption by private players, blockchain has developed another iteration of a closed network. While in its earlier form blockchain was an open or shall we say a public system; in the hands of private parties it has ceased to be such (Figure 10.2). In fact, far from being the open network that it was under the bitcoin regime, each of the systems that have been developing blockchain has veered it towards a private source. Take, for example, the case of Ripple and Bitcoin; although both are versions of digital currency, Bitcoin is a public system, where the data is available for all to purvey; on the other hand, Ripple is a closed

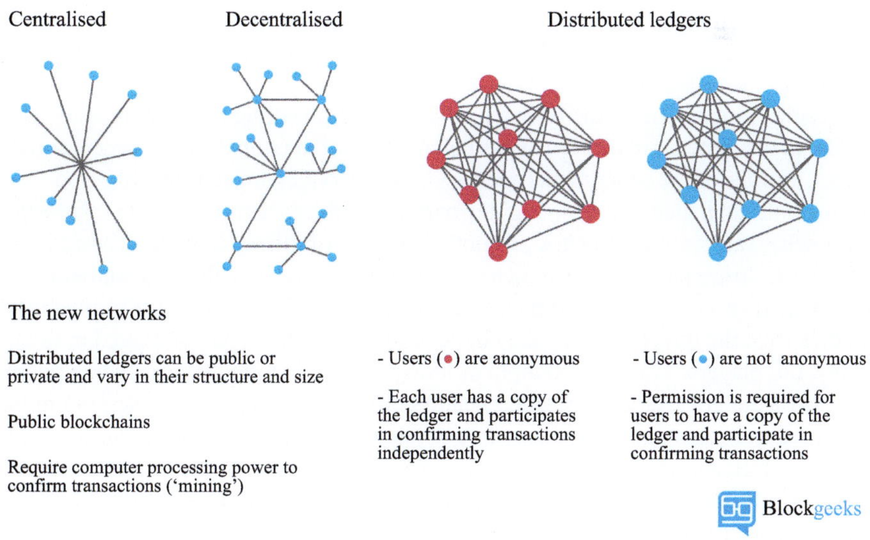

Figure 10.2 Centralised and decentralised ledgers, public and private blockchains (Source: Blockgeeks)

network. Despite having associations with many banks across the world, Ripple can allow money transfers only between its partner networks and has little interoperability. This can be easily explained with the help of another example of payment systems. Apple Pay is a propriety of Apple Computers, while Google Pay is a network for Android-run devices. While Google is also available on Apple's platform, and much like bitcoin, becomes an open-source system, Apple has restricted ability, by being only available for Apple users. Although this provides much more security to Apple, with a limited base its accessibility remains a problem. Thus, being a private player, it restricts the use case of blockchain as the information on the block is only available to a few players like with our Excel spreadsheet. On the other hand, in the case of bitcoin being an open or a public network, nearly anyone can access the information base [6].

10.4 Trust economies

Trust has become an important component of sharing economies. As technology permeates every strata of our society, trust gains much importance than profit models or business plans. In fact, the success of sharing economies is based on the trust between the players and a measure of security provided to the users. Thus, it is important that businesses and governments alike focus on the issue of trust and security, more so as they push digital. Blockchain is an unlikely champion of this trust system. Built as a ledger system to trade in Bitcoins, it has come a long way since. Although the infamy of the bitcoin network has damaged some of its reputation, the technology is being put to various uses to ensure that the new systems do not suffer from the drawbacks of the old. The chapter highlights the evolution of blockchain, and its use cases, while focusing on the trust and security aspect. It also lays to fore the drawbacks and limitations of technology, suggesting a way forward.

Besides money one of the primary objectives of installing blockchain is the trust deficit that the system creates. Trust has become one of the important currencies on the Internet, as none of the benefits that accrue via web services are tangible. It is important that trust is established between counter-parties for the system to operate efficiently. That is why, ratings and websites promoting the use of ratings have flourished in the recent past. Importantly, with a general discontent towards the government, this measure of trust is shifting from the government to people. People trust ratings more than they trust the government as they believe that the system may be rigged in favour of a few big players, Take the example of food joints and rating services, while people do not trust the government-provided health certificates much, the whole system has now turned to companies allowing people to rate restaurants based on services, quality of food and cleanliness and thus services like Zomato have become an important source for information. Similarly, in the case of sharing economy, people have started belaying more trust in companies like AirBnB and Bla Bla car than the traditional services. Blockchain may not replace the rating economy but is a good source of trust setting for people. With transparency at the core of its functions, it helps in establishing

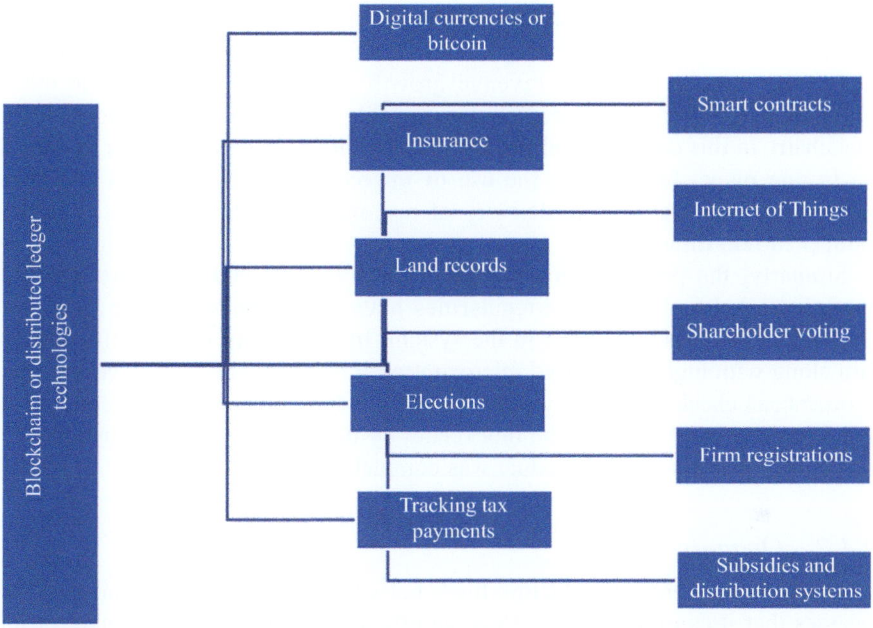

Figure 10.3 Beyond bitcoin: use cases of blockchain

trust between various sources, thereby creating something that the market needs and relies on direly [7].

While the application of blockchain has been for developed countries, developing countries like India can benefit from it as well. With Internet services becoming more pervasive across the region and people becoming more tech-friendly, blockchain can be boon for the developing and under-developed economies, one that ensures a system of checks and balances. Blockchain can curb the instances of corruption, help in the eradication of corruption. As the whole system is recorded block by block, the system can ensure transparency, which does not bode well for corrupt or secretive practices. More important, as each record is final, it shall also delve a blow to the legality problems in cases (Figure 10.3).

10.4.1 Production

Nothing can be complete without gauging the impact of blockchain and production, if blockchain is not able to make an impact of production, it would be futile to scale this technology for further uses. Thus, it becomes paramount that blockchain is used for production. Much like other processes, it can be used for improving supply chain efficiencies and ensuring trust on the customer side.

In the case of supply chain management, most companies using blockchain are set to become more efficient by ensuring little leakages from the system. The importance of supply chain management can be gauged from a survey conducted by Deloitte. According to the survey, 79% of the companies with high-performing supply chains reported higher revenue growth than the average. On the other hand, only 8% of the companies with less capable supply chain could do this [8]. Blockchain, in this case, can help in tracking of data across the network to get all data in one place. In addition, the use of interconnected blockchains can help evolve a system where a late delivery of raw material can be intimated to the producer in real time.

Similarly, the process can be used at the consumer end to ensure that the product is tested and that any irregularities have been reported by the company ensuring that there is more trust in the system. In the case of counterfeits, blockchain along with highly enhanced microcontroller chips can be used to ensure that customers can check the veracity of the products. More important, any testing done on any product can be put on the blockchain by the company for the customer to ensure that the testing of his product was complete.

10.4.2 Organisation

Blockchain is efficient in production lines, but what makes it more efficient is the processes that it can incorporate. Thus, an efficient organisation can make for an efficient production process. In this case, collective decision-making and smart contracts can really help companies ensure transparency and scalability.

10.4.2.1 Smart contracts

One of the primary examples of this trust can be smart contracts or credit notes that are dispensed off by banks as a measure of credit guarantee. These contracts can be turned into smart contracts with the inclusion of all counterparties under a blockchain managed contract. Take the case of credit notes and shipping. At present, there exists a lot of ambiguity in the case of the shipping industry and credit disbursal. The asymmetry of information can lead to frauds which further leads to the siphoning of money from banks. If a smart contract is ensured between counterparties including banks, insurance and shipping and traders, this ambiguity can be reduced to a great extent. For instance, if a trader in Country X were to receive an order from one in Country Y, then the one in X can take this order to Bank A in her own country and get credit for the production and export of such products. Meanwhile, if Y were to extend some credit, it will go to Bank B in her own country, and both the banks can cross verify this information on the new block that is created. Meanwhile, as banks are connected. X cannot take credit from any other Bank other than A, if that fulfils its requirements as banks share this information on their own blockchain to reduce fraud cases. More important, as both banks verify each other's transactions the loan disbursal process becomes utterly simplified, and the shipping company can be made a party to the contract. Now, the insurance provider can also come into play, as the whole system is connected each of the parties can draw up an insurance contract based on the services provided, with the transparency that already is existent in the system. In the end, the payment can be released once the shipping company delivers the order to country Y, thereby insuring a fast and efficient system. Thus, blockchain can cut down the uncertainties as different entities with different chains interact with each other seamlessly over one order [9].

10.4.2.2 Shareholder voting

One of the key problems for companies has been to get shareholders to become a part of the process. While the big fish do get their chance to vote on company decisions, the small shareholders rarely get a say. The process is tedious and cumbersome which makes it difficult for the shareholders to become holders of any value. As there is ambiguity about voting, for all to be present at a company location is virtually impossible. Blockchain may solve this dilemma, much like insurance, creating a separate block for each shareholder, and then asking them to vote in that block can be an easy solution. Not only will this provide shareholders the say, it would also make them equal participants in the company process. So, blockchain not only becomes transparent but also egalitarian as it allows each shareholder equal rights in the voting process, irrespective of their relative shareholding. Importantly, it becomes the whole company process more democratic, not only can shareholders be part of this decision, but board meetings can also be convened similarly, with voting taking place in the same fashion.

Already under implementation in countries like Ghana, blockchain voting provides a measure of importance to shareholders making them a part of the company decisions. While the share markets in developing countries have developed, many are still not aware of their rights. More important, there is ambiguity in conversion of shares from materialised to Demat form. A blockchain like system can ensure that government has data for share transfers and sales and purchases from one entity to another to the end of time and that too without lags. More important, it can ensure whether companies are being represented rightfully with as little measures of the discrepancy.

10.4.2.3 Firm registrations

Not only firm's shareholding, blockchain can also aid in registrations and tracking of ghost firms. One of the biggest problems for tax department has been handling of firms' data, as the department has no means to track which of the firms are just fly-by-night operator used for offshoring or money laundering. With blockchain all this can be made easy, as each director is assigned a director identification number, the system can track what kind of firms a person is operating and how many of these enterprises are actually filing returns and conducting some business. This would also make closure and opening of firms much easier for the government, while also aiding start-ups to file returns and in bankruptcy. More important, it would contribute to the ease of doing business by creating a transparent system, aided with privacy of data and fast-tracking of processes.

The Volkswagen scandal

In 2014, governments and regulators uncovered a new scandal related to Volkswagen vehicles. The company deliberately was cheating regulators with devices that would not reveal the true pollution level emitted by the cars. So, under the lab testing phase, the car would show different pollution levels from those that were on road. Once uncovered, the scandal led to mass recalls with the company having to apologise to people for the betrayal of trust. More so, investigations revealed that while many had warned the company about this mishap, some from the top management ignored such memos.

Blockchain, in this case, would have avoided instances of such scandal and company could have saved face. As the system is immutable, any instance as such would have been recording fixing liability to the top of the chain. Instead, what ensued after the scandal is a process of cover-ups where notes on the issue were found missing or misplaced. Thus, blockchain in this case could have been a saviour for the organisation with each testing procedure recorded in blocks, creating data which could not have been erased by the company.

10.4.3 Services

Just as in the case of production and organisation, blockchain can ensure a healthy delivery of services, where monetisation of services and their delivery become easier.

10.4.3.1 IoT and healthcare

According to a report by IEEE, blockchain can also be used to protect the Internet of Things (IoT) applications. IoT may have been the buzzword for a few years now, but with increased hacking attempts there have been concerns about its application. With the present infrastructure, IoT systems are so vulnerable that if a hacker were to get access to even one of the devices, he/she could bring down the whole network. Blockchain can eliminate this uncertainty with regard to the IoT devices. As each block is protected by a separate code, even if the hacker gets access to one block, the other devices remain secure as the system prevents him/her from accessing any further information. Thus, the chain is protected. This may be a boon not only for IoT systems but also for blockchain. As technology can ensure that systems across businesses and homes stay secure, a replication of this can also ensure a fast digitalisation of the healthcare industry.

Similarly, blockchain can ensure that health systems are not vulnerable to hacking attempts. So, instances like hacking of healthcare institutions can be stopped. In this case, blockchain can ensure that hacking attempts do not go beyond one system or machine thereby preventing a shutdown of the whole system. More important, with agencies like UK's National Health Service talking about digitalisation on a mass scale, blockchain can limit the extent of hacking. This can, and has, also been implemented in public utilities, where smart grids can be protected by blockchain applications, ensuring that public utilities are not affected. As technology pervades every aspect of our lives, it becomes so much more critical to maintain the security aspect. Blockchain to a certain extent can provide such relief.

10.4.3.2 Smart grids/energy

In the case of smart grids, the utility of blockchain goes beyond ensuring efficient delivery of services. Here, blockchain can also ensure a monetisation model. Smart grids work by ensuring that most problems are reported in real time and bills and all are generated by the government. But in the case of the new green energy revolution, it can do much more. For instance, in certain cases not only can the people buy energy from the government, also they can supply energy to the government. In case they are generating more than what they can use and then the government can dispense of this energy to those who need it. By ensuring this model, it can make green energy alternatives attractive and ensure that most people turn green. Once the solar technology is efficient enough, a blockchain will ensure a contract between the consumer and the producer or supplier of energy. Thus, if the supplier takes any quota of energy from the consumer in this case, he/she will compensate accordingly and the data being on the blockchain would ensure it is transparent and rid of flaws.

10.4.4 Finance

Finance has been one of the first and most important domains of blockchain adoption. The first iteration of blockchain was in digital currencies and banks and financial institutions have gone on to expand their use of blockchain-forming consortiums to most effectively use the technology.

10.4.4.1 Banking and finance

Banking has been one of the last domains of technology adoption. It took the finance world 20 years to adopt the Internet, while mobile phone banking came in much later. To look at a much recent example, and one that is closer to home, India's Apple Pay moment came two years after Apple had launched it and almost a decade since the government started online transfers using the National Electronic Fund Transfer (NEFT), Real-Time Gross Settlement (RTGS) and Instant Money Payment System (IMPS). But once verified by finance, a technology gains much wider application across different groups and opens several possibilities for a trickledown effect to initiate [10].

Blockchain has been one such technology which has come in vogue as the banking sector has been ready to adopt it. Although it has taken almost a decade for this technology, it was used as a source of bitcoin to be adopted [11]. Banking and the world of finance have given blockchain a new meaning. The technology that forms the backbone of bitcoin dispensation system is soon becoming one of the most overused phenomena in banking. Importantly, with the trust belayed on it blockchain has found applications beyond the realm of banking [12]. This is evident from the amount of tech start-ups related to blockchain that have emerged over the years, and how funding amount of these has gone up more than ten times in the last five years.

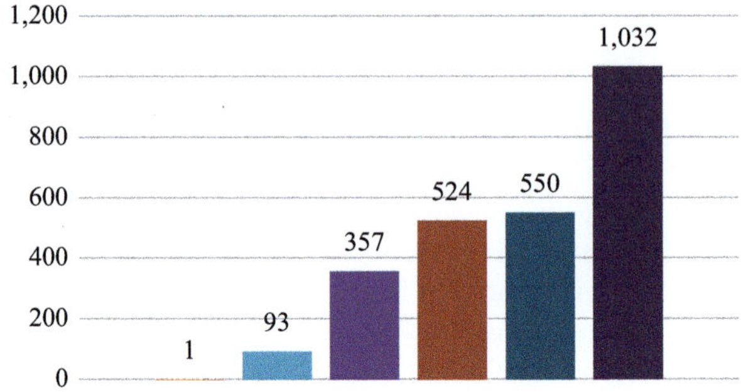

Figure 10.4 Funding and investment of blockchain start-up companies worldwide from 2012 to 2017 (in million US dollars) (Source: Statista)

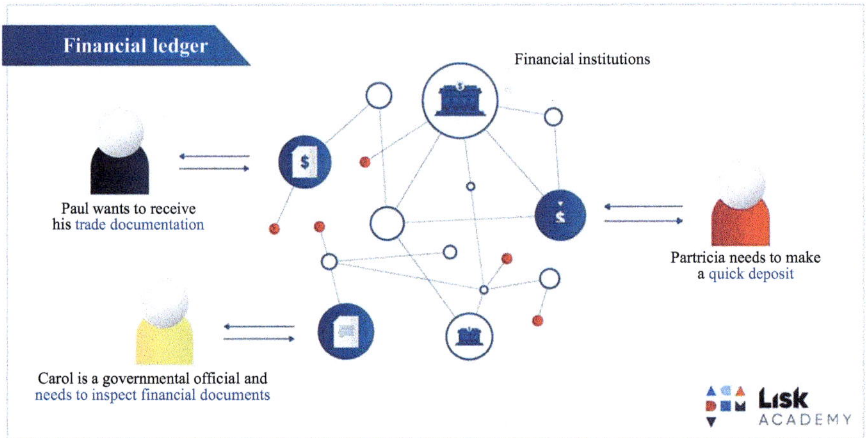

Figure 10.5 Blockchain can revolutionise banking (Source: Forbes)

Statista data shows that while there Blockchain companies garnered $93 million in 2013, this increased to $1.032 billion in 2017 (Figure 10.4).

One of the primary benefits of blockchain is that its use cases go beyond the remit of digital currencies and stretch from the area of transactions to insurance. A Santander InnoVentures report points out that Blockchain will save $20 billion in costs for financial institutions. Morgan Stanley highlights that it can reduce costs by as much as 50% as compared to traditional channels [19].

Thus, blockchain can be a saviour for the banking and financial institution which has been reeling under slow transactions and asymmetric information systems. Take the case of banking industry and the need for a central clearing house.

While banking has worked well for years with a central clearing house system, and it has been one of the primary functions of the central bank, it can be devoid of such functions and be the regulator in the industry if blockchain or distributed ledger technology is allowed to function (Figure 10.5). In the case of blockchain and an interconnected one, banks would not be required to make a separate bill of transactions for a central clearing house as all data and transactions would be recorded by one and shall be visible to another. The company or the government developing this system can, of course, charge for such services but what the system would provide is a way for banks to operate without any operational delays and lags. Besides, with the central function of a central clearing house cut-down, banks would not require an in-house team to check such transactions drastically reducing employee costs and banking costs, both for the banks and general public. So, a transfer that used to take hours or even minutes can be ratified within seconds. In addition, given the security, which still needs to be worked on, only chunks of data are available to be tampered while the entire system remains shielded from the purview of hackers [13].

10.4.4.2 Insurance

Not only shipping blockchain can also prove to be a boon for the insurance industry in general. The system by ensuring transparency and speed allows for the creation of niche insurance markets that can work for limited time transactions. Take the case of delay insurance on flights. The easier case in this example can be the purchase of insurance before getting on to a flight. While as you purchase this insurance, and the amount, the insurance company can decide on the payment amount and the premium to be decided based on past data and the case of flight delays, as you purchase that insurance it builds up a contract specifically for you, that allows for the payment of X amount if the flight is delayed over two hours. While the contract is enforced, it becomes immutable as none of the parties can

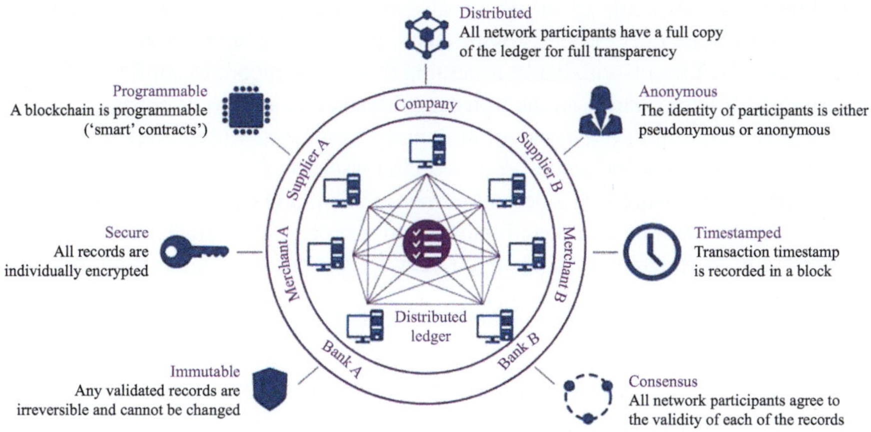

Figure 10.6 Blockchain in insurance (Source: KPMG Global)

change it afterwards. If the flight does turn up late, the insurance company can any case check it against the public database on flights and remit the amount to the customer immediately. Such niche faster payment products can be created for any of the offerings/services that can be cross verified, from rail travel to even accident insurance while travelling via an Uber from one place to another, eventually opening up new avenues for the insurance companies and customers as technologies interact with each other, especially as sharing economy becomes much more prevalent [14].

This is one place where countries like India can remove ambiguity with regard to insurance payments (Figure 10.6). Once the system is mechanised, insurance payouts can be faster, and with interactions of chains, it can reduce the litigation burden. As insurance cases account for a major chunk of court's load, discrepancies in the payouts. So, auto, life and health insurance cases can be settled at a faster rate and without much hassle [15].

10.4.5 Governance

Just as in the case of finance, governing institutions have found many use cases of blockchain. Ledgers are how the government works and thus blockchain can help governments utilise this technology to improve its recording systems.

10.4.5.1 Land records

Another common form of practice where blockchain can prove to be consequential is the maintenance of land records. Blockchain can ease the process of registry, as the registration of land records can be done online. As the records are immutable any transfer of registry or land would also reflect in records each time such documentation is done. Besides, it being an open blockchain it would allow everyone to have access to the resource. So, in the case of our spread sheet example, it would mean that each registration would reflect in one block, with any changes becoming the part of the same block, thereby creating a chain of events for people to follow through. Not only would this reduce the time of litigation, also it would ease the process of sale and purchase of land, as only property with clear titles can be sold. More so, it would reduce instances of fraud, with banks accessing this public record to confirm purchases and disburse off loans, based on the transfer of deeds and system of entries.

Not only would this system reduce ambiguity, combined with drone mapping also it can finalise land titles and settled disputed land issues with the help of registrations. The system would ensure that sale and purchase of land is easier and seamless. The process of registry can be completed online, and one can also convert their properties from leasehold to freehold without much delays. The process would save money in court fees and unnecessary litigation [16].

10.4.5.2 Elections

The shareholder voting may still be liable to some uncertainties, voting in elections is much easier. Each voter at present carries voter id or some form of identification which allows them to vote in elections. This identification, whichever it might be would be unique to that person, so it creates a separate store of data for that information. Blockchain as a public network can make use of such sorting to create a separate channel for each of the voters. As voting is considered a secret ballot, the security within blockchain can ensure that each line of data from voters is coded, so that even if the election commission gets to know that the person has voted, it does not get to know how and with what preference a choice has been made. While governments are in the process of instituting a mechanism for soldiers serving in other countries or territories to vote this way, this can also pave the way for governments to allow citizens to vote from the privacy of their homes without many problems. Moreover, it can ensure more participation on behalf of voters.

Much like shareholders, voting in developing countries is also a problematic issue, especially for migrant labour. It becomes extremely difficult to trace the voting process, blockchain combined with registry and land leasing agreements can ensure that the voter ID card is as mobile as the person. So, a person from New Delhi residing in Bengaluru can vote in elections in Bengaluru, and the system can ensure that once he/she has voted in Bengaluru, he/she cannot come back and vote in New Delhi.

10.4.5.3 Tracking tax payments

Much like with companies, an individual can be treated. While each individual is assigned a unique ID, the tax department can track a person's transactions to account for her money income. A case in point can be the cashless system, as more services turn to the cashless economy, there can be a point whereby the IT department can track transactions related to a person in a secure locker. As such process would be user-verified, these transactions can then be paired with income data, to look for the discrepancies in reporting of taxable income. More important, such a system can also serve as a measure of tracking how money is spent via transactions and how much tax an individual has deposited. While the backbone of the whole system is transparency, there is also a need for privacy, as people will not be comfortable showing their spending as a matter of public record.

A major area of concern for developing economies is of tax base. While tax collections have increased, tax base remains the same for these economies, due to tax evasion practices. More important, as the process becomes more participative, the tax department can track investments and spending of a person to pare the data with payments to ensure tax is paid fully and there is no tax evasion. Once, income tax collections increase the government can check by reducing sales taxes, which will help the poor and marginalised sections.

10.4.5.4 Subsidies and distribution system

More than the taxes, blockchain can help funnel the subsidy money whether it is in form of goods, services or cash transactions. The ledger system can help create a systematic record of payments in the form of subsidies with the two-way system ensuring that the right person gets to enjoy the services and not the middle men. As the system requires authentication by both parties it would ensure trust and transparency from both. Using authentication IDs and a linking of system, the system can generate a perfect outreach programme, where the trickle-down of benefits is complete and absolute. For instance, if say the government were to disburse grains, they can be received by the distributor, who shall have to verify the quantity, and the same cycle shall repeat when it reaches the beneficiary ensuring that everybody gets their due share.

With myriad schemes launched by the government, one of the major issues has been the tracking of benefits under such a scheme. The tracking of data at each point of time would ensure that benefits like mid-day meal reach the general populace, with rounds of check at each step, there would be little chance of wrong doing.

Demonetising gains

The Indian government carried out, by far, the largest note exchange exercise in its history on 8 November 2016. While there are many critics and supporters of this move, there has been no doubt that the government and its institutions were ill-prepared for such a big step on such a gigantic scale. Perhaps, this is the reason why even after exchanging Rs 15.36 lakh crore worth of currency, there has been little tracking of the money that has been received by the central bank. In fact, cash exchange at banks, which was allowed at Rs 4,000 per day, had been the biggest cause of worry. People employed unscrupulous means to get their currency whitewashed and some with the collusion of banks were easily able to get away with. While the government did promote digitalisation as a way forward, it was not able to use it to the full extent.

Blockchain, in this instance, would have helped the government achieve efficiency and plugged the leakages in the system. As all data would have been available across bank servers, if any person would have gone to exchange money from another bank after exchanging Rs 4,000 from one, would be recorded in the system or would not be allowed. More important, in such a scenario any use of government Ids by banks would also need to be cross verified by the consumer to double-check each entry. Besides coding of notes could have helped the central bank recheck the notes.

10.5 Utilising the blockchain

But there are certain limitations to the system that need be addressed before deploying it fully on a global or a national scale. The security concerns need to be addressed along with privacy concerns. Importantly, it becomes of utmost importance to decide, who owns the blockchain.

One of the reasons for the success of bitcoin and failure of other blockchain systems has been the issue of control. While bitcoin was a public control system, others were private limiting their ability to interact with others. There is a need to

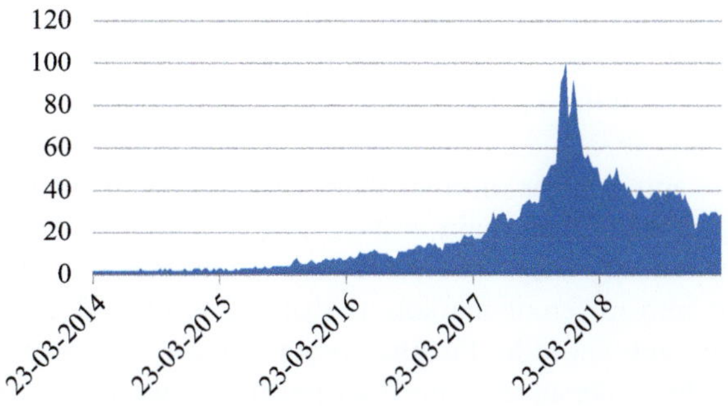

Figure 10.7 Peaked interest? Is blockchain losing relevance. Google Trends of blockchain and related word searches. (Source: Google Trends)

understand that for blockchains to work perfectly there must be a system that can traverse the lines of profitability and corporate motives. Even in terms of private systems, separate systems need to interact with each other with ease for a market to be created as such. That does not mean that private channels are doomed. While they may work excellently for small jobs, for scalability and pan-world coverage, there must be either an interaction of chains with each other or a system which is entirely public and decentralised.

Although security is one of the primary features of blockchain, it can only ensure the security of the system and not of each node. Thus, it is important that the system is made secure. In terms of privacy, there is a problem as well. While the system offers privacy, there is a problem that public chains have to balance privacy with transparency. The system may offer more transparency but the price for it would be a sacrifice of privacy, which many people may not be willing to give up [19].

One of the basic problems with the system is that people need to understand technology, and they need to be aware of its use. Blockchain would only be successful if people understand the use cases of technology and use it to their advantage. People need to be aware of the blockchain in the form of smart contracts, like digital currencies. Technology needs to percolate down to the lower strata of society for them to use it to their advantage. In terms of subsidies, if people are not aware of their use of technology, they will not be able to take advantage of this technology or repose their faith in the system (Figure 10.7).

Another major issue with blockchain is that of scalability. The system has worked well over limited users but that may not be the case with large number of users and across different blockchains. The heavy amount of traffic that a large-scale implementation entails is something that no blockchain is ready for [17].

The only way the government can ensure access to technology, privacy, security, is by pushing its use. The government can do so by promoting the use of technology to the lower strata of society and by stitching together a blockchain system. It needs to create a system, whereby, each and every party which part of the system is can access such information. It needs to ensure that trust can be established otherwise the system will not be scalable or will not be able to function the way it should. The linking of the system is important to ensure that system works efficiently.

But all these applications of blockchain will not be complete if the technology is kept out in isolation. Blockchain in itself is a great system to ensure that digital currencies work the way they should, but one of the major drawbacks for bitcoin from expanding into different markets is that it has limited application amongst people with high tech aptitude. For the common man, bitcoin is a scary word, as neither does he/she understand it nor does he/she know how to use it. The blockchain can remain a similar phenomenon if it is not allowed to interact with other technologies and made user-understandable if not user-friendly.

Any discussion on blockchain will not be complete if the technology is not allowed to interact with artificial intelligence or machine-learning algorithms. The first case in this point can be that of tax collections, as it is virtually impossible for each block to be created after each transaction, it is important that AI creates an assembly system where all transactions are clubbed under one block and under one head. Similarly, for niche product markets to be explored, a new category of products needs to interact well with blockchain systems to work perfectly.

One of the problems with implementing technology is that it should be accessible to all. The problem with higher technology platforms is that they are out of the reach of the general population and are, thus, limited in their applicability. There is a fear that despite being an efficient system that it is, it will not be able to reach masses as they would not understand the technology and would be left out of the process. Thus, the government needs to address the issue of technology percolation if it wants blockchain to enhance transparency and provide security.

Most important, in the case of developing economies, tech knowledge needs to be enhanced if the government is to implement such programmes. Thus, for countries like India, it would be prudent if the economy goes to teach its general populace on how to use technology. As the technology is used more, more faith would be reposed in the new systems, enhancing a system of trust.

One of the classic case examples of this selection bias is women missing out on the blockchain bandwagon. While the participation rates for women engineers have gone up in recent years, and there has been a surge in women in engineering education, most of this has translated in legacy sectors [18]. Even if we look into the field of cryptocurrencies and blockchain development, the number of women involved in the development of such is limited. As far as the interaction of technology is concerned, in most developing economies, women are found on a lower threshold in the use of technology than men. So, their awareness of programmes and technology use cases remains limited. If blockchain is to have any significant impact, it cannot ignore the missing half. Thus, there is a need for women both in the development of blockchains and access to technology.

10.6 Conclusion

This begs the question of whether blockchain can be the technology of the future. Amidst a landscape that is fast evolving can blockchain maintain its intrigue and relevance. But with anything smart, blockchain can come into relevance. As we integrate more systems with smart technologies, blockchain can be the link that maintains security and provides trust. The future of blockchain lies in the future of smart. As more grids get connected, as more IoT devices are used and as more trade and services happen on a global scale, blockchain can be the safe way of providing transmission.

And, this has started to happen too. Blockchain has come to occupy an important position in the privacy and security debate presenting an irony for the current generation. So, while the millennials are happy to post their personal lives on Facebook, Twitter and Instagram, they are also turning to services that are secure and guarantee privacy. Amidst a space of privacy scandals and hacking attempts where website data is being uploaded on social platforms, people have started turning to blockchain-enabled mails and messaging platforms. It will not be surprising if the governments and companies do the same.

A prime example of this can be the Blackberry revolution, which coupled with messaging also provided privacy and security backing. Blockchain backed start-ups and enterprises can start a similar revolution. But in this case, not one company, but the technology would be leading the charge. Smart traffic systems and smart grids can be one of the few uses that are not detailed in the chapter of this technology. More important, any ledger system or database management can have the convenience of blockchain. But scalability like before can be an issue. Much like with Blackberry where scalable technology meant a death knell for the service, blockchain may suffer a similar fate. Thus, for blockchain to be truly revolutionary, scalability along with pricing would be the prime factor for its growth. Blockchain may not be the technology that dazzles the common folk, but like encryption, it can ensure that many other technologies can make a change the society requires or needs.

While Blockchain was hailed as the technology of the future, especially after digital currencies started coming in vogue, the interest around the technology may be peaking. Although funding for blockchain-related start-ups has increased – as per Statista this increased to $1.032 billion in 2017 against $93 million in 2013 – the number of searches related to technology has decreased from its peak in 2017.

But the technology has found use cases beyond bitcoin and digital currencies, as technology is becoming more pervasive, permeating each aspect of our lives trust and security have assumed primary importance, more so when we are conducting business across the globe. Blockchain is a way of establishing this trust on the Internet while also creating an integrated systems architecture that is secure. Thus, blockchain can build a trust economy, based on the tenets of security. New systems such as smart contracts, IoT and insurance can all use blockchain to create new products in the market and ensure that there is little asymmetric information.

One of the basic problems in economics has been asymmetric information, where some people have more knowledge about a product than others which often leads to ambiguity in pricing and market adjustment. With blockchain, some of this ambiguity can be curbed. With this purpose the chapter has divided the functions of blockchain within two domains, categorising the use of technology for government and non-government functions, with trust as the primary component of each.

In case of smart contracts, the technology can ensure that all services involved in a contract system and related parties work in tandem to secure the contract. So, a contract would ensure that each party involved in the process honours the contract in a time-bound manner. This is also true of insurance contracts and the sector, where the claim process is a long and arduous process. Subsequently, the system can also be used to provide benefits for tracking of tax, land and other processes, while also ensuring that consumer's privacy remains paramount.

Blockchain may have been the technology of the future, but scalability and limited application were hindering its progress. The technology works in the form of a spreadsheet that may be accessible to all, but not editable for everyone, with each piece of information protected by a code [19,20]. This may have been beneficial for the system for cross-checking of information and in terms of security, but with limited users it can prove to be a costly affair.

While there is no doubt that blockchain is all set to become the technology of the future, it would not go far and beyond till it becomes scalable for enterprises and governments. That is if the governments and enterprises are not able to link blockchain together, the use cases may remain limited. Thus, it becomes paramount that the government moves towards a society where blockchains can be linked to derive the maximum benefit.

A major issue of concern is that of leaving out a sizeable population from the benefits of blockchain. Until citizens are not aware of the use cases of blockchain, there is a good chance that they will be left out of the process. Technology adoption has to reach the bottom of the pyramid and the oppressed classes for it to be truly revolutionary; citizen awareness is important about the adoption of technology so that even if they do not understand the nitty-gritty they do understand the how and why of it. This integration and equality have to translate not only between the rich and poor, but between male and female as well. If blockchain is not to be as exclusionary, as other technologies have been, education of it must be inclusive. With the world in need of integration, blockchain seems to be just the solution it needs, but it will require a few fixes and scalability issues for it to become a technology for the future.

References

[1] Schwab, K. (2017). The Fourth Industrial Revolution, New Delhi, India: Penguin.

[2] Morris, D. Z. (2016). Leaderless, Blockchain-Based Venture Capital Fund Raises $100 Million, and Counting. Retrieved from Fortune [Retrieved 21 May 2016].

[3] Armstrong, S. (2016). Move over Bitcoin, the Blockchain Is Only Just Getting Started. Retrieved from Wired [Retrieved 7 Nov. 2016].

[4] Shah, R. (2018). How Can the Banking Sector Leverage Blockchain Technology? Retrieved from PostBox Communications [Retrieved 1 Mar. 2018].

[5] Kelly, J. (2016). Banks Adopting Blockchain 'Dramatically Faster' Than Expected: IBM. Retrieved from Reuters [Retrieved 28 Sep. 2016].

[6] Explainer. (2016). Permissioned Blockchains. Retrieved from Monax [Retrieved 20 Nov. 2016].

[7] Iansiti, M. and Lakhani, K. R. (2017). The Truth About Blockchain. Harvard University: Harvard Business Review.

[8] Inamullah, M. (2018). How Blockchain Will Transform the Manufacturing Industry. Medium.

[9] Institute for Development and Research in Banking Technology. (2017). Applications of Blockchain Technology. Mumbai: IDBRT.

[10] Dailyfintech. (2018). Blockchain May Finally Disrupt Payments from Micropayments to Credit Cards to SWIFT. Retrieved from Daily Fintech: dailyfintech.com [Retrieved 2 Oct. 2018].

[11] Raval, S. (2016). What Is a Decentralized Application? In Decentralized Applications: Harnessing Bitcoin's Blockchain Technology (pp. 1–2). Sebastopol, California: O'Reilly Media, Inc.

[12] *The Economist*. (2015). Blockchains: The Great Chain of Being Sure about Things. Retrieved from The Economist [Retrieved 31 Oct. 2015].

[13] Tapscott, D. and Tapscott, A. (2016). Here's Why Blockchains Will Change the World. Retrieved from Fortune [Retrieved 8 May 2016].

[14] Crosby, M., Nachiappan, Pattanayak, P., Verma, S., and Kalyanaraman, V. (2015). Blockchain Technology: Beyond Bitcoin. Berkeley, CA: University of California, Berkeley.

[15] BlockchainHub. (2018). Blockchains & Distributed Ledger Technologies. Retrieved from BlockchainHub [Retrieved 19 Jan. 2018].

[16] Brito, J. and Castillo, A. (2013). Bitcoin: A Primer for Policymakers. Fairfax, VA: George Mason University.

[17] Peck, M. (2017). Reinforcing the Links of Blockchain. IEEE.

[18] Singh, S. and Fenton, S. L. (2014). Women Engineers: A Comparative Study between India and Australia. International Journal of Advancements in Research & Technology, 3(7), 108–122.

[19] Gera, I. (2016). Tech Cafe: The Blockchain Blockbuster. Retrieved from Financial Express [Retrieved 25 May 2016].

[20] Gera, I. (2017). Ransomware: After Petya Attack, Eyes Turn to Blockchain for a Way Out. Retrieved from Financial Express [Retrieved 29 Jun. 2017].

Chapter 11

System integration for Industry 4.0

Nseabasi Peter Essien[1], Uduakobong-Aniebiat Okon[1], and Peace Asuqùo Frank[1]

11.1 Introduction

System integration is a process commonly implemented in the fields of engineering and information technology. It involves the combination of various computing systems and software packages in order to create a larger system, and this is what drives Industry 4.0 to work at its optimum. System integration increases the value of a system by creating new functionalities through the combination of subsystems and software applications. The world is currently experiencing a fourth iteration of the Industrial Revolution, Industry 4.0, which merges computers and automation to enhance efficiency in the manufacturing industry and also includes cyber-physical systems, the Internet of Things, and cloud computing. Industry 4.0 takes into account all kinds of technologies and machines, from smartphones and tablets to cars, whitegoods, web-enabled televisions, and more. Also, software development is not left out of his process for the effective and efficient development of software products. Software development and applications are increasingly spreading in all areas of human endeavors. It therefore means that, to meet the needs of the world population, Industry 4.0 principles must be applied. As mentioned in our previous work, the design of virtual learning system (VLS) software as a web-based application is not merely writing series of pages, linking them together, and presenting them as an application but good and efficient system integration strategy of free and open-source software using PHP scripts implementing VLS (middleware) layer, the web server layer using Apache, the relational database layer using MySQL, and the operating system layer using Linux (interchangeably with Window OS). The design is greatly achieved and the problems of MySQL address.

[1]Department of Vocational Education, University of Uyo, Uyo, Nigeria

11.2 Application of system integration in VLS database replication design

MySQL adapts master–slave scenario. But for master–master replication, updates can be done both ways, which our system needed; hence through the concept of master–slave algorithm of MySQL, master–master update algorithm was developed, translated into programmable code, and integrated into the VLS. We used MySQL UNISON and Rsync replication policies to evaluate the performance of our replication protocol. The evaluation compares throughput and CPU wait_time and it was shown that our system can handle rapid variations in client updates while maintaining the quality of service across the application.

11.3 Database replication

Database replication is the creation and maintenance of multiple copies of the same database. In most implementations of database replication, for example MySQL, one database server maintains the master copy of the database and the additional database servers maintain slave copies of the database. Database writes are sent to the master database server and are then replicated by the slave servers. Database reads are divided among all the database servers, which results in a large performance advantage due to load sharing [1].

In addition, database replication can also improve the availability because the slave database servers can be configured to take over the role if the master database becomes unavailable. Replication is the periodic electronic refreshing (copying) of a database from one computer server to another so that all users on the network constantly share the same level of information/data [2,3].

Similarly, database replication is quickly becoming a critical tool for providing higher availability, survivability, and high performance for database applications. However, to provide useful replication, one has to solve the nontrivial problem of maintaining data consistency between all the replicas [4]. In this work, we describe a complete and provable algorithm that provides global persistent consistent order in a TCP or Unicast environment. The algorithm builds a generic replication engine which runs outside the database and can be integrated with existing open-source databases, for example, MySQL and applications. The replication engine supports various open-source models, relaxing or enforcing the consistency constraints as needed by the application. We implemented the replication engine on top of virtual learning environment application program and provided experimental performance results.

11.3.1 Root of replication

Replication has been studied through the years in many different areas. Its roots span into two big areas: databases and distributed systems. These two computer science branches historically had different approaches to replication. The first big difference was the basic entity that had to be replicated: in the database world it was data, while in the distributed systems it was a process [5]. Another important difference was the purpose for

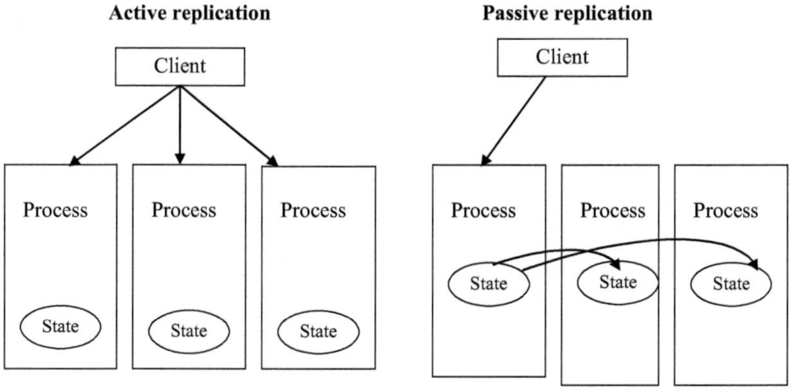

Figure 11.1 Active versus passive replication. Source: Content-aware component replication—JavaBeams Enterprise Case Study (2005)

performing replication. Databases were replicated mainly for scalability and the load distribution while in distribution systems replication was used to achieve fault tolerance [6]. Because the two areas had different purposes, different requirements were defined for replication algorithms; in particular, consistency guarantees have been defined.

11.3.1.1 Replication in a distributed system

In distributed system, replication is mainly used to improve fault tolerance. The entity being replicated is a process [7]; two replication strategies have been used in distributed systems: active and passive replication (Figure 11.1). In active replication, each client request is processed by all the servers while in passive replication there is only one server (called primary) that processes client requests. After processing a request, the primary server updates the state on the other (backup) servers and sends back the response to the client. If the primary fails, one of the backup servers takes its place.

11.3.1.2 Replication in databases

In database replication, algorithms are categorized into eager and lazy replication. In eager replication, we replicate each operation before returning the result to the client. On the other hand, in lazy replication, changes are not propagated immediately but delayed for a later time. This may be at a time interval or a transaction commit [8]. The main interest in this work is the distinct database replication algorithms use to perform updates. In some certain solutions, there is only one replica responsible for updating the data, while in some more advanced algorithms different replicas may update data. Also as we are interested in maintaining the data consistent, some form of concurrency control mechanism must be used. From the above, we can see that updates can be done either at single site or from multiple sites (Figure 11.2).

11.3.1.3 Replication in MySQL

MySQL replication works by having a master server where all the inserts, updates, and deletes (basically any writing done) and one or more slave servers that poll the

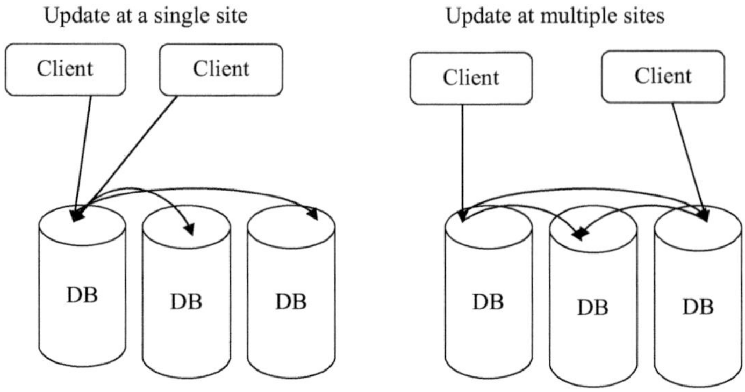

Figure 11.2 Single primary versus multiple primaries. Source: Replicated database service for application adapted from Robinson-ACM SIGOPS European Workshop 2005

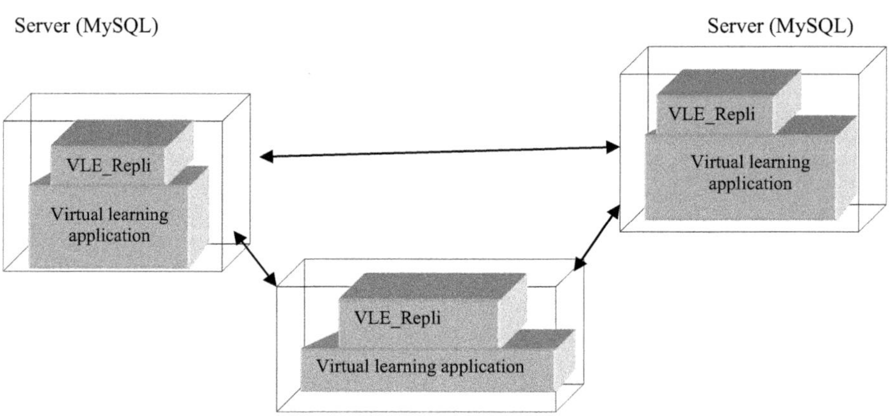

Figure 11.3 UML replication class diagram

master server to replicate the database (Figure 11.3). One can only issue select queries to the slave server. One can also have multiple master servers which are going to be covered in this work.

Database replication on community cluster scales well and is a viable alternative for using expensive multiprocessor and/or network storage hardware configuration [9]. The purpose and the motivation of this work is to design a smart replication framework that will adapt MySQL to **master–master** replication policy instead of **master–slave** replication policy. As MySQL is an open-source database, the language of implementation is PHP script.

In the system, the connection between the application and the database tiers is a set of schedulers (VLE_Repli), one per application that distributes the incoming requests to a cluster of database replicas. Each scheduler (VLE_Repli) upon receiving a query from the application server sends it using a read-one, write-all replication scheme to the replica set allocated to the application. Virtual learning environment system offers support replication through VLE_Repli service. The replications are propagated with invocations by passing the replication context. Replications are coordinated by VLE_Repli, which coordinates the virtual learning environment application. The application scheduler (VLE_Repli) provides consistent replication. The scheduler tags update with their field names of the tables they need to read and send them to the replicas. Where there is a proxy server, the scheduler communicates with a database proxy at each replica to implement the replication.

11.3.1.4 Replication algorithm (server–server replication)

In [10], Edemenang suggested that wait-and-signal operations are indivisible. This means the reader and the writer cannot operate at the same time. This is implemented during the replication of MySQL update of virtual learning database as it involves reading and writing (update) during the execution of queries.

Problem: Algorithms that will search table (database) with attributes or fields for a particular pattern. If the pattern is found, then it is replaced (replicated) by another given pattern (updated database).

Algorithm development: The need to replace (replicate) database content by another occurs very frequently in databases serving virtual learning environment because of teaching material, assignment, notice, update, etc. The underlying task is to update all occurrences on a line of a particular database pattern by another pattern depending on the one updated.

According to Schneider [11], every algorithm needs to adopt a particular programming language and technique for implementation. Base on this, therefore, the mechanism we are trying to implement can be thought of as positioning the start of the table (database attribute) pattern at the first field or attribute character in the database, the second field or attribute character in the database, and so on. At each positive, the degree of match between the pattern and the next must be determined. In this way, there will be no risk of missing a match. The central part of our database fields or attributes pattern searching strategy is that, while there are still other fields or attributes in the database to be replaced the search strategy continues.

(a) "Locate" database field or attribute pattern at the next position in the database table
(b) See if there is a complete match at the current field or attribute position.

From the way the pattern length *patlength* and the field or attribute text size (text lengths) interacts, we can conclude that there are:

Field or attribute text length-pat length (database length) +1 position at which the pattern can be placed in the table whenever we compare a field or attribute text character and a database table pattern one of two situations prevails, the pair of characters either match or mismatch. If there is a match then we are obliged to make an additional test to see if the complete match situation has been established. Once we see the problem in this light, the need for a separate matching loop disappears. Our central searching strategy has now evolved to:

While $i \leq$ (field or attribute) textlength—patlength (database length) + 1 do

(a) If pattern [j] = + * t [$i + j - i$], then

(a.1) increase j by one
(a.2) if a complete match then
Perform skip (the field or attribute is the same test the next field or attribute)
else
(a.1) reset pattern pointer j (Error message)
(a.2) move pattern to next text position by incrementing i

We have now reached the point where we can successfully search for and detect the fields or attribute pattern match in the database. The task that remains is to formulate the replication (editing copy) steps.

In the general case, we cannot expect that the input and edited lines will be of the same length because the original pattern and its replacement may be of different length. This suggests that it will probably be easiest to create a new copy of the common parts in producing the edited lines. Our replication steps will therefore involve a sequence of copying and pattern replacement steps with replacement taking place when the searching algorithm finds a complete match and copying prevailing otherwise. So, in producing the edited line, we must either copy from the original line or the replacement pattern. The copying from the original line will therefore need to precede hand in hand with the search. The question that must be answered is, how can we integrate the copying operation into the search? It is apparent that whenever there is a mismatch and a corresponding shift of pattern relative to the text, copying can take place. In the case where a character match is made, it is not so apparent what should be done. Examining this situation more carefully we see that it will only be increased when a complete match is made and so in the partial-match situation on copying will be needed. A complete-match signals the need to copy not from the original line but instead from the substitute pattern. So, in fact, the two copying situations that we must deal with are well defined:

(a) When a mismatch copy from the original line.
(b) When a complete match copy from a new pattern (one to be updated).

Let us consider the mismatch copy first. Once the proposal for the copy might be:

$$newtext[i] := txt[i]$$

This, however, does not take into account that a new line will grow at a different rate if the old and new patterns are different in length. The pattern position variable i will still be appropriate but a new variable k which can grow at a different rate will be needed for the edited line. The copy in the mismatch situation will then be of the form:

$$newtext[k] : txt[i]$$

When we encounter a complete match we need to copy in a complete pattern rather than a single character as this corresponds to the edit situation. The new pattern *newpattern* must be inserted directly after where the last character has been inserted in the edited line of text (i.e. after position k), since a number of characters need to be copied. The best way to do it will be a loop, that is,

```
for l:=1 to newpatlength do
      begin
   k:=k+1;
newtext[k]:newpattern[l]
      end
```

Once the pattern is copied, we will need to move past the text positions occupied by the old pattern. We can do this by incrementing i by the old pattern length

$$i: \ = i + patlength$$

At this point, we will also need to reset the pointer for the search pattern. We have now established an editing mechanism that will replace a pattern located anywhere and any number of times in the line to be edited. We have also set up a mechanism to copy from the original text line. Examining this mechanism closely, we see that it can fail to copy the last few characters from the text because of smaller value of the positioning index relative to the number of characters in the original text. We must, therefore, insert steps to copy these "leftover" characters. That is,

```
While i ≤ textlength do
      begin
   K:=k+1;
 Nextext[k]:txt[i];
   i:=i+1
      end
```

Once these editing requirements are incorporated into our pattern searching scheme, we will have the complete algorithm.

11.3.1.5 Algorithm description

1. Establish the database, the search pattern, and the replacement pattern and their associated lengths in characters.
2. Set initial values for the position in the old text, the new text, and the search pattern.
3. While all pattern positions in the database have not been examined do
   (a) If the current database and pattern characters match, then
      - (a.1) extend indices to next pattern/text character then
      - (a.2) if a complete match, then
         - (2.a) copy new pattern into current position in the edited line
         - (2.b) move past old pattern in the text
         - (2.c) reset pointer for the search pattern

Else
- (a′.1) copy current text character to the next position in the edited text,
- (a′.2) reset the search pattern pointer,
- (a′.3) move pattern to the next text position.

4. Copy the leftover characters in the original text line.
5. Return the edited line of text.

(Source: How to solve it by computer. Implementing with PHP Code iv VLE.)

11.3.1.6 MySQL algorithm policies

- The primary code is executed first.
- An acceptance test is evaluated.
- If the acceptance test evaluates to TRUE, then the next primary code segment following the current one is executed.

  Otherwise rollback to a state prior to the current state of the program.
- Repeat the execution of the standby spares until the acceptance test evaluates to TRUE or else the system fails

```
Procedure Update ()
      databaseTable (list of variable)
            Begin (*Content list of tables are copied into temporal variables*)
               Rp = 1(* initially perform primary *)
                  End Update; (* Update*)
```

The update operation creates a new environment for the running program so that recovery is possible using the temporary variables to restart the program at an earlier state. Thus, we have

```
Function Check (Tables)
         Test: variable match
            Begin
               If test
                  Then Begin (* successful – replicate *)
                     (* discard copies in temporaries *)
                        Check = TRUE
                        Rp = 0
   Else Begin (* failure*)
         (* restore copies for temporaries *)
            Check = fail or false
               Rp = rp + 1 (* increase rollback counter *)
            End
End check
```

Source: Edemenang (2000), Software Engineering Analysis and Verification, p. 434 [10]. Adopted from MySQL/PHP Database Application Published by M & T, USA.

Explanation of MySQL replication algorithm: We describe the replication algorithm in detail in this section. We begin by outlining the phases that the algorithm goes through in its operations and the states that each replica can be in and then go on to discuss each phase in turn.

Phase and state: At any moment, a system running MySQL replication algorithm is in one of three global phases, service, election, and recovery, as shown in Figure 11.4.

The system is normally in service phase, accepting and processing the update request from the client, and produces a temporal copy of the update.

```
Procedure Update ()
   databaseTable (list of variable)
         Begin (*Content list of tables are copied into temporal variables*)
            Rp = 1(* initially perform primary *)
               End Update; (* Update*)
```

On completion, the system moves to the **election** phase. During this phase, the replicas hold an election to assemble a majority and choose one of their numbers as master (this is as well depending on system setup). When the **election** succeeds, the replicas that participated in it are called active replicas, and then the algorithm moves on to the **recovery** phase.

The update operation creates a new environment for the running program so that recovery is possible using the temporary variables to restart the program at an earlier state. Thus, we have

```
Function Check (Tables)
   Test: variable match
      Begin
         If test
            Then Begin (* successful - replicate *)
               (* discard copies in temporaries *)
                  Check = TRUE
                  Rp = 0
```

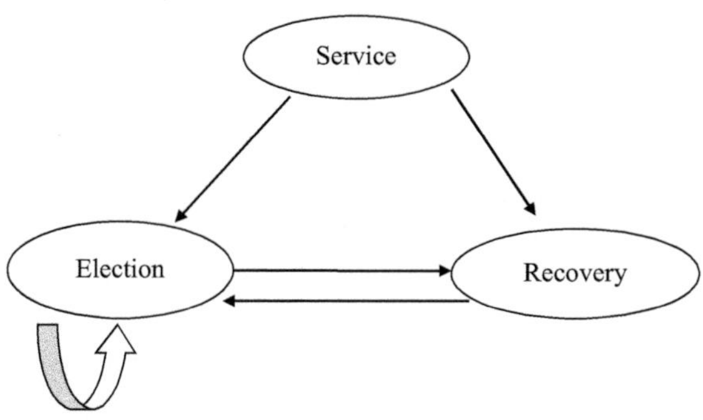

Figure 11.4 Phases in MySQL replication

During the recovery, the master directs the system in two interwoven tasks. First, the active replicas reconcile any difference among their copies of the data. Second, each active replica advances a set of variables it holds in stable storage so that inactive replicas can be identified as out of date during the next **election**. When these tasks have succeeded, the system re-enters the **service** phase.

Assuming during the reconciliation process, there are mismatches, failure notice is reported, the system rolls back to the **service** phase.

```
Else Begin (*failure*)
        (*restore copies for temporaries *)
          Check = fail or false
               Rp = rp + 1 (* increase rollback counter *)
            End
          End check
```

11.4 Virtual learning system replication policy

Through the concept of replication in MySQL scenario, VLS algorithm for master–master was developed with a system configuration that maintained consistent data items. During the update process, data objects are retrieved and modified. When "VLE_Repli" is invoked, the system searches for the object with the same structure in the database of the other node. If an object is the same, the VLE_Repli performs the update on the object. If not, an error message is returned and the process terminates. Details are as below:

- Retrieve the data object that requires modifications
- Update the object. For example: CSC 111; course notes, assignments, reference materials, etc.
- Select "update" button when done
- Note that in our program, "update" updates the object on the current node and triggers replication on the other nodes
- It performs checks on the data object to confirm the match
- If the attributes of the two objects are the same, replication takes place
- It repeats this in all the nodes
- If the object match could not be found on the other server, an error message is returned
- Replication stops

The VLS replication algorithm is shown below.

```
Procedure Update ()// perform update object (database and schemas modifications)
   Function Check (Tables) // trigger "update" button to carry out check on
the object
          (relation having the same database and schema in other
          connected node(s))
     Test: variable match // verify and confirm that all corresponding table
            relations are the same
```

Begin// start the verification and confirmations in all table
If test successful// Test for success
Then Begin (* replicate *)// if successful, update (replicate)
Else Begin (* failure*)// if not successful
Get schema from master // ensure that the right schema is obtained from the table
Loop Function Check (object)// go back and check the source of update
(* report system fail *)// report error notice and end the operation
End
End

There are two major advantages of VLS over that of MySQL. Update on other nodes takes place at real time. Whereas in MySQL policy, before update is carried out, the triage of **service, election**, and **recovery** must take place before the replication takes place. Another advantage of VLS replication is that replication occurs concurrently in all the masters connected, whereas in MySQL, only the master can update all the slaves connected to it which could lead to overworking the master.

11.4.1 Testing of experiment II (transition and analysis) replication testing

This experiment demonstrated MySQL replication (UNISON and RSYNC) and VLS replication design. The main idea was to overcome the master–slave adopted by UNISON and RSYNC. VLE_Repli adopted master–master replication but still maintaining the MySQL platform.

Replication throughput and CPU wait time were used to determine the superiority in performance using Pentium IV with low system configuration specification.

For each experiment, the activity of the CPU on throughput and wait time was monitored during the execution of client's updates. In this experiment, total update, update per second, number of tables updated, and CPU-time were invoked on the screen and written into the database table under Microsoft Window environment. Also in the previous experiment, data were collected from *VMSTST Linux utility* in plotting the graphs both using Microsoft excel and Lotus 1-2-3 in plotting the graphs.

The plates below show the screen captured and performance summary tables from different or various MySQL database modifications, and then demonstrate the update using (MySQL UNISON and Rsync) and VLS replication designed. VLS designed not only has shown higher performance in the update but also adapted the master–master replication.

This experiment demonstrated **MySQL** replication (UNISON and RSYNC) and VLS replication design. The main idea was to overcome the master–slave adopted by UNISON and Rsync. VLE_Repli adopted the master–master replication but still maintaining the MySQL platform.

Note that on the selection of RSYNC and UNISON which adopt master–slave replication, MySQL needs reconfiguration as below because it does both the update and replication. Only VLS does update and MySQL does the replication.

11.4.2 Configuration on master at command prompt

Create user dbname '@'%'identified by 'password';

Grant replication slave on *.* to 'dbname'

Flush tables with read block;

Show master status

Unlock tables;

11.4.3 Configuration on slave at command prompt

Stop slave;

Change master to

→ Master_host = ' '

→ Master_user = ' '

→ Master_password = ' '

→ Master_log_file = ' '

→ Master_log_pos = ' '

MySQL configuration file:

→ Retrieve all **MySQL.config** file in all the connecting systems

→ Retrieve **My.ini**

→ Call program line containing **Log-bin** = **Mysql-bin** and enable the program line

→ Also enable **server-id** = **1**

→ Save the configuration files again and exit

When an update is done on the master, it automatically replicates on the slave. Hence, updates are done on master and not slave in a master–slave scenario. But for the master–master replication, updates can be done both ways.

For each experiment, the activity of the CPU was monitored during the execution of the client's updates. In these experiments, total update, update per second, number of tables updated, and CPU-time were invoked on the screen and written into the database table.

The plates below show the screen capture and performance summary tables from different or various MySQL database modifications, and then demonstrate the update using (MySQL UNISON and RSYNC) and VLS replication designed. VLS designed not only has shown higher performance in update but also adapted the master–master replication.

11.4.4 Hardware and software requirement for replication

Our replication experimental algorithm consists of web application (virtual learning application), load balancer, and database server tiers. All these components use the same hardware. We use the Apache 1.3.31 web server and the MySQL 4.0.16 database server with **NewVLEARN** table. Several experiments were tested using Red Hat Fedora core 3 Linux operating system with the 2.6 kernel and WINDOW operating system was used interchangeably. All nodes were connected via 100 Mbps Ethernet LAN.

Table 11.1 Rsync, UNISON, and VLS performance summary (varying database update)

Session	Type	Total update	Update per sec.	CPU time
1	Rsync	3,569	60	59.48
	UNISON	3,569	59	60.49
	VLS	3,569	192	18.59
2	Rsync	256	60	4.27
	UNISON	256	59	4.24
	VLS	256	192	1.33
3	Rsync	138	60	2.30
	UNISON	138	59	2.34
	VLS	138	192	0.72
4	Rsync	382	60	6.40
	UNISON	382	59	6.50
	VLS	382	192	1.10

Table 11.2 Update per sec. (throughput) for Rsync, UNISON, and VLS (Rsync and UNISON are normal replication process of MySQL–master to slave), whereas VLS is enhanced—master to master

S/N	Update	Rsync	UNISON	VLS
1	Session 1	60	59	192
2	Session 2	60	59	192
3	Session 3	60	59	192
4	Session 4	60	59	192

Source: Field experiment – imported from program run/execution code

Database replication output: The result produces four graphs. For each experiment, the activity of the CPU on the throughput and response time was monitored during the execution of the client's update using ***VMSTST Linux utility*** directory for the two experiments conducted.

Throughput: Tables 11.1–11.3 show the complete set of results when running varying updates for three algorithms. These updates vary so that the system is in underload initially but becomes overloaded after sometimes depending on the number of client's update.

Figures 11.5–11.7 show various levels of performance (throughput, CPU wait time in both bar and line graph) at a varying session of update replication. From the performances above, it is clearly shown that the enhanced policy greatly improves the performance of the VLS.

Table 11.3 CPU wait time for Rsync, UNISON, and VLS (Rsync and UNISON are normal replication process of MySQL—master to slave), whereas VLS is enhanced—master to master

S/N	CPU wait time	Rsync	UNISON	VLS
1	Session 1	59.48	60.49	18.59
2	Session 2	4.27	4.24	1.33
3	Session 3	2.3	2.34	0.72
4	Session 4	6.4	6.5	1.1

Source: Field experiment – imported from program run/execution code

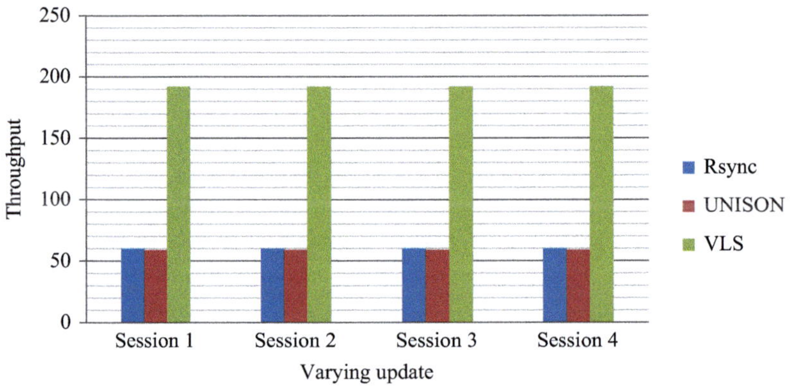

Figure 11.5 Demonstration of throughput performance (MySQL against enhanced)

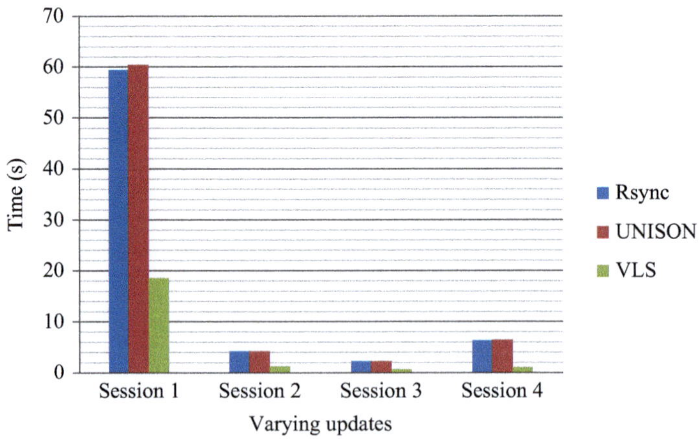

Figure 11.6 Demonstration of CPU wait time for Rsync, UNISON against enhanced VLS

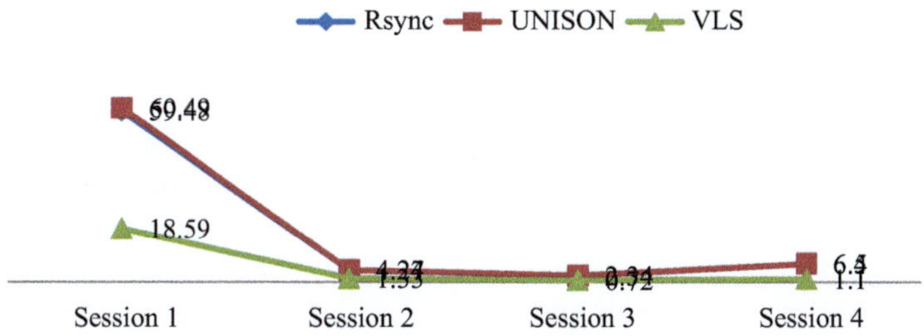

Figure 11.7 Demonstrations of CPU wait time performance in the line graph

11.5 Conclusion

The current trend in software development industrial revolution is that software developers have stop building software from scratch; rather software applications are now assembled from well-stocked catalogs of reusable software components to develop complex or large software applications through system integration. System integration, as defined earlier, is a process of bringing together the component subsystems into one system (an aggregation of the subsystems cooperating so that the system can deliver the overarching functionality) and ensuring that the subsystems function together as a system. The main advantage of Industry 4.0 in software development system integration is the need to improve the productivity and quality of their operations. The goal is to create system component reusability so that it can easily be expanded through integration with other subsystems to develop a more complex system to speed up the software development process.

Today, computer software development using software system integration is playing an important role in the development of all kinds of software applications for businesses in the world. System integration in all scientific inventions affects almost all aspects of our lives and our everyday activities. This is because there were many critical problems existing with the old-established software engineering models: low productivity and quality, and high cost and risk. The root cause comes from an attempt to develop software products as an independent entity, and this always led to the highlighted problems above. For solving these critical problems efficiently, a new software engineering model, component reuse, through system integration principle is presented. The essential difference between the old-established software engineering development process and system integration process is the software component reuse (open source). This has made revolutionary changes to almost all aspects in software engineering to efficiently handle software complexity, invisibility, changeability, and conformity, and solve the critical problems (low productivity and quality, high cost, and risk) existing with the old-established software engineering practice of developing software from the scratch. The problems addressed as demonstrated in the development of VLS where various algorithm components were integrated as a subsystem for building a complex system with less cost and time. The old-established software engineering process itself has become an obstacle rather than a driving force for software development in the twenty-first century: As pointed out by Jay and Jonathan Xiong [12], the unified process suffers from several weaknesses because it is only development. We are on the edge of a revolutionary shift of software development process, pioneered by the software system integration process, and likely to change our very attitudes in software systems modeling and engineering.

Through the implementation of system integration, it can be seen that second major problem with MySQL database is that it provides master–slave replication, but through system integration, it is possible to build VLS and improve on the efficiency for master–master replication to meet our purpose without building other systems.

References

[1] Gokul, S. and Ashvin, G. (2016). *Database replication policies for dynamic content application.* The proceedings of the *IEEE* International Conference on Data Engineering (ICDE), Canada

[2] Version Management with CVS. (2014). Retrieved July 17, 2018, from http//: www.evshome.org/doc/manual

[3] Holliday, J. and Agrawal, D. (2016). *Database replication using epidemic update.* Technical Report TRCS 001, University of California, Santa Barbara, 2016

[4] Amaza, C. (2016). Conflict-aware scheduling for dynamic content applications. In Proceedings of the Fifth USENIX Symposium on Internet Technologies and Systems

[5] Haney, M. (2017). From total order to database replication – self tuning and self administering. Cited from Microsoft Research. Retrieved May 27, 2018, from http://www.research.microsoft.com/research/dmx/AutoAdmin

[6] Karin, P. (2012). *Replication database services for world-wide application.* ACM SIGOPS European Workshop

[7] Bernstein, P. (2013). Concurrency control and recovery in database systems, cited from Addison-Wesley, 1999, Prentice Hall India

[8] Felder, P (2014). *Reconciling replication and transactions for the end-to-end replication of CORBA application.* Proceeding of CoopIS/DOA/ODBASE, pp. 19–22

[9] Cecchet, E (2014). C-JDBC: Flexible database clustering middleware. In Proceedings of *USENIX* 2004 Annual Technical Conference

[10] Edemenang, E. (2012). Software Engineering Analysis and Verification. Adopted from MySQL/PHP Database Application Published by M & T, USA. p. 434

[11] Schneider, D. (2017). Performance evaluation of algorithms in a share-nothing multiprocessor environment. Proceeding. ACM SIFMON Conference. 2007.

Retrieved. August 13th, 2019, from https://www.sitepoint.com/mysql-master-slavereplication-setting-up/

[12] Jay, X. and Jonathan, X. (2012). Complete Revolution in Software Engineering Based on Complexity of Science. Retrieved March 23, 2019 from http://www.academia.edu

Chapter 12

Additive manufacturing toward Industry 4.0

Puvanasvaran A. Perumal[1] and Kalvin Paran Untol[1]

Additive manufacturing (AM) is the process of creating an object by building it one layer at a time. It is something contrary to subtractive assembling, in which an item is made by removing at a strong square of material until the last item is finished. AM is the scientific name of the 3D printing practices used in the market where it differentiates the process from the material removal methods used in the manufacturing field. AM also eliminates many constraints imposed by conventional manufacturing. AM [1] is the “procedure of joining materials to make objects from 3D model information, for the most part, layer upon layer”. It is otherwise called rapid prototyping.

Utilizing AM [2], a design in the form of a computerized 3D solid model can be readily changed into a completed item without the utilization of extra apparatuses and cutting instruments. The adaptability of AM enables manufacturers to advance the structure for lean generation, which by its temperament eliminates squander. Kruth *et al.* state that rapid prototyping, for the most part, alludes to the procedures that produce molded parts by progressive creation or expansion of strong material, in that varying on a very basic level from shaping and material expulsion fabricating systems [3]. AM innovations can be utilized to make metallic parts. This leap forward in assembling innovation makes conceivable the manufacture of new shapes and geometrical highlights.

Although the manufacturing feasibility of sample parts with these processes has been the subject of a few investigations, the leap forward in assembling is yet to be trailed by achievement in the planning process [4]. In the age of Industry 4.0, most of the sectors will probably use AM with further improvements to the quality of produced parts.

12.1 AM in various industries

Initially seen as a process for concept modeling and rapid prototyping, AM has expanded over the last five years or so to include applications in many areas of our day-to-day lives [5]. From prototyping and tooling to direct part manufacturing in industrial sectors such as architectural, medical, dental, aerospace, automotive,

[1]Faculty of Manufacturing Engineering, Universiti Teknikal Malaysia Melaka, Melaka, Malaysia

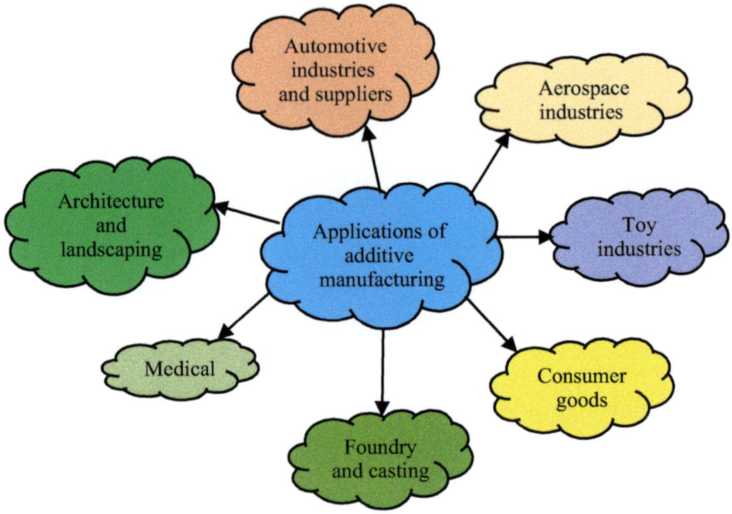

Figure 12.1 Applications of AM in the industries

furniture, and jewelry, new and innovative applications are constantly being developed. It can be said that AM belongs to the class of disruptive technologies, revolutionizing the way we think about design and manufacturing. From consumer goods produced in small batches to large-scale manufacture, the applications of AM are vast. Figure 12.1 shows the application of the AM in various industries.

12.1.1 Automotive industries and suppliers

Car manufacturers and suppliers have been among the first users of AM when it emerged in the late 1980s because they do a lot of design and redesign and they have been operating 3D CAD systems professionally for a long time. In this manner, the conveyance of impeccable informational indexes is not an issue for car original equipment manufacturer (OEM) and providers. These days, the number of prototypes increases because diversification makes to the development of product variations much more important. Model parts are utilized for in-house assessment just as to help exchanges with providers. The expanding significance of customization brings up the issue of direct AM of parts to abstain from tooling with its related expense in time and cash.

12.1.1.1 Car components: interior and exterior

Interior design especially adds to the character of the vehicle, frequently prompting and affecting the last purchasing choice. As opposed to the outside plan, it generally comprises numerous parts starting from scores of providers. Most of the parts are at last created by plastic infusion-lshaping. Therefore, the AM parts are mainly utilized for testing and introduction of vehicle ideas. Nevertheless, as series are becoming smaller and the number of variations increases, increasingly more AM parts are straightforwardly utilized in the final car. Albeit all AM procedures can be utilized to make interior car parts, laser sintering and polymerization are the top choices. Laser sintering commonly prompts legitimately usable parts, whereas stereolithography or polymer streamed parts are ordinarily utilized as ace parts for optional procedures.

Special editions of high-volume production cars often do not only have more powerful engines but also demonstrate their performance optically by exterior parts such as front and rear spoilers or side skirts.

12.1.2 Aerospace industries

On account of the small series and frequent special design requirements of their clients, the avionic business attempts to keep away from tools and ideally utilizes the tool-less AM processes. An achievement for the immediate creation of aviation interior parts was the presentation of a fire-resistant material for laser sintering, which currently is accessible for polymerization and expulsion forms too. Many of the interior pieces of planes do not vary particularly from parts made for vehicles. Consequently, test parts for car applications can be taken as tests for aviation applications. Creating metal and earthenware forms empower the immediate manufacture of specialized parts both for the cell and for the motor.

12.1.3 Toy industry

Even though toys are likewise "consumables," the toy business normally is addressed separately. It deals not only with the various plastic parts for children's toys but also with more and more customized models of cars, planes, and trains, even for adults. These models require fine subtleties and a touchy scaling, taking care of little subtleties uniquely in contrast to bigger ones. Contingent upon the scale, diverse AM forms are more reasonable than others are.

12.1.4 Consumer goods

Today, consumer goods do not just need to perform their expected functions, but they also must follow a certain trend. They should be centered around the necessities of a unique shopper gathering, including their preferred structure direction. Way of life items characterizes another up-and-coming business sector. As way of life changes quickly, it is prescribed to research patterns and test the market before generation. Consequently, models are required. Bowls, containers, lights, and other increasingly brightening things are favored items for creators who utilize the new opportunity of configuration provided by AM to defeat the geometrical limitations.

Considering AM, the three-dimensional perception of geodesic information is going to frame a developing specialty showcase. For this sort of models, the 3D printing procedure by Z-Corporation is valuable, because the parts can be continuously colored thus avoiding manual work. Considering AM, better approaches for geodesic showcases can be made. Models of a globe displaying the land portion and the sea portion of the earth with an exaggerated scaling of the topographic lines of the mountains and of the ocean floor in order to point out the details. To make the fine spike-molded subtleties superficially, stereolithography was picked. Subsequently, manual shading must be applied. AM enables the client to arrange his/her favored scaling. The closer an item moves toward the human individual, the more intensive becomes the interaction between both, and the more individual features are required. To make the parts required, mold- and die-bound manufacturing is no longer applicable. This opens points of interest for AM forms, which can make huge amounts of various parts in a single form, regardless of whether every one of them is distinctive as far as individual highlights.

12.1.5 Foundry and casting

Foundries use AM procedures to get models and tests of the later item (3D imaging) or to make centers and cavities for production. On the off chance that only a demonstrator part is required, all AM advancements can be applied. 3D printing is preferred if the part is not precisely stacked and a modest and rapidly accessible example is required. The application level is fast prototyping/strong imaging.

AM offers new driving forces primarily for sand throwing and venture throwing because the required lost centers and holes can be made rapidly and effectively. The improved unpredictability of the AM parts permits acknowledging geometries that cannot be made physically or by special tools. Because of the basic scaling of AM parts, the lost centers and pits can be streamlined effectively. For sand throwing applications, laser sintering or 3D printing of foundry sand is utilized, while lost centers for speculation throwing are made by the polymerization of resolvable or dissolve-capable thermoplastic waxes or gums. Lost examples are produced using polystyrene by laser sintering. Whenever penetrated with wax, 3D printing parts can be utilized also. As the centers and depressions are utilized for creation, the application level is quick assembling/direct tooling. Lost centers and examples can likewise be made by infusion of wax into silicon molds acquired from AM experts. Whenever completed appropriately, all AM innovations can be applied.

12.1.6 Medical

Humans are still individuals, who need individual treatment including customized aids such as implants, episthesis, orthosis (leg braces), and others. For a legitimate fit, the 3D information should be procured by restorative imaging procedures, for example, registered computed tomography or ultrasonic. A typical configuration for therapeutic pictures is DICOM. Extraordinary programming permits a reasonable limit determination and a 3D reproduction that gives the premise to a lot of standard triangle language (STL) information that can be prepared in any AM machine. On account of the great surface quality and the point-by-point generation, laser stereolithography and polymer flying are utilized ideally to make restorative models, for example, skulls and other human bone structures. Inside empty structures, for example, the frontal sinus or the filigree subcranial bone structure can be reproduced best by these procedures.

In any case, laser sintering, 3D printing, combined layer producing (expulsion, fused deposition modeling (FDM)), or layer overlay fabricating (LLM) convey medicinal models also and are as often as possible utilized for this reason.

For sintering and FDM, there are exceptional, affirmed materials for restorative utilize accessible that can be disinfected.

3D printing opens interesting opportunities for anaplastologists and different experts who make exclusively formed parts for conclusive throwing. The information of missing organs or parts of organs is taken from restorative imaging, 3D remaking, and consequent 3D demonstrating by programming, for example, sensable. This information is ideally founded on reflecting. The ideal counterfeit organ must be adjusted independently to every patient's circumstance. To abbreviate this system and to enable the anaplastologist to focus on their center aptitudes, the structured part is made by 3D printing.

12.1.7 Architecture and landscaping

Architects normally present their imaginative thoughts using scaled models. Since they are working with 3D configuration projects, the 3D information required for AM are straightforwardly accessible and the model making can be fundamentally improved by utilizing AM models or model components. The very complex, thin-walled freeform-shaped structure could hardly be made by traditional model making but needs to be built using AM. Laser sintering was picked to make a nitty-gritty model to guarantee a functional degree of strength that keeps it from being harmed by contacting. Although the block-formed components were sintered, they could likewise be made generally by processing along these lines making it a supposed mixture model. The model was basically utilized for an open introduction to the task.

Other than the displaying of single structures or parts of structures, a developing application is the creation of models of houses, towns, and scenes. A considerable lot of these require shading, either to call attention to the advantages of the development or to show milestones. The information can be acquired from a GIS or it very well may be separated from the Internet, for example, from Google Earth or 3D distribution center, and showed in 3D utilizing different AM forms. Among others, AM can convey 3D showcases of anyone's home, an old neighborhood, or companion's site, conspicuous structures, scaffolds and that is only the tip of the iceberg.

12.2 Different materials used in AM

Today, AM permits processing materials of every material class, specifically plastics, metals, and ceramics. Sintering of plastics and metals can be viewed as broadly utilized standard procedures, while for the expulsion of metal or clay-filled materials forms are still being worked on. The quantity of various materials inside every material class is still very restricted, although this number has expanded altogether in the course of the most recent years and grows continually due to international research activities. The purpose behind the set number of materials is that by and large a synchronous material and procedure advancement is required. Material for plastic laser sintering, for instance, must not exclusively be locally meltable, however simple to recoat which requires adjusted edges. Added substances and procedure subtleties, for example, protecting gas

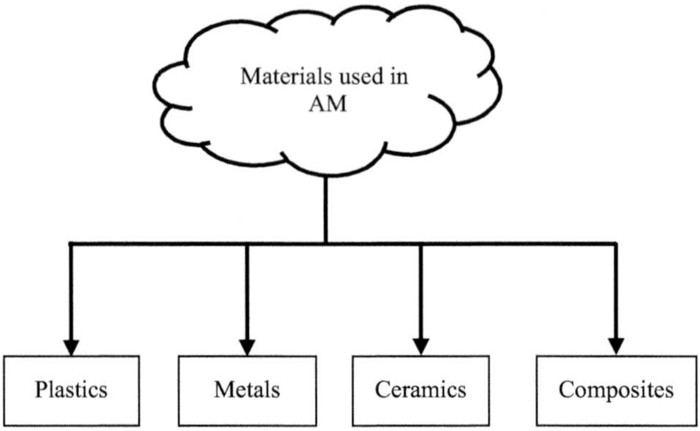

Figure 12.2 Types of materials used in AM

and preheating, smother neighborhood vanishing, oxidation, and other inter-process effects and interactions with the environment. This is one motivation behind why powder materials for laser sintering are not the same as powders for sinter covering, even though they are very comparable regarding their compound organization.

Mainly for materials for plastic laser sintering and stereolithography independent markets are developing. Metal powders are fundamentally the same as powders for laser covering and welding and along these lines have been outstanding for a long time. The client can pick among a wide assortment but must qualify the process that implies the advancement of the material information sheet. On the other hand, the materials that are released by the machine manufacturer can be utilized, accepting the limited number of materials and the value level. Some AM-related issues are progressively explained and happen by and large during assembling because now there is no long-time experience in AM. The most significant issues are maturing and UV stability for plastics and corrosion, deterioration, sedimentation, and oxidation for metal powders, as well as pores and inclusions for all AM processes. Figure 12.2 shows the types of material used in AM of different uses for different products.

12.2.1 Plastics

Plastics were the first group of materials to be processed by AM and they still provide the biggest part of materials. Materials for stereolithography are acrylic or epoxy gums that must help photograph polymerization. Today, the clingy and fragile materials of the mid-1990s are supplanted materials that mimic materials for plastic injection molding. This was accomplished by filling the tar with nanoparticles to build heat diversion temperature and mechanical security. Moreover, the variety of materials was expanded and now incorporates straightforward and non-straightforward, versatile, hardened, and a lot of increasingly different materials.

For plastic laser sintering, polyamides are the favored materials. Despite the fact that polyamides are one of the most well-known thermoplastic material families for infusion forming, which makes trust in this material, the AM-polyamides and the ones utilized for plastic infusion-shaping contrast fundamentally from one another. Warping and distortion were severe problems in the early days of AM, however, they are today reduced to a minimum due to preheating and improved scanning strategies. There is an expansive and progressively expanding assortment of polyamide-based powder materials for laser sintering available. This incorporates fire-resistant, aluminum filled qualities that can be sterilized. Improved mechanical properties are given by glass-filled powders, even though this term is confounding because circles and rice-grain-formed particles are utilized rather than strands to permit recoating. They give higher solidness contrasted with unfilled characteristics yet do not arrive at the properties that can be normal from fiber-filled infusion formed materials. Today, at least every level is represented by an AM material and procedure, aside from the degree of iridized materials. Polyimides are an extremely fascinating gathering of exceptionally strong and heat and chemical resistant polymers that would be truly attractive to use as AM materials.

12.2.2 Metals

The most frequently utilized techniques for AM of metals are sintering in the variation of selective laser melting and fusing. The material comes as powder with an essential molecule size of 20–30 μm. Since laser pillar distance across, layer thickness, and width of the track are in a similar size range, the checking structure is unmistakable on the top. The materials are fundamentally the same as the materials utilized for laser covering or welding with filler material. Albeit different marketed powders can be utilized, it should be contemplated that the capability of the material must be made or possibly assessed in-house. Then again, powders conveyed by the AM machine makers accompany material information sheets dependent on demonstrated parameters that consolidate advanced output procedures too.

For AM of metals, tempered steel, apparatus steel, CoCr-compounds, titanium, magnesium, aluminum as well as valuable metals such as gold and silver are accessible. The scope of metal materials is considerably more extensive, and their properties impersonate the materials that are utilized for conventional assembling far better than plastics.

12.2.3 Ceramics

AM utilizing ceramic materials is yet dependent on a specialty of layer fabricating advancements. Materials are accessible from the entire range of earthenware production, for example, aluminum oxide, Al_2O_3; silicon dioxide or silica, SiO_2; zirconium oxide or zirconia, ZrO_2; silicon carbide, SiC; and silicon nitride, Si_3N_4. Items are solid pottery, mainly with flow-through channels and high-temperature-loaded structures, for example, heat exchangers. Characterized large-scale porosities that help the joining of inserts are a remarkable selling purpose of resorbable bio-pottery. Small-scale porosities encourage the generation of microreactors.

12.2.4 Composites

Composites in the sense of lightweight strengthened structures are scarcely known in AM. They comprise of more than one material and along these lines can likewise be viewed as reviewed materials. Composites are ordinarily used to make lightweight items that are uniform in structure and isotropic, or if nothing else, isotropic under characterized edges of burden.

In general, layer laminate manufacturing can create composite parts with coordinated strands or textures, if these fortifications are accessible as prepregs or level semi-completed materials that can be integrated into the process. A specially adapted process for making reinforced curved parts from ceramic fiber (SiC) avoids cutting the fibers. It can place the layers under characterized yet various edges to adjust the structure to the normal burden. Also, the part can have a (somewhat) bended surface to make basic components and to maintain a strategic distance from stair steps parallel to the territory of the heap.

12.3 Global evolution of AM

There is, at present, a lot of dialog around the 3D printing industry (or AM). It is regularly packaged together with mechanical autonomy, digitization, and huge information in the "Industry 4.0" or "fourth industrial revolution" vision of the manufacturing plant of things to come. Such is the intensity of the media's advanced dreams for the 3D printed future. However, the overall population is ignorant that 3D printing is not new. The innovation has existed for a very long while—the primary business frameworks were available in the late 1980s. Figure 12.3 shows the evolution of AM from a global perspective.

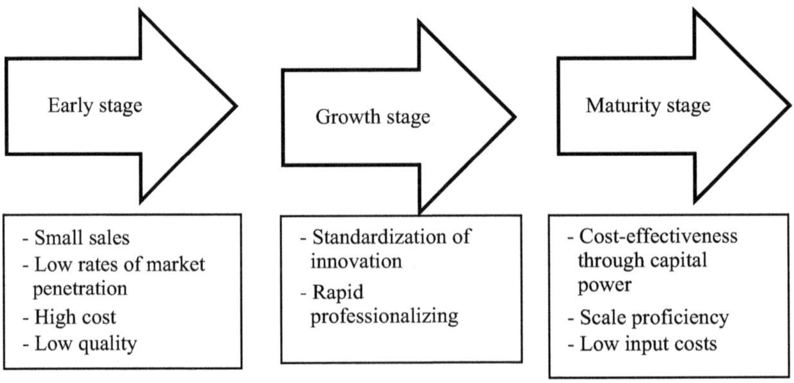

Figure 12.3 Evolution of AM

12.3.1 Early stages

Contemporary system investigation proposes that there are four phases in the existence cycle of ventures—introduction, growth, maturity, and decline. The beginning period is described by small sales, low rates of market penetration, high costs, and low quality. AM invested quite some time in this phase with boundaries to the development including industry-wide advancement hampered by patent restrictions and difficulties reducing the cost of machines.

12.3.2 Growth stages

The AM sector is now in the growth stage. This stage sees quickening market infiltration as costs fall. The 2018 Wohler's Report tells us that worldwide revenues from AM products in 2017 totaled \$3.133 billion, a 17.4% increase from \$2.669 billion in 2016. Few would dispute that the industry is growing, and at a quick pace. Enormous yearly increments in income keep on drawing in enthusiasm from different divisions (finance, manufacturing).

For some businesses, the growth stage results from the standardization of innovation (facilitating deals and dissemination and creating economies of scale underway). AM covers numerous procedures (material extrusion, material jetting, binder jetting, sheet lamination, vat photopolymerization, powder bed fusion, directed energy deposition) and materials (polymers, metals, ceramics, composites, and others). For instance, the creation of metal dental implants is driven by the powder bed fusion process (selective laser melting) utilizing titanium.

During this growth stage, the AM business is rapidly professionalizing. This does not imply that the business was not proficient during the early stage, rather than the size and scope of the firm activities is increasing and requires the input of supportive services such as human resources and business management. Numerous AM firms are controlled by specialists who established the organizations. Obviously, those organizations are bolstered by entrepreneurs. Now we are now increasingly seeing the influx of business and management talent from other sectors, filling important roles in senior management, supply chain operations, and consulting.

12.3.3 Maturity stages

While components of maturity can be seen in AM, for example, in applications such as spare parts and prototyping, the general business divisions are still immovably in the growth stage. The tipping point will probably be more extensive use of AM advancements in mass manufacturing. In any case, it appears to be impossible for AM to fit in with existing models of industry advancement. For instance, the maturity stage is portrayed by cost-effectiveness through capital power, scale proficiency, and low input costs. This is unlikely to apply to AM, as the matter of fact the utility and application of AM in manufacturing is counter to these understandings of the logic of design and production. AM will not supplant customary types of manufacturing (injection molding, subtractive methods), it will be complementary. Furthermore, based on a business model that takes a wider view of value and costs savings.

The maturity stage additionally generally observes production shifting, first to newly industrialized and then to developing countries. AM can possibly drastically change how items are made, and where they are made. It will be quite a while before the AM industry matures. We must rethink what industry "maturity" is for the AM segment, and its effect on manufacturing and for those watching the advancement of the part. There is as of now little definitive data accessible on the AM part. As AM arrives at maturity, bigger and freely possessed firms will develop. This will give more information on gear, materials, and services sales, a deeper understanding into the benefit.

In principle, the AM business has so far pursued a conventional industry improvement direction. Nonetheless, as it arrives at maturity, AM will probably challenge the models. Traditional understandings of the various components required for businesses to develop should change, especially as the expanding selection of AM will likewise change the advancement of those ventures wherein it is applied.

12.4 Future direction of AM

AM is a new technology that is very close to becoming a legitimate manufacturing method. At this moment, if a part was designed, for example, for milling, it does not make sense to 3D print it (higher cost, less quality). A new set of tools is required to explore what you can get from AM, a design method is needed to generate the most optimal, lightweight design. This is not possible to do manually. In the future, you will see computer procedures that allow you to only set the framework and let the problem be solved for you. This method goes far beyond optimization.

Generative design uses artificial intelligence to generate the best possible object and provides a design that is most optimal for printing. It shortens the design process from weeks or months to hours. The great part about this new software is that no experts are needed anymore. In this way, a basic understanding will still be required in the future, but engineering design will be accessible to a broader range of people.

Looking at the rise of generative design, it is no surprise that 90% of revenue in the 3D printing software segment comes from the design software. This is exactly where the value will be in the future. Once 3D printing is well established and the manufacturing methods are all equal, the person that designs the best product will win.

12.5 Conclusion

In conclusion, AM opens new opportunities for design and manufacturing cross-wise over various enterprises. Contrasted with traditional techniques, increasingly complex structures and geometries can be accomplished using customized design, greater efficiencies, higher performance, and better environmental sustainability. AM plays an important role in industries. AM technologies will soon be leading to the next major industrial revolution. AM plays a key role in Industry 4.0, saving time and costs, being decisive for process efficiency and reducing its complexity, allowing for rapid prototyping and highly decentralized production processes. Therefore, the innovation is seeing expanded reception past prototyping and tooling into the end and extra part generation. Therefore, AM has a significant task to carry out in the scope of assembling techniques. Companies can deploy to evolve their products in response to market demands. Importantly, as the innovation keeps on improving, AM changes from a problematic innovation utilized distinctly by trailblazers to a typical strategy for center creation.

References

[1] Mellor, S., Hao, L., and Zhang, D. (2016). Additive manufacturing: a framework for implementation. *International Journal of Production Economics*, 149, 194–201. doi: 10.1016/j.ijpe.2013.07.008

[2] Huang, S. H., Liu, P., Mokasdar, A., and Hou, L. (2012). Additive manufacturing and its societal impact: a literature review. *The International Journal of Advanced Manufacturing Technology*, 67(5–8), 1191–1203.

[3] Kruth, J.-P., Leu, M., and Nakagawa, T. (1998). Progress in additive manufacturing and rapid prototyping. *CIRP Annals*, 47(2), 525–540.

[4] Vayre, B., Vignat, F., and Villeneuve, F. (2012). Designing for additive manufacturing. *Procedia CIRP*, 3, 632–637. doi: 10.1016/j.procir.2012.07.108

[5] Gebhardt, A. (2011). Understanding Additive Manufacturing: Rapid Prototyping, Rapid Tooling, Rapid Manufacturing. Cincinnati, OH: Hanser Publications.

Chapter 13

Cloud computing in Industrial Revolution 4.0

Mohankumar Palaniswamy[1] *and Leong Wai Yie*[2]

Industries also evolved much since 1700. There have been four industrial revolutions since the 1700s. The first industrial revolution in 1780 was about steam engines, textile industries, and mechanical engineering. The second in 1840 was about steel industries. The third in 1900 was about electricity and automobiles, whereas, the fourth industrial revolution was about the IT industry, and it is generally accepted that the fourth industrial revolution has just begun [1]. As such, the term "Industry 4.0" was pinned by the German government in 2011 [2]. Industry 4.0 or Fourth Industrial Revolution is all about the Internet of Things and services (IoTS), cyber-physical systems (CPSs), and interaction and exchange of data through the Internet or cloud computing.

During 1960, one system can perform only one task at a time. Multiple systems needed to run multiple tasks simultaneously [3]. Now moving forward to the present 2020, the single system can perform multiple tasks within a few seconds. Such technology is achieved by scientific advancements like the Internet, web services, Internet of things, and cloud computing. Like the industrial revolutions, there have been several improvements and developments in computing, processing, and accessing the stored data. The evolution of computing is provided in Figure 13.1.

13.1 Cloud computing

"Cloud computing is the on-demand delivery of IT resources like data, storage, and power over the internet instead of buying, owning, and maintaining physical data centres and servers" [4]. In simple, cloud computing is the distribution or supply of reserve from one to another using the Internet [5] (Figure 13.2). Every small, medium, large enterprises, organizations, companies, industries, and factories use cloud computing for different purposes such as data storage, data backup, data analytics, email, virtual machines (VMs), and web applications are to name a few [4].

[1]School of Engineering, Taylor's University, Selangor Darul Ehsan, Malaysia
[2]Faculty of Engineering and Built Environment, MAHSA University, Selangor Darul Ehsan, Malaysia

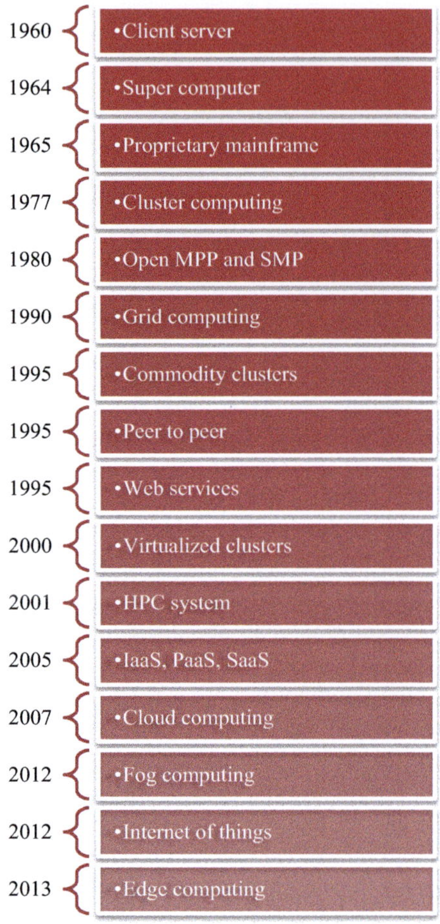

Figure 13.1 Evolution of computing technology [3]

13.1.1 Benefits of cloud computing

Every industry, organization, and company uses cloud computing. Even individuals, with or without knowing, use cloud computing. Cloud computing stands behind online service, email service, streaming, games, cloud storage, etc. The overall benefits of cloud computing include the following features. Agility—quick access to the resources, install and setup new technologies within few minutes. Elasticity—scaling the resources based on need. Wastage of resources or delay in decision-making is restricted. Cost savings—a large amount can be saved in capital expenses and redirected toward variable expenses. Expandable—business can be easily expanded to new terrestrial to reach end users. Productivity—time requirement of hardware and software setup in a new site can be dismissed. Performance—cloud-computing services reduce the network latency than traditional local network

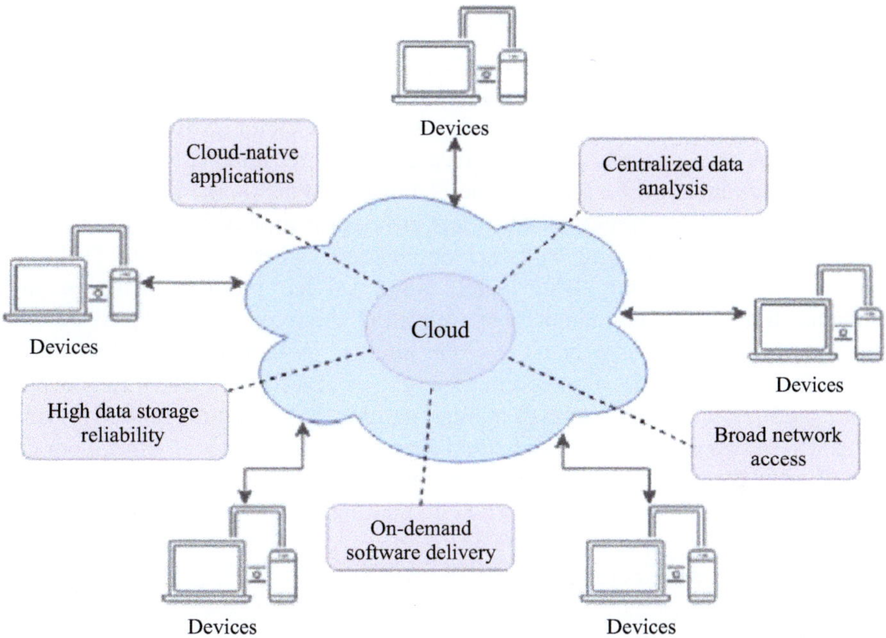

Figure 13.2 Cloud-based network [6]

setup. Reliable—cloud computing undergoes regular maintenance such as data backup and data mirroring so that user data can be recovered even during disaster. Security—cloud providers offer set of policies and controls that safeguards the user data from threats [4,5].

13.1.2 Types of cloud computing

Cloud services are provided to consumers based on different cloud-computing architecture. The three major cloud computing architectures are:

Public cloud—Public clouds are provided and operated by a third-party provider. They provide their resources and the resources of their partners, for example, Microsoft Azure. They provide hardware- and software-related IT support. A user can access those support using applications or web browser.

Private cloud—Private clouds are provided, operated, and maintained by a single organization. The resources are present in the organization's headquarters. It can be accessed only by its branches located at different geographical places.

Hybrid cloud—Hybrid clouds are the combination of public and private clouds. For example, when an organization downloads data from its private cloud and share data to its private cloud from public cloud, then it is called as hybrid cloud.

13.1.3 Types of cloud services

Cloud-computing services are broadly classified into three main categories. They are Infrastructure as a Service (IaaS), Platform as a Service (PaaS), and Serverless software as a Service (SaaS). IaaS is the most basic cloud service. It offers IT infrastructures such as servers, VMs, dedicated hardware, storage, and networks. It provides a high level of flexibility in the IT business. PaaS is the service which is targeted toward IT developers. PaaS offers developers an environment for developing, testing, and managing software applications where the onsite IT infrastructure does not matter. With SaaS, cloud service provider manages and maintains all the hardware- and software-related IT infrastructures and provides the end user a fully developed product over the Internet. With technological advancement in the recent decade, cloud computing got more advanced and travel toward the end user through other forms of computing like fog computing and edge computing. These two are the subsets of cloud computing (Figure 13.3). Since both the fog and edge computing utilize an intermediate level of processing and storage, consumers use it interchangeably. In fog computing, local area network acts as a gateway, whereas, computing is done on smart devices by programmable automation controllers in edge computing [8].

13.2 Fog computing

Unlike the centralized cloud computing, a decentralized computing where the resources are stored in logical locations between data birthplace and cloud is referred as fog computing [10]. Fog computing reduces the bandwidth amount needed when compared to the cloud computing by creating a low latency network (Figure 13.4). Fog computing is used in self-driving cars where the data is sent to the manufacturer for diagnostics and maintenance, smart cities, and electrical grids where the utility systems need to run efficiently, real-time analytics where decision-making takes place. The drawback of fog computing is that it heavily relies on data transport, which in turn requires a broadband or high-speed Internet access [11].

13.3 Edge computing

Any type of computing that is done at or very close to the source is termed as edge computing. It also reduces latency and bandwidth use [12]. A classic example to explain edge computing would be the camera equipped with a motion detection sensor. This type of camera does internal computing, records, and sends the footage to the server only when there is a motion detected. It uses less bandwidth and storage. Whereas, a camera without motion detection sensor records video for a very long time, uses large bandwidth, and more storage. Edge computing has low server usage, latency, and added functionality. Edge computing has a drawback too. It is more prone to malicious attack and it needs extra local hardware. A comparison between cloud, fog, and edge on their features is presented in Table 13.1.

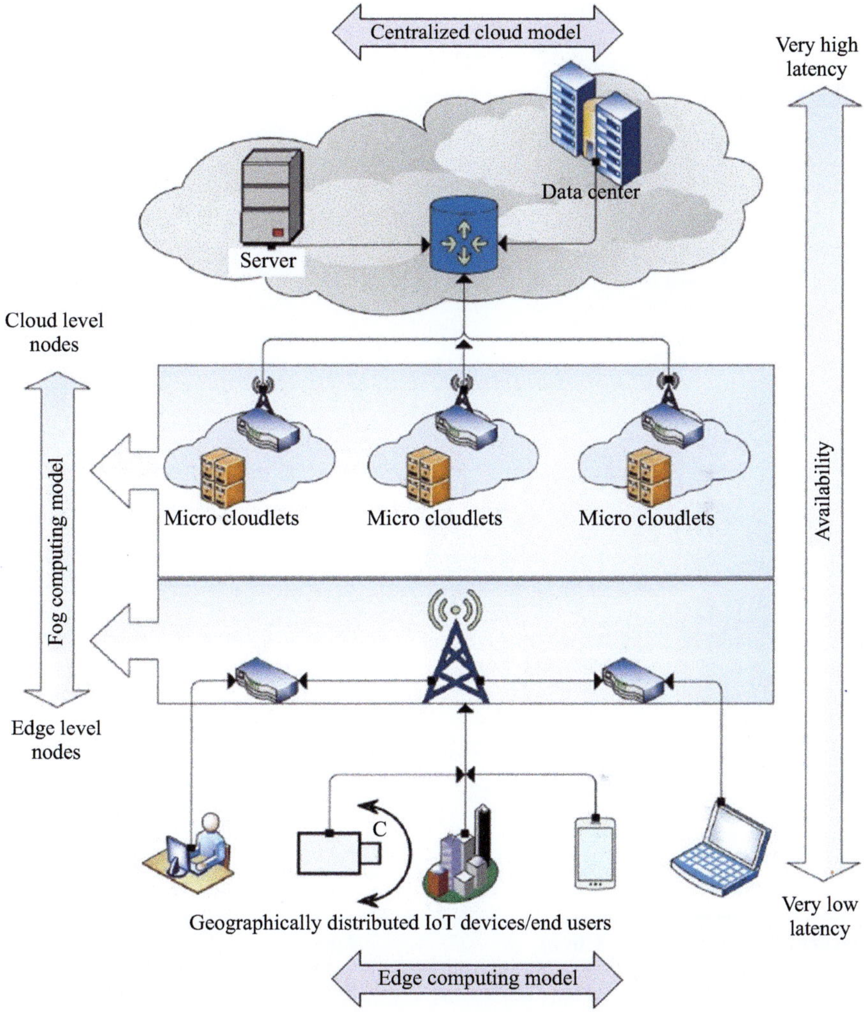

Figure 13.3 Types of computing [9]

13.4 Security and privacy issues

Rapid growth in technology has opened a new dimension in data-storage technology. In recent times, data storage has leveled up from the physical form to virtual form [13,14]. There are some disputes concerning the virtual data centers about its privacy, security, performance, backups, and cost. This section describes about the security issues and solutions to cloud, fog, and edge computing. Some common attacks on computing are denial of service—sending a huge number of messages to the network for authentication from an invalid return address to prevent others

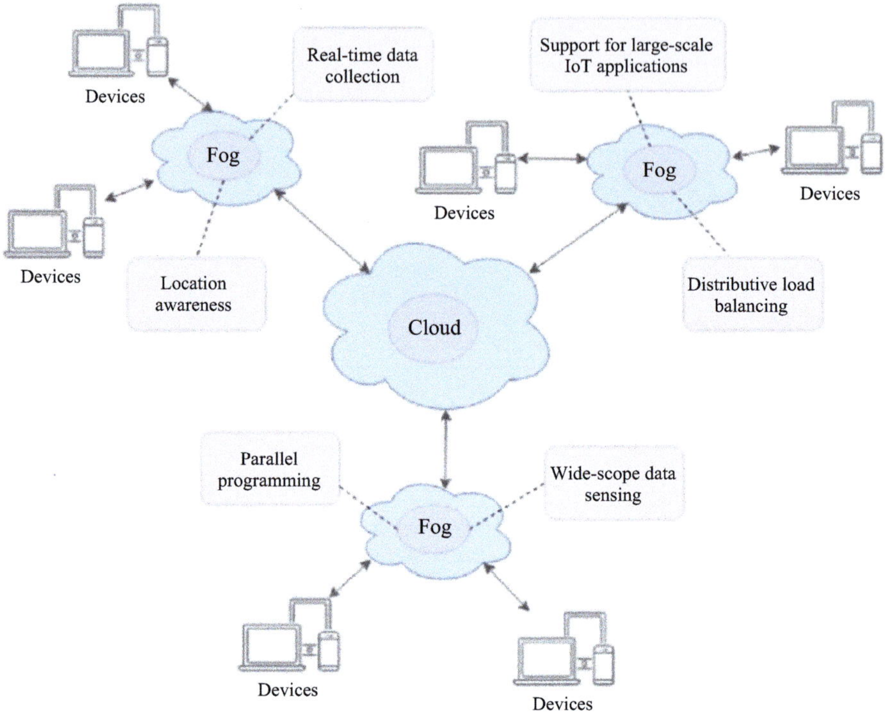

Figure 13.4 Fog-based network [6]

Table 13.1 Comparison between cloud, fog, and edge [9]

Features	Cloud	Fog	Edge
Server availability	Few	High	Less
Coverage	Global	Distributed and localized	Specific
Location identification	No	Yes	Yes
Real-time response	Low	High	High
Storage	High	Average	Limited
Working environment	Cloud service provider-owned place	Home, malls, streets	Specific provider (mobile, camera)
Architecture	Centralized	Distributed	Distributed
Number of users	Internet-connected users	Local users	Specific users
Service providers	Google, Microsoft Azure, Amazon	Cisco, Intel	Cellular network companies

Table 13.2 Security issues and solutions

Computing		Security issues	Solutions
Cloud computing	IaaS	Unauthorized control	Monitor network
		Data theft	Implement firewall
		Inability to monitor activities	Segment network
		Monitoring VM from another machine	
	PaaS	Absence of secured software in service provider	Encapsulation of access control
		Recovery and backup from system failure	Securing the operating system
		Inadequate provisions	Authorization enforcement
		Legacy applications	
	SaaS	Inefficient authorization implementation	Encryption
		Data loss	Recovery facilities
			Backup of data
Fog computing		Limited network	Encryption
		Virtualization	Trusted platform module
		Fog node issues	VM monitor
		Multitenancy	USB with decoy
Edge computing		Weak credentials	Secure all edge nodes
		Limited network	Encryption
		Unlimited updates	Continuous monitoring
		Insecure communications	User behavior profile
			Intrusion detection system

accessing the network, injection of cloud malware—injecting malicious or VM into the server, side-channel attack—reverse-engineering the device's cryptography, session hijack—intercepting and hijacking a user's session, Eavesdropping—an illicit user hide its presence and monitor authorized user, data tampering—tampering the data during communication or while in storage, and communication interception—illicit user exploiting the communication between two legitimate users [15–18]. A more detailed view on the security issues and solutions is provided in Table 13.2.

13.5 Cloud computing in Industrial Revolution 4.0

Cloud-based platform plays a major role in Industrial Revolution 4.0 (IR 4.0) by interconnecting the Internet and manufacturing department. Cloud can integrate several CPS and digital services to manufacture physical parts or components within an Internet-based environment. Rudolph and Emmelmann [19], in 2017, put forward a concept of using cloud computing in additive manufacturing as part of IR 4.0. They developed a framework and integrated algorithm where data preparation, additive manufacturing, post-processing, and delivery are the key components.

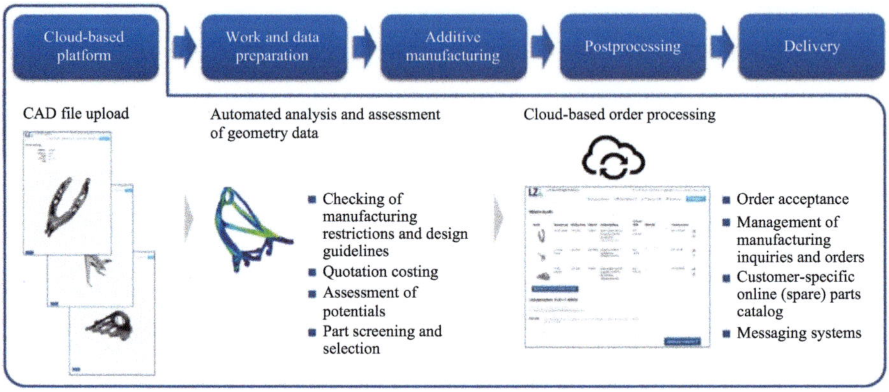

Figure 13.5 Cloud-based additive manufacturing [19]

According to the framework (Figure 13.5), a client can upload the geometry of the required part through online in STL or CAD format, which is considered as a default format for additive manufacturing [20]. The designs, volume, dimensions, material, number of pieces, guidelines, and requirements of the uploaded geometry are all carefully examined by the manufacturer. Material cost, production cost, pre- and post-processing were also considered. Pre-processing includes data preparation and machine setup. Post-processing includes heat treatment, wire cutting, sand-blasting, and the removal of support structures. Based on all these, an automated quotation or an offer is made. Client can either accept or reject it. Implementing this framework in manufacturing industries will provide a significant increase in the effectiveness of order processing. This framework does also support feature for client to view and track the status of their order through online.

In 2019, O'Donovan *et al.* [21] compared the latency and reliability performances of cyber-physical interfaces in engineering applications Industry 4.0 using cloud and fog computing. Their study focused on decentralized intelligence, near real-time performance, industrial data privacy, openness and interoperability. They used entry-level and standard configurations for cloud and fog CPS in their study (Figure 13.6). A test computer was set up to host JMeter, which was configured with parameters to send, receive, and measure transmissions happening between each CPS. For this study, they set up a temporary local network using Huawei router with two primary devices—(a) PC with JMeter, and (b) Raspberry Pi with Openscoring engine to enable real-time scoring. After network setup, using Amazon Web Services, cloud-based cyber-physical interface was constructed. It consisted of one virtual CPU, 1 GB memory, and Linux operating system. Fog-based cyber-physical interface was constructed with Raspberry Pi3 model B with 64-bit ARMv8 1.2 GHz processor, 1 GB memory, inbuilt Wi-Fi capabilities, and Linux operating system. They found that fog's decentralized, local autonomous topology may provide greater consistency, privacy, reliability, and security for Industry 4.0 applications than the cloud. The difference in latency between cloud and fog was found to be

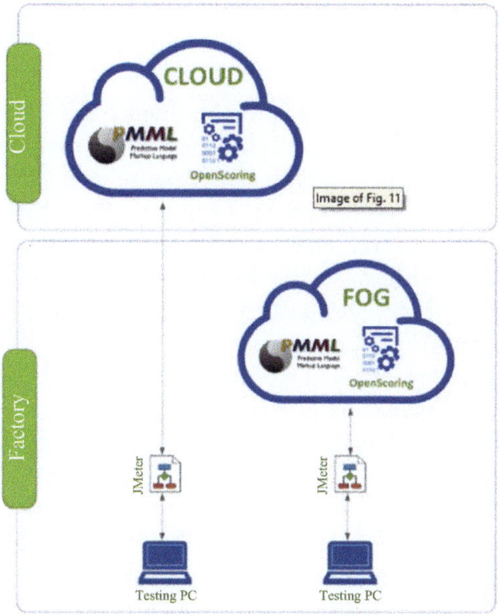

Figure 13.6 Setup for cloud and fog CPS [21]

ranging from 67.7% to 99.4%. Also, the fog interface showed 0% failure rate and cloud interface with 0.11%, 1.42%, and 6.6% under different level of stress. They concluded that engineering applications requiring raw performance can take advantage of cloud computing, whereas applications requiring consistent and reliable real-time execution can benefit from fog computing.

A recent systematic literature review [22] on 77 papers found a positive relationship between cloud computing and supply chain. More specifically, cloud computing has a strong impact on supply chain integration, especially in the formation and physical flows. It also concluded that more research is needed in design, finance, purchase, and warehouse integrations in the future to have a better understanding in cloud computing and supply chain integration relationships.

Khayer *et al.* [23] conducted a study on cloud computing adaptation and its impact on small and medium enterprises (SMEs) performance using dual-stage analytical approach by combining structural equation modeling and artificial neural network. Their study revealed that service quality, perceived risks, top management supports, cloud providers' influence, server location, computer efficacy, and resistance to change have a strong effect on adopting cloud computing in SMEs. They also found that cloud computing has a positive impact on the SMEs performance.

Ooi *et al.* [24] conducted a brief research on how cloud computing can land in innovativeness and firm performance in manufacturing industries by proposing 11 hypotheses. (1) Performance expectancy has a positive influence on innovativeness, (2) effort expectancy has a positive influence on firm performance, (3) effort expectancy has a positive influence on performance expectancy, (4) top management support has a positive influence on firm performance, (5) top management support has a positive influence on effort expectancy, (6) firm size has a positive influence on innovativeness, (7) firm size has a positive influence on firm performance, (8) firm size has a positive influence on effort expectancy, (9) absorptive capacity has a positive influence on innovativeness, (10) absorptive capacity has a positive influence on firm performance, and (11) innovativeness has a positive influence on firm performance. They too used structural equation modeling, artificial neural network, and partial least squares to conduct research. Research findings revealed that 7 of 11 hypotheses were supported. Performance expectancy, firm size, and absorptive capacity have a positive significant influence on innovativeness and firm performance.

Certain SMEs have implemented advanced planning and scheduling (APS) systems in their manufacturing units. However, APS systems lack flexibility. Due to this, Liu *et al.* [25] developed a cloud-based advanced planning and scheduling (CAPS) system for the automotive parts manufacturing industry. CAPS includes five major modules: industry case template, user interface, input basic data, cloud-based simulation planning, and output. Applications of CAPS in an automotive assembly demonstrated high planning quality, low implementation, and low maintenance cost.

13.6 Cloud computing in the communication sector

Implementing quality of service (QoS) in cloud-based communication demands three major issues to be solved [26]. They are (1) coexistence of different wireless protocols, (2) interoperability between communication systems, and (3) engineering to allow adaptive factory operation. To overcome these hurdles, Kunst *et al.* [27] proposed a resource-sharing architecture particularly in the domain of Industry 4.0. Their CPS architecture had two additional layers, namely networking layer and resources broker. Networking layer consists of network technologies to access the cloud. Resources broker is for managing and controlling the network resources. Kunnst *et al.* used MATLAB® and system evaluation methodology from WiMAX to evaluate and simulate the coexistence of three wireless Internet technologies: 5G, 4G, and IEEE 802.11. The total simulation consisted of three different traffics. They are (1) 60% of HTTP for supervisory control and data acquisition, (2) 20% of VoIP-based manipulator fine tuning, and (3) 20% of video-based production quality control. Results showed that both delay and jitter QoS metrics were below the threshold.

Cloud computing helps not only in manufacturing and data storing but also in communications [39–45]. While running a business, there is always a need for travel and meeting. Due to high travel expenses, an alternate way has to be found. A study by Radwan *et al.* [28] presented a new approach that converts current 2D video calls into 3D video calls using cloud computing. Their new approach begins with (1) developing a cloud infrastructure to handle video communication using OpenStack cloud infrastructure [29], (2) set up video call using webRTC technology [30], (3) capture video using a video input device, and (4) 3D video is created using image processing techniques which can be watched using 3D glasses. Radwan *et al.* used

VM to test the performance of 3D call and compare it with traditional peer-to-peer video call. Experimental results showed that their approach provided a better CPU performance when compared with peer-to-peer 3D video call. Adding to that, their new approach has a better peak signal-to-noise ratio value by 0.991 dB compared to peer to peer.

13.7 Cloud computing in healthcare sector

Like Industry 4.0 for manufacturing sector, by incorporating Internet of Things (IoT), big data, and cloud computing, healthcare sector is progressing to healthcare 4.0 or eHealth [31]. Healthcare industries use cloud technology to design and customize treatments for their patients. Healthcare industries need to be up to date because different diseases can develop from time to time. In recent times, the usage of IoT to interconnect medical resources and provide efficient healthcare services to patients has been increased [32]. Healthcare services generate large data. Those data need to be stored, processed, and analyzed. Thus, stakeholders prefer to use cloud computing in healthcare sectors [33]. But it has been accompanied with a problem. A huge amount of IoT data from health services needed to be handled with more execution time such as waiting and turnaround time delay and large resource utilization [34,35]. Later, in 2018, Elhoseny *et al.* [36] proposed a hybrid model of IoT and cloud computing to manage big data in health services. The architecture of the proposed hybrid model contains four main components. They are stakeholders' device, stakeholders' request, cloud broker, and network administrator. To enhance the VMs selection, they used three optimizers, namely genetic algorithm, particle swarm optimizer, and parallel particle swarm optimizer. A set of experiments were conducted regarding execution time, data processing, and system efficiency among those three optimizers. Their results showed that the proposed hybrid model outperformed the state-of-the-art models and system efficiency was also significantly improved by 5.2%.

A literature review study [37] was done on determining the factors influencing the acceptance of cloud-computing implementation in healthcare sector. They reviewed 55 articles and obtained 21 factors based on the thematic analysis method. They are compatibility, top management support, relative advantage, security, complexity, external pressure, IT knowledge, cost, trust, trialability, regulations and government support, innovativeness, external expertise, sharing and collaboration, user experiences, awareness, firm size, social influence, task, vendor support, and business continuity.

13.8 Scholarly articles in cloud computing

All over the world several researchers do research in IR 4.0, its components, and publish their results. Those results get cited by other researchers. The ratio of annual total citations received and the expected total citations based on the average in a particular field is called as Field Weighted Citation Impact (FWCI). Based on the NRF issue report on analysis on research level of the five major platform technologies related to the IR 4.0 between 2012 to 2016, the following graphs (refer Figures 13.7 to 13.10) are provided: Scholarly Output (SO) of Asia Pacific countries and their FWCI, Total Scholarly Output (TSO) on industrial components and their FWCI, SO on cloud computing between 2012 to 2016 and its FWCI, and top five countries on SO of cloud computing and their citation per publication (C/P).

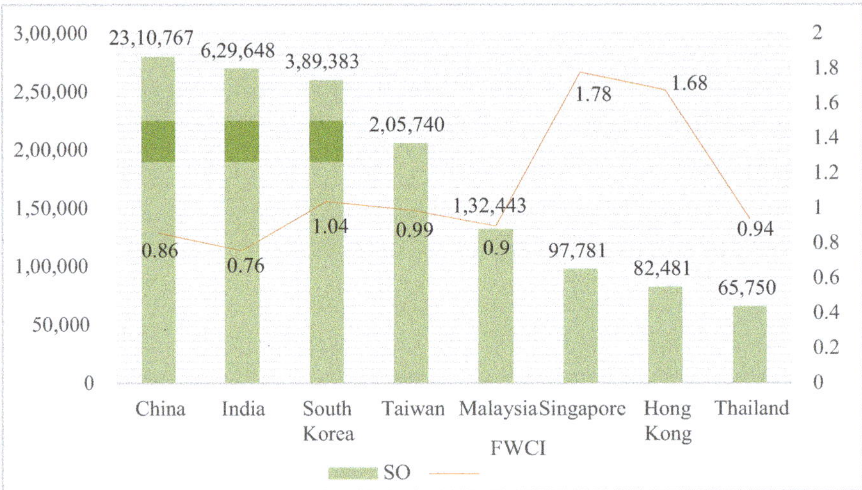

Figure 13.7 SO of Asia Pacific countries and their FWCI

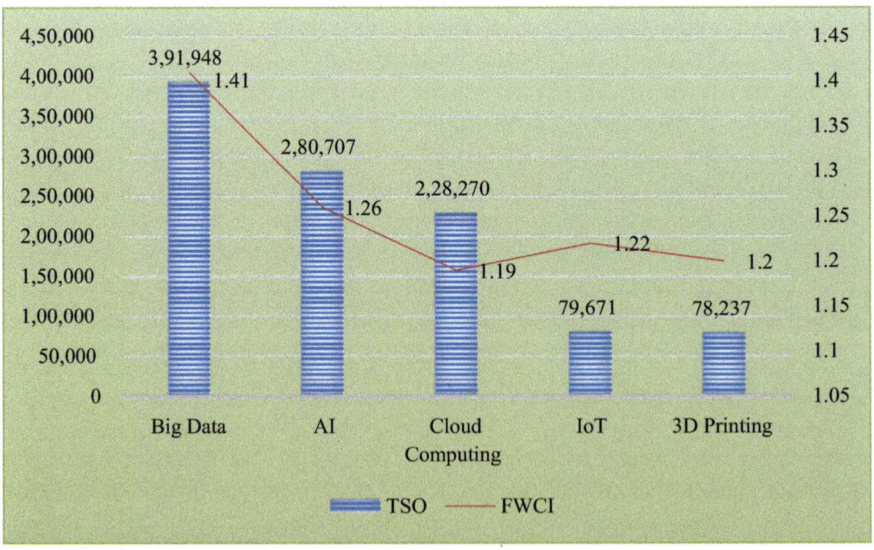

Figure 13.8 TSO on industrial components and their FWCI

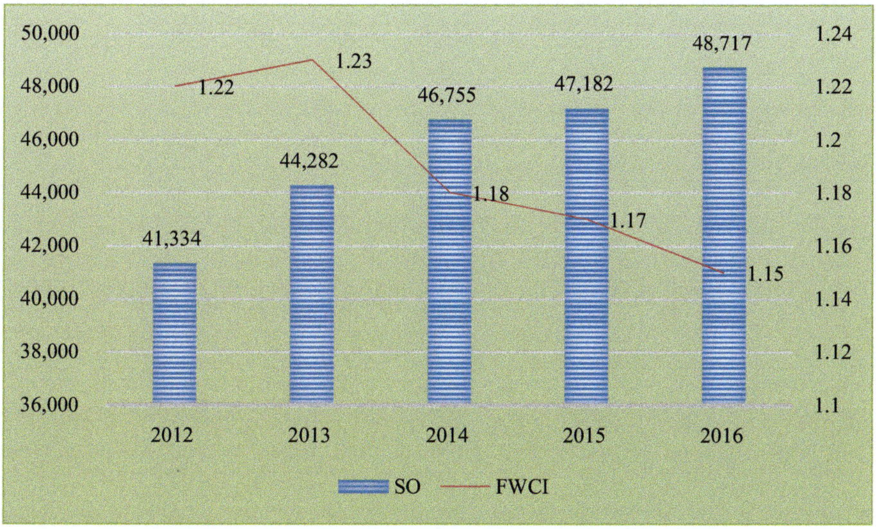

Figure 13.9 SO on cloud computing and its FWCI

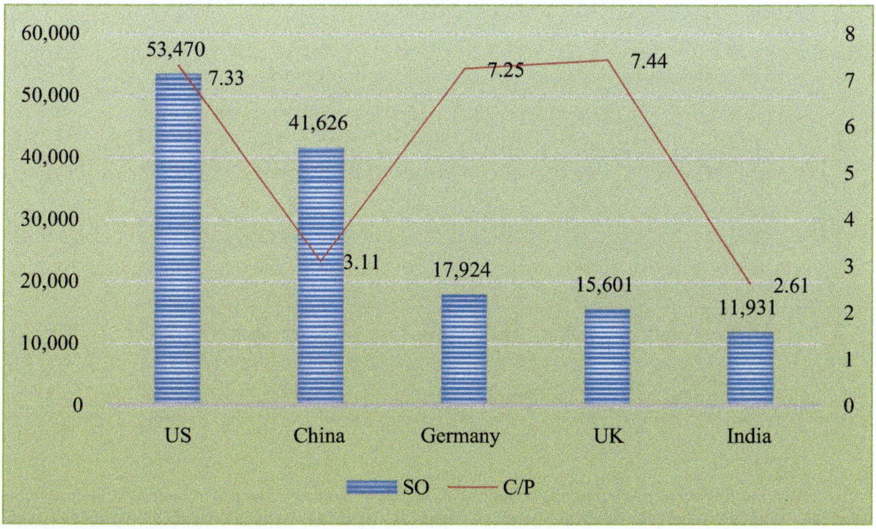

Figure 13.10 Top countries in SO on cloud computing and their C/P

13.9 Conclusion

The concept of industrial revolution was proposed in 2011. Even in this 2020, that concept is still popular but not thoroughly implemented globally. Governments and private organizations are slowly understanding the theory of cloud computing. The ruling governments must provide simplified conditions that are both government-oriented and private oriented to increase the number of cloud service providers so that the quality of service will be improved [38]. With more awareness and support, Industrial Revolution 4.0 can be achieved soon.

References

[1] J. Morgan, "What is the Fourth Industrial Revolution?," Forbes, 19 February 2016. [Online]. Available: https://www.forbes.com/sites/jacobmorgan/2016/02/19/what-is-the-4th-industrial-revolution/#5738154ff392. [Accessed 24 January 2020].

[2] A. Lele, "Industry 4.0," in *Disruptive Technologies for the Militaries and Security*, vol. 132, Singapore, Springer, 2019, pp. 205–215.

[3] S. S. Gill, S. Tuli, M. Xu, *et al.*, "Transformative Effects of IoT, Blockchain and Artificial Intelligence on Cloud Computing: Evolution, Vision, Trends and Open Challenges," *Internet of Things*, vol. 8, p. 100118, 2019.

[4] Amazon, "AWS," [Online]. Available: https://aws.amazon.com/what-is-cloud-computing/. [Accessed 24 February 2020].

[5] Microsoft, "What is cloud computing?," Microsoft Azure, [Online]. Available: https://azure.microsoft.com/en-in/overview/what-is-cloud-computing/. [Accessed 24 February 2020].

[6] M. J. Baucas and P. Spachos, "Using cloud and Fog Computing for Large Scale IoT-Based Urban Sound Classification," *Simulation Modelling Practice and Theory*, vol. 101, p. 102013, 2020.

[7] Microsoft, "What is cloud computing?," Microsoft Azure, [Online]. Available: https://azure.microsoft.com/en-in/overview/what-is-cloud-computing/. [Accessed 24 February 2020].

[8] S. Parikh, D. Dave, R. Patel and N. Doshi, "Security and Privacy Issues in Cloud, Fog and Edge Computing," in *The 10th International Conference on Emerging Ubiquitous Systems and Pervasive Networks*, 2019.

[9] M. R. Anawar, S. Wang, M. A. Zia, A. K. Jadoon, U. Akram and S. Raza, "Fog Computing: An Overview of Big IoT Data Analytics," *Wireless Communications and Mobile Computing*, vol. 2018, p. 22, 2018.

[10] J. DeMuro, "What is fog computing?," Techradar, 18 December 2019. [Online]. Available: https://www.techradar.com/in/news/what-is-fog-computing. [Accessed 26 February 2020].

[11] B. Butler, "What is fog computing? Connecting the cloud to things," Networkworld, 17 January 2018. [Online]. Available: https://www.networkworld.com/article/3243111/what-is-fog-computing-connecting-the-cloud-to-things.html. [Accessed 26 February 2020].

[12] Cloudflare, "What is edge computing?," Cloudflare, [Online]. Available: https://www.cloudflare.com/learning/serverless/glossary/what-is-edge-computing/. [Accessed 26 February 2020].

[13] M. B. Ali, T. Wood-Harper and M. Mohamad, "Benefits and Challenges of Cloud Computing Adoption and Usage in Higher Education: A Systematic

Literature Review," *International Journal of Enterprise Information Systems*, vol. 14, pp. 64–77, 2018.

[14] E. Yadegaridehkordi, M. H. N. B. Md Nasir, N. F. B. Mohd Noor, L. Shuib and N. Badie, "Predicting the Adoption of Cloud-Based Technology Using Fuzzy Analytic Hierarchy Process and Structural Equation Modelling Approaches," *Applied Soft Computing*, vol. 66, pp. 77–89, 2018.

[15] A. Singh and M. Shrivastava, "Overview of Attacks on Cloud Computing," *International Journal of Engineering and Innovative Technology*, vol. 1, pp. 321–323, 2012.

[16] P. Kumar, N. Zaidi and T. Choudhury, "Fog Computing: Common Security Issues and Proposed Countermeasures," in *2016 International Conference System Modeling & Advancement in Research Trends*, Moradabad, 2016.

[17] S. Khan, S. Parkinson and Y. Qin, "Fog Computing Security: A Review of Current Applications and Security Solutions," *Journal of Cloud Computing*, vol. 19, 2017.

[18] Maniah, E. Abdurachman, F. L. Gaol and B. Soewito, "Survey on Threats and Risks in the Cloud Computing Environment," in *The Fifth Information Systems International Conference*, Surabaya, 2019.

[19] J.-P. Rudolph and C. Emmelmann, "A Cloud-Based Platform for Automated Order Processing in Additive Manufacturing," in *The 50th CIRP Conference on Manufacturing Systems*, Taiwan, 2017.

[20] C. C. Kai, G. G. K. Jacob and T. Mei, "Interface between CAD and Rapid Prototyping systems. Part 1: A Study of Existing Interfaces," *The International Journal of Advanced Manufacturing Technology*, vol. 13, pp. 566–570, 1997.

[21] P. O'Donovan, C. Gallagher, K. Leahy and D. T. O'Sullivan, "A Comparison of Fog and Cloud Computing Cyber-Physical Interfaces for Industry 4.0 Real-Time Embedded Machine Learning Engineering Applications," *Computers in Industry*, vol. 110, pp. 12–35, 2019.

[22] L. Novais, J. M. Maqueira and A. Ortiz-Bas, "A Systematic Literature Review of Cloud Computing Use in Supply Chain Integration," *Computers & Industrial Engineering*, vol. 129, pp. 296–314, 2019.

[23] A. Khayer, M. S. Talukder, Y. Bao and M. N. Hossain, "Cloud Computing Adoption and Its Impact on SMEs' Performance for Cloud Supported Operations: A Dual-Stage Analytical Approach," *Technology in Society*, vol. 60, p. 101225, 2020.

[24] K. B. Ooi, V. H. Lee, G. W.-H. Tan, T.-S. Hew and J.-J. Hew, "Cloud Computing in Manufacturing: The Next Industrial Revolution in Malaysia?," *Expert Systems with Applications*, vol. 93, pp. 376–394, 2017.

[25] J.-L. Liu, L.-C. Wang and P.-C. Chu, "Development of a Cloud-Based Advanced Planning and Scheduling System for Automotive Parts Manufacturing Industry," in *29th International Conference on Flexible Automation and Intelligent Manufacturing*, Limerick, 2019.

[26] M. Wollschlaeger, T. Sauter and J. Jasperneite, "The Future of Industrial Communication: Automation Networks in the Era of the Internet of Things and Industry 4.0," *IEEE Industrial Electronics Magazine*, vol. 11, pp. 17–27, 2017.

[27] R. Kunst, L. Avila, A. Binotto, E. Pignaton, S. Bampi and J. Rochol, "Improving Devices Communication in Industry 4.0 Wireless Networks," *Engineering Applications of Artificial Intelligence*, vol. 83, pp. 1–12, 2019.

[28] N. Radwan, M. Abdelhalim and A. Abdelraouf, "Implement 3D Video Call Using Cloud Computing Infrastructure," *Ain Shams Engineering Journal*, vol. 11, no. 2, pp. 363–375, 2020.

[29] R. Kamboj and A. Arya, "Openstack: Open Source Cloud Computing IaaS Platform," *International Journal of Advanced Research in Computer Science Software Engineering*, vol. 4, pp. 1200–1202, 2014.

[30] J. Cui and Z. Lin, "Research and Implementation of WebRTC Signaling via WebSocket-Based for Real-Time Multimedia Communications," in *5th International Conference on Computer Sciences and Automation Engineering*, 2016.

[31] G. Aceto, V. Persico and A. Pescape, "A Survey on Information and Communication Technologies for Industry 4.0: State-of-the-Art, Taxonomies, Perspectives, and Challenges," *IEEE Communications Surveys & Tutorials*, vol. 21, pp. 3467–3501, 2019.

[32] A. Darwish, A. E. Hassanien, M. Elhoseny, A. K. Sangaiah and K. Muhammad, "The Impact of the Hybrid Platform of Internet of Things and Cloud Computing on Healthcare Systems: Opportunities, Challenges, and Open Problems," *Journal of Ambient Intelligence and Humanized Computing*, vol. 10, pp. 4151–4166, 2019.

[33] E. Pacini, C. Mateos and C. G. Garino, "Dynamic Scheduling Based on Particle Swarm Optimization for Cloud-Based Scientific Experiments," *CLEI Electronic Journal*, vol. 17, p. 12, 2014.

[34] K. Muhammad, M. Sajjad, M. Y. Lee and S. W. Baik, "Efficient Visual Attention Driven Framework for Key Frames Extraction from Hysteroscopy Videos," *Biomedical Signal Processing and Control*, vol. 33, pp. 161–168, 2017.

[35] N. Alassaf, B. Alkazemi and A. Gutub, "Applicable Light-Weight Cryptography to Secure Medical Data in IoT Systems," *Journal of Research in Engineering and Applied Sciences*, vol. 2, pp. 50–58, 2017.

[36] M. Elhoseny, A. Abdelaziz, A. S. Salama, A. Riad, K. Muhammad and A. K. Sangaiah, "A Hybrid Model of Internet of Things and Cloud Computing to Manage Big Data in Health Services Applications," *Future Generation Computer Systems*, vol. 86, pp. 1383–1394, 2018.

[37] M. T. Amron, R. Ibrahim, N. A. Abu Bakar and S. Chuprat, "Determining Factors Influencing the Acceptance of Cloud Computing Implementation," in *The Fifth Information Systems International Conference*, Surabaya, 2019.

[38] O. Ali and V. Osmanaj, "The Role of Government Regulations in the Adoption of Cloud Computing: A Case Study of Local Government," *Computer Law & Security Review*, vol. 36, p. 105396, 2020.

[39] W. Y. Leong, and D. P. Mandic, "Towards Adaptive Blind Extraction of Post-Nonlinearly Mixed Signals," *2006 16th IEEE Signal Processing Society Workshop on Machine Learning for Signal Processing*, IEEE, pp. 91–96, 2006.

[40] S. Huang, D. H. Zhang, W. Y. Leong, H. L. Chan, K. M. Goh and J. B. Zhang, "Detecting tool breakage using accelerometer in ball-nose end milling," *2008 10th International Conference on Control, Automation, Robotics and Vision*, IEEE, pp. 927–933, 2008.

[41] W. Y. Leong, J. Homer and D. P. Mandic, "An Implementation of Nonlinear Multiuser Detection in Rayleigh Fading Channel," *EURASIP Journal on Wireless Communications and Networking (EURASIP JWCN)*, vol. 2006, pp. 1–9, 45647, United States.

[42] W. Y. Leong and J. Homer, "Blind Multiuser Receiver in Rayleigh Fading Channel," *Australian Communications Theory Workshop (AusCTW'05)*, pp. 145–150, Brisbane, 2005.

[43] E. B. T. Dennis, R. S. Tung, W. Y. Leong, and C. M. T. Joel, "Sleep Disorder Detection and Identification," *Procedia engineering*, Elsevier, no. 41, pp. 289–295, 2012.

[44] W. Y. Leong, "Implementing Blind Source Separation in Signal Processing and Telecommunications," Thesis, University of Queensland, Australia, 2005.

[45] P. Mohankumaran and W. Y. Leong, "3D Modelling with CT and MRI Images of a Scoliotic Vertebrae," *Journal of Engineering Science and Technology EURECA*, pp.188–198, 2015.

Chapter 14

Cybersecurity in Industry 4.0 context: background, issues, and future directions

Haqi Khalid[1], Shaiful Jahari Hashim[1], Sharifah Mumtazah Syed Ahmad[1], Fazirulhisyam Hashim[1] and Muhammad Akmal Chaudary[2]

A new revolution called Industry 4.0 (I4.0) is emerging and trending, in which industrial systems comprised of numerous sensors, actuators, and intelligent elements are interfaced and integrated into the smart factories with Internet communication technologies. I4.0 is currently driven by disruptive innovations that promise to provide opportunities for new value creations in all major market sectors. Cybersecurity is a common requirement in any Internet technology, thus it remains a major challenge to adopters of I4.0. This chapter provides a brief overview of a number of key components, principles, and paradigms of I4.0 technologies pertaining to cybersecurity. In addition, this chapter introduces industry-relevant cybersecurity vulnerabilities, risks, threats, and countermeasures with high-profile attack examples (e.g. BlackEnergy, Stuxnet) to help readers to appreciate and understand the state of the art. Finally, the chapter attempts to highlight the open issues and future directions of the system components in the context of cybersecurity for I4.0.

14.1 Introduction: background and motivation

With regard to Industry 4.0 (I4.0) paradigms, an increasing number of companies are connecting plants and factories to the Internet, which are also called Industrial Internet, to improve their effectiveness and efficiency. Cybersecurity issues are one of the most important challenges to address in the Industrial Internet of Things (IIoT) devices [1,2].

I4.0 is the current trend of manufacturing automation technologies, mainly including enabled technologies such as cyber-physical systems (CPSs), the Internet of

[1]Department of Computer and Communication Systems Engineering, Faculty of Engineering, Universiti Putra Malaysia, Serdang, Malaysia
[2]Department of Electrical Engineering, College of Engineering, Ajman University, Ajman, United Arab Emirates

Things (IoT), and cloud computing [3,4]. By adopting a new IoT paradigm, people can create a smart world with the major developments of microelectronics and communication technology [5]. The IoT industry also includes applications (also known as the IIoT) [5]. IIoT has adopted IoT to improve productivity, efficiency, security, intelligence, and reliability in the context of the interconnection system [6].

I4.0 was coined in Germany and was used to identify the new proposal for the future of Germany's economic policy for the first time in 2011 [8]. In today's industries, there is a high demand for connectivity among components, while maintaining basic needs, such as business continuity and high availability. This requires new, intelligent production processes that are better adapted to the production needs and processes with a more efficient allocation of resources. Figure 14.1 illustrates the revolution of I4.0, which is the timeline of the evolution of manufacturing and the industrial sector in general.

Thus, came a new industrial revolution known as the Fourth Industrial Revolution. I4.0 is composed of three main stages: first, digital records are obtained through industrial asset sensors that collect data in close imitation of human feelings and ideas. It is called fusion of the sensors. Second, the analytical capacity of the collected data is implemented through analysis and visualizing of the data with sensors. Many different background operations are performed, from signal processing to optimizing, visualizing, cognitive calculations, and high-efficiency calculations. An industrial cloud supports the service system to manage the huge volume of data. Third, aggregate information must be converted into meaningful results, including additive production, autonomous robots, digital design, and simulation, to translate information into action. Raw data are processed with application for data analysis in the industrial cloud and then transformed into practical knowledge [9].

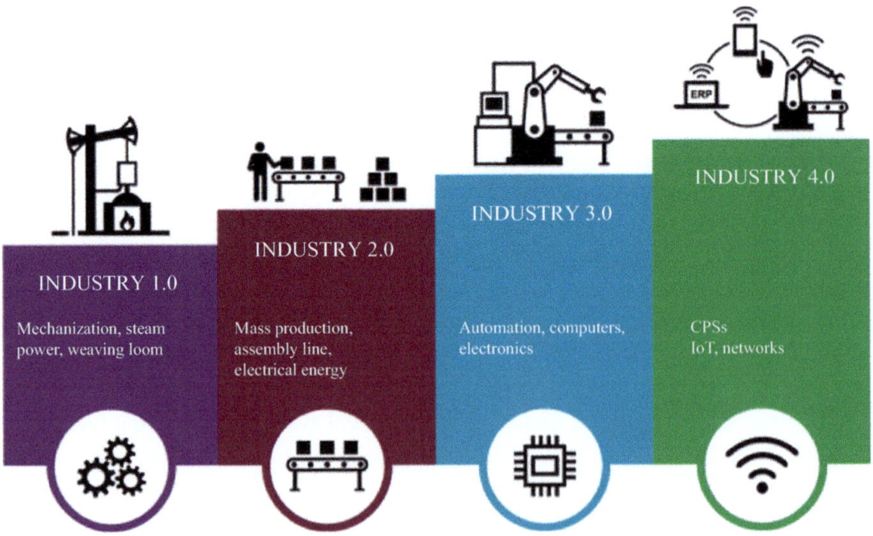

Figure 14.1 Industrial Revolution (Source [7])

The physical entry terminals of I4.0 have IoT inserted into them and are therefore ultimately vulnerable to cyberattacks. Although this additional connectivity contributes to increasing the productivity, it creates a treat for cybercrimes in the network. The sensitivity of such networks is well known to cybercriminals in which successful attacks may create devastating impacts such as—lost incomes, a fall in profitability, irreparable fire damage or a devastating threat to people and assets. Developing a comprehensive strategy for cyber risk is critical to value chains as they combine operative technology and IT-I4.0's very driving force. The current internet technology is, however, plagued by the cyber difficulties which present significant challenges and obstacles to the adopters of I4.0 faces traditional cybersecurity problems together with its unique self-imbibed security challenges. In the lack of an appropriate solution to these challenges, I4.0 may not achieve its true potential [10].

For this reason, this chapter starts by introducing the background and motivation behind the cybersecurity of I4.0 in Section 14.1. Secondly, cybersecurity characterization is outlined in Section 14.2. The major security of cybersecurity principles presented in Section 14.3. Section 14.4 presents I4.0 system components. This is followed by the open issues which are discussed in Section 14.5. And, future directions for industries, researchers, and developers are highlighted in Section 14.6. Section 14.7 concludes the chapter.

14.2 I4.0 cybersecurity characterizations

The system assets (system components) of the Industrial IoT first should be determined before addressing the security threats. An asset is a valuable and sensitive resource in a company's industry. The main components of any IoT system are system hardware, software, services, and service data (including the buildings, machinery, etc.). After the industrial assets have been identified, mostly involved in cybersecurity concerns (i.e., in the contexts of I4.0), the following advantages can be obtained:

1. Defining the inherent vulnerabilities of the systems affecting their safety;
2. Defining the cyber threats to the systems;
3. Identifying the risks related to cyberattacks;
4. Countermeasures to address the problems of cybersecurity.

14.2.1 Cybersecurity vulnerabilities

Vulnerabilities are design weaknesses that allow an intruder to execute the commands, access unauthorized data, and/or to carry out denial-of-service (DoS) attacks [11,12]. Vulnerabilities are found in various components of the I4.0. Several vulnerabilities exist: communication and network protocols infrastructure, application servers, database servers, human machine interfaces, program logic controllers, remote terminal units in each supervisory control and data acquisition (SCDA) system component.

These might include system hardware or software failures, policy weaknesses, system procedures, and users' weaknesses [13]. The vulnerabilities could be identified as remote access, software and local area networks (LAN) [14], and both virtual cloud resource and IT systems machines can be associated. There are various types of vulnerabilities affecting CPS and industrial control systems (ICSs); those unknown (i.e. zero-day vulnerabilities) are very common [14–17]; they are placed between different components of each interface where there is an exchange of information [18].

Evidence indicates that IoT devices are at risk for botnets (e.g. number of Internet-connected devices), as the security of many producers is not a priority. This is due to the common use of default passwords and open ports; there is a lack of an integrated system to automatically receive firmware updates; and firmware is often forgotten once installed (owners do not know if their device is used for malicious purposes or the need to update the firmware). Jansen [7] identifies the following reasons for the components of most industrial devices:

- Devices run for weeks or months in many plants without any security and anti-virus tools;
- Various ICS network controllers can be disturbed by malformed network traffic or even high traffic volumes since they were designed at a time when cybersecurity was not a matter of concern;
- Several ICS networks have multiple ways of entering cybersecurity threats, bypassing existing measures of cybersecurity (e.g. laptops running in and out of systems or USB sticks running on multiple computers, without having malware properly checked);
- Nearly all ICS networks are still being implemented as a large, flat, and unrelated network without physical or virtual isolation (allows malware spread even to the remote sites).

To address these problems, companies should conduct a process of vulnerability evaluation to identify and assess the potential system vulnerabilities. The National Institute of Standards and Technology (NIST) [11] defines a vulnerability assessment as a systematic evaluation of ICS or product, identifying security weaknesses, providing data from which to predict the effectiveness of the security action proposed, and confirmation of the suitability of such actions following their implementation. Vulnerability assessments are typically carried out using a network or host-based method, using automated scanning tools for systems and vulnerabilities discovery, testing, analysis, and reporting. The technical, physical, and governance-based vulnerabilities can also be identified by manual methods.

14.2.2 Cybersecurity threats

A threat is an action that benefits from and has a negative impact on the security weaknesses in the system. Threats may arise from two major sources: people and nature [19]. The threats to computers could be severe, including terrible earthquakes, hurricanes, fires, floods, or others. Few protections against natural disasters can be implemented and nobody can prevent them. Disaster recovery schemes such as supports and contingency plans are the best way to protect the systems against natural threats. Human threats are the causes from people such as malicious internal threats [20] (someone has authorized access) or external threats [21] (persons or organizations that work outside the network) seeking to exploit and disrupt a system. The following are the categories of human threats:

- Unstructured threats comprising mostly of inexperienced people who use easily accessible hacking tools.
- Structured threats can support, develop, and utilize scripts and codes as people know about system vulnerabilities. An example of a structured threat is the advanced persistent threats (APT). APT offers an advanced network attack aimed at high-quality information in companies and government organizations, for manufacturing, financial industry, and national defense [9,22]. Three dimensions can characterize the attacks against the interconnected physical systems [9]:

1. Type of attacker (i.e. insider or external);
2. The aims and objectives of the attacker (i.e. large-scale destruction or a specific target);
3. Mode of attack (e.g. active or passive).

In relation to the I4.0 contexts, the German Federal Office for Information Security (BSI) lists the following key categories of cyber threat [15].

- External direct access attacks;
- Indirect attacks on the service provider IT systems that have granted external access;
- Unknown attack vectors without the ability to detect vulnerabilities (or zero-day exploits);
- Nontargeted malware that infects components and affects their functionality;
- Intrusion into nearby networks or network areas (e.g., the current office network).

In the attack mode, active attacks are designed to alter the system resource or affect system operations, (including distributed denial-of-service (DDoS) attacks and compromised-key assaults), whereas passive attacks are intended to learn from, and/or use the system information, instead of changing the system.

Unauthorized access to information systems and confidential data may generally occur when cyber threats are successful. This may result in unauthorized use, disclosure, breakdown, modification, or destruction of critical data and/or data interfaces, and a DoS by network and computer (in the worst case, it could lead to reduced system availability).

14.2.3 Cybersecurity risks

The risk refers to "the level of influence on the operations of an organization (including mission, functioning, image, or reputation), the company's assets or individuals arising in the event of an information system operation given the potential impact and potential risk of threats." Safety risks related to information systems are based on the loss of information or information systems confidentiality, integrity, or availability [9,11]. Confidentiality, in particular, means that unauthorized users can keep the information secret; in this way, data can be disclosed due to the lack of confidentiality.

Integrity, which means protecting the confidentiality of data or resources, could lead to unauthorized modification unless it is preserved. In this case, there is a risk that records, data, and data are corrupted or modified. Availability makes a system accessible and usable on request. Consequently, a DoS can occur if there is a guarantee of the availability of the system, causing a lack of productivity of the physical system. The risks are highly caused vis-à-vis the exposure and attacks as described in Sections 14.2.3.1 and 14.2.3.2.

14.2.3.1 Exposure

Exposure is a system configuration problem or mistake that enables an attacker to carry out information-collection activities. Resilience to physical attacks is one of the most challenging issues in IoT. Devices can be left unattended in most IoT applications and are likely to be directly accessible to attackers. Such exposure allows an attacker to capture the device, remove secrets, change its programing, or replace them with a malicious device which is under the attacker's control [2,14].

14.2.3.2 Attacks

Attacks are taken using various techniques and tools that harm a system or interrupt normal operation by exploiting vulnerabilities. Attackers perform attacks for any achievement or benefit. The cost of the attack is called effort measurement in terms of expertise, resources, and motivation of the attacker [23]. Attack actors are people that pose a digital world threat. They may include hackers, criminals, and governments [22].

In many ways, there can be attacks, including active network attacks for traffic monitoring of sensitive information, passive attacks, such as unprotected traffic network communications and authentication information weak encryption, close-in, inside operation, and so on. Common types of cyberattacks are:

1. DoS: This type of attack attempts to prevent the desired users from using a machine or network resource. The majority of IoT devices are vulnerable to resource energetic attacks due to low memory and limited computing resources.
2. Physical attacks: This type of hardware component attack tampers. Due to the unattended and distributed IoT nature, most devices are typically in outdoor environments that are very likely to be physically attacked.
3. Privacy attacks: IoT privacy protection has become an increasing challenge as large amounts of information are easily accessible by remote access mechanisms. The most frequent attacks are:
4. Password-based attacks: Intruders are attempting to duplicate a valid user password. The attempt can be made in two ways: (1) the dictionary attack, which tries possible combinations of letters and numbers to devise passwords; (2) the attack with brute force uses cracking tools to find out all the possible combinations of passwords.

- Tracking: The unique device identification number is used to track the user movements. Tracking a user's location allows them to be identified in anonymous situations.
- Eavesdropping: It involves listening to two parties' conversation.
- Cyberespionage: It includes spying or obtaining secret information from individuals, organizations, or governments using cracking and malicious software.
- Data mining: It allows attackers to discover data in certain databases.

5. Recognition attacks: These include unauthorized system, service, and vulnerability discovery and mapping. Recognition attacks include scanning network ports, packet sniffers, traffic analysis, and sending IP address queries.
6. Access attacks: Unauthorized individuals have the right to access networks and devices. Two different kinds of access attacks are available: the first is physical access, whereby a physical device is available for the intruder. The second is remote access to IP-linked devices.
7. Destructive attack: Space is used to disrupt and destroy the lives and property on a large scale. Terrorism and vengeance attacks are examples of destructive attacks.
8. Supervisory Control and Data Acquisition (SCADA) Attacks: SCADA is vulnerable to many cyberattacks like any other TCP/IP system. The system can be attacked in one way or another:
   - Shut the system down, using DoS. For example, BlackEnergy3 (BE3) was involved in Ukraine's 2015 cyberattacks resulting in power outages. BE3 was used in the lead up to the attack to collect information on the ICS environment and was probably used to compromise user credentials by the network operators, although it did not have a direct role for cutting out power. Unlike previous incidents involving variants of Black Energy, BE3 was delivered to the Ukrainian energy companies via spear-phishing e-mails and weaponized Microsoft Word documents [8].
   - Take control of the system with Trojans or viruses. For example, in 2008, a virus named Stuxnet was attacked at Iran's nuclear facility in Natanz [22].
9. Cybercrimes: The Internet and smart objects are used for the use of users and data, such as IP theft, ID theft, brand theft, and fraud, for material purposes.

14.2.4 Cybersecurity countermeasures

Countermeasures are a set of actions, equipment, procedures, and techniques to address the threats, vulnerabilities, or attacks by preventing or removing threats, minimizing, or reporting the damaging effects [11,15]. The following three high-level approaches to guarantee industrial control systems security are described by Jansen in [7]:

1. Harden the area, meaning isolating the plant network from the office network, using firewalls and demilitarized zone where appropriate;

2. Deep defenses, using various layers of defense across the network to stop and contain malware that breaks the perimeter.
3. Remote access, one of the most common ways in which firewalls are penetrated. In this case, the virtual private network is recommended in a separate demilitarized area to isolate the remote users.

Furthermore, the security controls implemented should be continually updated (installing new security patch) at the level of the device, at the level of the network (updating new threat firewall signatures) and at the level of the plant/industry (monitoring and analyzing of the actual log sources). In other words, effective industrial systems should protect against an unauthorized access or change of information for authorized users whether in storing, trading or transit and service denial, including the measures required to identify, document, and address the threats under the Committee on National Security Systems (CNSS) [24,25].

Kankanhalli *et al.* [26] report that the prevention efforts include the deployment of the state-of-the-art security software or industrial asset security checks such as advanced access control, intrusion detection, firewall, monitoring systems, and generation of exception reports. Deterrent efforts include the development of security policy and guidelines, training for users, and the development of experienced industrial system auditors. The 50 most important categories of security countermeasures commonly adopted to evaluate the appropriateness of industrials system security are listed in Table 14.1.

14.3 I4.0 security principles

To succeed in implementing efficient industrial IoT security, certain principles must be assured for security to be guaranteed in the entire Industrial IoT system. According to the National Cyber Security Strategy (NCSS) [27], and the CNSS [24], the following six principles must be addressed to ensure an efficient IoT environment since IoT is rapidly growing.

14.3.1 Confidentiality

Confidentiality is an important feature of Industrial IoT security, but in certain situations when data are published, it may not be mandatory. This enables the information and data not to be exposed from within or outside the system to any unauthorized person or party. Data and information confidentiality is maintained by using encryption algorithms on the data stored and transmitted and by reducing the access to data locations [28]. For example, patient information, business data, and/or military information, as well as security credentials and secret keys, must be kept hidden from unauthorized entities.

14.3.2 Integrity

Integrity is, in most cases, a compulsory security feature to provide IoT users with reliable services. There are different requirements for integrity for different Industrial

Table 14.1 A summary of important security countermeasures

IT-related countermeasures	Software	Operator entrance log
		System recovery
		Multiuser system
		Scanner
		Automatic debug and test
		Access control to the program source
		Verification of the system modified
		Covert channels and Trojan code
	Network	Antivirus software
		Encryption
		User authentication
		Instruction detection systems
		Firewalls
		Alternative circuit
		Digital signatures
		Limitation of connection time
	Hardware	Remote mirroring
		Surveillance system use
		Entrance limitation
		Emergency power source (UPS)
		Periodical disk checking
	Data	Information backup
		Data access controls
		Authentication
		User access rights, authorization
		Enforced path
		Event logging
		Information-handling procedures
		Management of the removable media

(Continues)

Table 14.1 (Continued)

		Disposal of the media
Non-IT-related countermeasures	Risk transference	Security service provider
		Security outsourcing
		First-party insurance
		Third-party/public liability insurance
	Personnel	Confidentiality agreement
		Invalid account removal
		Information security consultant
		Security audit irregularly
		Security education and training
		Operational procedures training
		Incident report procedures
	Regulation and legality (including risk transference) security policy	Information security policy
		Security in job responsibilities
		Business continuity management
		Compliance with legal requirements privacy of personal information
		Intellectual property rights
	Physical facilities and environment	Lightning protector
		Air conditioner
		Fireproof installations
		Waterproof installations
		Quakeproof installations

IoT systems. It also is capable of maintaining the data in its current form and preventing unauthorized manipulation. In other words, the data must be kept away from the external and insider who try to modify it.

So, A destination gets wrong data and believes it is accurate. For example, a remote patient monitoring system has an extremely high integrity check for random errors because of information sensitivity. Data loss or manipulation can take place because of a communication which can lead to death [10].

14.3.3 Availability

This generally allows the system to deliver services and production of products on time. The availability of each subsystem means that it can function correctly and works on time and when necessary [28]. In other words, availability guarantees the proper functioning of all industrial IoT subsystems by avoiding all types of corruption, including hardware and software failure, power failures, and DoS attacks. Similarly, a device user (or the device itself), whenever necessary, must be able to access the services. To provide services even in the presence of malicious entities or adverse situations, different software and hardware components must be robust on IoT devices. For example, fire monitoring or health monitoring systems would probably be more available than roadside sensors.

14.3.4 Authenticity

Authentication of ubiquitous IoT connectivity is based on the nature of industrial IoT environments in which communication between equipment and equipment machine-to-machine (M2M), between man and device, and/or between man and human would be possible. Various authentication requirements require various systems solutions.

Strong solutions, such as authentication of bank cards or banking systems, are required. The authorizations property allows only authorized entities (any authenticated entity) to carry out certain network operations. However, the most important task is to be international, i.e. e-Passport, while others must be local [29].

14.3.5 Nonrepudiation

In cases where the user or device cannot deny an action, the nonrepudiation feature provides certain evidence. Nonrepudiation may not be considered a major security feature for most Industrial IoT. It can, for instance, be applicable in certain contexts where payment systems cannot be denied by users or providers. In the event of an incident of repudiation, an entity would be traced back to its actions by an accountability process that can help to verify what happened inside its story, and who was actually responsible [10].

14.3.6 Privacy

Privacy is the right of an entity to determine whether it interacts with its environment and the level to which that entity is willing to share information about itself with others. Industrial IoT's main privacy objectives are:

- **Device privacy:** It depends on physical privacy and switching privacy. In cases of device theft or loss or resilience to side-channel attacks, sensitive information can be released from the device.
- **Storage privacy:** Two factors should be taken into consideration to protect the privacy of data stored on devices:

1. Devices should store possible amounts of data necessary.
2. Regulation should be extended to provide user data protection after end-of-device life (Wipe) if the device is stolen, lost, or not in use.

- **Communication privacy:** It depends on a device's availability and device integrity and trustworthiness. To derogate from the communication of data privacy, IoT devices should communicate only where necessary.
- **Processing privacy:** It depends on the device and the integrity of communication. Without the knowledge of the data owner, data should not be disclosed or retained by third parties.
- **Privacy identity:** The identity of any device should only be identified by an authorized person/device.
- **Privacy location:** Only the approved entity should be found in the geographical position of the relevant device (human/device).

14.4 I4.0 system components

The term "Industry 4.0" is recent and is associated with the development of Information and Communication Technologies (ICT), in particular, the integration of ICT, in productive processes [30]. Several incidents in these sectors have been reported so far. Although these incidents are regarded as isolated, yet they can be alarming. New cybersecurity risks arise through the interconnection of critical infrastructures that must be identified, explored, and dealt with before too early [7]. This section will go through most of the popular system components in I4.0. In I4.0, cybersecurity experts must now deal with the problems of cybersecurity [7].

The Standards Overview (ISO 27000 series, IEC 62541 OPC UA) and Times Sensitive Nets (IEEE 17222016), which address these challenges OPC UA, presented the overall report [31]. Similarly, a new IoT cyber risk evaluation design principles in relation to the I4.0 presented [32] to allow for a quick and up-to-date overview of the existing and emerging I4.0 IoT development. In this way, many researchers are motivated to implement the techniques in industrial matters. This section will, however, focus on the applications that are being integrated and implemented into the industry, such as authentication and security approaches.

14.4.1 Cloud computing

According to the NIST [11], there are four types of cloud deployment models, which are discussed in detail in the following sections.

14.4.1.1 Public cloud

This cloud infrastructure is made open for everyone, including storage, applications, and other services and users only pay for the time that the service is used for pay-per-use. Public clouds are however less secure since all users can use applications and data. Example: Google App Engine, smart cloud IBM. Public cloud benefits: availability and reliability, pay-as-you-go, and self-service freedom [33,34].

14.4.1.2 Private cloud

This cloud infrastructure is implemented within a single organization and is open to a limited group of users. The organization is controlling its resources and applications. So, it enhanced private cloud security. Example: Ubuntu Enterprise Cloud, Ubuntu PVC. Private cloud benefits: high-level security, use of existing resources, and full customization control [33].

14.4.1.3 Community cloud

Community cloud involves the distribution in the same community of computing infrastructure between two organizations. Cloud infrastructure and computer resources are exclusive to two or more organizations with common privacy, security, and regulatory elements. This can help reduce the cost of capital for its implementation as the costs are allocated among the organizations. Example: "Gov Cloud" by Google. Community cloud benefits [33]: lower costs than private cloud, limited users, high security.

14.4.1.4 Fog computing

Fog computing is considered, above all, an extension of the cloud computing on the edge of the network that enables new IoT and I4.0 applications and services [35]. However, in I4.0, the use of intelligent CPSs utilizing IoT, mobile, and cloud computers make the factories increasingly intelligent and efficient. However, the works of the authors in [36,37] researched the deployment of fog computing systems to minimize the installation costs under limitations of the maximum demand capacity, maximum latency, coverage, and maximum device capacity.

The fog-based architecture of I4.0 applications was studied to reduce the overall battery restriction of IoT nodes' energy consumption [35,38,39]. There are currently a few studies which together with I4.0 introduce fog computing. In this section, these research works are presented. Recently, the I4.0 Fog-based architecture to minimize IoT node's energy consumption, which is usually battery restricted, is proposed in [35]. Also, Navantia's IAR architecture has been designated in [37], using cloud and fog computing to reduce latency response and discharge traditional cloud-based systems.

Further, a smart computer system framework with fog and edge devices in a logistics center was planned for use in [36]. Then a DMGA with both MA and GA merits was proposed. In addition, an industrial CPS based on fog computing can be used to implement and operationalize productive machine-learning models to inform real-time decision-making in the factory [39]. Likewise, authors in [38] present the Industrial Cyber-Physical System fog computation to use and integrate readymade models and compare a fog-physical interface to the reliability and consistency of traditional cloud interfaces.

14.4.2 Big data

With a growing number of data sources and the drop-in price of IT storage and computer resources in recent years, it has now become feasible to collect and analyze large quantities of data. The development of these methods improves the operational efficiency. The work of [40] developed an integrated big data analytics framework for data-driven risk management where the sensor data for healthcare applications were emphasized to enhance system scalability, security and efficiency [41–43].

Under the designation "big data" in I4.0, a lot was done in various disciplines, such as sales prediction, production planning, or the mining and clustering of the user relationships [40]. For instance, it was introduced in the three areas of CPS, digital manufacturing, I4.0 progress, and Big Data analytics [44]. Similarly, an approach which integrates heterogeneous data sources, also allows for automated big data risk assessment in I4.0 presented in [40]. In the context of I4.0, big data challenges and opportunities are shown in [42]. Data collection, processing, integration, modeling, storage, security, confidentiality, analysis, lifecycle, and presentation are detailed. In addition, secure architecture for the IIoT is proposed in [43], the key management security mechanism for the protection of big data in the industry uses 4.0 to storage and process saleable sensor data (big data) for healthcare applications. A survey has been carried out to focus on this critical crossroads. And in cyber-physical and Big Data systems, the detailed methods are introduced to help the readers understand their role in I4.0 [41].

14.4.3 Interoperability and transparency

Systems' ability to exchange the data with unambiguous meaning is interoperability. However, the literature defines interoperability in different ways and new models have recently been proposed for interoperability assessments [45]. Though, a focus on interoperability, one of the principles of I4.0, while standardization is indispensable for communication, should also be helped for interpretation of the significance of communication by machines and systems and here is interoperability: the meaning of the contents of the data being exchanged in I4.0 [45–50]. Recently, [46] showed the interrelationship between Industrial Internet Reference Architecture (IIRA) and Reference Architecture Model for industry 4.0 (RAMI 4.0) and how certain testbeds deployed under IIC – particularly the Infosys AE and IDT testbeds.

Also, in [47], a promising approach for optimizing CPS interoperability was investigated called formal concept analysis in I4.0. Interoperability literature was examined to find the automated approaches addressing semantic interoperability, especially in large-scale dynamic CPSs [45]. In addition, there was a general indication for the functional mapping and an interoperability modeling link between the industrial internet reference architecture (IIRA) digital twin concept (RAMI 4.0 AS) was marked with the RAMI 4 [49]. Machine-to-machine integration within I4.0 proposed in [48] should allow the machine to communicate and machines must achieve levels of interoperability that are consistent with system needs. There is a popular application used in data transparency design such as the following points.

14.4.3.1 Application program interface

An application program interface (API) is a set of routines, protocols, and tools to build software applications. In general, an API indicates the interaction between software components. In addition, graphical user interface parts are programmed with APIs. A decent API enables the development of a program with all design blocks. Then a programmer assembles the blocks. For operating systems, apps, or websites, there are many varying kinds of APIs. Windows, among others, has numerous API sets that are used by hardware and applications—it is the API that allows it to work when copying and pasting text from one application to another. API lists Google Maps, Twitter, YouTube, Flickr, and Amazon Product Advertising as some of the most popular APIs.

14.4.3.2 GraphQL

GraphQL is a query language that allows API users to describe the data they want, allowing API developers to focus on data relations and business rules rather than worries about the different payloads of the API. In this description, a GraphQL client can request the exact information in a single application. These types are used by GraphQL to guarantee that customers ask only to make clear and valuable errors [51]. Since the public release of GraphQL, many GraphQL clients and server-side query solver runtime has been implemented in software systems and programming languages.

Like any software program, these implementations can be subject to functional and performance problems. GraphQL is a powerful query language that does a lot correctly. GraphQL provides a highly elegant methodology when properly implemented for data recovery, backend stability, and increased query efficiency. GraphQL also allows you to query exactly whenever you want. This is excellent to use an API, but also has complex security implications. Instead of asking for legitimate and useful information, the malicious actor can submit an expensive and nested query to overload your server, database or network, or everything else. Thus, it opens to a DoS attack without the right protections [52].

However, authorization is a mechanism by which the system determines the level of access to the systems controlled by a specific (authenticated) user. A database management system may be developed to enable certain persons to collect the information from a database, but not to change the stored data in a database, while enabling other persons to change the information, for example, which could be or may not be related to a web-based scenario. Authentication can be processed with the context of queries. The basic idea is that the application can be injected with arbitrary code which is then transmitted to the request, so that the authorization by the developing company can be controlled very thoroughly.

This is a very secure system that is probably far stronger than other SQL injection and Langsec problem solutions [53]. However, a simple but expressive technique called deviation testing that automatically searches for anomalies in the way a schema is served proposed in [51]. The technique uses a set of deviation rules to input the existing test case and automatically generates variations from the testing case. The purpose of deviation testing is to increase the test coverage and to support developers in their GraphQL implementations and APIs in finding possible bugs.

A new conceptual framework is provided in [54] for web-based data access, the framework widely used in different applications. The client can obtain the data from the server side through this architecture, reducing the number of network applications and avoiding excessive data collection or capture. Graphic databases were presented by the authors in [55] as a viable storage for clinical data and graphs as a filter mechanism for enforcing local data policies, which guarantees privacy in medical data transactions.

14.4.4 Blockchain (distributed ledgers)

Blockchain technology is one of the possible foundations for providing a stable protocol for independent business activities [56]. Blockchain is another I4.0 technology that has become very popular in other fields such as finance and has the potential to provide a higher level of transparency, security, trust, and efficiency in the supply chain and enable the use of smart contracts [57]. The scope of this section on blockchain technology is in the context of I4.0.

Many researchers have recently implemented a decentralized blockchain solution that preserves privacy and audits it to ensure secure communication between agents and help in developing adequate solutions to change working conditions [56–58]. Furthermore, a blockchain is used to receive, validate, and make available inventory data collected from unmanned aerial vehicles (UAVs) to stakeholders and organize a secure interaction between independent agents [59–61]. Similarly, overhead and delay of packets decreasing and BC scalability increased in comparison to the various baselines, to enable the development of resilient, fault-tolerant IoT, and middleware blockchain applications have been highlighted in [62–65]. Yet, in [56], the authors have organized a communication system among agents on a pair using the decentralized blockchain ethereum technology and smart contracts.

Also, a communications architecture that includes both a blockchain and smart contracts together with a UAV development for RFID-based inventory and traceability applications is presented in [60]. Additionally, in [59], a BCT approach to the treatment of the trustworthiness of companies in the form of smart contracts as a text is developed. In addition, a model that utilizes blockchain and multi-agent systems in a network of entities to help represent an entity and to support the decisions by providing additional knowledge is developed in [57]. To ensure the enforcement of fine-grained access control policies, the blockchain-based mutual authentication system, BSeIn, was also presented in [58].

In recent years [66], a blockchain-based IoT architecture has been introduced that provides light and decentralized security and privacy. Also, authors in [67] proposed a way of managing IoT devices using the computer platform Ethereum. The work of [68], a hierarchical structure and distributed confidence to preserve security and privacy in the BC, is the basis for the proposed framework and makes it better adapted to meet the specific IoT requirements. In [64], it is proposed that a lightweight scalable BC for IoT, an IoT-friendly consensus algorithm, which eliminates the need to resolve the puzzle before an improvement block is added to the BC.

The structure and functionality of the PlaTIBART presentations in [63], is a platform for the development, deployment, execution, management, and testing of ITBA for the trans-active IoT blockchain applications with repeated testing. In addition, current blockchain protocols designed for IoT networks in [65] are state-of-the-art. An efficient decentralized mechanism for authentication called trust bubbles proposed is in [69]. In addition, work in [70] analyzed the current trends in research on BC approaches and technology utilization in an IoT environment. Likewise, an architecture to enable secure firmware updates on a distributed-trust IoT network using blockchain presented in [62].

14.4.5 Software-defined network

Software-defined network (SDN) is a technology which can efficiently manage the whole network and convert the complex network architectural into a simple and manageable one [71]. SDN means that a network control unit is physically separated from the forwarding device in which a control plane manages several devices [72]. SDN will probably be one of the most important technologies in I4.0, allowing programmatic monitoring of the network architecture, enabling network access change on demand [73]. Various researches have shown that SDN or software-defined IIoT is impacting approaches and a number of researches are conducted using SDN-enabled networks to create highly scalable and flexible networks while adapting them to changes in the network environment [74]. In addition, the data time was reduced and user experience in latency-sensitive applications improved. Moreover, firewalls limit the exposure to industrial control devices so that security risks are minimized [73,75,76]. To manage physical devices and provide the interface for information exchanges, work in [77], proposes an IIoT software-defined architecture.

Controller(s) scalability problems in the SDN architecture presented [78], it organizes discussion of scalability control plane into two broad approaches: approaches relating to topology and approaches relating to mechanisms. A multi-attribute secure communication model is designed in [79], a SDN for the IIoT environment. Besides, the authors of [80] proposed a meta-heuristic evolutionary algorithm (PACSA-MSCP) that is based on a parallel version of max-min ant colony leveraged by an asynchronous simulated annealing algorithm.

By conducting an IIoT study and research-related areas, SDN and EC for IIoT adaptive transmission architecture are proposed in [81]. In [82], the methodology for implementing the industrial IIoT energy-efficient wireless sensor nodes with a view to the data center's three-tier IIoT architecture backbone is proposed. The data-driven approach, which utilizes anomaly detection invariants, is developed in the water treatment testbed shown to identify and detect the integrity attacks in [83].

Moreover, work in [71] presents several security threats resolved through SDN and new threats arising from the implementation of SDN. The work of the authors in [84] Hy-LP – a hybrid protocol and framework for IIoT networks. Hy-LP enables seamless and lightweight interactions with and between smart IIoT devices, in inter- and intra-domain deployments. The software-defined IIoT architecture in [85] proposes that the adaptive mode can be selected in fog computing using computing mode-based CMS and ASTP-based execution sequences. Compound approach to protect the IIoT command attack is presented in [86] by developing a lightweight authentication scheme to ensure a sender's identity.

A SDN-based testbed and a combined cybersecurity-resiliency ontology were proposed by the author of [87] to be used in the requirements for the capture of the virtual network design stages. A SDIN architecture, which addresses the existing disadvantages of the IIoT, for instance, resource use, data processing, and system compatibility, is proposed in [88]. A heterogeneous, hierarchic, and multi-tier architecture for communication, and edge-driven intelligent data distribution was proposed in [75]. In addition, work in [72] introduced the notion of SDCM for promoting cloud-based production and other 4.0 industry pillars through agility, flexibility, and adaptability while minimizing different complex problems.

In [74], software-defined infrastructure is then used in a network environment in I4.0. In addition, ISD 4.0 researchers are examining SDN in [89], as it can be used intelligently to automate multiple tasks, such as user management, routing, surveillance, control, security and configuration, not limited by others. Likewise, a SDN-enabled MEC-assisted network architecture incorporating various types of access technologies is proposed in [76] to provide low-speed and high-reliability communication. Furthermore, work in [73] introduced the new SDN Firewall, which applies the standard architecture automatically, without affecting the flexibility of the network.

14.4.6 Multi-factor authentication

In order to provide a secure environment in I4.0, cybersecurity applications play an important role in industries to prevent lack of security properties and overcome the common attacks (e.g. DoS, replay attack, MiTM). Some of the authentication techniques that used in industrial IoT are discussed as following:

14.4.6.1 Kerberos

Kerberos is a protocol for authenticating client and server through an insecure network connection, where they can authenticate mutually [90]. It is a technology that makes authentication possible and secure in open and distributed networks. Kerberos has recently been paid attention to the security robustness provided by many researchers [91] and secured communications to keep attackers from having access to many computers. It was therefore resistant to attacks and efficient time and space to meet IoT demands [92].

However, the invention is concerned with a multistage security system and methodology designed to mitigate unauthorized data access in the proposed industrial control environment [92]. In addition, in [90], an authentication protocol for biometric-Kerberos targets for m-commerce applications is proposed. The protocol engineering framework was also introduced by [93] to allow for structured, formal, and interoperable definition of software and hardware, architecture, design and development, configuration, documentation, and maintenance. In addition, authors in [91] described the design and implementation by using three-level authentication Kerberos to ensure the smart home system authentication through IoT. In [92], the protocol used AS and the KDC to add Expanded-Kerberos to it, an efficient and secure protocol is proposed for protecting the privacy and sensitive data of users. Furthermore, a third-party authentication scheme is presented [79], which is used by Kerberos.

Consequently, users may check the authenticity and integrity of their saved data in the cloud using Kerberos in the designed scheme. In addition, a CoAP framework, which solves a fine-grain access control problem, proposed in [94], makes use of ideas for other access control systems, such as Kerberos and RADIUS, and integrates both with the CoAP protocol to achieve an audio access control framework for IoT. In the same context of network security, a double authentication mechanism has been proposed with the Kerberos system [95].

14.4.6.2 Two-factor authentication

Two-factor authentication generally requires that a user provides a knowledge factor or two or more (e.g. something the user knows, such as a password, answer a question), an inherent factor (e.g. a user's knowledge, such as a fingerprint, retinal scan, other biometric data) and a possession factor (e.g. something the user has, such as key, token) [96]. This mechanism is justified in terms of security and usability while maintaining high performance for all operations and ensuring end-to-end authentication among IoT devices/applications [97–102]. In addition, it has ensured strong confidentiality that the adversaries can never trace any vehicle. In [103–105], several essential security features have been ensured, and lower computing, communication, and storage costs are involved. This section generally relates to authentication systems and, more specifically, systems and processes to provide authentication for two-factor systems in different types of infrastructure and operating environments in critical infrastructure.

Nevertheless, in [97], a new user-friendly two-factor authentication system (DoubleSec) for mobile devices was also developed in the design of a security architecture that combines and makes use of best practices in cryptography and security protocols. Too many systems, methods, and computer-readable media for two-factor authentication of a robust system in the operating infrastructure are provided in [96]. Similarly, a three-way handshake for two-way authentication between sensor nodes and the establishment of a secure communication channel is proposed and implemented in [98].

In addition, in [99], authors presented a two-factor authentication scheme by using QR connection, PIN, and the possession of a smartphone. A recent efficient lightweight OTP scheme for two-factor authentication between equipment, applications, and IoT communications, based on IDE (IBE ECC), was proposed in [100]. Furthermore, a new cloud-based IIoT deployment biometrics-based user authentication scheme (BP2UA) was introduced in [106]. In the two-factor research field for industrial wireless sensor networks (WSNs), the authors in [107] have taken a substantial step forward in breaking the vicious break-fix break-fast cycle.

A new two-factor authentication protocol on PUF noise factor for IoT devices, as designed in [104], is also preserved in regard to privacy. A two-factor multi-server architecture authentication-and-key-agreement scheme incorporating the elliptic curve cryptosystem (ECC) is presented in [108]. In addition to two authentication factor IoT, the authors of [101] presented an infrastructure of a biometric end-to-end security solution. The authentication scheme of 2FLIP was later proposed in [102], using two main methods: CA decentralization and Biological password 2FA decentralization. Furthermore, the [109] paper proposed SoundAuth, a 2FA mechanism that can reliably detect the co-location of both users.

14.4.6.3 Three-factor authentication

Several schemes have been proposed for the authentication and key agreements for industry, IoT, and WSNs. However, the data collected in real time in the IIoT environment are transferred to the public channel and security and privacy issues are raised in this environment [110]. It provides security for WSN, user untraceability, and computational efficiency for mobile services [111–113]. By using three-factor authentications, the user authentication and the production of the session key can increase the system security and additional functionality [105,110,114–116].

In the industrial environment, there are a few works. A security-enhanced authentication and key agreement scheme to overcome these security weaknesses using biometric information and an elliptic curve cryptosystem proposed in [113]. Also, in [112], authors proposed a three-factor authentication and key agreement scheme, which provides initiator untracability. Likewise, the three-factor IoT Environments authentication for WSNs, using biometrics to manage biometric information, adopts the fuzzy commitment scheme and corrects error codes [111]. A lightweight remote user authentication protocol [114], using a gateway node-based architecture for IoT environment, requires that the user first registers via the gateway node.

In addition, by using the three-factor idea in [110], the new user authenticated key agreements scheme, only authorized users can access the services of the specified IoT sensing devices installed in the IIoT environment. In another paper, three authentication factor protocols proposed in [115] were mainly reviewed and analyzed. The enhanced 3-AKE protocol in [116] proved its security under an enhanced model of security. The authors of [105] are also introducing a three-factor industrial WSN user authentication protocol.

14.5 Open issues

The fourth industrial revolution, I4.0, is the integration into the IIoT, which is to see the digital transformation of manufacturing. The use of many techniques in I4.0 could lead to massive problems, which can only make the services insignificant. Some of the challenges faced are explained in the following sections.

14.5.1 Fog computing issues

Fog computing is primarily an extension of cloud computing at the edge of the network that enables new industry-related applications and services 4.0. This new computing paradigm is not yet completely defined, and several challenges such as the following must be considered.

- ***DoS***: DoS or DDoS are the steps by which a system or application cannot be made available. I4.0 relies on a large number of interconnected systems and processes; in these environments, DoS attacks represent a very significant threat. In addition, as cloud computing is becoming popular and widely used in the smart factory concept, more hackers are likely to find new ways to exploit system vulnerabilities, such as the application of DoS. A company can suffer high cost damage from a DoS cyberattack. This attack causes material damage (servers and sensors must be replaced or returned to normal operation; network reprogramming is needed; systems should be redesigned), and it also causes financial and operational damage (service disruption, complex resuming protocols, new training for machine operators). An approach to the DoS attacks must be proposed.
- ***Industrial augmented reality (IAR):*** IAR solutions are expected to provide dynamic on-demand, fast reactions from remote servers in an I4.0 factory. Such information can also vary in size of payload, as IAR devices can request several megabytes of video from just a few kilobytes of text content. Therefore, IAR systems not only need to provide augmented reality (AR) functionality, but also have to be able to exchange and manage content as fast as possible [37].

14.5.2 Big data issues

Big data is an important issue for the researchers in I4.0 and offers opportunities in every area of research. Big data with different systems has many challenges which are described below:

- ***Data transformation:*** Industry data is generated mainly by machine data. Programmable logic controller (PLC) receives and generates electrical and physical data. As heterogeneous data is generated in manufacturing, that needs to be converted into a suitable interoperability form to conclude or predict machine failure on the basis of big archives. For communication between processes, data must be transformed and cleaned in different situations in the same format. To remotely control robotic machines online, data needs to be converted into visual forms for machine operators. In I4.0, the use of smart technology needs to transform data into a format consistent with different smart devices. It is a challenge for I4.0 to keep all these data before and after the transformation [42].

- ***Data integration and modeling:*** Interoperability is the key automation principle in I4.0, where several types of devices communicate. It is a challenge to integrate the various types of data generated in a single platform for fast production. For real-time processes, communication, and collection of data for integration with remote machines and CPSs is required. Data integration is a delay process in the distributed storage system. Data integration, where a real-time action is required, is important for remotely managed and operated machines. The design of industrial 4.0 and industrial Internet applications requires consistent models of data, which are reliable, scalable, and secure. From data collection to the end-users, all attributes must be specified. Industry 4.0 is challenging to control, operate automated processes, and estimate product cost, integrate, and model industrial big data [41].
- ***Real-time access***: I4.0 focuses on real-time access to industrial big data. Real-time access is necessary for cyber-physical sensor systems, actuators, and other devices. Real-time access to industrial big data is needed to handle it in the shortest possible time for smart tolerance of faults and failure detection. The network bandwidth must be rapidly and unloaded. When physical devices are remotely controlled, if delayed, then it causes a problem to the next physical devices, because all actuators operate in a predefined time period sequence [42].
- ***Security and privacy:*** The two main concerns in the world of big data are security and privacy. I4.0's quickly increasing volume of heterogeneous data and the shift to cloud increases the security risk. Data from external threats need to be secured. Due to the remote control of physical devices, industrial data risk is higher than the other normal data system. The hacker can be control over the physical machinery because all things are remotely controlled in I4.0 via the web portal. The key problems for I4.0 are therefore data authentication and privacy [42].
- ***Data analytics:*** Since the last decade, big data analysis has been the key field of data science. It is a blend of IT, mathematics, statistics, and machine learning. The data extraction from industrial big data forced entrepreneurs to take this into account for future industry planning and decision-making. The core areas related to the industrial Big Data analytics are fraud analysis, recommendation systems, industrial fault detection, processes mining, management, machine data, transport, and analysis of the market. Different physical devices require fast and real-time analysis of heterogeneous data creation. Incompleteness is a problem with real-time analysis and requires algorithms such that before analysis data is preprocessed. The scalability of analysis is also a challenge, as data is increased with production [42–44].

14.5.3 Interoperability issues

This section considers interoperability as one of the requirements for I4.0 to promote "smart interoperability" for the devices. This factor is not completely accomplished yet, so some problems still need to be addressed:

- ***Integration***: The cooperative manufacturing systems include a large number of information systems distributed in relation to physical machines through the large complex-networked architectures, such as Cooperative Enterprise Information Systems (CEIS) which can access a wide range of information and must interact to achieve their purpose. The architects and developers of CEIS have a difficult problem to deal with interoperability. There is a growing demand for integrating such systems tightly with organizational and manufacturing work so that these information systems can be fully, directly and immediately exploited by the intra and inter-enterprise processes [47].
- ***Supply chains:*** The sparse collection of sensors and actuators is used to optimize their value chain both by production lines and supply chains. Devices, therefore, play a central part in these manufacturing lines, which are increasingly complex and require extreme (if not impossible) human monitoring and control. This issue needs to be addressed properly [50].

14.5.4 GraphQL issues

A GraphQL wrapper needs to implement necessary API authentication mechanisms. We provide the support for the (a) API key and basic authentication and (b) OAuth 2 and OpenID Connect in GraphQL wrappers.

- **API keys and basic authentication:** Authentication viewers provide a means of passing authentication information to users (API keys, username, and password). Viewer types are specific GraphQL (Input), which wrap all other GraphQL (Input) object types that need authentication in their resolve functions. The viewers define the mandatory arguments APIKey and password for all enclosed child types and propagate their values to resolution functions. In GraphQL queries, viewers place sensitive credentials that demand dedicated security mechanisms (i.e. transport encryption) [117].
- **OAuth 2 and OpenID Connect:** OAuth 2 is an authorization framework where users can authenticate and authorize certain actions of an application on a third-party OAUTH services server. Applications are registered on the OAUTH server, and users are forwarded to the server afterward. The users authenticate with the OAUTH server, which returns the application access tokens. The tokens are used to interact with protected resources. OpenID Connect is a layer above OAuth 2 which prescribes the JSON Web Token to be employed for GraphQL wrappers, equivalent to OAuth 2. The above-described flow is independent of a GraphQL wrapper, but it relies on the application to obtain the tokens necessary for the application (e.g. the server that hosts the GraphQL wrapper). The GraphQL wrapper resolve functions must (a) be able

to access token from an application, and (b) send these tokens to target REST (-like) API requests, typically including them in an Authorization header [117].

14.5.5 Blockchain issues

With the emergence of a new revolution, industrial processes face major changes: I4.0. The conditions are fulfilled to accelerate and improve the concepts of this new revolution through blockchain technology implemented in this environment [57]. But the BC is still an emergent technology and faces many challenges such as the following:

- ***Quality of service:*** The inherent trust of the QoS values is not a good idea in real-world applications. There are also a range of challenges and risks for centralized authorities, including scalability, high overhead maintenance, management of DDoS, single point of failure, scams, and trust. The central government also has a range of challenges and risks. Further, EaaS IoT, SoA, and Cloud technologies expand geographically distributed services over the internet exponentially. The management of attribute-based digital certificates is operational and inexpensive as QoS values are dynamic and need updating and verification in real time [57].
- ***Privacy protection and anonymity:*** A simple authentication password, as discussed above, cannot protect the privacy of the real identities of the terminals, because a malicious cloud can trivially identify a terminal based on daily activity via interactions [58].
- ***Fine-grain access control:*** The integration of certain strategies, such as the user role mapping table and the authorization table of roles, will generally lead to a fine grains access control. However, cloud attacks are vulnerable to the approaches based on the use of these tables, as the cloud relies on these approaches [58].
- ***Integrity and confidentiality:*** A local cloud database records/stores the access history of terminals as discussed earlier. There is therefore a risk that unauthorized access and modification of such databases may occur. Traceability and auditability of access records are therefore hard to achieve [58].
- ***Multi-agent communication:*** Autonomous agents are increasing significantly in business processes, transaction costs, and risks connected with information transfer errors among IT systems. Cost is also rising since the control over the agents is centralized, as the failure of the system or the capture of intruders leads to work being paralyzed.

14.5.6 SDN issues

The assumption is that production machine "Pre-Industry 4.0" has very small Internet connectivity only; therefore, they have very limited use and do not meet the security requirements required for Internet connectivity [89].

- ***Security:*** I4.0 investigates SDN since it can intelligently automate various tasks such as user administrators, routing, control, security, and configuration, but not just for them. These tasks can be performed also using traditional, non-

SDN-based proprietary networking devices, such as switches and routers. The proprietary networking devices are static in a certain area of the network, necessary to configure each one of them individually, complex (the hardware part of the device contains billions of gateways and more than 6,000 standard documents are being implemented in the software part of the systems and applications) and not programmable (contains no open/standard APIs), and it is difficult to use [89].

- ***Real-time monitoring:*** Industries must be able to perform numerous calculations, transmit data at high speed, and manage high performance and reliable data easily through a huge amount of data processing and complex operations. Big data monitoring infrastructure is necessary [74].

14.5.7 Kerberos issues

It is an authentication protocol that enables the client and server to authenticate one another over an untrusted network connection [90]. Security, which means the preservation of the physical environment, for example, people, infrastructure, machinery, and equipment, is almost the same for the industrial IoT industry [118]. Industrial IoT's purpose is to provide a fast and up-to-date overview of the current IoT developments in the I4.0, including CPSs, the IoT, cloud computing and cognitive computing [32]. This section presents the current and emerging technologies of the Kerberos. The authentication scheme of Kerberos was widely used in many technologies, such as WSN, Smart Home, cloud computing, and Internet, from our comparison shown in Table 14.2. So, some problems, like the following, remain in the scheme.

- ***DoS:*** The DOS attacks were not resolved by Kerberos. Therefore, DoS is vulnerable to the online part of the scheme. So, the Kerberos used in Internet networks urgently needed to prevent DoS attacks [119].
- ***Time synchronization***: Each network principle must have a clock that is loosely synchronized to other principles so that Kerberos time costs are highly determined and taken into account [119].
- ***Computation cost:*** ECC is a primitive cryptographic operation that is used for encryption of user information but that is known with high computation during hashing, pairing, and multiplication calculation. In curve cryptography. Therefore, calculation costs must be efficient [95].
- ***File transfer:*** With elliptic curve cryptographing communication costs, the transfer of files or data between entities takes longer, which means that the transfer of data is slow, so the mechanism has to help increase the transfer of data [95].
- ***Privacy:*** The implementation of a Kerberos Protocol Intrusion Detection System may allow attackers to manipulate user data and personal information. The modification can then be intercepted and used to maliciously personify the user and to provide user privacy with extra methods [120].
- ***Big number of devices:*** The burden of an underlying network infrastructure is increased by a large number of intelligent systems deployed in SG-IIoT. To

Table 14.2 Kerberos requirements and techniques comparison

Reference	Problem	Technique	Kerberos requirements					Limitation
			Mutual authentication	**Confidentiality**	**Integrity**	**Single sign-on**	**Trusted third-party**	
[121]	Access to PLC and control systems at using an alphanumeric string, these type passwords are highly prone to attack or discovery.	A multi-tiered security framework.	A	A	A	N/A	A	Not fully considered the Kerberos authentication in the proposed architecture.
[90]	If attackers impersonate the genuine users, mobile devices are incapable of identifying the authentication.	A biometric Kerberos-based user identity authentication scheme.	A	N/A	A	A	A	No technical details about the use and protection of biometrics.
[93]	Focus instead of introducing the improvements and extensions to the protocol on correction errors detected.	Object–process methodology with Kerberos.	N/A	N/A	A	N/A	N/A	Low rate of adoption and need to move from the common standard UML de-facto.
[91]	Internet users can easily remotely monitor and control domestic devices from anywhere and anytime.	GPRS module and RF modules are used with three-level Kerberos authentications.	N/A	N/A	A	N/A	N/A	System needs improvement to overcome the delay.

(Continues)

[79]	The huge data volume generated by various smart devices used in SG-IIoT makes the underlying network infrastructure more costly.	SDN-IIoT communication model, an attribute-based encryption scheme, and Kerberos.	A	N/A	A	N/A	A	Need to design a secure communication framework for the distributed SG IIoT environment using the SDN backbone.
[122]	Problem of vulnerable password guessing attacks caused by dictionary attacks, replay attacks in the authentication process, and man-in-the-middle attacks in wireless LAN.	Kerberos extensible authentication protocol (K+EAP).	A	N/A	N/A	N/A	N/A	High computation overhead.
[120]	Transmission across the network with plain text is vulnerable in terms of security and privacy because information sent across the network may be intercepted and subsequently used to impersonate the user maliciously	Intrusion detection system to monitor Kerberos.	A	A	A	N/A	N/A	Helps to lack the privacy preserving
[119]	For the home delivery company, direct imitation of the typical cross-realm network authentically does not apply.	The crowdsourcing delivery model is based on Kerberos.	A	A	N/A	A	A	Vulnerability to DOS, time synchronization

(Continues)

Table 14.2 (Continued)

Reference	Problem	Technique	Kerberos requirements					Limitation
			Mutual authentication	Confidentiality	Integrity	Single sign-on	Trusted third-party	
[95]	Network attackers usually access a wide variety of computers by exploiting their attack armies' vulnerabilities.	Cryptography of elliptical curve, Kerberos system, and RSA.	N/A	A	N/A	A	A	Communication cost, files transfer.
[92]	The protocols currently used are not secured. The credentials are therefore shared through the unsecured channel such as the Internet and can therefore be extremely vulnerable to attacks.	Separating the modules of Kerberos and distributing it on multiple nodes autonomously.	A	A	N/A	N/A	A	Not efficient in changing mode with lack of privacy protection.

A* = Available, N/A* = Not-available

achieve a trusted real-time data flow, the communications infrastructure must have a custom network management paradigm that handles interoperability between different smart devices used on SG-systems [79].

- ***Single sign on/one-time password:*** Kerberos is an important feature. Only once must a customer type a password with single sign-on (SSO). Customer tickets such as the ticket granting ticket (TGT) and ticket granting server (SGT) are stored on a client's computer in a cache. Our comparison shows that most existing systems do not support OTP/SSO and therefore that things must be taken into account of the IoT [119].
- ***Confidentiality:*** Confidentiality is one of the Kerberos requirements. In any authentication scheme that depends on Kerberos, it is very important to consider it. Many methods in security and privacy have been proposed lately, but the confidentiality of the network environment remains an important challenge [119].

14.5.8 Two-factor authentication issues

Two-factor authentication is a popular practice for authenticating a user before allowing access to the proper system. The two-factor authentication usually requires that a user provides two or two different knowledge factors (e.g. something a user knew, such as password, answer to question), a factor inherent (e.g. something the user has, such as fingerprint, retinal scanning, other biometric data) [96]. The enhanced level of security for two-factor authentication, access control for numerous cyber assets in the infrastructure industries (e.g. computer systems, data bases, equipment) remains relatively basic. Table 14.3 illustrates the two-factor requirements for authentication and comparison of techniques.

- ***Vulnerable to attacks:*** Many existing two authentication schemes provide a secure authentication system in the literature. However, not all of the proposed systems fulfill the security properties of IoT and Industrial environments still vulnerable to many active attacks (Sinkhole, Sybil, Replay. MiTMs, Forward Secrecy, password guessing attack, DoS, physical attack, and impersonation attacks).
- ***Privacy preserving:*** Preserving privacy means preserving personal information concerning individual users (e.g. user identity, personal keys). Privacy is important for ensuring unlinkability, anonymity, and untraceability in two-factor authentication schemes. However, this technique still lacks confidentiality, and an effective system is needed to address this challenge.
- ***Mutual authentication:*** It refers to two entities that simultaneously authenticate each other. In IoT, Industrial, it is not sufficient only to authorize the sender to the receiver, but the sender should also ensure that the possible recipient of sensitive data is identified and authorized. The authentication should therefore probably be mutually authenticated to ensure the security of the domain. The existing systems do not yet include mutual authentication; therefore, it must be taken into account in the upcoming proposals.
- ***Time synchronization:*** Wide time synchronization of the network would therefore be necessary, introducing added complexity and consumption of resources [98]. And time stamp use can even be a serious problem in more

Table 14.3 Two-factor authentication requirements and techniques comparison

Security prosperities	[97]	[96]	[98]	[100]	[123]	[99]	[102]	[101]	[104]	[105]	[107]	[106]	[108]	[109]	[103]
Resilience to the impersonation attack	N/A	N/A	A	N/A	N/A	A	A	A	A	A	A	A	A	N/A	A
Resilience to the password guessing attack	A	N/A	N/A	A	N/A	A	N/A	N/A	A	A	A	A	A	N/A	A
Resilient-to-physical-observation	N/A	A	N/A	N/A	N/A	A	N/A	A	A	N/A	A	N/A	A	N/A	N/A
Trusted third-party	A	N/A	N/A	N/A	N/A	A	N/A	N/A	N/A	A	N/A	A	N/A	A	N/A
Unlinkable	N/A	N/A	N/A	N/A	N/A	A	A	N/A	N/A	N/A	N/A	N/A	N/A	N/A	A
Authentication	A	A	A	A	A	A	A	A	A	A	A	A	A	A	A
Anonymity and untraceability	N/A	N/A	N/A	A	N/A	N/A	A	N/A	N/A	A	A	A	A	N/A	A
Resist to DoS	N/A	N/A	A	N/A	N/A	N/A	A	N/A	N/A	N/A	N/A	A	A	N/A	A
Resist to MiTM attack	N/A	N/A	A	A	N/A	N/A	N/A	N/A	N/A	N/A	N/A	A	A	A	A
Integrity	A	A	A	A	A	A	A	A	A	N/A	A	N/A	N/A	N/A	N/A
Privacy preserving	A	N/A	N/A	N/A	N/A	N/A	A	N/A	A	N/A	N/A	A	N/A	N/A	A
Privacy protection	A	A	N/A	N/A	N/A	N/A	N/A	N/A	A	A	N/A	N/A	N/A	A	N/A
Secure system key update	N/A	N/A	N/A	N/A	N/A	N/A	A	N/A	N/A	A	A	N/A	A	N/A	A
Prevents clock synchronization problem	N/A	N/A	A	N/A	A	N/A	N/A	N/A	A	N/A	A	N/A	N/A	A	A
Device security	A	A	A	A	A	A	A	A	A	A	A	A	A	A	A
Mutual authentication	A	N/A	A	N/A	N/A	N/A	N/A	N/A	A	A	A	A	A	N/A	A
Resist replay attack	N/A	N/A	A	A	N/A	A	A	A	A	A	A	A	A	N/A	A
Suitable for IoT applications	N/A	N/A	A	A	N/A	A	A	A	A	A	A	A	A	N/A	A
Suitable for industrial	A	A	A	A	A	A	A	A	A	A	A	A	A	A	A
Forward secrecy	N/A	A	N/A	N/A	N/A	N/A	N/A	N/A	N/A	N/A	N/A	N/A	A	N/A	A
Formal verification/Analysis	N/A	N/A	A	N/A	N/A	N/A	A	N/A	A	N/A	N/A	A	A	N/A	N/A
Computation cost	N/A	N/A	A	A	N/A	N/A	A	A	A	A	N/A	A	A	N/A	A
Sinkhole attack	N/A	N/A	A	N/A	N/A	N/A	N/A	N/A	N/A	N/A	N/A	N/A	A	N/A	N/A
Sybil attack	N/A	N/A	A	N/A	N/A	N/A	N/A	N/A	N/A	N/A	N/A	N/A	A	N/A	N/A

powerful environments, because a synchronization failure makes a protocol otherwise secure vulnerable to attack. As a result, many of the existing systems do not still use time synchronization and need to be addressed.

- ***Computation cost:*** In literature, the truth of using various methods to enhance privacy and security may increase the system's computing costs. The authors, therefore, need to consider calculations as they may lead to a vulnerability of system for attacks, such as DoS and other user data to be handled to achieve high transport densities or malicious parties.
- ***Trusted third-party:*** Adversaries who have access to the trusted device can attempt to extract data that are stored on the device in a trusted third-party. Adversaries may also attempt to extract sensor data from an unacceptable device. Therefore an alternative trusted party should be used to protect the privacy of users [103].
- ***Formal verification/analysis:*** The scheme is not formally ensured by a lack of verification of proposed schemes. Due to the solidity of the formal control tools, schemes have to ensure that they are correct. Therefore, two-factor authentications require formal verification/analysis to secure system security.

14.5.9 Three-factor authentication issues

Authentication by three factors is an extensive idea to two authentication factors and generally: what you are aware of (i.e. password), what you have (i.e. smart card, phone) and who you are [116]. But it was determined that three-factor authentication ideas would overcome the two-factor authentication challenges. This idea is still in the early stages, however, and may have problems to face. In this section, some of the problems are explained. Table 14.4 shows three requirements for the authentication factor and comparison of techniques.

- ***Vulnerable to attacks:*** The weakness of the existing 3FA systems is obviously vulnerable for unused considerable attacks in the literature comparison. Although their scheme has a lot to offer, it still has some weaknesses, like deviation, insider password, impersonation, attack tracking, privileged insider attacks, attack at the password, biometric updates, device detection attack, parallel session attack, and dictionary attack. Security property, therefore, is important for protecting the scheme and new models for overcoming these vulnerabilities are necessary.
- ***Forward secrecy:*** It is a feature of the key agreement protocols that ensures that your session key is not compromised even if the server's private key is compromised [113]. It is sufficient to achieve perfect secrecy on 3FA systems so that he/she cannot use long term private keys to compute previously established session keys. However, most of the systems currently in place lack PFS and an efficient design.
- ***Privacy preserving:*** From the literature point of view, we noted that trace-ability and connectivity were suffering from the previous methods proposed. So, it is important and necessary to improve the schemes so as to ensure every user's privacy (untraceability, unlinkability, and anonymity).

Table 14.4 Three-factor authentication requirements and techniques comparison

		[112]	[113]	[111]	[114]	[116]	[110]	[115]	[105]
Attacks	Impersonation attacks	N/A	A	A	A	A	A	N/A	A
	Stolen smart card attack	N/A	A	A	A	A	A	A	A
	Offline password guessing attack	N/A	A	N/A	N/A	A	A	N/A	N/A
	Attack in the password and biometrics update phase	N/A	N/A	N/A	A	N/A	A	N/A	N/A
	Privileged insider attack	N/A	N/A	A	N/A	A	A	N/A	N/A
	Man-in-the-middle attack	N/A	A	N/A	A	A	A	N/A	N/A
	Sensing device-capturing attack	N/A	N/A	N/A	N/A	A	A	N/A	N/A
	Replay attack	N/A	N/A	A	A	A	A	N/A	A
	DoS attack	N/A	N/A	N/A	A	A	A	N/A	N/A
	Dictionary attack	N/A	N/A	N/A	A	N/A	N/A	N/A	N/A
	Parallel session attack	N/A	N/A	A	A	N/A	N/A	N/A	N/A
Properties	User anonymity and untraceability	A	A	A	A	A	A	A	A
	Mutual authentication	A	A	A	A	A	A	N/A	A
	Friendly password update	A	A	A	A	N/A	A	N/A	A
	Perfect forward secrecy	N/A	A	N/A	N/A	A	N/A	A	N/A
	Suitable IoT/Industrial			A	A	A	A	A	A
	Known session key security	N/A	N/A	A	N/A	A	N/A	A	A
	Biometric template privacy	A	A	N/A	A	A	A	N/A	N/A
	Revocability	N/A	N/A	N/A	N/A	A	A	N/A	N/A
	Usage of ECC	N/A	A	N/A	N/A	A	N/A	N/A	N/A
	Avoiding clock synchronization problem	N/A	N/A	A	N/A	A	N/A	N/A	A
	Privacy preserving	N/A	N/A	A	N/A	N/A	N/A	N/A	N/A
	Formal verification/Analysis	A	A	N/A	A	A	A	N/A	N/A

- ***Revocability:*** When a user has been compromised and removed from the system, Gateway removes the identity of the entry user from its database. If an unsubscribed user attempts to access the gateway nodes, verification will not be passed. However, we note that the current scheme is vulnerable to revocability and has not been implemented. A scheme is needed urgently to prevent revocation to pass a check if the user tries to login again.
- ***Clock synchronization problem:*** In many ways, clock synchronization was common and this use can lead to a serious threat. And it could further increase the cost of the calculation. This problem is avoided by an improvement or a proposed model.

14.6 Future directions

This section outlines brief directions from the literature to mitigate developers, researchers, industry/factories' problems in the various fields of I4.0. The directions for the future are classified by nature and are summarized in Table 14.5.

14.6.1 Directions to the developer/designer

Due to the development of I4.0, the future directions help to develop solutions by developers and designers with technical problems in the use of various I4.0 devices. However, leveraging big data prototype can be used in practical use case scenarios to further examine its performance and investigate its potential. In this regard, data mining and machine-learning approaches could be used to learn the patterns automatically [43]. To achieve more security components and services, an innovative platform must be used in future industrial developments, let alone I4.0 [74].

In addition to temporal filtering, the Transparent Firewall implements an Access Control system based on the OPC UA standard and conducted security tests in order to test the viability of the proposal [73]. Clearly, further research is needed to develop such solutions that are sufficiently transparent and reliable to be trusted in automation systems and operational technology, but this is an interesting opportunity to develop a solution that is applicable across standardization domains [45]. Also, a lightweight blockchain (LSB) prototype implementation to obtain its performance in real-world settings should be developed [64]. An application of BC to solve the problem of data exchange and trading that must be implemented should be designed [124].

14.6.2 Directions to researchers

This section provides researchers with research directions in the specific areas of I4.0. In I4.0, future research on wider problem areas should address more efficient algorithms with advanced techniques, such as coevolution, multiobjective optimization, and decomposition [36]. Research should address the challenges identified in applications that are difficult or impossible to deal with in the cybersecurity analysis procedure model [126].

Table 14.5 Summary of the research future directions in I4.0

Directions to developer/designer	Big data prototype in practical use case scenarios.	[43]
	Innovation platform for security components.	[74]
	Develop access control server on the OPC UA.	[73]
	Develop interfaces that are transparent and reliable.	[45]
	Design an application of BC for data exchange.	[124]
	Develop a prototype implementation of lightweight scalable BC.	[64]
	Develop network management tools.	[63]
	Prototype of the SDN-based sensor node.	[125]
	Design with software-defined IIoT.	[77]
	Enhancing the system's flexibility to multi-interact.	[82]
	Implement Markov fingerprinting with Kerberos.	[120]
	Prototype protocols (e.g. Kerberos) to the offline real life.	[119]
	Set up a testbed for memory requirements.	[114]
Directions to researchers	Large-scale problem instances.	[36]
	Procedural model for a cybersecurity analysis.	[126]
	Detect cyberattacks targeting I4.0.	[127]
	Blockchain with IoT-based smart contracts.	[60]
	Integrating an efficient framework into business process management (BPM) systems.	[59]
	Implementation of blockchain mechanism in a simulator.	[62]
	Induce new types of cybersecurity for I4.0.	[62]
	Lacked discovery information for client applications.	[73]
	Support using decentralized AAA servers.	[94]
	Improving and enriching OPL.	[93]
	Improve three-level Kerberos authentication.	[91]
	Security and privacy protection requirements of IoT.	[105]
	A fully decentralized VANET authentication.	[102]
	Security schemes for IoT applications.	[111]
Directions to industries/factories	Provide low-latency responses for I4.0.	[37]
	Transparent and reliable to be trusted in automation systems.	[45]
	Limited computing and energy resources of some industrial devices.	[57]
	Deploy a lightweight, robust, and scalable OTP scheme on a real IoT.	[100]
	Use two-factor authentication for passwords or factory set accounts.	[128]

The algorithm can be used to successfully detect cyberattacks targeting I4.0 systems and can be coupled with active response mechanisms in order to stop real cyberattacks [127]. More research should focus on additional experiments and the implementation of a specific blockchain with IoT-based smart contracts [60]. Additionally, research must identify how to manage certain industrial devices' limited computing and energy resources, which become a critical problem if the blockchain is used in industrial systems as a result of the mining process [57].

Also, an efficient framework should be integrated into business process management (BPM) system to automate the evaluation and verification of the trustworthiness in service selection and composition [59], in addition, the implementation of blockchain mechanism in a simulator along with suitable algorithms and a mathematical model is suitable [62]. In future research, the ability of the simplified software-defined cloud architecture to induce new kinds of cybersecurity in I4.0 systems should also be explored. Regarding spatial filtering, the standard OPC Discovery Service has lacked discovery information for client applications, so further research needs to be conducted to automatically generate access lists [73]. A researcher should study the mechanism that fully supports the use of decentralized AAA servers [94]. Another study can be conducted to enhance the OPL, to make it more natural to human readers and appeal to it [93].

A researcher should also improve three-level Kerberos authentication so that it can be faster [91]. Additional studies should also analyze and extract the IoT-based applications specific security and privacy requirements [105]. A completely decentralized VANET authentication scheme needs to be addressed while maintaining the relevant efficiency [102]. Since WSN is important in a number of smart environments, such as smart grid, smart healthcare, and intelligent transport systems, researchers need to examine the IoT security systems [111].

14.6.3 Directions to industries/factories

This section presents important directions for industries and factories in a variety of research areas. Most directions are related to security, privacy, and security communication technologies of I4.0. Further studies should provide low latency response by mixing elements both of the computing paradigms which could enable future IAR applications for I4.0 to be developed in real time [37]. More research is needed for solutions to be transparent and reliable enough to rely on automation and operational technology, but this is an important opportunity to develop a solution that can be applied throughout standardization [45].

In an integrated system such as how certain industrial appliances manage the limited computing and energy resources, this becomes a critical problem when a blockchain is applied to industrial systems as a result of the mining process [57]. Further research must conduct to design a lightweight, robust, and scalable one-time password (OTP) scheme on a real IoT platform [100]. For privileged root access and remote access, two-factor authentications are required. This eliminates weak passwords or user-defined accounts for access [128].

14.7 Conclusion

This book section examines cybersecurity problems in I4.0 contexts using a structured approach to the literature review by analyzing the quality of the content of recent studies. The book chapter's evaluation is focused, in particular, on five analytical areas: (1) the definition of system vulnerabilities, cyber threats, risks, exposure, attacks, and countermeasures in the context of I4.0 scenarios; (2) the identification of major objectives for security and privacy; (3) the definition of techniques in I4.0, integrated in these fields; (4) classify the issues in I4.0; and (5) categorize the future directions within the environment.

The need for cybersecurity in I4.0 will expand in the future not only into several infrastructures but also into the consumer area with industrial IoT where privacy and physical protection are of more significance, especially where adversaries can access them physically. In this section, we provided an overview and background on cybersecurity of I4.0. This chapter also, provide an open issues and future directions related to cybersecurity of I4.0.

References

[1] M. Lezzi, M. Lazoi, and A. Corallo, "Computers in industry cybersecurity for Industry 4.0 in the current literature: a reference framework," *Comput. Ind.*, vol. 103, pp. 97–110, 2018.

[2] E. B. a B. W. World, "Cybersecurity for Industry 4.0 Cybersecurity implications for government, industry and homeland security," in *Cybersecurity implications for government, industry and homeland security*, 2018.

[3] Y. Liao, F. Deschamps, E. De Freitas, and R. Loures, "Past, present and future of Industry 4.0 – a systematic literature review and research agenda proposal," *Int. J. Prod. Res.*, vol. 7543, no. November, p. 0, 2017.

[4] E. Sisinni, A. Saifullah, S. Han, U. Jennehag, and M. Gidlund, "Industrial internet of things: challenges, opportunities, and directions," *IEEE Trans. Ind. Informatics*, vol. 14, no. 11, pp. 4724–4734, 2018.

[5] K. K. R. Choo, S. Gritzalis, and J. H. Park, "Cryptographic solutions for industrial internet-of-things: research challenges and opportunities," *IEEE Trans. Ind. Informatics*, vol. 14, no. 8, pp. 3567–3569, 2018.

[6] J. D. Contreras, J. I. Garcia, J. D. Pastrana, and U. Valle, "Developing of Industry 4.0 applications," *Int. J. Online Eng.*, vol. 13, no. 10, pp. 30–47, 2017.

[7] N. Benias and A. P. Markopoulos, "A review on the readiness level and cyber-security challenges in Industry 4.0," in *South-East Europe Design Automation, Computer Engineering, Computer Networks and Social Media Conference, SEEDA-CECNSM 2017*, 2017.

[8] M. Piccarozzi, B. Aquilani, and C. Gatti, "Industry 4.0 in management studies: a systematic literature review," *Sustain.*, vol. 10, no. 10, pp. 1–24, 2018.

[9] M. Lezzi, M. Lazoi, and A. Corallo, "Cybersecurity for Industry 4.0 in the current literature: a reference framework," *Comput. Ind.*, vol. 103, pp. 97–110, 2018.

[10] L. Thames and D. Schaefer, "Cybersecurity for Industry 4.0," 2017, pp. 1–33.

[11] K. Stouffer, S. Lightman, V. Pillitteri, M. Abrams, and A. Hahn, "The NIST definition of cloud computing," *Natl. Inst. Stand. Technol.*, p. 255, 2014.

[12] X. Fan, K. Fan, Y. Wang, and R. Zhou, "Overview of cyber-security of industrial control system," *2015 Int. Conf. Cyber Secur. Smart Cities, Ind. Control Syst. Commun. SSIC 2015 – Proc.*, pp. 1–7, 2015.

[13] K. David and M. Sc, "A guide to computer network security," *Choice Rev. Online*, vol. 51, no. 04, pp. 51-2126-51–2126, 2013.

[14] K. Kobara, "Cyber physical security for industrial control systems and IoT," *IEICE Trans. Inf. Syst. E99D*, no. 4, pp. 787–795, 2016.

[15] H. Flatt and S. Schriegel, "Analysis of the Cyber-Security of Industry 4.0 Technologies based on RAMI 4.0 and Identification of Requirements," in *2016 IEEE 21st International Conference on Emerging Technologies and Factory Automation (ETFA)*, 2016.

[16] L. Urquhart and D. Mcauley, "Avoiding the internet of insecure industrial things," *Comput. Law Secur. Rev. Int. J. Technol. Law Pract.*, vol. 34, no. 3, pp. 450–466, 2018.

[17] I. C. Systems and P. D. Mcauley, "Abstract: Keywords: Part I: Introduction Part II: From Exploration to Consumption," in *TILTing Perspectives 2017: Regulating a Connected World, Tilburg, Netherlands*, 2017, no. May, pp. 67–72.

[18] H. He, C. Maple, T. Watson, and A. Tiwari, "The Security Challenges in the IoT Enabled Cyber-Physical Systems and Opportunities for Evolutionary Computing & Other Computational Intelligence," in *2016 IEEE Congress on Evolutionary Computation (CEC)*, 2016, pp. 1015–1021.

[19] K. Dahbur, "A survey of risks, threats and vulnerabilities in cloud computing," 2011.

[20] A. Duncan, S. Creese, and M. Goldsmith, "Insider Attacks in Cloud Computing," in *2012 IEEE 11th International Conference on Trust*, Security and Privacy in Computing and Communications, 2012.

[21] P. Baybutt, "Assessing risks from threats to process plants: threat and vulnerability analysis," *Process Saf. Prog.*, no. December, pp. 269–275, 2002.

[22] M. Abomhara and G. M. Koien, "Cyber security and the internet of things: vulnerabilities, threats, intruders and attacks," *J. Cyber Secur. Mobil.*, vol. 4, no. 1, pp. 65–88, 2015.

[23] E. Bertino, L. Martino, F. Paci, and A. Squicciarini, Security for web services and service-oriented architectures. Springer Science & Business Media, 2009.

[24] C. I. No and R. May, "CNSS, National Information Assurance (IA) Glossary (CNSS Instruction no. 4009)," *http://www.cnss.gov/instructions.html*, 2006. [Online]. Available: https://www.hsdl.org/?abstract&did=7447.

[25] Q. J. Yeh and A. J. T. Chang, "Threats and countermeasures for information system security: A cross-industry study," *Inf. Manag.*, vol. 44, no. 5, pp. 480–491, 2007.
[26] A. Kankanhalli, H. H. Teo, B. C. Y. Tan, and K. K. Wei, "An integrative study of information systems security effectiveness," *Int. J. Inf. Manage.*, vol. 23, no. 2, pp. 139–154, 2003.
[27] H. A. M. Luiijf, K. Besseling, M. Spoelstra, and P. de Graaf, Ten national cyber security strategies: a comparison. International Workshop on Critical Information Infrastructures Security. Springer, Berlin, Heidelberg, 2011.
[28] G. A. Fink, T. W. Edgar, T. R. Rice, D. G. MacDonald, and C. E. Crawford, *Security and Privacy in Cyber-Physical Systems*. Elsevier Inc., 2016.
[29] B. Review, "Book review secrets & lies: digital security in a networked world," *Int. Hydrogr. Rev.*, vol. 2, no. 1, pp. 103–104, 2001.
[30] T. Pereira, L. Barreto, and A. Amaral, "Network and information security challenges within Industry 4.0 paradigm," *Procedia Manuf.*, vol. 13, pp. 1253–1260, 2017.
[31] M.-A. Kaufhold, C. Reuter, T. Ludwig, and S. Scholl, "Interoperability and Security Challenges of Industrie 4.0," in *Informatik 2017*, 2017, pp. 973–985.
[32] R. Petar, D. De Roure, J. R. C. Nurse, and R. Nicolescu, "Integration of cyber security frameworks, models and approaches for building design principles for the Internet-of-Things in Industry 4.0 research plan research methodology initiatives," in *Living in the Internet of Things: Cybersecurity of the IoT – 2018*, 2018, pp. 1–6.
[33] M. Haris and R. Z. Khan, "A Systematic Review on Cloud Computing," *Int. J. Comput. Sci. Eng.*, no. December, 2018.
[34] W. Gunawan, "Measuring information security and cybersecurity on private cloud computing," *J. Theor. Appl. Inf. Technol.*, vol. 96, no. 1, pp. 156–168, 2019.
[35] G. Peralta, M. Iglesias-Urkia, M. Barcelo, R. Gomez, A. Moran, and J. Bilbao, "Fog computing based efficient IoT scheme for the Industry 4.0," in *Proceedings of the 2017 IEEE International Workshop of Electronics, Control, Measurement, Signals and their Application to Mechatronics, ECMSM 2017*, 2017.
[36] C. C. Lin and J. W. Yang, "Cost-efficient deployment of fog computing systems at logistics centers in Industry 4.0," *IEEE Trans. Ind. Informatics*, vol. 14, no. 10, pp. 4603–4611, 2018.
[37] T. M. Fernández-Caramés, P. Fraga-Lamas, M. Suárez-Albela, and M. Vilar-Montesinos, "A fog computing and cloudlet based augmented reality system for the Industry 4.0 shipyard," *Sensors (Switzerland)*, vol. 18, no. 6, pp. 1–18, 2018.
[38] P. O'Donovan, C. Gallagher, K. Bruton, and D. T. J. O'Sullivan, "A fog computing industrial cyber-physical system for embedded low-latency machine learning Industry 4.0 applications," *Manuf. Lett.*, vol. 15, pp. 139–142, 2018.
[39] P. O'Donovan, "An industrial analytics methodology and fog computing cyber-physical system for Industry 4.0 embedded machine learning applications," 2018.

[40] T. Niesen, C. Houy, P. Fettke, and P. Loos, "Towards an integrative big data analysis framework for data-driven risk management in Industry 4.0," in *Proceedings of the Annual Hawaii International Conference on System Sciences*, 2016, vol. 2016–March, pp. 5065–5074.

[41] L. Da Xu and L. Duan, "Big data for cyber physical systems in Industry 4.0: a survey," *Enterp. Inf. Syst.*, vol. 13, no. 2, pp. 148–169, 2019.

[42] M. Khan, X. Wu, X. Xu, and W. Dou, "Big Data Challenges and Opportunities in the Big Data Challenges and Opportunities in the Hype," *IEEE Int. Conf.*, no. May, pp. 1–6, 2017.

[43] G. Manogaran, C. Thota, and D. Lopez, "Big Data Security Intelligence for Healthcare Industry 4.0," in *Cybersecurity for Industry 4.0.*, 2017, pp. 103–126.

[44] W. Lidong and W. Guanghui, "Big data in cyber-physical systems, digital manufacturing and Industry 4.0," *Int. J. Eng. Manuf.*, vol. 6, no. 4, pp. 1–8, 2016.

[45] J. Nilsson and F. Sandin, "Semantic Interoperability in Industry 4.0: Survey of Recent Developments and Outlook," in *Proceedings – IEEE 16th International Conference on Industrial Informatics, INDIN 2018*, 2018, pp. 127–132.

[46] M. Pai, "Interoperability between IIC architecture & Industry 4.0 reference architecture for industrial assets," *Infosys*, pp. 1–12, 2016.

[47] D. Morozov, M. Lezoche, H. Panetto, D. Morozov, M. Lezoche, and H. Panetto, "FCA modelling for CPS interoperability optimization in Industry 4.0," in *7th International Conference on Information Society and Technology, ICIST 2017*, 2017, pp. 55–59.

[48] R. M. De Salles, F. A Coda, J. R Silva, D. J. D. S. Filho, P. E. Miyagi, and F. Junqueira, "Requirements analysis for machine to machine integration within Industry 4.0," in *2018 13th IEEE International Conference on Industry Applications*, 2018.

[49] G. Pedone and I. Mezgár, "Model similarity evidence and interoperability affinity in cloud-ready Industry 4.0 technologies," *Comput. Ind.*, vol. 100, no. May, pp. 278–286, 2018.

[50] F. Lelli, "Interoperability of the time of Industry 4.0 and the Internet of Things," *Futur. Internet*, vol. 11, no. 2, p. 36, 2019.

[51] D. M. Vargas, A. F. Blanco, A. C. Vidaurre, *et al.*, "Deviation Testing: A Test Case Generation Technique for GraphQL APIs," pp. 1–9, 2018.

[52] M. Stoiber, "Securing Your GraphQL API from Malicious Queries," *https://blog.apollographql.com/securing-your-graphql-api-from-malicious-queries-16130a324a6b*, 2018.

[53] K. Sandoval, "Security Points to Consider Before Implementing GraphQL," *https://nordicapis.com/security-points-to-consider-before-implementing-graphql/*, 2017.

[54] Y. Guo, F. Deng, and X. Yang, "Design and implementation of real-time management system architecture based on GraphQL," *IOP Conf. Ser. Mater. Sci. Eng.*, vol. 466, no. 1, 2018.

[55] G. Mathew and Z. Obradovic, "Constraint graphs as security filters for privacy assurance in medical transactions," *ACM Trans. Embed. Comput. Syst.*, pp. 502–504, 2012.

[56] A. Kapitonov, S. Lonshakov, A. Krupenkin, and I. Berman, "Blockchain-based protocol of autonomous business activity for multi-agent systems consisting of UAVs," *2017 Work. Res. Educ. Dev. Unmanned Aer. Syst. RED-UAS 2017*, pp. 84–89, 2017.
[57] P. Pinheiro and R. Barbosa, "Highlights of practical applications of agents, multi-agent systems, and complexity: the PAAMS collection," in *Highlights Prac. Appl. Agents, Multi-Agent Syst.*, vol. 887, 2018, pp. 149–160.
[58] C. Lin, D. He, X. Huang, K. K. R. Choo, and A. V. Vasilakos, "BSeIn: a blockchain-based secure mutual authentication with fine-grained access control system for Industry 4.0," *J. Netw. Comput. Appl.*, vol. 116, no. March, pp. 42–52, 2018.
[59] W. Viriyasitavat, L. Da Xu, Z. Bi, and A. Sapsomboon, "Blockchain-based business process management (BPM) framework for service composition in Industry 4.0," *J. Intell. Manuf.*, pp. 1–12, 2018.
[60] T. M. Fernández-Caramés, O. Blanco-Novoa, M. Suárez-Albela, and P. Fraga-Lamas, "An UAV and Blockchain-based System for Industry 4.0 Inventory and Traceability Applications," in *International Electronic Conference on Sensors and Applications*, 2018, vol. 15.
[61] A. Kapitonov, I. Berman, S. Lonshakov, and A. Krupenkin, "Blockchain Based Protocol for Economical Communication in Industry 4.0," in *2018 Crypto Valley Conference on Blockchain Technology (CVCBT)*, 2018, pp. 41–44.
[62] G. Gabriel, R. Roy, and S. B. R. Kumar, "An Architecture to Enable Secure Firmware Updates on a Distributed-Trust IoT Network Using Blockchain," in *International Conference on Computer Networks and Communication Technologies*, 2018, pp. 671–679.
[63] M. A. Walker, A. Dubey, A. Laszka, and D. C. Schmidt, "PlaTIBART: a Platform for Transactive IoT Blockchain Applications with Repeatable Testing," in *Proceedings of the 4th Workshop on Middleware and Applications for the Internet of Things*, 2017, pp. 17–22.
[64] A. Dorri, S. S. Kanhere, R. Jurdak, and P. Gauravaram, "LSB: A Lightweight Scalable BlockChain for IoT Security and Privacy," *arXiv Prepr. arXiv1712.02969*, pp. 1–17, 2017.
[65] M. A. Ferrag, M. Derdour, M. Mukherjee, and A. Derhab, "Blockchain technologies for the Internet of Things: research issues and challenges," *IEEE Internet Things J.*, vol. PP, no. c, p. 1, 2018.
[66] A. Dorri, S. S. Kanhere, and R. Jurdak, "Blockchain in Internet of Things: Challenges and Solutions," *arXiv Prepr. arXiv1608.05187*, 2016.
[67] S. Huh, S. Cho, and S. Kim, "Managing IoT Devices using Blockchain Platform," in *2017 19th International Conference on Advanced Communication Technology (ICACT)*, 2017, pp. 464–467.
[68] A. Dorri, S. S. Kanhere, R. Jurdak, and P. Gauravaram, "Blockchain for IoT Security and Privacy: The Case Study of a Smart Home," in *2017 IEEE International Conference on Pervasive Computing and Communications Workshops (PerCom Workshops)*, 2017, pp. 618–623.

[69] M. T. Hammi, B. Hammi, P. Bellot, and A. Serhrouchni, "Bubbles of Trust: a decentralized Blockchain-based authentication system for IoT," *Comput. Secur.*, 2018.

[70] A. Panarello, N. Tapas, G. Merlino, F. Longo, and A. Puliafito, "Blockchain and IoT integration: a systematic survey," *Sensors*, vol. 18, no. 8, p. 2575, 2018.

[71] J. CEKIC and P. DUJANOVIC, "Software defined networking architecture, security and energy efficiency: a survey," *IEEE Commun. Surv. Tutor.*, vol. 19, no. 1, pp. 3–15, 2017.

[72] L. Thames and D. Schaefer, "Software-defined cloud manufacturing for Industry 4.0," *Procedia CIRP*, vol. 52, pp. 12–17, 2016.

[73] A. Tsuchiya, F. Fraile, I. Koshijima, A. Ortiz, and R. Poler, "Software defined networking firewall for Industry 4.0 manufacturing systems," *J. Ind. Eng. Manag.*, vol. 11, no. 2, pp. 318–333, 2018.

[74] Y. W. Ma, Y. C. Chen, and J. L. Chen, "SDN-enabled network virtualization for Industry 4.0 based on IoTs and cloud computing," in *International Conference on Advanced Communication Technology, ICACT*, 2017, pp. 199–202.

[75] M. C. Lucas-Estan, T. P.Raptis, M. Sepulcre, A. Passarella, C. Regueiro, and O. Lazaro, "A software defined hierarchical communication and data management architecture for Industry 4.0," in *2018 14th Annual Conference on Wireless On-demand Network Systems and Services (WONS)*, 2018, vol. 5.

[76] J. Liu, J. Wan, B. Zeng, Q. Wang, H. Song, and M. Qiu, "A scalable and quick-response software defined vehicular network assisted by mobile edge computing," *IEEE Commun. Mag.*, vol. 55, no. 7, pp. 94–100, 2017.

[77] J. Wan, S. Tang, Z. Shu, *et al.*, "Software-defined industrial Internet of Things in the context of Industry 4.0," *IEEE Sens. J.*, vol. 16, no. 20, pp. 7373–7380, 2016.

[78] M. Karakus and A. Durresi, "A survey: control plane scalability issues and approaches in software-defined networking (SDN)," *Comput. Networks*, vol. 112, pp. 279–293, 2017.

[79] R. Chaudhary, G. S. Aujla, S. Garg, N. Kumar, and J. J. P. C. Rodrigues, "SDN-enabled multi-attribute-based secure communication for smart grid in IIoT environment," *IEEE Trans. Ind. Informatics*, vol. 14, no. 6, pp. 2629–2640, 2018.

[80] H. R. Faragardi, M. Vahabi, H. Fotouhi, T. Nolte, and T. Fahringer, "An efficient placement of sinks and SDN controller nodes for optimizing the design cost of industrial IoT systems," *Softw. – Pract. Exp.*, vol. 48, no. 10, pp. 1893–1919, 2018.

[81] X. Li, D. Li, J. Wan, C. Liu, and M. Imran, "Adaptive transmission optimization in SDN-based industrial internet of things with edge computing," *IEEE Internet Things J.*, vol. 5, no. 3, pp. 1351–1360, 2018.

[82] W. P. Nwadiugwu and D.-S. Kim, "Energy-efficient Sensors in Data Centers for Industrial Internet of Things (IIoT)," in *2018 3rd International Conference On Internet of Things: Smart Innovation and Usages (IoT-SIU)*, 2018, pp. 1–6.

[83] S. Madhawa, P. Balakrishnan, and U. Arumugam, "Employing invariants for anomaly detection in software defined networking based industrial internet of things," *J. Intell. Fuzzy Syst.*, vol. 35, no. 2, pp. 1267–1279, 2018.

[84] G. Hatzivasilis, K. Fysarakis, O. Soultatos, I. Askoxylakis, I. Papaefstathiou, and G. Demetriou, "The industrial Internet of Things as an enabler for a circular economy Hy-LP: a novel IIoT protocol, evaluated on a wind park's SDN/NFV-enabled 5G industrial network," *Comput. Commun.*, vol. 119, no. February, 2018, pp. 127–137, 2018.

[85] J. Wang and D. Li, "Adaptive computing optimization in software-defined network-based industrial internet of things with fog computing," *Sensors (Switzerland)*, vol. 18, no. 8, 2018.

[86] P. Chi and M. Wang, "A Lightweight Compound Defense Framework Against Injection Attacks in IIoT," in *2018 IEEE Conference on Dependable and Secure Computing (DSC)*, 2018, pp. 1–8.

[87] R. F. Babiceanu and R. Seker, "Cyber resilience protection for industrial internet of things: a software-defined networking approach," *Comput. Ind.*, vol. 104, pp. 47–58, 2019.

[88] F. Xu, H. Y. B, S. Cui, and C. Zhao, "Software defined industrial network: architecture and edge offloading strategy," in *International Conference on Communications and Networking in China*, 2019, vol. 209, no. 61431008, pp. 46–56.

[89] R. Khondoker, P. Larbig, D. Scheuermann, F. Weber, and K. Bayarou, "Addressing Industry 4.0 Security by Software-Defined Networking," in *Guide to Security in SDN and NFV*, 2017, pp. 229–251.

[90] F. Han, M. Alkhathami, and R. Van Schyndel, "Biometric-Kerberos authentication scheme for secure mobile computing services," in *2013 6th International Congress on Image and Signal Processing (CISP)*, 2013, vol. 03, no. Cisp, pp. 1694–1698.

[91] P. P. Gaikwad, J. P. Gabhane, and S. S. Golait, "3-Level Secure Kerberos Authentication for Smart Home Systems Using IoT," in *2015 1st International Conference on Next Generation Computing Technologies (NGCT)*, 2015, no. September, pp. 262–268.

[92] N. B. Korade, B. Veeramallu, A. R. More, V. S. Potey, and S. G. Langhe, "Modified Kerberos for IoT by dynamic expansion," *Int. J. Pure Appl. Math.*, vol. 118, no. 24, pp. 1–13, 2018.

[93] Y. Mordecai and D. Dori, "Model-Based Protocol Engineering: Specifying Kerberos with Object-Process Methodology," in *2014 IEEE 28th Convention of Electrical & Electronics Engineers in Israel (IEEEI)*, 2014, pp. 1–5.

[94] P. P. Pereira, J. Eliasson, and J. Delsing, "An authentication and access control framework for CoAP-based Internet of Things," in *IECON Proceedings (Industrial Electronics Conference)*, 2014, no. D, pp. 5293–5299.

[95] D. Selvamani and V. Selvi, "A two-fold authentication mechanism for network security," *Asian J. Eng. Appl. Technol.*, vol. 7, no. 2, pp. 86–90, 2018.

[96] A. Wright, T. X. Us, and R. H. Ca, "Two-factor authentcation background," 2014.

[97] M. Rennhard, M. Tschannen, and T. Christen, "SecureSafe: A Highly Secure Online Data Safe Industrial Use Case," 2012.

[98] M. N. Zurich, "Optimization of Two-way Authentication Protocol in Internet of Things," University of Zurich, 2014.

[99] A. Borisov, "A novel approach for user authentication to industrial components using QR codes," in *Proceedings – International Computer Software and Applications Conference*, 2015, vol. 3, pp. 61–66.

[100] V. L. Shivraj, M. A. Rajan, M. Singh, and P. Balamuralidhar, "One Time Password Authentication Scheme Based on Elliptic Curves for Internet of Things (IoT)," in *2015 5th National Symposium on Information Technology: Towards New Smart World (NSITNSW)*, 2015, no. c, pp. 1–6.

[101] M. S. Hossain, G. Muhammad, S. M. M. Rahman, W. Abdul, A. Alelaiwi, and A. Alamri, "Toward end-to-end biomet rics-based security for IoT infrastructure," *IEEE Wirel. Commun.*, vol. 23, no. 5, pp. 44–51, 2016.

[102] F. Wang, Y. Xu, H. Zhang, Y. Zhang, and L. Zhu, "2FLIP: a two-factor lightweight privacy-preserving authentication scheme for VANET," *IEEE Trans. Veh. Technol.*, vol. 65, no. 2, pp. 896–911, 2016.

[103] P. Gope, A. K. Das, N. Kumar, and Y. Cheng, "Lightweight and physically secure anonymous mutual authentication protocol for real-time data access in industrial wireless sensor networks," *IEEE Trans. Ind. Informatics*, vol. PP, no. X, pp. 1–1, 2019.

[104] P. Gope, B. Sikdar, and S. Member, "Lightweight and privacy-preserving two-factor authentication scheme for IoT devices," *IEEE Internet Things J.*, vol. PP, no. X, p. 1, 2018.

[105] X. Li, J. Peng, J. Niu, F. Wu, J. Liao, and K. K. R. Choo, "A robust and energy efficient authentication protocol for industrial internet of things," *IEEE Internet Things J.*, vol. 5, no. 3, pp. 1606–1615, 2018.

[106] A. K. Das, M. Wazid, N. Kumar, A. V. Vasilakos, and J. J. P. C. Rodrigues, "Biometrics-based privacy-preserving user authentication scheme for cloud-based industrial Internet of Things deployment," *IEEE Internet Things J.*, vol. 5, no. 6, pp. 4900–4913, 2018.

[107] D. Wang, W. Li, and P. Wang, "Measuring two-factor authentication schemes for real-time data access in industrial wireless sensor networks," *IEEE Trans. Ind. Informatics*, vol. 14, no. 9, pp. 4081–4092, 2018.

[108] G. Xu, S. Qiu, H. Ahmad, *et al.*, "A multi-server two-factor authentication scheme with un-traceability using elliptic curve cryptography," *Sensors*, vol. 18, no. 7, pp. 2394, 2018.

[109] M. Wang, W. T. Zhu, S. Yan, and Q. Wang, "SoundAuth: secure zero-effort two-factor authentication based on audio signals," *2018 IEEE Conf. Commun. Netw. Secur. CNS 2018*, pp. 1–9, 2018.

[110] J. Srinivas, A. K. Das, M. Wazid, and N. Kumar, "Anonymous lightweight chaotic map-based authenticated key agreement protocol for industrial internet of things," *IEEE Trans. Dependable Secur. Comput.*, vol. PP, no. c, p. 1, 2018.

[111] S. Kumari, K.-K. R. Choo, F. Wu, X. Li, A. K. Sangaiah, and J. Niu, "A three-factor anonymous authentication scheme for wireless sensor networks in internet of things environments," *J. Netw. Comput. Appl.*, vol. 103, pp. 194–204, 2017.

[112] N. T. Nguyen, H. D. Le, and C. C. Chang, "Provably secure and efficient three-factor authenticated key agreement scheme with untraceability," *Int. J. Netw. Secur.*, vol. 18, no. 2, pp. 335–344, 2016.

[113] Y. H. Park and Y. H. Park, "Three-factor user authentication and key agreement using elliptic curve cryptosystem in wireless sensor networks," *Sensors (Switzerland)*, vol. 16, no. 12, 2016.

[114] P. K. Dhillon and S. Kalra, "A lightweight biometrics based remote user authentication scheme for IoT services," *J. Inf. Secur. Appl.*, vol. 34, pp. 255–270, 2017.

[115] P. Wang, F. Wei, B. Li, Y. Zhao, and W. Li, "Cryptanalysis and security enhancement of three authentication schemes in wireless sensor networks," *Wirel. Commun. Mob. Comput.*, vol. 2018, pp. 1–11, 2018.

[116] X. Jia, D. He, L. Li, and K. K. R. Choo, "Signature-based three-factor authenticated key exchange for internet of things applications," *Multimed. Tools Appl.*, vol. 77, no. 14, pp. 18355–18382, 2018.

[117] E. W. B, A. Cha, and J. A. Laredo, "Generating GraphQL-Wrappers for REST (-like) APIs," in *Web Engineering*, vol. 1, Springer International Publishing, 2018, pp. 65–83.

[118] M. Waidner and M. Kasper, "Security in industrie 4.0 – challenges and solutions for the fourth industrial revolution," in *2016 Design, Automation & Test in Europe Conference & Exhibition (DATE)*, 2016, pp. 1303–1308.

[119] H. Li, Y. Niu, J. Yi, and H. Li, "Securing offline delivery services by using kerberos authentication," *IEEE Access*, vol. 6, pp. 40735–40746, 2018.

[120] F. Al-Ayed, C. Hu, and H. Liu, "An efficient practice of privacy implementation: Kerberos and Markov chain to secure file transfer sessions," *Int. J. Netw. Secur.*, vol. 20, no. 4, p. 7, 2018.

[121] B. Alan Batke, R. Joseph Baier, and D. Michael Callaghan, "System and methodology providing mult-ter security for network pendust," 2012.

[122] Y. Ma and H. Ning, "The improvement of wireless LAN security authentication mechanism based on Kerberos," *2018 Int. Conf. Electron. Technol. ICET 2018*, pp. 392–397, 2018.

[123] A. Sinclair and C. Heights, "Industrial protocol system authentication and firewall," 2015.

[124] A. P. Id and N. Tapas, *Blockchain and IoT Integration: A Systematic Survey*. 2018.

[125] H. I. Kobo, A. M. Abu-Mahfouz, and G. P. Hancke, "A survey on software-defined wireless sensor networks: challenges and design requirements," *IEEE Access*, vol. 5, pp. 1872–1899, 2017.

[126] H. Flatt, S. Schriegel, J. Jasperneite, H. Trsek, and H. Adamczyk, "Analysis of the cyber-security of Industry 4.0 technologies based on RAMI 4.0 and identification of requirements," in *IEEE International Conference on*

Emerging Technologies and Factory Automation, ETFA, 2016, vol. 2016–November.

[127] L. Thames and D. Schaefer, "Cybersecurity for Industry 4.0 and advanced manufacturing environments with ensemble intelligence," in *Cybersecurity for Industry 4.0.*, Springer, Cham, 2017, pp. 243–265.

[128] D. T. Sullivan, "Survey of malware threats and recommendations to improve cybersecurity for industrial control systems version 1.0," 2015.

Chapter 15

IoT-based data acquisition monitoring system for solar photovoltaic panel

Ranjit Singh Sarban Singh[1], Muhammad Izzat bin Nurdin[1], Wong Yan Chiew[1] and Tan Chee Fai[2]

Internet of things (IoT) is known as the extension of Internet connectivity for physical devices which is used daily. With this advantage, IoT technology is being used widely for many applications, and one that widely being implemented is for the solar system. As the solar system technology has advanced toward an effective real-time monitoring, elements such as data acquisition and monitoring are very important to deeply understand the installed solar system. To effectively monitor the installed solar system in real-time condition, it is also important to individually monitor each component such as solar photovoltaic panel, charge controller, inverter system and others.

Hence, this research explains about the IoT-based data acquisition monitoring system for solar photovoltaic panel for a solar system. The IoT-based data acquisition monitoring system for solar photovoltaic panel consists of four units of thermocouple (TC) sensors integrated with MAX31855 amplifier, one unit of INA 219 DC current/voltage sensor and Raspberry Pi Zero Wireless device as device manager. The proposed IoT-based data acquisition monitoring system is developed to sense, measure and calculate the current, voltage, power and temperature. This information is stored into the SD card of the Raspberry Pi Zero Wireless which is known as the device manager. The stored current, voltage, power and temperature information in SD card Raspberry Pi Zero Wireless is then transferred to the cloud storage wirelessly. This information is then extracted into self-developed website. The information can be easily viewed at the website and helps the consumers to monitor the installed solar system, especially the performances of solar photovoltaic panel. Monitoring the solar photovoltaic panel in real time using the

[1]Advanced Sensors and Embedded Control Research Group (ASECS), Centre for Telecommunication, Research and Innovation (CeTRI), Department of Computer Engineering, Fakulti Kejuruteraan Elektronik dan Kejuruteraan Komputer (FKEKK), Universiti Teknikal Malaysia Melaka (UTeM), Hang Tuah Jaya, Melaka, Malaysia

[2]Department of Mechatronics & Biomedical Engineering, Lee Kong Chian Faculty of Engineering & Science, Universiti Tunku Abdul Rahman, Selangor, Malaysia

IoT-based data acquisition monitoring system can effectively facilitate a system-level maintenance and immediate fault-detection can be performed.

15.1 System design and development

This chapter explains about the methodological process of the proposed project system design, integration and installation. This project methodology is divided into two sections: (a) hardware system design and development and (b) software design and development.

In the hardware system design and development, the project methodology is explained starting from the conceptual idea till the hardware system is realised. And, for the software design and development, an IoT architecture is implemented to gather required information and cloud-based user interface is designed and developed to monitor all the information or produced output from the developed hardware system.

15.1.1 Conceptual TC sensors placement, system design, integration and installation

Figure 15.1 shows the top view of the conceptual TC sensor placement, system design, integration and installation on a solar panel module. The overall system also consists of a DC-to-AC power inverter, battery as energy storage system (ESS), solar charge controller (SCC) and load to test the system functionability.

Referring to Figure 15.1 and Table 15.1, four units of TC sensors are integrated with MAX 31855 amplifier, which is to improve the sensitivity and measure the solar panel module temperature. The voltage and current sensor, which is also known as DC current sensor, is integrated at the output solar panel module, especially at the positive wire of solar panel module to sense and measure the voltage and current output from the solar panel module. Raspberry Pi Zero Wireless, which is known as single-board system, integrates all the four TC sensors integrated with

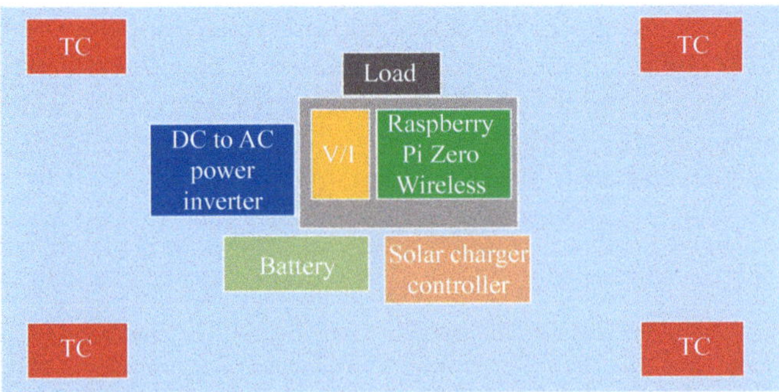

Figure 15.1 Conceptual TC sensors placement, system design, integration and installation

Table 15.1 List of components for system development

Items	Quantity
Thermocouple (TC) Sensor	4
Voltage/Current (V/I) Sensor (INA 219)	1
Raspberry Pi Zero Wireless	1
Battery	1
Solar Charger Controller	1
DC to AC Power Inverter	1
Load (Lamp)	1
Solar Panel Module	1

MAX 31855 amplifier and DC current sensor (voltage and current sensor). The temperature data, voltage and current data from the integrated sensors are recorded onto the Raspberry Pi Zero Wireless micro SD card before this data is sent to the cloud storage system.

Figure 15.2 shows the overall system architecture design which has four TC sensors integrated with MAX 31855 amplifier and INA 219 DC current sensor. Other devices connected are SCC, DC-to-AC power inverter, battery – ESS. Table 15.2 shows the wiring colours for the system shown in Figure 15.2.

Table 15.3 represents the port numbering for the SCC model Z10 shown in Figure 15.2. The port numbering at SCC model Z10 is to allow the SCC connectivity to the battery – ESS and load.

In this section, the detailed explanation of each component connectivity is described, referring to Figure 15.2. The +12 V port B output of the solar panel module is input into port B at INA 219 DC current sensor and the output port A at the INA 219 DC sensor is connected to input port A battery – ESS. The −12 V port C of battery – ESS is connected to the negative port C solar panel module.

Table 15.4 explains the INA 219 voltage/current sensor output connection to the Raspberry Pi Zero Wireless. The port numbering at the INA 219 voltage/current sensor is referred to the ports at the Raspberry Pi Zero Wireless. Therefore, the port 4 – VCC and port 34 GND at the INA voltage/current sensor is connected to port 4 – 3.3 V and port 34 – GND at the Raspberry Pi Zero Wireless. The port 3 – SDA (serial information) and port 5 serial clock line (SCL) at the INA 219 voltage/current sensor is connected to port 3 – GPIO 02 and port 5 – GPIO 03 at the Raspberry Pi Zero Wireless.

15.1.2 Hardware: sensors system design, integration and installation

Table 15.5 explains about the TC sensors (TC1, TC2, TC3 and TC4) integrated with MAX31855 Amplifier output ports connectivity to the ports at Raspberry Pi Zero Wireless. Referring to Figure 15.2, TC1, TC2, TC3 and TC4 TC sensors are

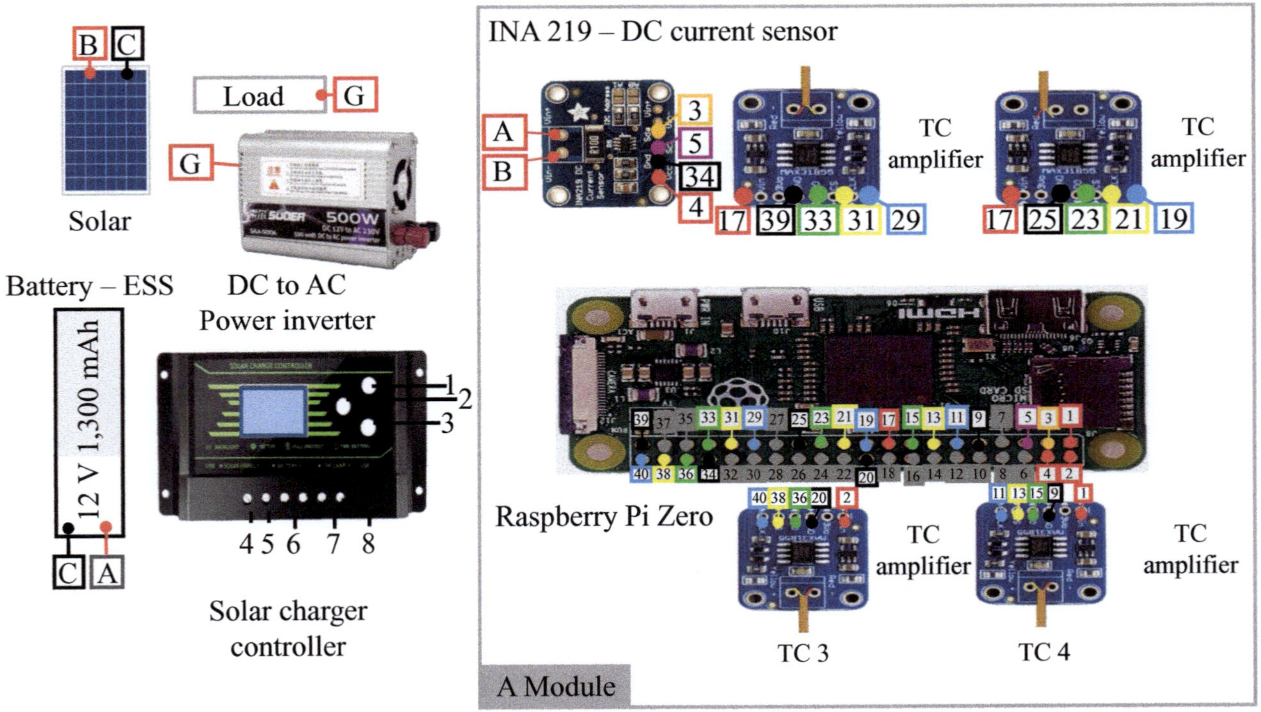

Figure 15.2 TC sensors integrated with MAX31855

Table 15.2 Wiring colours

Wiring colour	Description
	VCC – Power supply 3.3 V, 5 V, 12 V
	GND – Ground
	DO – Digital output
	CS – Chip select
	CLK – Clock
	SCL – Serial data
	SDA – Serial clock

Table 15.3 Solar charge controller model Z10

Ports	Name	Item	Name
1	Plus	5	Solar panel terminal
2	Set up	6	Battery terminal
3	Minus	7	Local terminal
4	USB 5 V	8	USB 5 V

Table 15.4 Connection of INA 219 DC current/voltage sensor

INA 219 DC current/voltage sensor (port description)	Raspberry Pi Zero Wireless – ports	Port description
VCC – Port 4	4	3.3 V
GND – Port 34	34	GND
SDA (serial information) – Port 3	3	GPIO 02
SCL (serial clock line) – Port 5	5	GPIO 03

connected at the input ports of MAX31855 Amplifier. According to Table 15.5, ports 33, 31, 29 of TC1 MAX31855 Amplifier are connected to ports 33 (GPIO 22), 31 (GPIO 27), and 29 (GPIO 17) at the Raspberry Pi Zero Wireless. TC2 MAX31855 Amplifier 23, 21 and 29 ports are connected to ports 23(GPIO 11), 21 (GPIO 09) and 29 (GPIO 10) ports at the Raspberry Pi Zero Wireless. TC3 MAX31855 Amplifier 23, 21 and 29 ports are connected to ports 36 (GPIO 23), 38 (GPIO 24) and 40 (GPIO 18) ports at the Raspberry Pi Zero Wireless. TC4 MAX31855 Amplifier 23, 21 and 29 ports are connected to ports 15 (GPIO 07), 13 (GPIO 12) and 9 (GPIO 08) ports at the Raspberry Pi Zero Wireless.

Table 15.5 Connection of TC sensor integrated with MAX3185 amplifier

Components	Amplifier MAX31855 – ports	Ports description	Raspberry Pi Zero Wireless – ports	Ports – GPIO
TC 1	33	DO (Data output)	33	GPIO 22
	31	CS (Chip select)	31	GPIO 27
	29	CLK (Clock)	29	GPIO 17
TC 2	23	DO (Data output)	23	GPIO 11
	21	CS (Chip select)	21	GPIO 09
	19	CLK (Clock)	19	GPIO 10
TC 3	36	DO (Data output)	36	GPIO 23
	38	CS (Chip select)	38	GPIO 24
	40	CLK (Clock)	40	GPIO 18
TC 4	15	DO (Data output)	15	GPIO 07
	13	CS (Chip select)	13	GPIO 12
	9	CLK (Clock)	9	GPIO 08
All MAX 31855	17, 1, 2	V_{in}	17, 1, 2	3.3 V
	39, 25, 20, 9	GND	39, 25, 20, 9	GND

Table 15.5 presents the connectivity summary for the proposed system in Figure 15.2. This information is used to develop the actual hardware for which the development methodology is discussed in the following sections.

15.2 Software design and development

This section discusses about the software design and development for the proposed hardware system and a graphical user interface (GUI). Two types of software are developed for the proposed hardware system. One is the embedded software design developed for the physical hardware system. The embedded software design is a program or also known as the operating system for the proposed hardware to operate as desired. It also connects all the other sub-components to the Raspberry Pi Zero Wireless board. The other software development is known as website

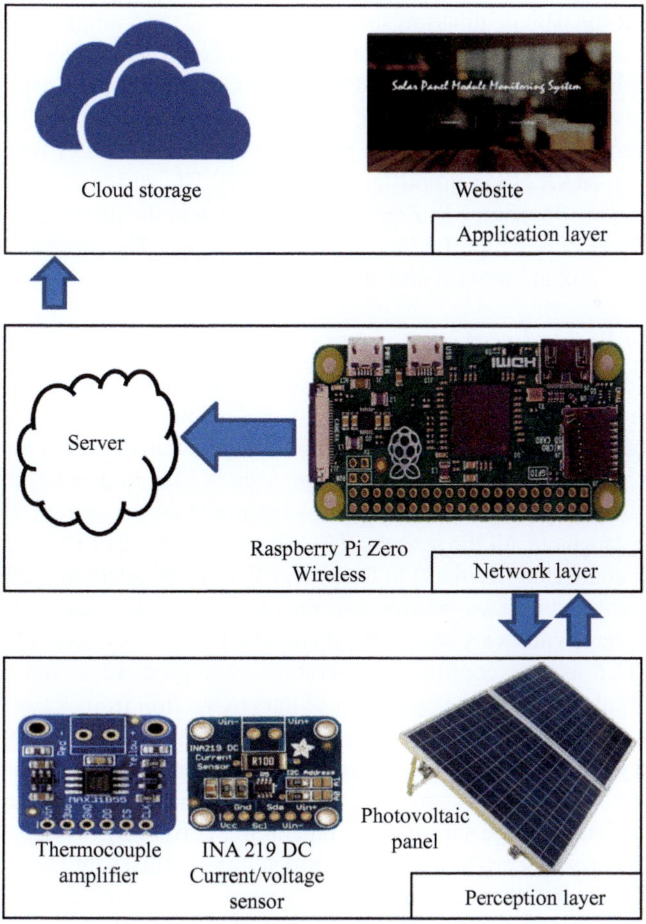

Figure 15.3 Embedded software design integration – Raspberry Pi Zero Wireless hardware system and website

development which is to create GUI and to create engagement between the users and proposed hardware system

15.2.1 Embedded software design integration – Raspberry Pi Zero Wireless hardware system and website

The embedded software is an important aspect to create a bridging connection between the developed hardware and developed GUI. Proper design of the embedded software for both hardware operation and website GUI can extract the required information such as temperature, current and voltage of a solar photovoltaic panel. Hence, the three-layer IoT architecture as shown in Figure 15.3 is employed for this research. The three layers are: (a) perception, (b) network and (c) application. The three-layer IoT architectures are generally used frameworks to provide an appropriate control at device, data storage and process management control.

Perception layer is the first layer shown in Figure 15.3. The perception layer explains about the data acquisition sensors, TC MAX31855 amplifier and INA 219 DC current/voltage. All sensors are connected and controlled by a device manager, which is Raspberry Pi Zero Wireless. The network adapter in the Raspberry Pi Zero Wireless is used to create a network connectivity to the network layer for data sending. Referring to Figure 15.3, the perception layer consists of four TC sensors integrated with MAX31855 amplifier, one INA219 DC current/voltage sensor. At the network layer, Raspberry Pi Zero Wireless connects the proposed system to the Internet to secure connection to the cloud storage server. The network connectivity must be at 2.4 GHz at strength and stable.

15.2.2 Embedded software design

This section explains about the embedded software design developed for the sensors and hardware system operation in Figure 15.2. Referring to Figure 15.4, the designed embedded software for the Raspberry Pi Zero Wireless initialised the Raspbian operating system. Then, phyton programming language is used to develop the embedded software for the Raspberry Pi Zero Wireless input and output system. Using phyton programming language, all the sensors (TC sensors and INA 219 DC current/voltage) are internally connected (as input) to Raspberry Pi Zero Wireless to perform the temperature, current and voltage sensing and measurement. The measured temperature, current and voltage values are saved into Raspberry Pi Zero Wireless SD card storage.

Figure 15.5 is the continuous operation of Figure 15.4; the operation in Figure 15.5 explains about the operation of data extraction from Raspberry Pi Zero Wireless SD card storage system to the cloud storage system and website. Initially the Raspberry Pi Zero Wireless system initialised the connectivity to the cloud storage server. Upon stabilising the connectivity between Raspberry Pi Zero Wireless and cloud storage server, the temperature, current and voltage information saved in Raspberry Pi Zero Wireless SD card storage is then sent and saved into the cloud storage server. Otherwise, the connectivity is terminated, and re-initialisation is required. The stored temperature, current and voltage information at the cloud

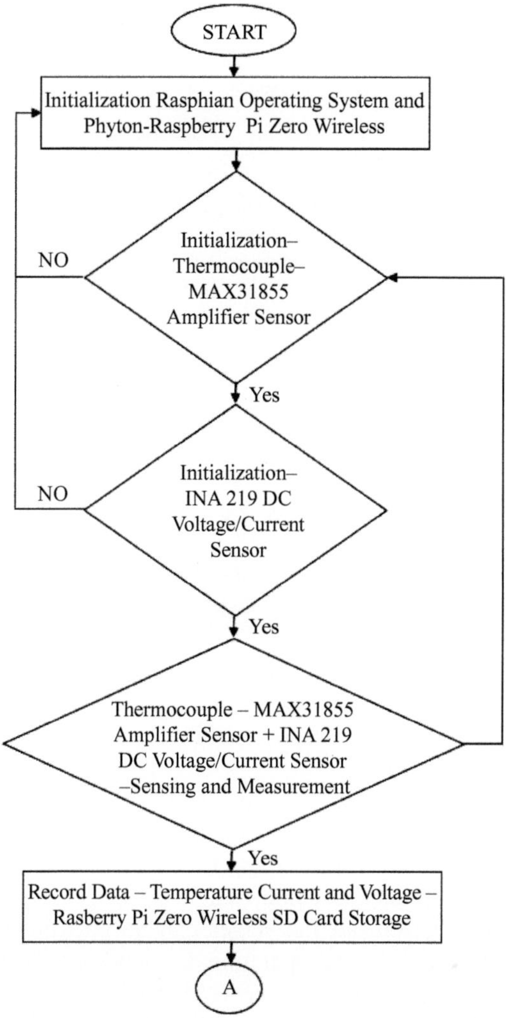

Figure 15.4 Embedded software design – sensors and hardware system operation

storage server is then sent to the website for viewing. At the website, consumers can individually monitor each one solar photovoltaic panel performances. Any abnormality can be easily detected, and consumers can lodge report to the maintenance team to perform the maintenance or replacement.

This section explains about the methodological design and developed hardware of IoT-based data acquisition monitoring system to analyse the solar photovoltaic panel performance. The process of design and development is initiated with a conceptual idea, and then expand it to actual hardware design and development. Upon completion of actual hardware development, embedded software design for

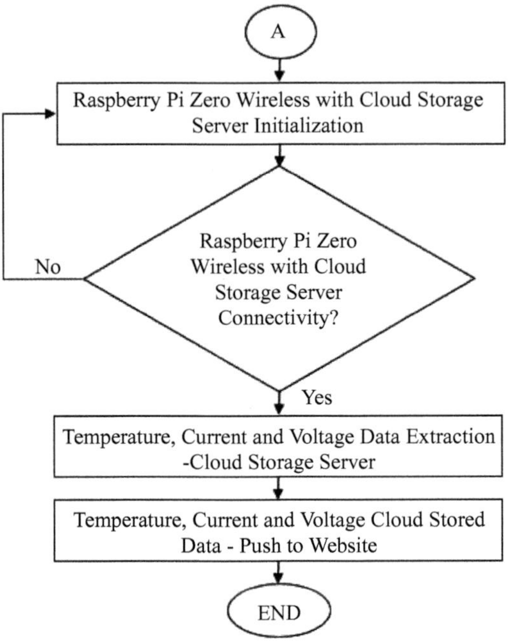

Figure 15.5 Embedded software design – Raspberry Pi Zero Wireless SD card storage to cloud and website storage system

hardware operation, cloud storage and website is initiated. First, embedded software operational program for hardware is developed, and then upon establishment of embedded software operational program, the embedded software for website is initiated. All this development is conducted in stages because each stage must operate as it is desired before all the stages can be integrated into one system. Finally, after validating each stage operational, all the stages are combined, and the overall system operation and functionality is analysed and validated based on the recorded results. Therefore, in the next section, the results and discussion of the overall system operation and functionality is presented.

15.3 Results and discussion

This section presents the captured results based on the developed IoT-based data acquisition monitoring system to analyse the solar photovoltaic panel performance.

15.3.1 Hardware system design and development

Table 15.6 presents the numbered components in Figure 15.6. Figure 15.6 is showing the arrangement of the components to develop the IoT-based data acquisition monitoring system.

Table 15.6 Components numbering for hardware system design and development

Numbering	Components
1	Solar panel module
2	DC-to-AC power inverter
3	Load (Lamp)
4	MAX31855 TC amplifier sensors
5	Solar charger controller
6	Rechargeable battery 12 V
7	TC sensor

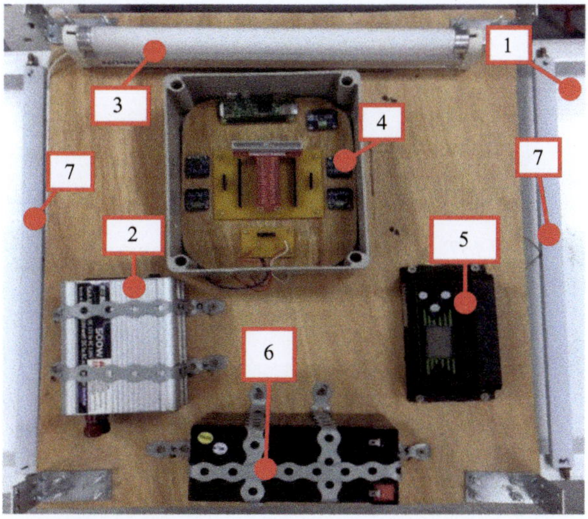

Figure 15.6 Complete numbering and arrangement of hardware design and development without wires connection

Figure 15.7 shows the developed IoT-based data acquisition monitoring system with wires connected. The wires connectivity in Figure 15.7 can be referred to Figure 15.2, Tables 15.2–15.5.

As for solar photovoltaic panel connectivity and INA 219 DC current/voltage sensor to the SCC, Figure 15.2 is referred. The wires connectivity if referred to Tables 15.3 and 15.4. Also, MAX31855 TC amplifier which is integrated to TC sensors and Raspberry Pi Zero Wireless wire connectivity can be referred to Table 15.5.

Figure 15.8 shows the completely developed IoT-based data acquisition monitoring system to analyse the solar photovoltaic panel. The completed hardware shown in Figure 15.8 is inclusive of all the wires connected and is ready for field testing.

Figure 15.7 Developed IoT-based data acquisition monitoring system hardware – wires connection

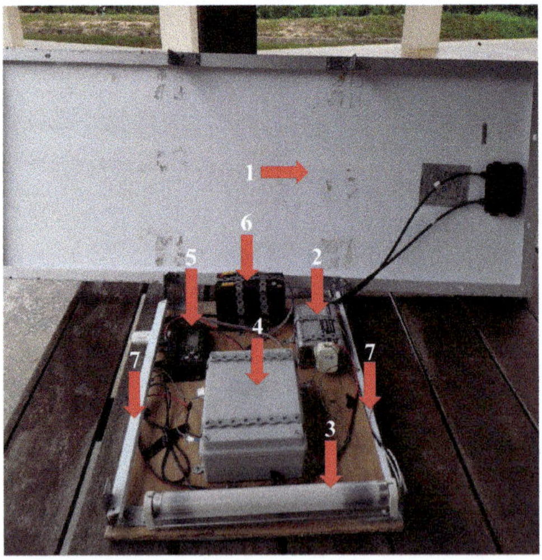

Figure 15.8 Completed IoT-based data acquisition monitoring system integrated with solar photovoltaic panel

Table 15.7 List of components for completed IoT-based data acquisition monitoring system

Number	Hardware
1	Solar panel
2	DC-to-AC power inverter
3	Lamp (Load)
4	System module
5	Solar charger
6	Rechargeable battery 12 V – ESS
7	Thermocouple

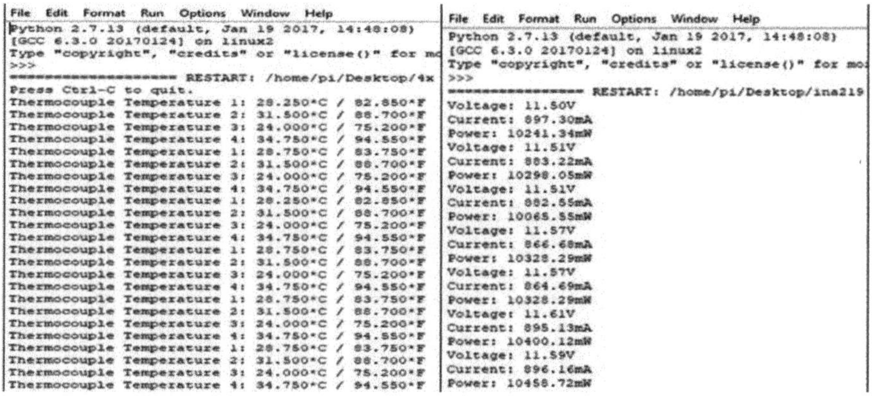

Section A Section B

Figure 15.9 Section A – temperature and Section B – current–voltage

15.3.2 Embedded software design and development

Figure 15.9 shows captured results of the TC sensors and INA 219 DC current/voltage sensor using the Phyton shell in Raspbian Raspberry Pi Zero Wireless operating system. There are two sections, Section A and Section B, shown in Figure 15.9. Section A presents the temperature sensed and reading of TC sensors, while Section B presents the current and voltage reading from INA 219 DC current/voltage sensor. The power is calculated using the current and voltage values in Section B.

Figure 15.10 and Table 15.8 show the temperature reading captured on 1 May 2019 for all the four TC sensors. Based on the plotted reading in Figure 15.10, there are three changing phases. The first phase is from 8 a.m. to 1 p.m. The temperature has slowly increased based on the temperature increment in the weather condition. Second phase is from 1 p.m. to 4 p.m., and the temperature is constant because the weather temperature has reached its peak temperature. And, the last phase is from 4 p.m. to 6 p.m. where the temperature slowly decreases due to the sun-setting

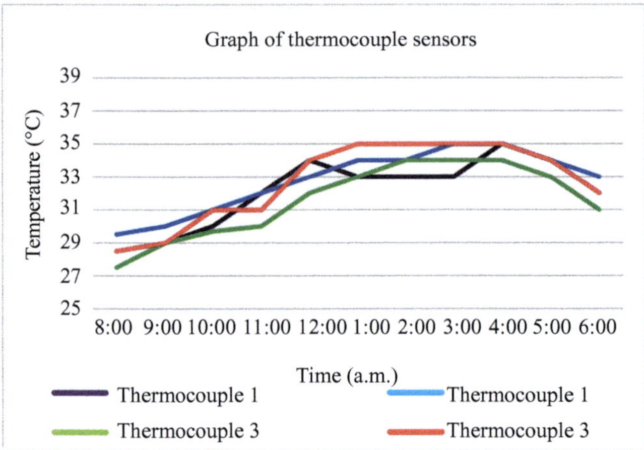

Figure 15.10 TC sensors temperature reading

condition. The temperature analysis is an important factor to understand the maximum output current and voltage when the sun is producing the highest temperature possible. This is analysed in the following section.

Figures 15.11 and 15.12 and Table 15.9 show the sensed and measure values of the current and voltage using the INA 219 DC current/voltage sensor. Figure 15.13 and Table 15.9 show the calculated power using (15.1) using the embedded software programming. Based on the plotted data in Figure 15.11, the measured current value is between 0.8 A and 2.92 A. It shows that in the morning the solar photovoltaic system produces low current due to low sunlight intensity; while the sunlight intensity increases, the solar photovoltaic panel current also increases. The current decreases again when the sun starts to set in the evening. This shows that the INA 219 DC current/voltage sensor can effectively sense and measure the current and voltage output of a solar photovoltaic panel. Finally, the power is calculated using (15.1).

$$\text{Voltage (V)} \times \text{current (A)} = \text{power (W)} \tag{15.1}$$

15.3.3 Cloud/database monitoring system

Figure 15.14 shows the main database interface which contains information about temperature, current, voltage and power. Referring to Figure 15.14, the section highlighted with blue box is known as localhost section which is to create the parameters name such as temperature, current, voltage and power. The section highlighted with red box as shown in Figure 15.14 is tables created to store the values of temperature, current, volte and power.

Figure 15.15 shows the temperature database section which contains a table to record the time, temperature of sensor 1 (Temp 1), temperature of sensor 2 (Temp 2), temperature of sensor 3 (Temp 3) and temperature of sensor 4 (Temp 4).

Table 15.8 TC sensors temperature reading

Time (h)	Thermocouple 1 (°C)	Thermocouple 2 (°C)	Thermocouple 3 (°C)	Thermocouple 4 (°C)
8:00 a.m.	28.5	29.5	27.5	28.5
9:00 a.m.	29	30	29	29
10:00 a.m.	30	31	29.7	31
11:00 a.m.	32	32	30	31
12:00 p.m.	34	33	32	34
1:00 p.m.	33	34	33	35
2:00 p.m.	33	34	34	35
3:00 p.m.	33	35	34	35
4:00 p.m.	35	35	34	35
5:00 p.m.	34	34	33	34
6:00 p.m.	32	33	31	32

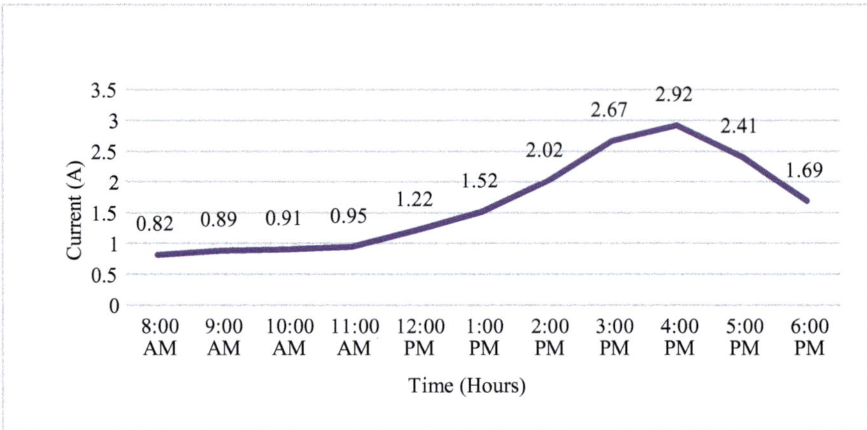

Figure 15.11 Current – INA 219 DC current/voltage sensor

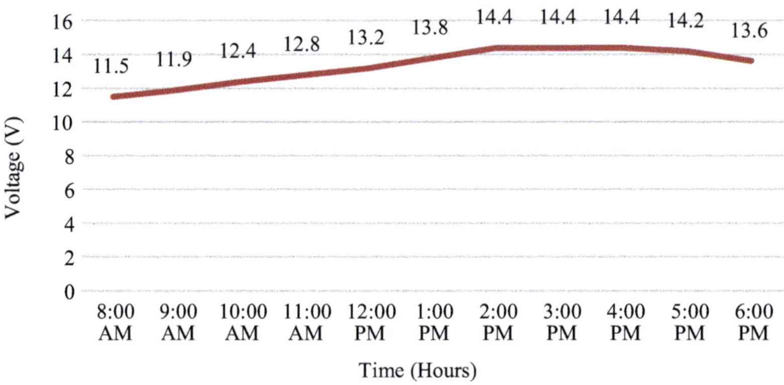

Figure 15.12 Voltage – INA 219 DC current/voltage sensor

The temperature reading recorded in temperature database presented in Figure 15.15 is extracted from Figure 15.10 and Table 15.8, which is stored in the SD card of Raspberry Pi Zero Wireless system shown in Figure 15.8.

Figure 15.16 shows the measured current, voltage and calculated power using (15.1) in the current, voltage and power database section. The current and voltage is sensed and measured by the INA 219 DC current/voltage sensor attached to the developed IoT-based data acquisition monitoring system, while the power is calculated using (15.1). The current, voltage and power recorded into the database are

Table 15.9 Measured voltage, current and calculated power

Time (h)	Voltage (V)	Current (A)	Power (W)
8:00 a.m.	11.5	0.82	9.20
9:00 a.m.	11.9	0.89	10.6
10:00 a.m.	12.4	0.91	11.3
11:00 a.m.	12.8	0.95	12.2
12:00 p.m.	13.2	1.22	16.1
1:00 p.m.	13.8	1.52	21.0
2:00 p.m.	14.4	2.02	29.0
3:00 p.m.	14.4	2.67	38.5
4:00 p.m.	14.4	2.92	42.2
5:00 p.m.	14.2	2.41	34.1
6:00 p.m.	13.6	1.69	22.9

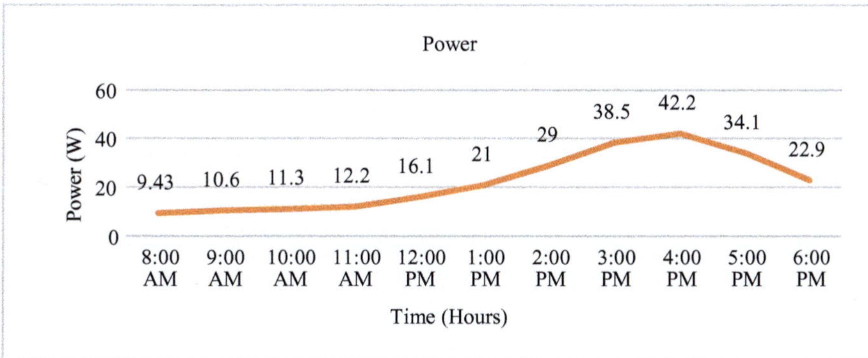

Figure 15.13 Calculated power – Equation (15.1)

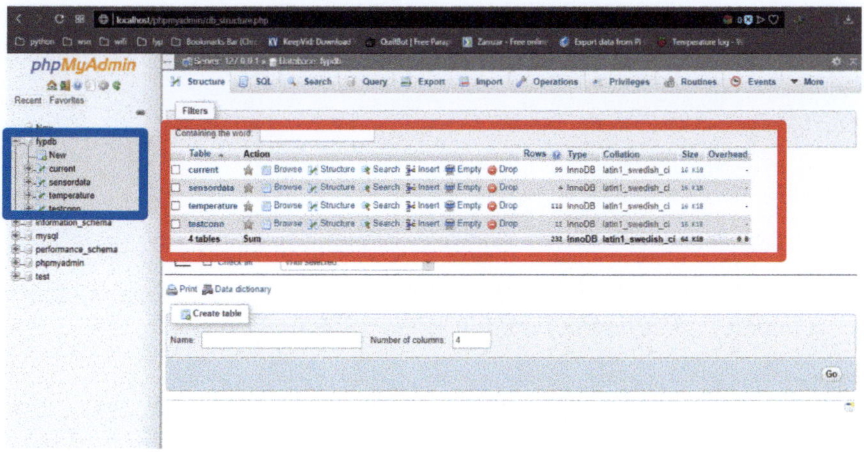

Figure 15.14 Temperature, current, voltage and power – localhost database

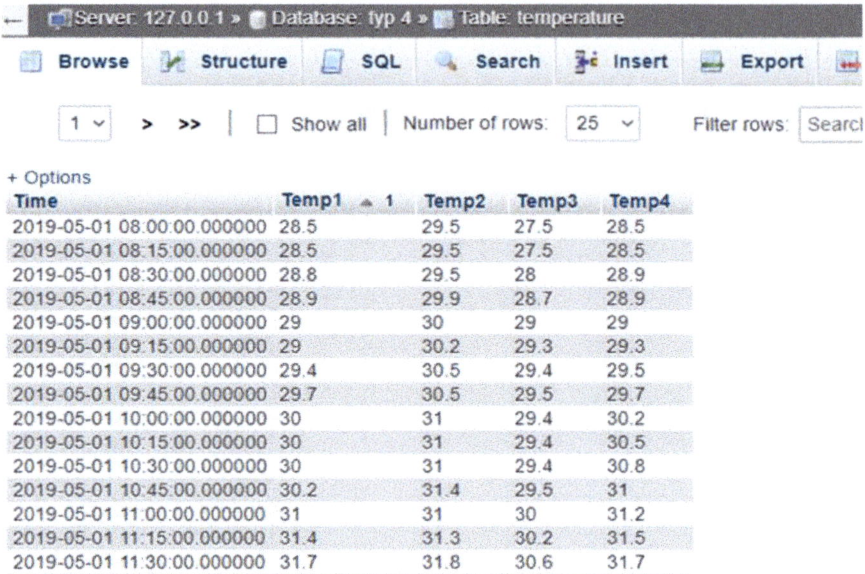

Time	Temp1	Temp2	Temp3	Temp4
2019-05-01 08:00:00.000000	28.5	29.5	27.5	28.5
2019-05-01 08:15:00.000000	28.5	29.5	27.5	28.5
2019-05-01 08:30:00.000000	28.8	29.5	28	28.9
2019-05-01 08:45:00.000000	28.9	29.9	28.7	28.9
2019-05-01 09:00:00.000000	29	30	29	29
2019-05-01 09:15:00.000000	29	30.2	29.3	29.3
2019-05-01 09:30:00.000000	29.4	30.5	29.4	29.5
2019-05-01 09:45:00.000000	29.7	30.5	29.5	29.7
2019-05-01 10:00:00.000000	30	31	29.4	30.2
2019-05-01 10:15:00.000000	30	31	29.4	30.5
2019-05-01 10:30:00.000000	30	31	29.4	30.8
2019-05-01 10:45:00.000000	30.2	31.4	29.5	31
2019-05-01 11:00:00.000000	31	31	30	31.2
2019-05-01 11:15:00.000000	31.4	31.3	30.2	31.5
2019-05-01 11:30:00.000000	31.7	31.8	30.6	31.7

Figure 15.15 Temperature sensors temperature database

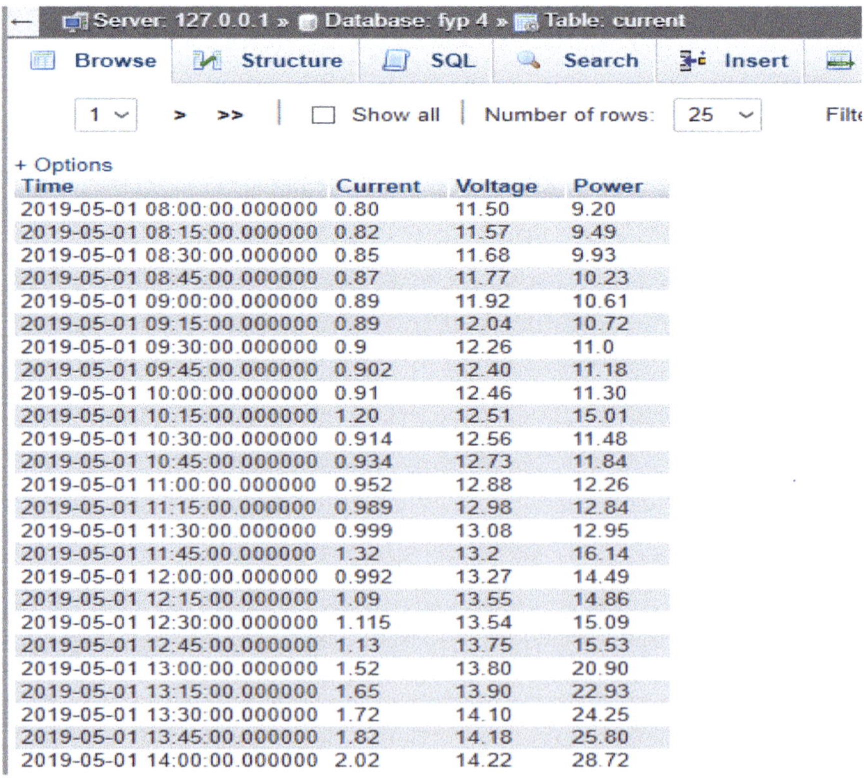

Time	Current	Voltage	Power
2019-05-01 08:00:00.000000	0.80	11.50	9.20
2019-05-01 08:15:00.000000	0.82	11.57	9.49
2019-05-01 08:30:00.000000	0.85	11.68	9.93
2019-05-01 08:45:00.000000	0.87	11.77	10.23
2019-05-01 09:00:00.000000	0.89	11.92	10.61
2019-05-01 09:15:00.000000	0.89	12.04	10.72
2019-05-01 09:30:00.000000	0.9	12.26	11.0
2019-05-01 09:45:00.000000	0.902	12.40	11.18
2019-05-01 10:00:00.000000	0.91	12.46	11.30
2019-05-01 10:15:00.000000	1.20	12.51	15.01
2019-05-01 10:30:00.000000	0.914	12.56	11.48
2019-05-01 10:45:00.000000	0.934	12.73	11.84
2019-05-01 11:00:00.000000	0.952	12.88	12.26
2019-05-01 11:15:00.000000	0.989	12.98	12.84
2019-05-01 11:30:00.000000	0.999	13.08	12.95
2019-05-01 11:45:00.000000	1.32	13.2	16.14
2019-05-01 12:00:00.000000	0.992	13.27	14.49
2019-05-01 12:15:00.000000	1.09	13.55	14.86
2019-05-01 12:30:00.000000	1.115	13.54	15.09
2019-05-01 12:45:00.000000	1.13	13.75	15.53
2019-05-01 13:00:00.000000	1.52	13.80	20.90
2019-05-01 13:15:00.000000	1.65	13.90	22.93
2019-05-01 13:30:00.000000	1.72	14.10	24.25
2019-05-01 13:45:00.000000	1.82	14.18	25.80
2019-05-01 14:00:00.000000	2.02	14.22	28.72

Figure 15.16 Voltage–current–power INA 219 DC current/voltage sensor

extracted from Figures 15.11–15.13 and Table 15.9. These values are stored in the SD card of Raspberry Pi Zero Wireless System, as shown in Figure 15.8.

15.3.4 IoT-based data acquisition monitoring system webpage –localhost

Figure 15.17 shows the GUI webpage for the IoT-based data acquisition monitoring system for analysing solar photovoltaic panel's current, voltage, power and temperature. The GUI webpage is used to present the sensed, measured and recorded information of current, voltage, power and temperature of a solar photovoltaic panel. All the current, voltage, power and temperature information shown in Figures 15.10–15.13 and Tables 15.8 and 15.9 are stored in the SD card of Raspberry Pi Zero Wireless system, as shown in Figure 15.8. Then, this information is wirelessly transferred to the localhost of a computer where the localhost acts as a server to store this information. The localhost also is equipped with the webpage as shown in Figure 15.17, which allows the consumers to monitor or check on the solar photovoltaic panel performances. The main page of the webpage shows the temperature and current/voltage section which can be easily accessed. The consumers can click on the temperature or current/voltage section to view the temperature of the solar photovoltaic panel, and consumers can also click on the current/voltage section to view on the current, voltage and power of the solar photovoltaic panel.

The overall temperature data for all the four TCs sensors are shown in Figures 15.18 and 15.19, while the individual temperature data are presented in Figures 15.20 and 15.21. The consumers can always decide to check the individual

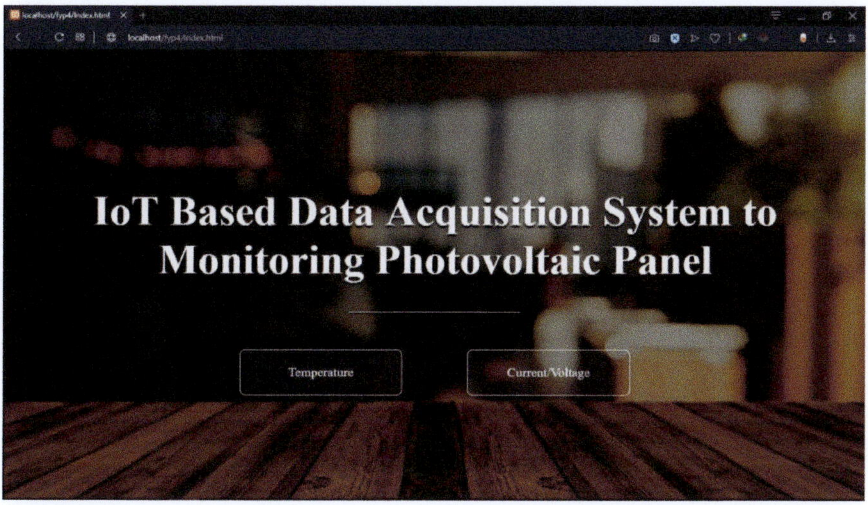

Figure 15.17 Webpage interface to monitor the photovoltaic panel

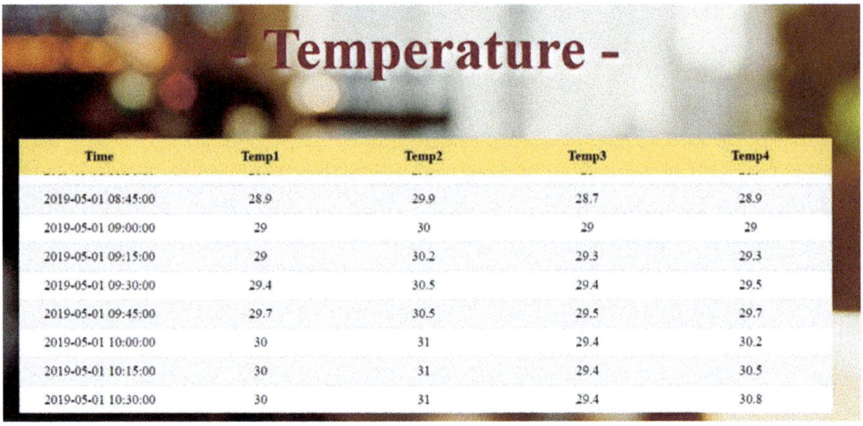

- Temperature -

Time	Temp1	Temp2	Temp3	Temp4
2019-05-01 08:45:00	28.9	29.9	28.7	28.9
2019-05-01 09:00:00	29	30	29	29
2019-05-01 09:15:00	29	30.2	29.3	29.3
2019-05-01 09:30:00	29.4	30.5	29.4	29.5
2019-05-01 09:45:00	29.7	30.5	29.5	29.7
2019-05-01 10:00:00	30	31	29.4	30.2
2019-05-01 10:15:00	30	31	29.4	30.5
2019-05-01 10:30:00	30	31	29.4	30.8

Figure 15.18 Temperature data for four TC sensors

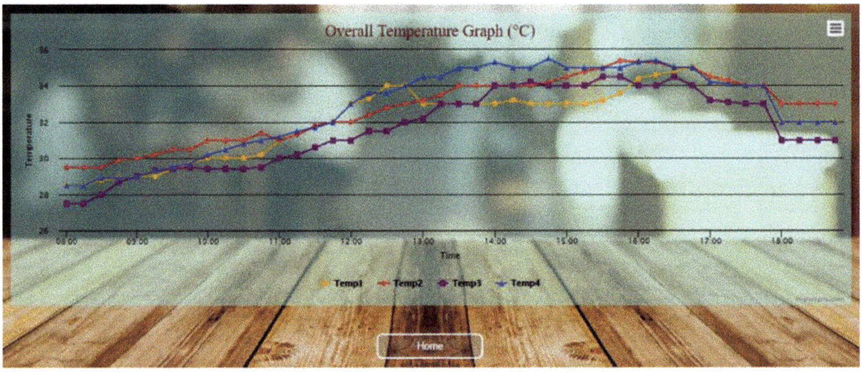

Figure 15.19 TC sensors (Temp 1, 2, 3 and 4) temperature data

temperature or consumers can look at the overall temperature of the solar photovoltaic panel.

The second section is the current/voltage section as shown in Figure 15.17. Consumers can click on the current/voltage section to check or view the current and voltage performance of a solar photovoltaic panel. The information presented in Figures 15.22–15.24 is the information stored into the SD card of Raspberry Pi Zero Wireless, as shown in Figure 15.8. Based on the information available on the webpage, consumers can determine the solar photovoltaic panel performances and overall can understand the installed system performances.

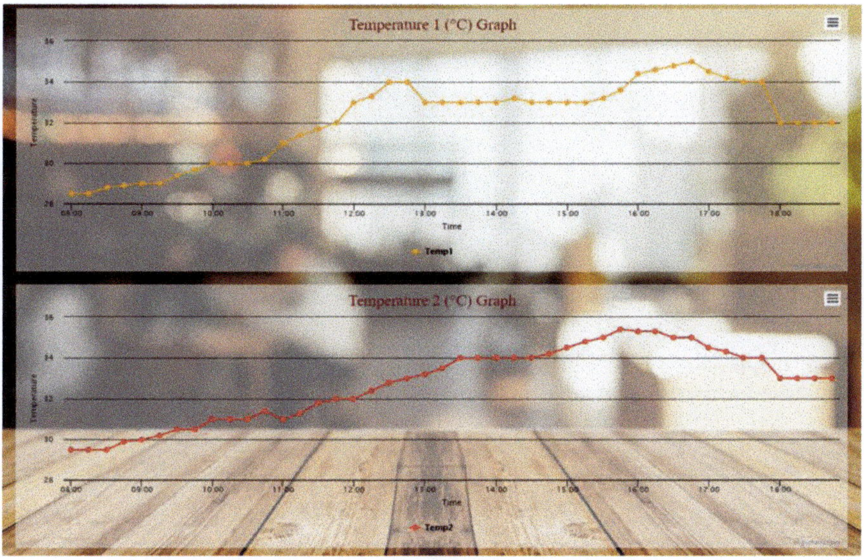

Figure 15.20 Temperature data for TC sensors 1 and 2

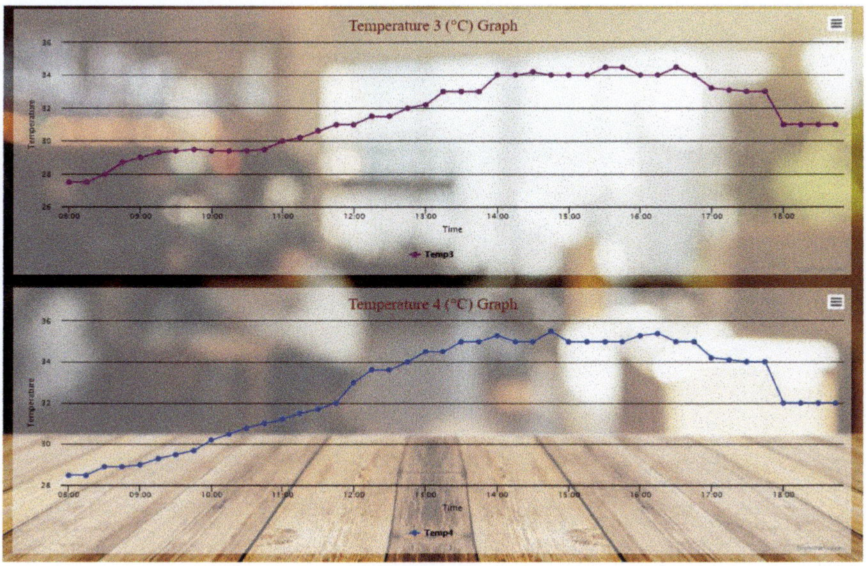

Figure 15.21 Temperature data for TC sensors 3 and 4

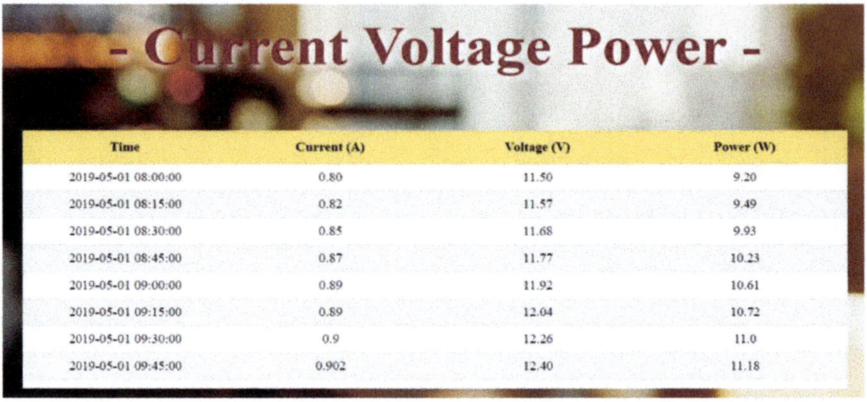

- Current Voltage Power -

Time	Current (A)	Voltage (V)	Power (W)
2019-05-01 08:00:00	0.80	11.50	9.20
2019-05-01 08:15:00	0.82	11.57	9.49
2019-05-01 08:30:00	0.85	11.68	9.93
2019-05-01 08:45:00	0.87	11.77	10.23
2019-05-01 09:00:00	0.89	11.92	10.61
2019-05-01 09:15:00	0.89	12.04	10.72
2019-05-01 09:30:00	0.9	12.26	11.0
2019-05-01 09:45:00	0.902	12.40	11.18

Figure 15.22 Current–voltage–power information

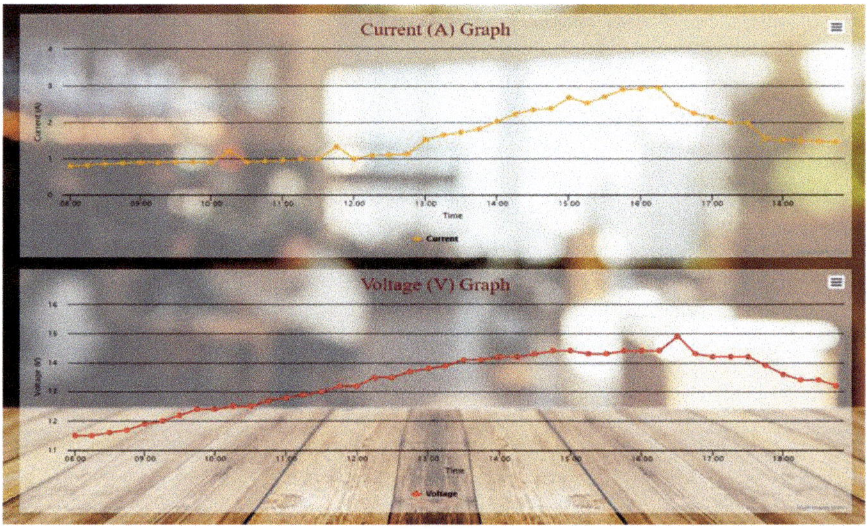

Figure 15.23 Current–voltage information

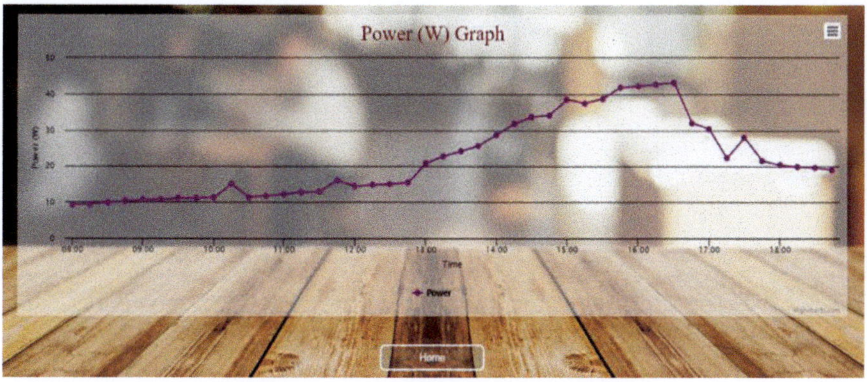

Figure 15.24 Power information

15.4 Conclusion

The proposed IoT-based data acquisition monitoring system to analyse the solar photovoltaic panel has been successfully developed, tested and validated its functionality and operation. The entire developed hardware and software integration explained the entire features of the IoT-based data acquisition monitoring system.

Raspberry Pi Zero Wireless used in the system is one of the tiny modern processors that can be used to control the operation for collecting data from the integrated sensors. Localhost database is used as the platform for gathering the data from the Raspberry Pi Zero Wireless and the collected data is present in webpages that are friendly to the consumers. The Adafruit TC sensors and INA 219 DC voltage/current sensor is used to sense and measure the temperature, current and voltage of each respective solar photovoltaic panel. The values of temperature, current and voltage are recorded into the SD card storage of Raspberry Pi Zero Wireless. In the nutshell, the presented system also provides accurate and real-time information about the temperature, current and voltage of a solar photovoltaic panel installed. In other words, any incorrect information recorded and retrieved from the Raspberry Pi Zero Wireless could immediately indicate the system's inconsistency.

Acknowledgement

The author(s) wish to thank the Ministry of Higher Education of Malaysia (MOHE), Advanced Sensors and Embedded Control Research Group (ASECS), Centre for Telecommunication Research & Innovation (CeTRI), Fakulti Kejuruteraan Elektronik dan Kejuruteraan Komputer (FKEKK), Universiti Teknikal Malaysia Melaka (UTeM), Melaka, Malaysia.

Chapter 16

Internet of Things (IoT) application for the development of building intelligent energy management system

Boon Tuan Tee[1] and Md Eirfan Safwan Md Jasman[1]

Buildings are a major consumer of electricity. They account for 60% of the total consumption of electricity worldwide. For instance, in Malaysia, 70% of the electricity usage during the year is because of buildings. Even in other nations, buildings are known for the high consumption of electricity. However, exhaustive data and figures are not completely available because of limited information and indicators. Hence, there has been considerable emphasis on analysing and formulating means for the effective management of electricity in buildings by means of effectual energy management systems. Basically, Internet of Things (IoT) is a platform which links devices through the Internet, allows them to talk to each other as well as to human beings, thus facilitating the accomplishment of requisite context-specific goals such as condition monitoring (for example, air-conditioning and mechanical ventilation (ACMV) fault detection), energy savings (for example, scheduling ACMV system on the basis of occupancy), and predictive maintenance (for example, air filter servicing in heating, ventilation, and air conditioner (HVAC)) [1]. As IoT is likely to transform the future of energy management system, this chapter intends to offer a synopsis of an IoT application in formulating an intelligent energy management system for buildings.

16.1 Introduction

IoT pertains to the idea of pervasive terminal devices and facilities, imbibed into the central intelligence of mobile devices, sensors, building automation systems, industrial systems, video monitoring systems, smart home installations, vehicles, personal portable wireless terminals and other smart gadgets, by means of various wireless or wired short-distance or long-distance communication network for attaining application integration as well as interoperability [2]. By utilising proper information security systems within the intranet (i.e. the network), extranet (the private network) or the Internet ecosystem, it offers safe, organised, personal and

[1]Fakulti Kejuruteraan Mekanikal, Universiti Teknikal Malaysia Melaka, Melaka, Malaysia

real-time location tracking, online monitoring, commanding, alarm linkage, process control, plan management, process security dimension, security, statistical reporting, online upgrade, decision support, service and management functions [2].

Setting up an IoT-driven data centre information system as the hub is crucial for addressing the issues related to building energy management. It should be favourable towards setting up a post-decision-support mechanism, improve the foundation of modern management and offer a base for prospective planning and decision-making. This can be fused with other information systems that assume the role of data aggregation and help in assisting management capacity and efficacy. IoT technology-based monitoring mechanisms for the indoor environment of buildings, as well as energy consumption, bring together all energy consumption equipment for making the construction energy-efficient. The technology that forms its core is the sensor network and computer information processing, aimed at constructing a cutting-edge, potent information acquisition and processing platform. The mechanism is apt for various prevailing and new constructions. It is the most appropriate system for transferring data in the indoor environment of buildings and monitoring mechanisms for the energy consumption [2].

16.2 Building indoor environment

When it comes to indoor environment quality (IEQ) in buildings, it incorporates indoor air quality (IAQ), thermal comfort, acoustics, and lighting [1]. As specified by prior studies and standards, it is acknowledged that poor IEQ has an unfavourable impact on occupational health as well as productivity, especially when it comes to kids and the elderly. Indoor living environs encompass many kinds of workplaces and spaces such as hospices, offices, public service centres, libraries, schools, leisure spaces and vehicle cabins. Usually, the huge number of inhabitants, the time spent inside and the greater density of occupants rationalise the need to formulate automatic supervision mechanisms for delivering a wholesome and productive workplace for teachers, students and school staff [3].

Buildings account for 60% of the worldwide energy consumption and 30% of the CO_2 releases. Thermal comfort assurance pertains to a considerable proportion of the stated energy consumption. The espousal of personalised conditioning mechanisms is apparently a consistent approach for enhancing user satisfactoriness with environment thermal conditions as thermal comfort is a complex topic with many interrelated factors which have to be construed [3].

The basic notion of IoT is the ubiquitous presence of many kinds of devices possessing collaboration and communication skills between them to attain a mutual objective. The IoT would not just enhance day-to-day life behaviours and tasks in various fields like assisted living environments, smart homes, and smart health, but it will also render pioneering data and computational means for formulating innovative software solutions. It would play a crucial role in monitoring the indoor environment of buildings. The swift rise of Information and Communication Technologies (ICTs) and the spread of IoT provide considerable prospects for the advancement of indoor environment information mechanisms. Nevertheless, there are still some challenges involved in accomplishing security, safety and privacy when it comes to building applications [3].

With regard to monitoring systems, remote information devices are tools which offer real-time data on air humidity, temperature, quality, illumination CO_2 concentration and noise levels. They also suggest which action to take when the user encounters non-optimal conditions. One can install such devices at home or in the workplace for attaining precise and timely data on the internal microclimate, connecting the person's subjective state with the objective conditions of the environs. For instance, for assessing the impacts of air quality and temperature on the sense of sleepiness in the course of a meeting, the noise levels which trigger disorders or reduce productivity should be gauged, the temperature should be adjusted and the place and time for ventilation of the environments should be determined. Certain controllers collect input from different lighting, temperature, location and contact sensors to let the user know in real time about any event to take place at home, such as the opening of a door, flooding in the cellar or switching off of an appliance [4].

Studies have been conducted on indoor air quality sensing for different public areas such as subways, department stores, schools, and offices. A system developed by Paulos *et al.* can measure and monitor office air quality by studying work efficiency and office indoor air environment. The overall employee work efficiency improved when the system was controlled through a wireless sensor network that was connected to mobile devices [5]. Kanjo, Lohani and Acharya formulated an environmental information monitoring system which makes use of precautions like indoor fine dust reduction, through mobile wireless LAN. It was observed by the authors that employee satisfactory level has improved with the work environment. [5]. Hwang and Yoe [6] conducted a monitoring and analysis of the indoor environment information via closed-circuit television and public environment information with the help of an application programming interface. Furthermore, they formulated an indoor environmental control system that has its basis on automatic situation recognition. Wang *et al.* [7] and Pötsch *et al.* [8] formulated an indoor environmental monitoring system that is based on a wireless sensor. This can be used for green buildings and the LoRaWAN stack, respectively. The system can visualise the gathered measurement position data and indoor environment, and it can be used for distribution of the temperature sensor to different locations within the target space. Moreover, the temperature is communicated in every space through a step colour chart. More specifically, the authors computed the distance from a window. The authors also put in place sensors at three levels beyond the horizontal point. Their system envisions the gathered data as a three-dimensional space chart based on the spatial distribution [5].

For the investigation of the indoor air quality monitoring system, researchers partitioned the fine dust concentration measurement values on the floor plan of the space and distributed them into multiple spaces. They were then expressed in two or three dimensions. Furthermore, the system possesses a simple structure for intuitively obtaining the indoor environmental condition, which enables a comparison of the dust concentration based on the space. In another development, Salamone *et al.* conducted studies that made use of simpler self-developed experimental tools. The open-source smart lamp was installed in an actual office environment. The reliability of IoT equipment was also examined [5]. Salamone *et al.* performed an assessment of the ventilation efficiency based on the ventilation method of indoor space through the computational fluid dynamics (CFD) technique [5].

16.3 Building energy management

With several new energy certifications and regulations, the requirement to decrease energy usage and emissions of greenhouse gases as well as the financial need to cope with rising energy expenses, the efficiency of energy consumption has become a significant aspect with respect to the management of the building energy usage. Building energy management involves the arrangement and operation of energy generation and energy usage units. The goals are focusing on resource preservation, environment protection and savings of costs, while the consumers have stable access to the energy they require. It is also linked closely to the environmental management, generation management, logistics and other customary business operations.

With the progress in recent technology, a new low-cost sensor generation connected to the World Wide Web is now available to decrease the building's energy usage. These systems can be employed in all kinds of buildings and are capable of intelligently monitoring environmental variables of indoor areas or energy usage of electrical outlets or individual equipment, providing consumers with real-time data on the smartphone which can be accessed from anywhere. The system also includes sending SMS or email notifications, together with eventual alerts or equipment breakdown warnings [2].

The combination of sensing and automatic trigger systems with Internet tracking is used to optimise energy usage. One of the adoptions is to integrate IoT technology into lighting devices and be able to connect with the power supply company so as to balance power production and energy usage efficiently. Such a proposed system can also offer consumers the opportunity to remotely manage their devices or manage them centrally via a cloud-based service and enable advanced operations like scheduling [2].

Other models associated with building energy management are green buildings and smart buildings which are also the topics currently significant in engineering and architecture. Among both the concepts, green buildings can save water, save energy, protect the climate and enable the users to ensure sustainable progress, while the smart buildings can be seen as green buildings with superior energy performance achieved by integrating IoT, ICT and other advanced technologies. For instance, lighting controls are generally used to reduce energy costs. Lighting management or advanced load management can decrease the lighting requirement when energy is highly expensive. Manual dimmers, which enable occupants to regulate light levels according to their needs, have become more inexpensive.

Lighting controls have proved to reduce the lighting power consumption by at least 35% in the new buildings and by 50% in the conventional buildings [2].

Besides reading data, some of the systems enable real-time functionalities, using a smartphone, on household appliances by managing the on–off schedule or regulating activity levels as required. Objects can be considered self-recognisable and gain intelligence because of their ability to convey information concerning them and gain access to collective information of the other devices. For instance, smart thermostats monitor the departure and arrival of the occupants to regulate the temperature when required and smart lighting switches on and off automatically after a long absence for security reasons. All objects play a proactive role by communicating with the Internet. The purpose of IoT is to represent the real world into the electronic world, providing electronic identity to real-world places and things. Places and articles equipped with radio frequency identification (RFID) labels or QR codes currently communicate data to mobile devices like smartphones or to the network [4].

In particular, there has been an increased interest in how IoT can be used for efficient energy consumption. By installing networks of smart, connected devices that constantly collect, assess and share data, IoT can provide deep insights into how huge buildings, factories, facilities and utilities consume energy. Such insights, in turn, can enable to develop energy management system strategies that lead to resource optimisation and more and more autonomic improvement.

16.4 IoT approach for data and information collection

Information and data gathering are done to collect the building's data. It includes the indoor condition and occupant comfort. By combining IoT and the data collection system, this data can be remotely accessed. This is because of the IoT characteristics that transfer real-world information into the electronic world. The data which is stored in the server digitally can be accessed and read anytime for multiple purposes, for instance, analysis and monitoring.

The integration of the data gathering system and IoT has increased the reliability and effectiveness of the information and the data gathered in this way. The quicker the process of gathering and assessing the data, the quicker the trigger can be generated to optimise the condition of the building.

16.4.1 Indoor environment monitoring

Indoor environment monitoring refers to a system that indicates the environment of the indoor region of the building. The optimal condition of the indoor environment is a condition in which the renters feel healthy and comfortable. These conditions are light intensity, the acoustic, IAQ and thermal comfort of the building. These are the aspects considered important by the scholars in measuring the comfort level of the occupants [1,9].

16.4.1.1 IoT workflow

To get the overall idea of the integration of IoT and the monitoring system, let us first have a look at the system workflow. Figure 16.1 displays indoor monitoring system workflow where this system is combined with IoT. The system workflow can be partitioned into five stages: sensors, microprocessor, gateway, database and end-user.

First stage (sensor): The sensor is the heart of the system. Usually, the kind of sensor chosen decides which system is proposed to be developed. Several kinds of sensors are available. Generally, electrical sensors are used in the system integrated with the IoT framework. This kind of sensor is selected since it becomes possible to transmit the data gathered by the sensor to the IoT framework. As the platform itself carries the connotation of conveying something physical to the electronic world, the best choice is the electrical sensor. Figure 16.2 illustrates some of the available sensors that are likely to be used in the indoor conditions monitoring system.

Second stage (microprocessor/microcontroller): Even though we got the sensor reading, we cannot directly read data from the sensor. Also, certain sensors need power supply for an external source to function. Hence, we require the microprocessor/ microcontroller for this purpose. Microprocessor/microcontroller acts as a power source and an interpreter for the sensor. Even though electrical signals are produced by

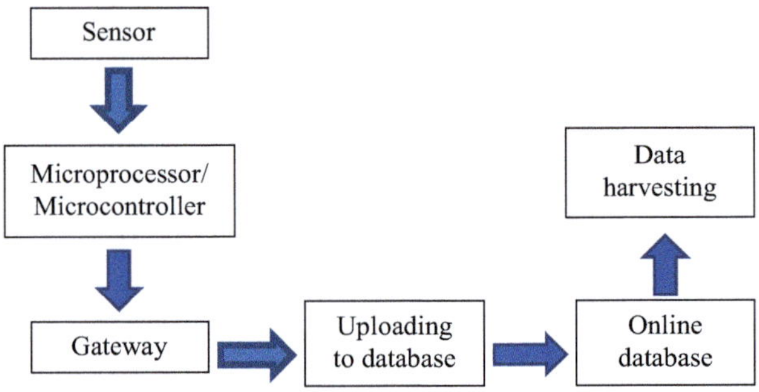

Figure 16.1 General workflow of the indoor environment monitoring system integrated into the IoT system

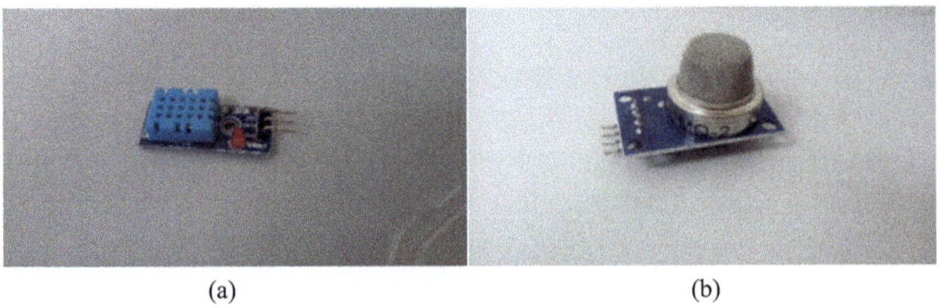

Figure 16.2 Two examples of sensor in the market: (a) DHT11 temperature and humidity sensor and (b) MQ-2 gas sensor

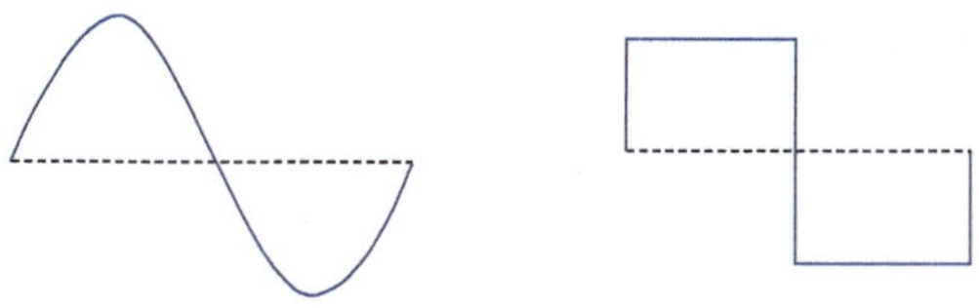

Figure 16.3 On the left is the analogue signal and on the right is the digital signal

Figure 16.4 Mega 2560, one of the available microcontrollers in the market

the electrical sensor, they can be classified into two signals: analogue signal and digital signal. The difference between the two signals is that the digital signal is a discrete one having 0 and 1, whereas the analogue signal varies constantly with time. Figure 16.3 illustrates the digital and the analogue signals with respect to time.

A majority of the microcontrollers/microprocessors can only interpret the digital signals, and so microcontrollers such as the Arduino UNO have a converter on their board. As the analogue signal can be converted into a digital signal, this type of sensor can be used in the digital data collection system. The microcontroller will gather and decipher the sensor signal and convert the data into a form such as a numerical value that can be stored (Figure 16.4). There are many differences between a microcontroller and a microprocessor with respect to factors such as weakness and strength. Therefore, the selection of a microprocessor or a micro-controller is done depending on the system design to be used.

Third stage (gateway): The gateway refers to the 'gate' which is also a path for the information and data that had been gathered and deciphered by the microprocessor/microcontroller to be sent and stored in the online database. The gateway is generally in the form of router which links directly the system with the online database. Based on

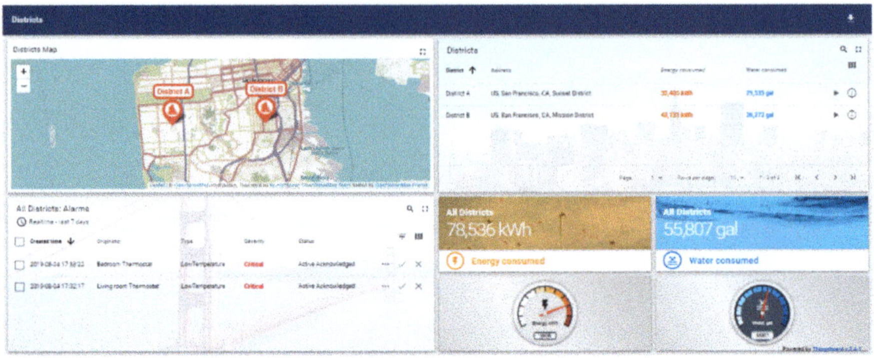

Figure 16.5 Example of the online database [10]

the design, the designer of the system has the independence to decide whether to connect through a computer or connect directly to the router. In the design where the system is directly connected through the router, it requires a wireless antennae set. This set of wireless antennae will enable the transmission of the data wirelessly to the online database.

Fourth stage (online database): The database that is online is also called a cloud or a server. It is the place where the entire information obtained from the sensor is being stored. The server has high potential though it is up to the abilities of the designer to decide the database layout. This server can be used as storing medium interactively to show the real-time reading with graphical presentation. Not only does this storage method provide uniqueness of data representation, it also facilitates access to the information and the database from any place having Internet availability (Figure 16.5).

Fifth stage (data harvesting): The data stored on the server is now available to the user. It is not necessary for the user to go near the sensor to access information and analyse it. The data stored in the server can be examined by the person authorised to monitor the building and its surroundings. The best part of using the IoT framework is that the monitoring of the information can be carried out by the system automatically. It is a smart system designed in a way that it alerts the user if the value of a parameter exceeds that of the defined criteria.

16.4.1.2 Importance of indoor environment monitoring

The significance of supervising the indoor area satisfactorily has been globally recognised. Some of the recent studies discuss the reasons for the significance of monitoring the indoor area. Table 16.1 lists certain studies that have been conducted recently.

In a previous study, Mishra *et al.*, 2016 [2,11] emphasised that 40% of global energy requirements came from buildings. But in an attempt to make the building ‘green’, we must take care of the occupants’ comfort. Mishra mentioned that the principal characteristic of the comfort of the occupants is their thermal comfort.

Table 16.1 Classification of the past studies

Author	Thermal comfort	IAQ	Visual comfort	Acoustic
Mishra *et al.* (2016) [2]	X			
Li and Jin (2018) [3]		X		
Junaid *et al.* (2018) [4]		X		
Asif *et al.* (2018) [5]	X	X		
D'Oca *et al.* (2018) [6]	X	X	X	
Rabiyanti *et al.* (2017) [7]				X

The thermal comfort includes temperature, flow of air and humidity, which are found to have a direct impact on the productivity and perception of the occupants.

Further, in 2018, Li and Jin [12] in their research, emphasised on the significance of supervising the quality of indoor air as it can prove to be a cause of severe illnesses. The model designed in the research comprises two functions: assessment and prediction of the level of IAQ. By employing a hybrid optimisation design, the contamination level of the air is automatically predicted while the evaluation is done to determine the quality of air. The design is proven to be reliable and more effective in a large city which is highly polluted.

Another matter of alarm for IAQ is the particulate matter (PM), also called dust. This dust is so tiny that it intrudes on a person's lung and cause health-related issues [13]. Junaid *et al.* [13] also mentioned that prolonged exposure to this unhygienic condition can cause chronic respiratory diseases and can also cause infections.

Interestingly, the concentration of PM can also indicate room occupancy [14]. Jeon [15] found a way to associate the concentration of PM with the room occupancy. In the research by employing Arduino framework, SEN0177 (PM sensor) and the IoT platform, the number of room occupiers can be found out even without being physically present there. Also, the system can recognise the occupants' activities. It would help exceedingly as the activities of the occupants also affect the atmosphere and the comfort of the room.

Rabiyanti *et al.* stated that the acoustic characteristics of the room also affect a person's comfort [16]. As per Rabianti, the acoustic level must not be higher than 55 dBA for the environment to be conducive for the process of learning.

Meanwhile, D'Oca conducted a questionnaire in 2018 which indicated the significance of visual comfort to attain a comfortable indoor environment [17]. D'Oca also mentioned that the regulation of the lighting conditions in the room helps in decreasing the extra glare and enhancing productivity.

While D'Oca used the manual approach of conducting a questionnaire, Zhao *et al.* (2017) used a web-based questionnaire along with a timestamp [18]. The results of this method are more dependable as the environment of the room is also monitored. The availability of the timestamp enables to analyse the change or reaction to the change in the situation. The system also produces estimates of the demands of change in the temperature level from the reactions of the occupants.

The monitoring of the quality of the indoor atmosphere is significant as it affects the occupant and, in case of low quality, makes the occupier uncomfortable. Hence, by combining IoT and indoor monitoring system, a reliable and effective assessment can be performed and corrective steps can be taken, thereby achieving a better indoor environment.

16.4.2 Energy performance assessment

The assessment of energy performance is a system that assesses the usage of energy in a building. It includes cooling, heating, and electrical power. In Malaysia, the primary usage of energy to be assessed is the usage of electrical energy. The smart meters issued by Malaysia energy supply company, Tenaga Nasional Berhad (TNB) recently are an example system that combines the assessment of energy performance with IoT.

Rented rooms and hostels that are equipped with split meters attached to each room implement systems similar to those discussed above (Figure 16.6). Therefore, the energy used by the individual unit can be measured. It will decrease the burden of the occupants as they have to pay only for the amount of energy they use. The installation of split meters will reduce the amount of energy used by the occupants as they will be able to save money by using less energy.

A smart building such as the office of Suruhanjaya Tenaga has superior energy performance assessment compared to the metered buildings. The system classifies the usage according to the floor and the purpose of the use of the energy. It provides an accurate outline of the energy performance assessment. In this building, the systems are integrated with IoT which sends the data daily to Singapore where the experts analyse it so as to retain the award of green-mark platinum structure. This is the advantage of using IoT, and it enables the data to be accessed from anywhere.

Several tools are present in the market that facilitate the users to monitor, regulate and optimise the consumption of energy. A ground-breaking energy-aware IT ecosystem had been proposed by Fotopoulos *et al.* [19], which offers personalised energy monitoring and awareness facilities that match with the activities of the

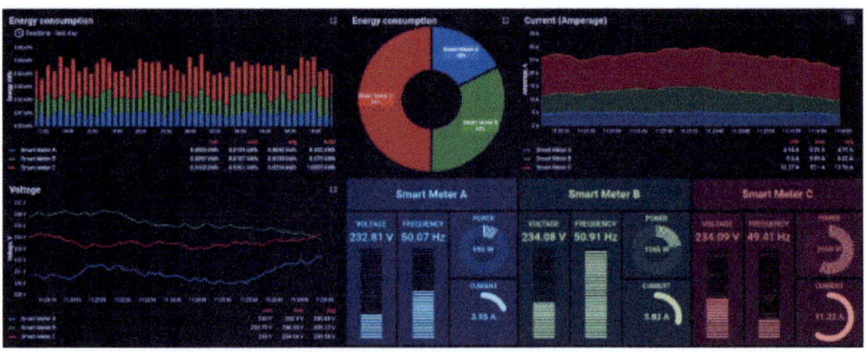

Figure 16.6 Example of meter for energy consumption monitoring of rented room which utilises IoT [10]

occupants. In addition, there are products from prominent firms implementing IoT framework which enables the management of the data related to the consumption of energy. A few of the available products from these firms are as follows:

Siemens Synco™ *is capable of functioning as a regulation system for building computerised system. It can regulate the HVAC of the building. The system is also able for multi-site monitoring. Lastly, it claims to be capable of monitoring the billing and energy consumption remotely* [20].

Schneider electric BMS solution [21] *is a framework for automatic building. Not only does the system manage the energy profile, it can also regulate and guide the user on power, fire, HVAC, and access control.*

Honeywell building management system [22] *is an open management system which is attuned to almost all of the open self-sufficient building systems. The system is capable of monitoring steam, electrical power, oil, and gas and water usage.*

Cylon active energy [23] *can adjust to any kind of building irrespective of the metering solution or building energy management system installed. It is a system for building energy monitoring with the capability to monitor and regulate several sites. It is also able to assess and recommend a concrete solution for the usage of energy, which is a strong advantage of the system.*

DEXCell energy management [24] *is a hardware-neutral and cloud-based system. It integrates advanced monitoring, assessment, alert signals and information storage in a scalable and easy-to-use SaaS solution.*

eSight energy management platform [25] *uses IoT which displays the crucial energy data using adjustable dashboards that shows important information at a glance and supports engagement. eSight's project tracking component assesses energy-saving opportunities, monitors the progress and verifies the savings using an all-in-one intuitive dashboard. It is also capable of creating and circulating bills solving the problems of the occupants of the building.*

Predictive energy optimisation [26] *is a product of the software platform of the Building IQ, whose aim is to improve the energy usage efficiency of huge and complicated structures.*

Enerit systematic energy management software [27] *puts forward best practices of energy usage management, provides complete ISO 50001 coverage and aligns with Energy STAR and Statement of Energy Performance (SEP).*

There are many more systems available in the market that provides tools for management and monitoring the energy usage. Some of the aforementioned tools are energy monitoring systems that are integrated with the building automation system with an ability to regulate the building with the aim to improve the performance of energy consumption in the building. Also, there are other modelling tools that provide help and suggestions to optimise energy consumption.

Hence, we discussed several available systems that use IoT in their energy performance evaluation. Usually, these systems analyse the consumption trend and devise a way to optimise the energy consumption. An effective system can decrease up to 40% energy consumption in a huge building. Automation of the building also

Table 16.2 Difference of data collection system with and without the integration with IoT

	With IoT	**Without IoT**
Sensor	Electrical sensor	Mechanical or electrical sensor
Data collection	Can be done remotely	Need to take directly from the device
Data-sharing	Fast and reliable	Slower
Storage	Large storage capacity	Moderate storage capacity
Run time	Can be run continuously without being physically at the data collection site	Need to be present when initiating the system and gradually monitor to collect the data

enables regulation of the building air conditioning system to function as efficiently as possible.

Table 16.2 shows the difference in data collection system with and without the integration with IoT. From the table, it can be inferred that the integration of the information collection system with IoT improves the efficiency and reliability of the process of the data gathering.

16.5 Conclusion

The emergence of the IoT platform shows a considerable progress in transforming the sphere of building energy usage management system. Nonetheless, like any novel and complex technology, there is a risk of slow transformation and being unable to reap the highest potential benefit. In the adoption of IoT technology (like smart MEMS devices and wearable devices), one of the challenges for the improvement of the energy consumption of the buildings is the dynamic complexity of its environment. It is a well-known fact that a structure has complex dynamics (thermal comfort, thermal load, occupancy of the occupants, etc.). This complexity makes it difficult for the IoT devices to understand such dynamics in order to regulate and improve the building's performance. In this regard, further examination that focuses on comfort and air quality modelling, prediction and occupancy detection, for instance, using IoT technology, should be carried out. An innovative framework is also required to integrate the IoT devices with the current building systems for sensors and controls.

As a huge amount of data is produced from IoT devices, intelligent hardware and processing devices are needed to interpret and use the data. This requirement is even greater when we are dealing with real-time computations. Another challenge is the privacy issue where cyber security measures must be taken to solve the issue.

In general, the use of IoT technology has played an important role in the advanced design and adoption in the building energy management system. This adoption has been proved to be substantially beneficial in betterment of the system and further studies are still needed to deal with the other concerns as discussed above.

References

[1] W. Tushar, N. Wijerathne, W.T. Li *et al.*, "IoT for green building management", IEEE Signal Processing Magazine.

[2] W. Chuyuan and Y. Li, "Design of energy consumption monitoring and energy saving management system of intelligent building based on the Internet of things", 2011 International Conference on Electronics Communications and Control (ICECC), 2011.

[3] G. Marques and R. Pitarma, "An internet of things-based environmental quality management system to supervise the indoor laboratory conditions", *Applied Science*, vol. 9, 2019.

[4] M. Casini, "Internet of things for Energy efficiency of buildings", *International Scientific Journal Architecture and Engineering*, 978-1502819499.

[5] J. Kim and H. Hwangbo, "Sensor-based optimization model for air quality improvement in home IoT", *Sensors*, vol. 18, 2018.

[6] J. Hwang and H. Yoe, "Study of the ubiquitous hog farm system using wireless sensor networks for environmental monitoring and facilities control", *Sensors*, vol. 10, 10752–10777, 2010.

[7] W. Wang, N. Wang, E. Jafer, M. Hayes, B. O'Flynn, and C. O'Mathuna, "Autonomous wireless sensor network based building energy and environment monitoring system design," Proceedings of the 2010 International Conference on Environmental Science and Information Application Technology (ESIAT), Wuhan, China, 17–18 July 2010, pp. 367–372, 2010.

[8] A. Pötsch, F. Haslhofer, and A. Springer, "Advanced remote debugging of LoRa-enabled IoT sensor nodes," Proceedings of the Seventh International Conference on the Internet of Things, Linz, Austria, 22–25 October 2017, p. 23, 2017.

[9] J. H. Choi and K. Lee, "Investigation of the feasibility of POE methodology for a modern commercial office building," *Build. Environ.*, vol. 143, pp. 591–604, 2018.

[10] ThingsBoard, I., *ThingsBoard*, (2019, 8 6). Retrieved from ThingsBoard: https://thingsboard.io/smart-metering/

[11] A. K. Mishra, M. G. L. C. Loomans, and J. L. M. Hensen, "Thermal comfort of heterogeneous and dynamic indoor conditions — An overview," *Build. Environ.*, vol. 109, pp. 82–100, 2016.

[12] R. Li and Y. Jin, "The early-warning system based on hybrid optimization algorithm and fuzzy synthetic evaluation model," *Inf. Sci. (Ny).*, vol. 435, pp. 296–319, 2018.

[13] M. Junaid, J. H. Syed, N. A. Abbasi, M. Z. Hashmi, R. N. Malik, and D. S. Pei, "Status of indoor air pollution (IAP) through particulate matter (PM)

emissions and associated health concerns in South Asia," *Chemosphere*, vol. 191, pp. 651–663, 2018.

[14] A. Asif, M. Zeeshan, and M. Jahanzaib, "Indoor temperature, relative humidity and CO_2 levels assessment in academic buildings with different heating, ventilation and air-conditioning systems," *Build. Environ.*, vol. 133, pp. 83–90, 2018.

[15] Y. Jeon, C. Cho, J. Seo, *et al.*, "IoT-based occupancy detection system in indoor residential environments," *Build. Environ.,* vol. 132, pp. 181–204, 2018.

[16] Rabiyanti, I. Rahmaniar, and J. C. P. Putra, "Effect of acoustic and thermal comfort to support learning process in a university," Procedia Eng., vol. 170, pp. 280–285, 2017.

[17] S. D'Oca, A. L. Pisello, M. De Simone, V. M. Barthelmes, T. Hong, and S. P. Corgnati, "Human-building interaction at work: findings from an inter-disciplinary cross-country survey in Italy," *Build. Environ.*, vol. 132, pp. 147–159, 2018.

[18] Z. Zhao, M. Houchati, and A. Beitelmal, "An energy efficiency assessment of the thermal comfort in an office building," *Energy Procedia*, vol. 134, pp. 885–893, 2017.

[19] E. Fotopoulou, A. Zafeiropoulos, F. Terroso-Sáenz, *et al.*, "Providing personalized energy management and awareness services for energy efficiency in smart buildings," *Sensors (Switzerland)*, vol. 17, no. 9, 2017.

[20] Seimens, "Synco | Building automation and control systems | Siemens." [Online]. Available: https://new.siemens.com/global/en/products/buildings/automation/synco.html. [Accessed: 24-May-2019].

[21] SchneiderElectric, "Building management solutions | Schneider Electric." [Online]. Available: https://www.schneider-electric.com/en/work/solutions/building-management/. [Accessed: 24-May-2019].

[22] Honeywell, "HVAC and BMS | building management and intelligent building solutions." [Online]. Available: https://buildingsolutions.honeywell.com/en-US/solutions/hvacbuildingmanagement/Pages/default.aspx. [Accessed: 24-May-2019].

[23] Cylon, "Energy management services." [Online]. Available: https://www.cylon.com/about/cylon-active-energy.html. [Accessed: 24-May-2019].

[24] Dexma, "About DEXMA: welcome to the energy analytics company." [Online]. Available: https://www.dexma.com/about-dexma/. [Accessed: 24-May-2019].

[25] eSightenergy, "eSight Energy | About." [Online]. Available: https://www.esightenergy.com/about/. [Accessed: 24-May-2019].

[26] BuildingIQ, "Predictive energy optimization – BuildingIQ." [Online]. Available: https://buildingiq.com/products/predictive-control-old/buildingiq-platform/. [Accessed: 24-May-2019].

[27] Enerit, "Enerit software system for ISO 50001Enerit | Systematic Energy Management Software | ISO 50001 Software," *Enerit Ltd.*

Chapter 17

Expert fault diagnosis system for building air conditioning mechanical ventilation

Chee-Nian Tan[1], CheeFai Tan[2], Ranjit Singh Sarban Singh[3] and Matthias Rauterberg[4]

The rapid energy consumption around the world has resulted in the shortage of fuel supply and overconsumption of the energy sources, thus worsening the environmental impacts. The overall energy consumption of residential and commercial buildings in the developed countries has reached between 20% and 40% [1]. This figure has exceeded those in the industrial and transportation sectors. Growth in population, increasing demanding for building services and comfort levels and increasing time spent in the buildings indicate that energy demand will continue to increase in the future. For this reason, energy efficiency in buildings is the main goal for energy policies globally [1]. The consumption of energy in air conditioning and mechanical ventilation (ACMV) systems is particularly significant, namely 50% of building consumption and 20% of total consumption in the United States. Figure 17.2 analyses the available information concerning energy consumption in buildings, with a particular focus on cooling systems. It was found that ACMV consumes about 60% of the total energy consumption of a typical commercial building. The maintenance and servicing of air conditioning systems require highly skilled and experienced technicians and engineers. Experts from this field might not be available all the time when air conditioning units break down or when maintenance is needed. When air conditioning systems break down unexpectedly, high costs will normally be incurred. In addition, when air conditioning systems are not running as efficiently as they should, it could lead to costly electrical bills. Unfortunately, regular maintenance and inspection cannot be done as technical experts are not always readily available. In other words, there will be a loss of expert knowledge when human expertise is not available. The chapter describes the application of an

[1]Faculty of Mechanical Engineering, Technical University of Malaysia Melaka, Melaka, Malaysia
[2]Department of Mechatronics & Biomedical Engineering, Lee Kong Chian Faculty of Engineering & Science, Universiti Tunku Abdul Rahman, Selangor, Malaysia
[3]Centre for Telecommunication, Research and Innovation (CeTRI), Fakulti Kejuruteraan Elektronik dan Kejuruteraan Komputer, Universiti Teknikal Malaysia Melaka, Melaka, Malaysia
[4]Department of Industrial Design, Eindhoven University of Technology, Eindhoven, The Netherlands

expert system shell to diagnose the faulty problem of building air conditioning system. Human experts are expensive and heavily needed for building maintenance, services and maintenance of the air conditioning system. Hence, the main objective of the developed system, namely an expert system, is to diagnose the building air conditioning system. With the aid of an expert system, diagnosis process of building air conditioning system will be more standardised and more precise compared to the conventional way. The limited values for this developed expert fault diagnosis system are based on the building air conditioning system design and expert's experiences.

17.1 Introduction

Digitalisation in Construction 4.0 is giving a great impact on the construction industry in ways to improve their productivity. Potential in digitalisation in the industry of construction can be seen in line with Industry 4.0 from four main aspects: digital data, digital access, automation and connectivity [2–4,16–22]. Digital data is the collection of electronics and analysis of data to get every new insight into every link in the value chain and then put these new insights into good use whereas digital access covers the mobile access to the Internet and internal networks. Automation is the latest technology that creates autonomous and self-organising systems. Connectivity explores the possibilities to link up and synchronise hitherto separate activities (Roland [5]).

The fourth industrial revolution lies in the powerhouse of German manufacturing and widely adopted by nations such as China, India and other Asian countries via the Internet of Things and Internet of Services becoming integrated with the manufacturing environment ([6,7]). However, in Construction 4.0, construction business will have a strong global network to connect their transporting materials, running errands, cleaning up, rearranging the building site and looking for materials and equipment. This will bring a huge improvement in the industrial and construction processes within engineering, material usage, supply chains and product lifecycle management. It is therefore perfectly understandable that many businesses see a need for optimisation in construction 4.0 (Roland [5]).

For building construction and services which along the lines of construction 4.0, it is an upcoming trend that construction companies nowadays concentrate on the digitalization of planning, construction and logistics with building information modelling (BIM). BIM is a 3D modelling that can provide professional building design, construction, facility operations service and physical characteristics of places as shown in Figure 17.1 [8,9]. With the help of BIM technology, an accurate virtual model of a building is digitally constructed. It helps architects, engineers and constructors visualise what is to be built in a simulated environment to identify any potential design, construction or operational issues ([10]). The use of building information modelling (BIM) is compulsory by 2020 for every public infrastructure in Germany, Netherlands, Denmark, Finland and Norway. For Malaysia, led by the Construction Industry Development Board (CIDB) under the construction master

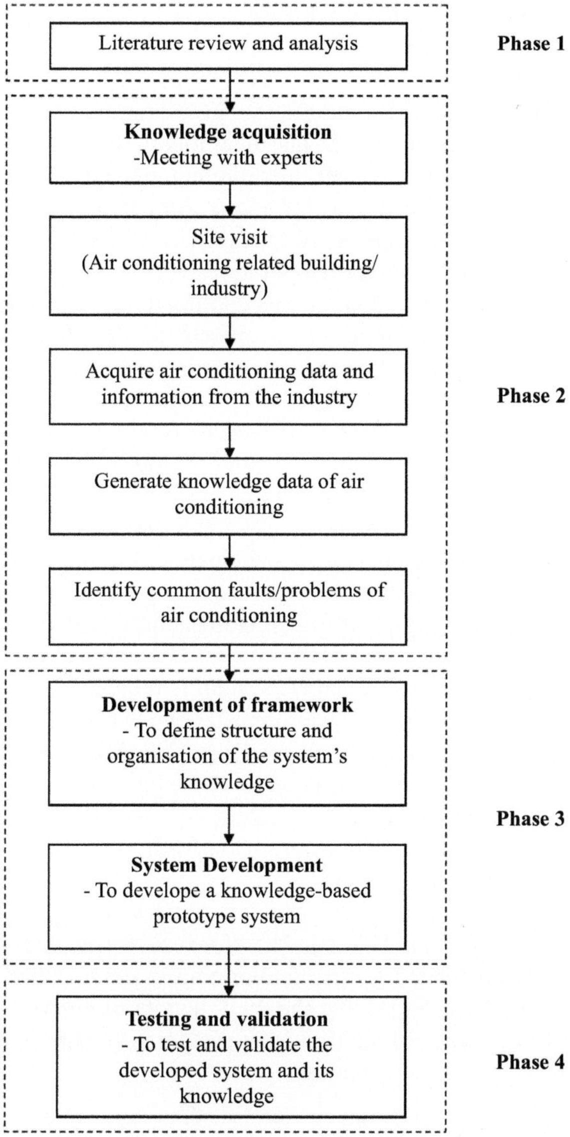

Figure 17.1 Caption text the architecture of the can be roughly sketched as consisting of a bottom sensor layer, a middle network layer and a top application layer at the bottom layer of the tags have found increasingly widespread applications in various business areas

plan 2016–20, it is hoped more emphasis on technology adoption across the project lifecycle will induce higher productivity.

On the other hand, energy is an imperative component for the development of a country as it is essential in various industries. Nowadays, non-renewable resources all over the world are diminishing at an alarming rate to meet the essential needs of mankind. According to a study [1], air conditioning consumes the most energy in buildings, households or even office workstations. Air conditioning in buildings and office workstations is a demanding trend as it provides a comfortable environment for occupants by reducing the indoor temperature.

As humans are highly dependent on this cooling device especially in countries with tropical weather, it usually needs to operate for extended periods. Hence, device failure or system breakdowns may occur anytime. When the air conditioning system breaks down, it can be very costly to hire technicians to carry out repair work. Therefore, the preventive action is better than repair works as the later cost more money and time. Thus, the expert system developed significantly reduces air conditioning maintenance cost, also it promotes proactive solutions.

Regular maintenance and repair of air conditioning systems are generally carried out by experienced technicians and engineers. Building ACMV experts are not available all the time to advise and review the possible references and data when the units break down [11]. In other words, there will be a loss of expert knowledge when human expertise is not available. Thus, to keep all the information and data of a field permanently, an expert system is required.

An expert system is a computer that emulates the behaviour of human experts within a well-defined, narrow domain of knowledge [12]. The expert system provides guidance and recommendations according to the situation based on engineering knowledge and experience. Besides, an expert system is one of the artificial intelligence technologies that was developed from research and it can simulate human cognitive skills for problem-solving [13].

A knowledge-based expert system is a computer software that can overcome problems with expert solutions [14,15]. Therefore, an expert system is developed with useful information related to air conditioning systems. With the help of an expert system, the time for diagnosing the main factors of air conditioning system breakdowns can be reduced. In addition, the system will also be able to provide recommendations based on the situation.

17.2 Overall system description

The knowledge-based system (KBS) for air conditioning fault diagnosis was developed based on the heuristic rules as well as the experience of air conditioning experts. There were four major phases involved in the development of KBS, as shown in Figure 17.1.

17.2.1 Literature review

The literature review is the first step that has to be done before commencing any research work. Knowledge and information related to the work done will be studied and noted. First, information journals, books and the Internet are collected. The information covered air conditioning systems, components and functions of air conditioning systems, a methodology of developing an expert system, information on expert system shell and Kappa-PC software. The information was needed to achieve the objectives of the research work.

17.2.2 Knowledge acquisition

The development of a KBS is a challenging stage as the required expert knowledge is heuristic and difficult to gather. The domain expert discussed in the section below is the main source of the project. There are two main sources for knowledge acquisition which are the (i) meeting with experts and (ii) manufacturer operating manuals.

17.2.2.1 Meeting with experts

One senior engineer (Mr Nizam), one facility management manager engineer (Mr Hisham), two technical ACMV specialist (Mr Wong and Mr Surah) and three senior technicians (Mr Nazri, Mr Amirul and Mr Shah) were the targeted domain experts who were involved in the knowledge acquisition process. The common faults and problems were obtained through discussions on ACMV problems and regular meetings with the domain experts. Besides, visiting several sites related to ACMV was also important to obtain site information. Sites visited include PJD Tower located at Jalan Tun Razak, Kuala Lumpur and Putri Park Hotel at Chow Kit, Kuala Lumpur.

17.2.2.2 Manufacturer operating manuals

Technical information about the air conditioning system and its operation manuals are normally stated clearly in specific manufacturer operating manuals. The manual contains detailed knowledge and technical knowledge of the domain. It also contains general procedures for fault diagnosis, machine start-up procedures, operating procedures, maintenance schedule and detailed specifications. Case studies were also conducted at the location stated in the manufacturer operating manuals to verify the validity of the expert system.

17.2.3 Design

By gathering and accumulating all the knowledge from the domain expert, it is crucial to select the knowledge representation technique. A prototype system is built to validate the research and provide guidance for future work. A complete and successful developed expert system must begin from step (a) to step (f) and details as below.

17.2.3.1 Selection of a prototype development tool

The development tool of KBS must have the ability to save time and fulfil all the criteria. The selection of a tool must fulfil the following conditions:

- Supportive of hybrid knowledge representation techniques.
- Varies with inference facility.
- Supportive of good interface facilities with external programs and systems.

The Kappa-PC expert system shell (1997) was chosen as the KBS tool as it fulfilled all the criteria above. Kappa-PC has strong object-oriented capabilities and allows production rules for the representation of knowledge. The knowledge base is to be constructed using heuristic information and additional calculations related to ACMV. It also provides a variety of user options.

17.2.3.2 Selection of knowledge representation technique

A knowledge representation technique that matches the air conditioning problems perfectly was required in the expert system. Hence, a rule-based system is required by the approach of:

$$If(X)Then(Y)$$

If the condition of the rule is satisfied, conclusion of the rule will be set as the result.

17.2.3.3 Selection of control technique

The backward chaining method is applied in this research project as the expert first considers the conclusions and then tries to prove it with all the supporting information.

17.2.3.4 Development of the prototype

An expert system shell was used to develop the system. The developed prototype is a model of the final system.

17.2.3.5 Development of interface

The expert system should provide an explanation facility that describes the system's reasoning to the user. The user interface is an important software module that provides explanation to the user. The user interface is developed in such a way that it is easy to access and user friendly.

17.2.3.6 Development of the product

The created system ought to refresh or update the system when new standards or procedures are accessible. Information regarding the maintenance and servicing of air conditioning units is stored in the knowledge base. The inference engine with the help of a Kappa-PC tool is responsible for scanning the information in the knowledge base by applying the 'if-then-rules' in order to get the best solution to tackle ACMV faults.

17.2.4 Testing and validation

The developed expert system is tested and evaluated to guarantee that software execution is relevant towards the objectives of this study. System validation and user acceptance are both important criteria for determining whether the expert system has achieved the main objective. System validation is important to determine whether the system performs the task satisfactorily whereas user acceptance deals with how well the system addresses the needs of the user.

Lastly, the developed system will be deployed into the work environment and ten technicians were asked to use the expert system to troubleshoot the ACMV system in the building. The objective of the field testing was to determine whether the expert system meets the actual fault diagnosis of problems and solutions.

17.2.5 Maintenance

System maintenance is important although it is the last phase of this research. It is required as most of the information in the knowledge base would need to be updated as time goes by. Companies may change new ACMV components or develop new operating procedures after a certain period of time and the expert system needs to be updated regularly. Anyway, this changing state requires proper modifications to the system.

17.3 System development

In this section, steps to develop an expert system using the Kappa-PC software are described. The Kappa-PC software is used to develop the expert system. The main feature of this system allows users to click and select their requirements from a window. The system will then suggest a solution to fulfil the need of the users.

The Kappa-PC application development system is a truly hybrid PC tool that combines critical technologies essential for the rapid development of low-cost, high-impact business applications (Kappa-PC). Solutions created in the Kappa-PC environment have shown a high return on investment in a wide range of PC applications including help desks, order processing, sales support, inventory control and manufacturing. Kappa-PC is suitable for developing and delivering powerful applications that become timely and cost-efficient solutions for critical problems.

Kappa-PC provides a wide range of tools for constructing and employing applications. The main window of the Kappa-PC system serves as a user interface to manage the development of an application.

The first step in this program is to create the class and the slot by using an Object Browser. A global instance is then created. When these steps are completed, the function of the system is created. Once the function is created, the next step is to create a rule for the system. After the rule is determined, the next goal is created for the system. After the goal and rule are created, the interface for the system is created. Once all steps have been accomplished, the system is tested in terms of its user interface. When the user interface passes the test and the system runs successfully, the program is considered complete. If the system results in failure, the program has to be rewritten.

The development of the system using Kappa-PC Software is divided into two major steps that are

(a) Development of the KBS
(b) Calculation and unit conversion

17.3.1 Development of the KBS

After the database of the system has been developed, it is ensured by the system development. Generally, the structure of the KBS consists of users, user interface, process information, database, knowledge base, inference engine and output.

17.3.1.1 Inference engine and knowledge base

The inference engine is an important part of an expert system which works based on rules. The inference engine scans and checks the condition given in the rules depending on the chaining process defined in the reasoning. The reasoning process is classified into two namely backward or forward chaining. Rules are scanned until one is found when constraint values match the user input. The scanning resumes and the results are deduced. The final results are reported to the user. The process continues until the final selection is made. Data is supplied to the interpreter via interactive dialogue by the user. The knowledge base of the proposed system consists of four separate groups of data concerning man, material, machine and method. Detailed information related to air conditioning is posted in the knowledge base. All the information is linked together by a hierarchical graph. Each element of the analytical model is represented as a frame or an object.

When a user enters the required information through the user interface, the inference engine is invoked. Then, the inference engine will try to search for the information from the knowledge base. In the KBS, the knowledge base and database work together. The inference engine will process the given information from the user interface based on the reasoning process defined in the reasoning database. Then, the inference engine will try to match the information with the data from the knowledge base and database. If the condition part of a rule in the knowledge base matches with the information given by the user, the conclusion of that rule is set as a result.

17.3.1.2 Making classes, subclasses and instances

Figure 17.2 shows the main menu of the Kappa-PC software. The window contains the Object Browser, Session, Edit Tools, KAL Interpreter, KAL View Debugger, Find Replace, Rule Relations, Rule Trace and Inference Browser.

The object browser allows the viewing and modification of the objects and their relationship in the application. It presents a graphical view of the object hierarchy. At the far left is the class root which is always a part of the knowledge base. Subclasses of the root (such as Menu, DDE, Image and KWindow) appear to its right, connected to the root by solid lines besides Global by dotted line. Any class objects with

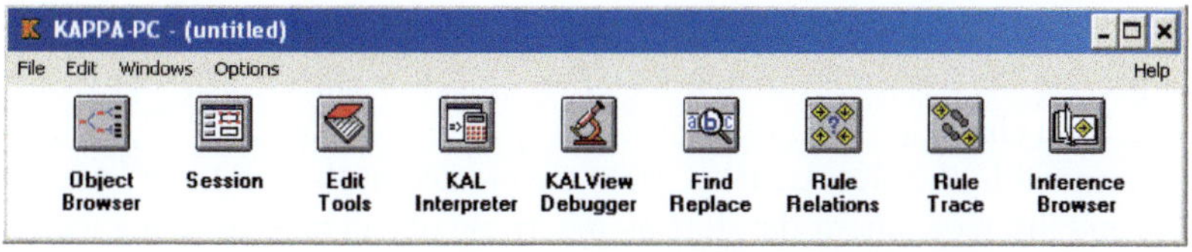

Figure 17.2 The main menu of the Kappa-PC software

Figure 17.3 Object browser

subclasses defined yet hidden are displayed with a box surrounding them; if the instances are hidden, then the objects are surrounded by an ellipse. Figure 17.3 shows the object browser of Kappa-PC.

In the Kappa-PC system, classes and instances are displayed on the object browser window. The classes and instances on the object browser window show a general overview of the way the knowledge base is structured, how it is organised into a hierarchy and how it forms the system's knowledge base.

Each of the class and instance objects contains slots and these are the places where information about things or concepts that are important in the knowledge domain is stored. In addition, each of the images that make up the user interface is represented by class and instance objects.

Objects can be either classes or instances within classes. A class is a more general object. It can be a group or collection such as ACMV. An instance is a more specific object. It is a particular item or event. Examples of specific objects include AHU, cooling towers, axial fans, chillers and others. To create a class, right-click on the parent class and select 'Add Subclass'. Next, a small window will appear and require the user to write the Class Name. Then, press the 'Ok' button and a subclass will appear on the hierarchy. Thirty-four classes had been created in this developed system.

To create an instance, right-click on the parent class and select 'Add Instance'. Next, window will appear and require the user to write the Instance Name. Then, the 'OK' button is pressed and an instance will appear on the hierarchy. A total of 325 instances were created in this developed system.

17.3.1.3 Slots and slot value for classes and instances

Slots represent the important properties of an object or class. Each slot in the class will describe a characteristic of the class. To specify the characteristic, value or text can be assigned to the slot. Usually, slots are created at the top-most level and the subclass and instance inherit the slots by inheriting object programming techniques.

17.3.1.4 Functions writing

KAL is the programming language used in the Kappa-PC expert system. It is an interpreted language designed for rapid prototyping. The developed system for air conditioning fault diagnosis is mainly divided into the main function and sub-functions such as the run function, conclude function, display function and reset function. The KAL programming language has been used for the development of the function. The function can be developed using the function editor.

Functions are the most flexible tool in KAL. Everything a user can do in Kappa-PC, from adding an object to activating a button, can be done easily, quickly and efficiently using the KAL function. Functions can be used to define the expression, or define other functions by combining standard functions.

Function is used to determine the action to be taken by a system when a user clicks on the button in the user interface. The function in Kappa-PC is in C-programming language. The function editor is used to define, examine and modify functions. The function editor has three boxes for developing application programs. Arguments contain the names of each function separated by spaces. The body contains the KAL expression defining the actions that the function which should follow. Comments contain static text information about the function. It has a menu bar for creating, editing and modifying the functions. It is activated by the editing tool window using the edit or new options. Some common functions that are created in this system are the *Reset* function and the *Back* function. Besides, there are few functions created for different purposes.

17.3.1.5 Main function

The main function includes the run function, sub-function, conclude function, display function and reset function. When the main function is executed, it will execute the sub-function step by step.

17.3.1.6 Conclude function

The conclude function is used to determine the fault diagnosis of the air conditioning system such as the causes and recommendations for a specific situation.

17.3.1.7 Display function

The display function is used to show the results to the user.

17.3.1.8 Reset function

The reset function will delete instances such as values, subclasses, slots and others in the object browser. In addition, the function also resets the entire slot value in the 'Global' instance.

17.3.2 Create a rule

Rules in Kappa-PC are basically a logical expression with *If* and *Then. If* defines the conditions for the inference engine to search for the rules that satisfy the user's input data. The use of 'And' and 'Or' is permitted within the IF condition list. *If* condition must be followed by a semicolon. *Then* will determine the goal that will be applied when the inference engine validates that the rules are true. If there is more than one conclusion, the lines within the THEN field must be surrounded by curly brackets and followed by a semicolon.

The rule editor is used to define, examine and modify the rules. The rule editor is activated from the editing tools. The rule editor contains five boxes which are the patterns, priority, if, then and comment. Patterns contain a list of all object patterns to be matched within this rule. If there are several patterns, they should be separated by spaces. Priority contains a number that the inference engine uses to resolve conflicts between the rules. The default priority is zero. If and then boxes contain the KAL expression. IF defines the condition of a rule and THEN defines the conclusion that the inference engine should perform if the rule is triggered. Comment refers to the comments about the rule. This comment is used for explanation purposes.

To create a rule for the system, the user needs to decide the reasoning method which can be forward chaining or backward chaining. The rule that proceeds from the premises towards the conclusion is called forward chaining. Meanwhile, backward chaining requires a predefined goal which can be written in the Goal Editor. The backward chaining process tries to satisfy the goal continuously.

To create a rule, the user needs to click on the 'Rule' tool in the Edit tool Window and select 'New'. The user will then be required to define the name of the rule. After the rule 'cfault1' is named, and the 'Ok' button is clicked, the Rule Editor Window will appear. On the Rule Editor Window, there are two boxes namely IF and THEN. To define the condition of a rule, the IF box is used. On the other hand, the THEN box is used to define the conclusion of a rule. The condition of rules consists of different sets of attribute values while the conclusion of rules consists of possible equipment type. In order to write up the rules of the condition, logic reasoning should be prepared as shown in Figures 17.4 and 17.5.

All of these rules are important when the inference browser is displayed after a session of the interface is conducted to determine the rule traces and their relations. A total of 38 rules were developed in this system.

17.3.3 Goal

Kappa-PC is rather restrictive in its use of goals. A goal generally refers to a slot value and whether or not its value has been inferred or set to a specific value. For example, the body of a valid goal may be:

KnownValue? (Person:Name);

In this expert system, the goal is generated through the backward chaining method. The inference engine will seek the rules based on a goal-driven method. An example of the goal editor is shown in Figure 17.6 with the expression which will generate the goal based on the determination of rules by the system inference engine.

Q: Cooling tower fan does not operate normally.

- ***Yes***
  - Q: Is the fan motor able to start?
    - → ***Yes***
      - AD: Check fan motor bearing and belt.
      - AD: Fan belt may be loosened or needs adjusting to the right position.
      - AD: Check the rotation.
    - → ***No***
      - AD: Return fan motor for full service.
- ***No***
  - Q: Is the fan operating?
    - → ***Yes***
      - AD: Call up electrician to check the fuse and wiring.
      - AD: Call up electrician to check the motor starter and overload.
    - → ***No***
      - AD: Call up electrician to check the motor. Repair or replace if defective.

Note: Q = Question, AD=Advice

Figure 17.4 The logic reasoning of rules (1)

Q: Compressor shuts off.

- → ***Yes***
  - Q: Compressor shuts off at high pressure?
  - ***Yes***
    - AD: Check the compressor pressure to determine if it is at a low or high pressure.
    - AD: The incoming pressure towards the compressor is high. Compressor will automatically stop operating when it detects over pressure. Over pressure can be due to contamination or dirt which accumulates at the incoming channel to the compressor. Flushing may need to be carried out.
  - → ***No***
    - AD: Check the compressor pressure to determine if it is at a low or high pressure.
    - AD: The incoming pressure is too low for it to run continuously. Low pressure can be due to the leakage of gas. Welding/Refill of refrigerant may be needed.
- → ***No***
  - Q: Is the electric circuit for the compressor tripped?
  - → ***Yes***
    - AD: Call up electrician to check fuse and wiring.
    - AD: Call up electrician to check overload.
  - → ***No***
    - AD: Call up electrician to check the compressor checking. Repair or replace if defective.

Note : Q = Question, AD=Advice

Figure 17.5 The logic reasoning of rules (2)

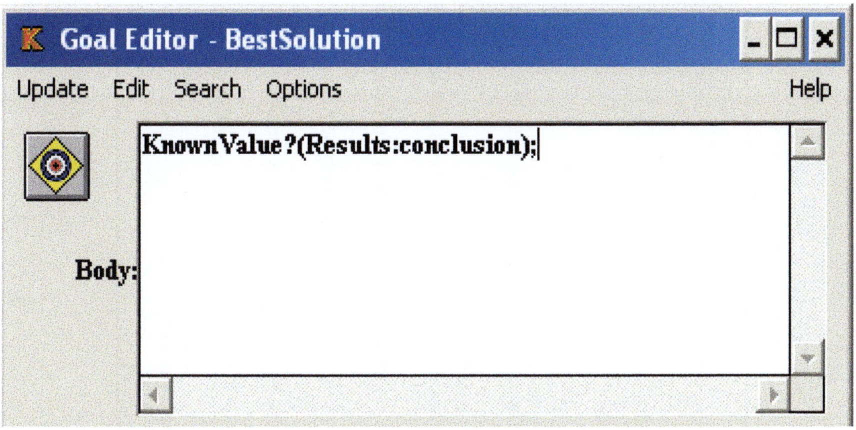

Figure 17.6 Example of a goal editor window

The generated goal will then be stored in the 'Conclusion' slot inside the class called 'Results'. Only one goal is created in this developed system.

17.3.4 User interface

There are two ways to work with images, namely by using the graphic development tools in the session windows and by using KAL graphic functions. The session window is where you create images that allow users to interact with the knowledge base. The session window is the main interface for the end-user of a Kappa-PC application. Kappa-PC allows the application developer to customise this window with different graphic choices and a display, creating an interface that simplifies the end user's task of interacting with the application.

The session window has two modes for two different purposes. The layout mode is used for manipulating graphic images through the mouse-and-menu interface. Meanwhile, the runtime mode is used to present the application interface to an end-user. A total number of twenty session windows were created in the developed system.

Individual graphic objects are known as images. Examples of images are line plots, bitmaps, state boxes and meters. Some of the images display information about the state of the application for the end users; some also allow the users to input information into the application. We can create, delete, hide, show, overlap, or move images at any time while an application is running. The function of each type of image is explained in Table 17.1.

To create text, for example, a user needs to select the text image on the image toolbar and click on the desired point where the user needs the text to be placed. Then, the user should double click on the text and a text option window will appear. At the title of the text option window, users can write the text. The user interfaces are linked to one another by using functions.

Table 17.1 Function of images in the session window layout mode

Images	Functions
Button	To create a button for users to click.
Text	To show the wordings in the window.
Edit	The space for users to type a value or display results.
Bitmap	To display a picture in the window.
Combo box	To create a dropdown menu for users to choose from.
Transcript image	This allows the user to display a file or a piece of text
State box	A state box is attached to a slot which has a set of allowable values. The actual value of the slot is the highlighted item in the state box.
Meter	A meter is again attached to a numeric slot and displays the value of the slot in meter form.
Slide bar	A slide bar is again attached to a slot and when the user slides the slide bar during the consultation, the value of the slot is changed accordingly.
Single box	The programmer can define a range of values for a slot. The user may select one of these options which becomes the value of the slot.
Multiple box	The programmer can define a range of values for a slot. The user may select a number of these options which become the value of a multi-valued slot.
Check box group	The programmer can define a range of values for a slot. The user may select a number of these options which become the value of a multi-valued slot.
Radio button group	The programmer can define a range of values for a slot. The user may select a number of these options which become the value of a multi-valued slot.
Check box	The programmer can define a range of values for a slot. The user may select one of these options which becomes the value of the named slot.

The same steps will be used to create the buttons 'EXIT' and 'HELP' in the title box. At the same time, type 'EXIT' and 'HELP' in the action box. After creating this, the session window will appear, as shown in Figure 17.7.

The title and action of buttons in the main session are shown in Table 17.2. Continue the same steps for other buttons that are present in the upcoming window session.

To modify the front session, the picture of the related topic can also be inserted. The same steps will be taken to modify the front session. Use the yellow palette of the user interface image that has just appeared or click on the 'Select' button in the session window. Next, click on 'Image' and choose 'Bitmap Option' from the sub-menu. A cross appears in the session window. Move this to a suitable place in the session window.

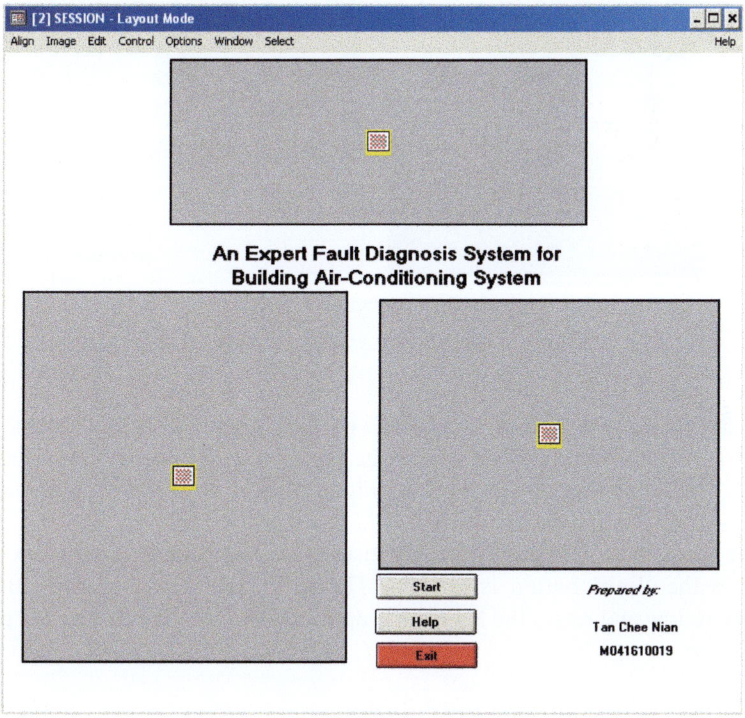

Figure 17.7 Front session with the button

Table 17.2 Title and action of buttons in the main session

Title	Action
Start	start
Help	help
Exit	exit

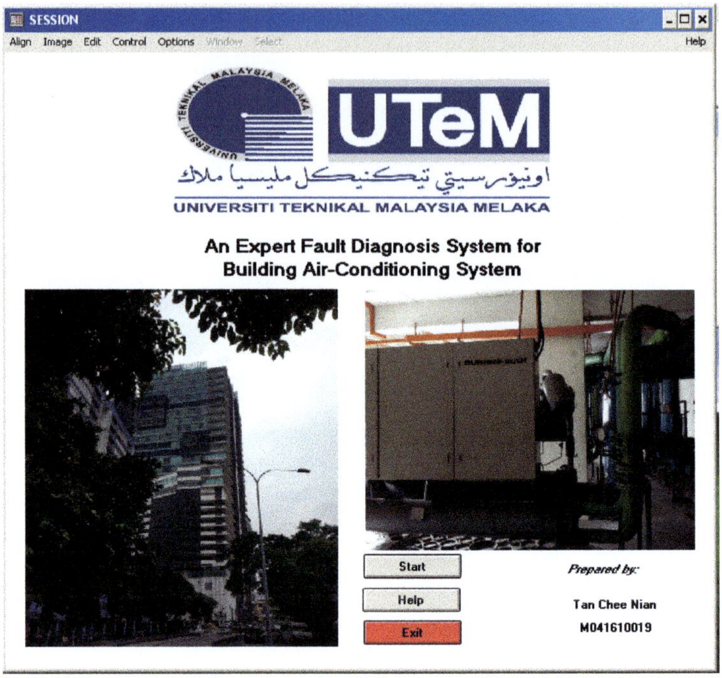

Figure 17.8 Front session with pictures

Now, make sure that the cursor is inside the purple-cornered box that has appeared and click twice. After that, create bitmap image boxes to insert the picture in the session window. The session window is shown in Figure 17.8. A complete list of the session in the developed system is shown in Table 17.3.

The 'Exit' button is used to let the user close the system. A window will pop out when the 'Exit' button is clicked. The user can choose to close the entire session window and retain the Kappa-PC application, close the Kappa-PC or cancel the function.

17.3.5 KAL view debugger window

The KAL view debugger as shown in Figure 17.9 is used to alert users about the errors in functions and methods. The KAL view debugger window is divided into three parts which include the source code window, the value window and the watch window. The source code window displays the source of the current function being examined. The value window is used to display the return value of each statement when stepping through the execution of the function. It also displays various messages. The watch window displays the values of the variables or slots being watched. It has a menu bar to access all the functions by which the KAL view debugger can be activated.

Table 17.3 Sessions in the developed system

Session	Topic
0	Main session
1	Fault diagnosis main session
2	Help
3	AHU
4	Cooling tower
5	Ducting
6	Axial fan
7	Water pump
8	Calculation (Cooling tower)
9	Chillers
10	Flushing method
12	AHU images
13	Cooling tower images
14	Ducting images
15	Axial fan images
16	Water pump images
17	Chillers images
18	Unit conversion
19	Calculation (AHU)

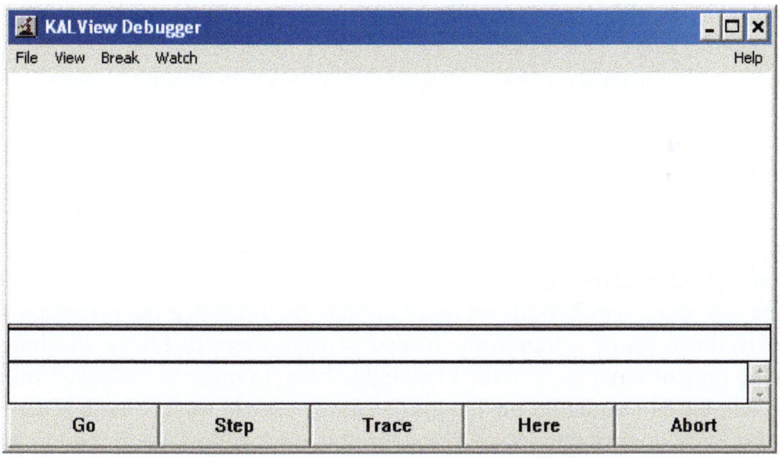

Figure 17.9 KAL view debugger window

17.3.6 Find and replace window

Find and replace window (Figure 17.10) is to replace the atom that appears in the knowledge base. In order to find an atom, the atom has to be entered in the Find field. As the Kappa-PC is case sensitive, the atom which has to be replaced should be typed properly. The atom which has to be replaced should be entered in the replace field.

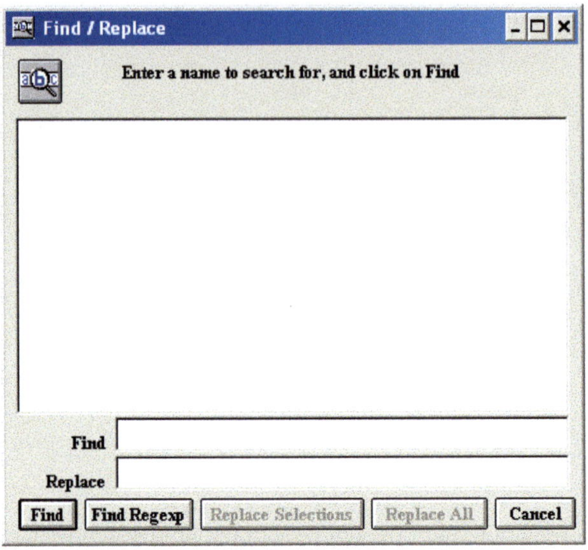

Figure 17.10 Find/Replace window

For the developed system, Find/Replace window has not been used to replace any atoms in the knowledge base. This is due to all objects are classified according to class, sub-class, instances and slots accordingly. There are no necessities to replace occurrences of the atoms as all objects have been arranged in a well-organised environment.

17.3.7 Rule window

Three tools including the rule relation window, rule trace window and inference browser are part of the rule window in Kappa-PC. The rule relation window in Figure 17.11 provides a graphical view of the relationship between rules. It can be revoked from the menu bar.

The rule trace window allows users to view the rules that the inference engine invoke in the form of a transcript. It also permits users to follow the impact of reasoning on particular slots in the knowledge base. In order to execute a rule trace during rule-based reasoning, tracing and breaking should be accomplished with the particular rules or slots before initiating the reasoning process. It helps the users to follow the reasoning process which can be either the forward or backward chaining process. It also permits users to follow the impact of reasoning on particular slots in the knowledge base.

In order to carry out the rule trace during rule-based reasoning, tracing and breaking should be accomplished with the particular rules or slots before initiating the reasoning process. It helps users to follow the backward chaining process. It has dropdown menus to accomplish the required function. Figure 17.12 shows the rule trace window.

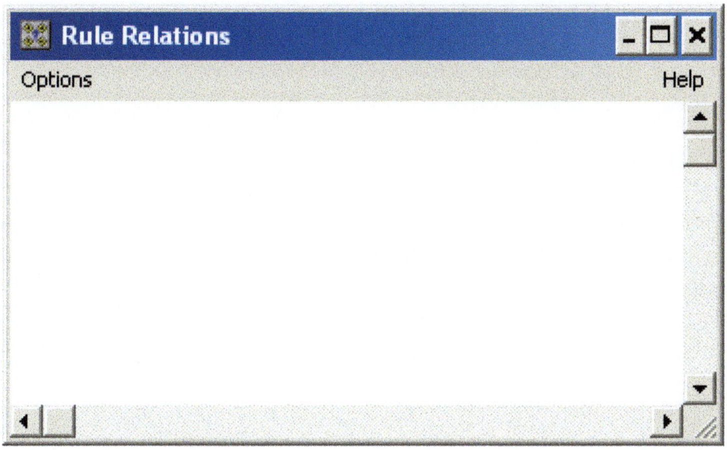

Figure 17.11 The rule relations

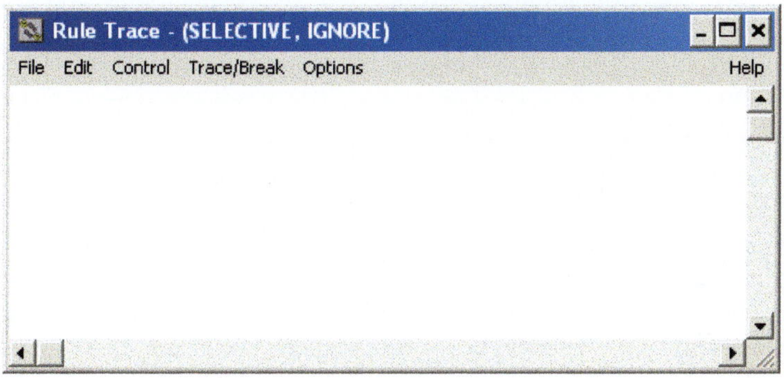

Figure 17.12 Rule trace window

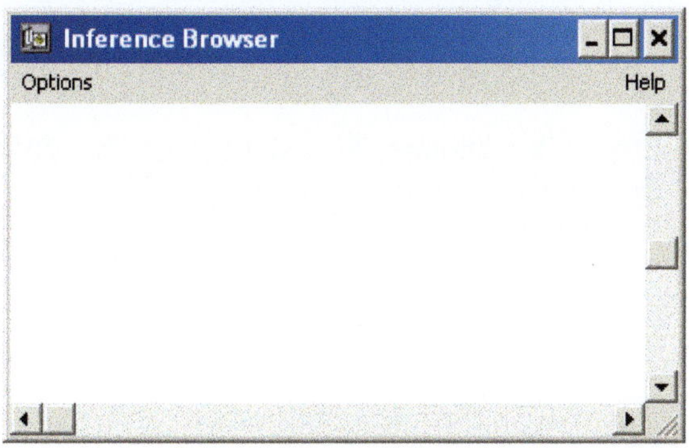

Figure 17.13 The inference browser window

The inference browser shown in Figure 17.13 displays the chaining relations among the rules in a graphical way. The inference browser is also used to trace the source of errors in the application of a KBS.

17.4 Summary

This chapter concludes the fundamental concepts and knowledge for system development of Kappa-PC software. Kappa-PC software utilises production rules, but the core value of knowledge representation is by means of the object. Kappa-PC objects can be in the form of classes, subclass or even instances. Overall, the developed expert system codifies the knowledge in the form of functions, rules, objects, goals and others in providing a wide range of applications.

References

[1] Chua, K., Chou, S., Yang, W. and Yan, J., 2013. Achieving Better Energy-Efficient Air Conditioning – A Review of Technologies and Strategies. *Applied Energy*, 104:87–104.

[2] Brodtmann, T., 2016. Why industry 4.0 is not just about industry. Available at <https://www.euractiv.com/section/digital/opinion/why-industry-4-0-is-not-just-about industry/> [Accessed on 25 Nov 2017].

[3] Mario, H., Tobias, P. and Boris, O., 2016. Design Principles for Industrie 4.0 Scenarios. 49th Hawaii International Conference on System Sciences, pp. 3928–3937.

[4] Malte, B., Niklas, F., Michael, K. and Marius, R., 2014. How Virtualization, Decentralization and Network Building Change the Manufacturing Landscape: An Industry 4.0 Perspective. World Academy of Science, Engineering and Technology, International *Journal of Mechanical, Aerospace, Industrial, Mechatronic and Manufacturing Engineering*, 8:37–44.

[5] Berger, R., 2016. Digitalization in the construction industry. Available at <https://www.rolandberger.com/.../tab_digitization_construction_industry_e_final.pdf> [Accessed on 27 November 2017].

[6] Einsiedler, I., 2013. Embedded Systeme fur Industrie 4.0. *Product Management*, 18:26–28.

[7] Hans, G.K., Peter, F., Thomas, F. and Michael, H., 2014. Industry 4.0. *Buisness & Information Systems Engineering.*

[8] Migilinskas, D., Popov, V., Juocevicius, V. and Ustinovichius, L., 2013. The Benefits, Obstacles and Problems of Practical Bim Implementation. *Procedia Engineering*, 57:767–774.

[9] Kalinichuk, S., 2015. Building Information Modeling Comprehensive Overview. *Journal of Systems Integration*, 25–34.

[10] Salman, A., 2011. Building Information Modeling (BIM): Trends, Benefits, Risks, and Challenges for the AEC Industry. *Leadership and Management in Engineering*, 11:32–38.

[11] Mansyur, R., Rahmat. R.A.O.K. and Ismail, A., 2013. A Prototype Rule-Based Expert System for Travel Demand Management. *UNIMAS E-Journal of Civil Engineering*, 4:34–39.

[12] Liebowitz, J., 1995. Expert Systems: A Short Introduction. *Engineering Fracture Mechanics,* 50:601–607.

[13] Negnevitsky, M., 2005. Artificial Intelligence: A Guide to Intelligent Systems, Harlow, England: Addison-Wesley.

[14] Dym, C.L., 1985. Expert Systems: New Approaches to Computer Aided Engineering. *J Eng Computer*, 1:9–25.

[15] Gemignani, M.C., Lakshmivarahan, S. and Wasserman, A.I., 1983. *Advances in Computers*, 22.

[16] W. Y. Leong, and D. P. Mandic, "Towards Adaptive Blind Extraction of Post-Nonlinearly Mixed Signals," *2006 16th IEEE Signal Processing Society Workshop on Machine Learning for Signal Processing*, IEEE, pp. 91–96, 2006.

[17] S. Huang, D. H. Zhang, W. Y. Leong, H. L. Chan, K. M. Goh and J. B. Zhang, "Detecting tool breakage using accelerometer in ball-nose end milling," *2008 10th International Conference on Control, Automation, Robotics and Vision*, IEEE, pp. 927–933, 2008.

[18] W. Y. Leong, J. Homer and D. P. Mandic, "An Implementation of Nonlinear Multiuser Detection in Rayleigh Fading Channel," *EURASIP Journal on Wireless Communications and Networking (EURASIP JWCN)*, vol. 2006, pp. 1–9, 45647, United States.

[19] W. Y. Leong and J. Homer, "Blind Multiuser Receiver in Rayleigh Fading Channel," *Australian Communications Theory Workshop (AusCTW'05)*, pp. 145–150, Brisbane, 2005.

[20] E. B. T. Dennis, R. S. Tung, W. Y. Leong, and C. M. T. Joel, "Sleep Disorder Detection and Identification," *Procedia engineering*, Elsevier, no. 41, pp. 289–295, 2012.

[21] W. Y. Leong, " HYPERLINK "http://scholar.google.com/scholar?cluster=466236349346697615&hl=en&oi=scholarr" Implementing Blind Source Separation in Signal Processing and Telecommunications," Thesis, University of Queensland, Australia, 2005.

[22] P. Mohankumaran and W. Y. Leong, "3D Modelling with CT and MRI Images of a Scoliotic Vertebrae," *Journal of Engineering Science and Technology EURECA*, pp.188–198, 2015.

Chapter 18

Lean green integration in manufacturing Industry 4.0

Puvanasvaran A. Perumal[1]

In recent years, programs intended to develop effective lean manufacturing systems have been implemented in many of the world's leading companies. Many of them have been highly successful in increasing efficiency, reducing costs, improving customer response time, and contributing to improved quality, greater profitability, and enhanced public image. Some companies have committed to reducing negative impacts of their operations on the environment. The "green" systems have created huge reductions in energy consumption, waste generation, and hazardous materials used. In addition, companies' images as socially responsible organizations are also highlighted.

Several research efforts indicate that lean companies show significant environmental improvements by being more resourceful and energy efficient. Some studies also show how lean and green systems share many of the same best practices to reduce their respective wastes. Yet, the consensus view is that these two systems tend to operate independently, administered by distinctly different personnel, even within the same manufacturing plant.

18.1 Green

The current definition of green is a notion of "helping to sustain the environment for future generations." Although ultimately true, urgencies at present demand more immediate and specific returns to launch projects in the near future. Currently, returns such as positive cash flow, reduced energy, material, and operating costs can make or break a company. Figure 18.1 shows the green building concept and its returns.

Figure 18.1 shows the six elements of green building concept in a mind-mapped manner.

[1]Department of Manufacturing Management, Faculty of Manufacturing Engineering, University Teknikal Malaysia Melaka, Malaysia

Figure 18.1 The green building concept mind mapping

18.2 Lean green

Green is a subset of sustainability (and, thus, green manufacturing is a subset of sustainable manufacturing). Lean green studies have been trying to fill the gap of insufficient tools and methods for environmental initiatives as compared to the sufficient tools and methods in production [1]. Ricoh defined sustainability in terms of development and progress as follows: "We are aiming to create a society whose environmental impact is below the level that the self-recovery capability of the natural environment can deal with." In addition, they give a simple example: "the reduction target of CO_2 emissions is generally based on the 1990 emission level, but in the future we need to limit emissions based on the estimated emission level that the self-recovery capability of the Earth could deal with." Hence, this definition points out the key elements that there is a strong social component in addition to the usual business and environmental emphasis (we usually define three "legs" to sustainability as economic benefits, environmental benefits, and social benefits), there is a comparison with a "sustainable level" that has natural roots (e.g., the ability of the environment to accommodate the inputs/impacts we make to it), and the need to adjust our level of impact to be consistent with that. Besides, Ricoh illustrated the need to adjust the target to ensure that we stay within acceptable levels of impact—that is, adjust our definition of the level of sustainability we are targeting.

This offers real challenges to engineers, especially with respect to manufacturing and production of goods. Currently, if we are not operating at a "sustainable level" (think Ricoh's example and the level of energy/material/water/other resources used for manufacturing or the impacts of manufacture) then we need to adjust our processes, systems, and enterprises to get to that level over time. This is not easy with the population growth and accompanying growth demand just "business as usual," which will increase unsustainable trends.

Therefore, we need to have a compounded reduction in our impact/use that considers both the increase in demand and difference between a sustainable and unsustainable level of consumption/impact. This mismatch is the so-called wedge pointed out by the excellent paper on stabilization wedges [2]. Green manufacturing deals with technologies and solutions that provide these wedges—help to "turn the super tanker" if you will and, ideally, with enough wedges we transition from business as usual to a sustainable level of impact/consumption. Hence, our premise is that this can be done to our competitive advantage and profitably. The rising strategy of integrating both lean and green manufacturing is a potential approach to acquire business goals of profit, efficiency and environment sustainability [3].

18.3 Integration of lean and green

Lean integrated with green relate to three sections: lean management system, lean waste reduction technique, and lean business results. Figure 18.2 illustrates the lean system model including the three related lean.

18.4 Lean green needs

Going green is an important trend in an era of environmental responsibility. Unlike lean manufacturing that focuses on ways to improve operations and cut wastes from the customer's perspective, green initiatives look at the ways to eliminate waste from the environment's perspective.

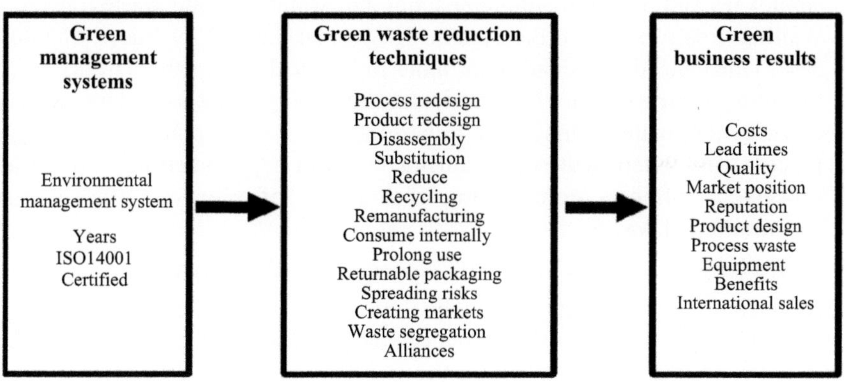

Figure 18.2 Parallel model for lean and green operation

With current tight credit market, high cost of raw materials and transportation, stiff global competition, and a weak dollar, lean and green manufacturing can provide the competitive advantage and profitability many manufacturers are looking for.

Besides, Taiichi Ohno's seven forms of waste are also one of the reasons why the integration is needed in green lean, which must be identified and eliminated in the pursuit of green office and business systems just as in production.

The seven forms of waste introduced by Ohno are as follows:

1. **C**orrection
2. **O**verproduction necessary
3. **M**otion
4. **M**aterial movement
5. **W**aiting
6. **I**nventory
7. **P**rocess.

COMMWIP is an acronym that has been developed to assist folks in remembering the seven forms of waste (note: the COMMWIP forms are the same as those introduced by Toyota's scientist, Ohno, but with different words and order). The good news is that tackling the seven forms of waste becomes the foundation for reducing the traditional safety and environmental wastes:

1. Correction equals defects, which equate to injuries and illnesses, along with wasted materials that create negative environmental impacts. Reducing defects reduces human and environmental exposures.
2. Overproduction may increase ergonomic risk to employees and result in outdated materials having to be disposed without being used.
3. Motions that do not add value often create stress and strain for employees. Reaching, bending, and twisting do not add value to the customer, but increase risk of injury to employees.
4. Material movement increases exposure to moving materials on the factory floor and, to a lesser degree, in the office. Reducing exposure reduces risk to employees and potential spills impacting air, water, or solid waste.
5. Waiting creates stress in the system whether on the factory floor or behind a desk. Unbalanced workloads contribute to physical and mental stress.
6. Inventory increases the risk of trip hazards and blind spots along with the potential that material may have to be scrapped at some point.
7. Process waste often creates non-value-added extra steps and errors in the system. Process is often a contributing root cause of injuries, illnesses, and environmental waste.

Nevertheless, it is important to remember that the globalization demands a greater competitiveness of the private and public companies, which should be focused not only in the subjects related to the efficiency of the production systems but also in those factors of the production which involve the preservation of the environment.

The White Paper describes the thinking behind the paradigm shift to going green and the increased business and profit it can generate. Like any change, or paradigm shift, there will be those who embrace it and those who resist it. Knowing the reasons for making the change toward a greener way of being, as seen in the business case, and then having a proven, systematic approach to going green, like the green value stream approach, provides the support needed to begin this change. Embracing the change early will reserve an organization's seat as a profitable leader in the increasingly global movement toward environmental sustainability.

18.5 Lean green benefits

There are many benefits in applying lean green to manufacturing. Many companies use lean green to achieve their mission and become more successful. Here, six benefits are discussed. However, the benefits in terms of economic business which increases competitiveness are highlighted. There are direct cost savings, increased customer loyalty and attraction, increased employee attraction and retention, ability to grow, innovation, and development of new technologies, and increased profit and shareholder value.

In business, lean green is applied for direct cost savings when doing business. Raw material, operational, and administrative costs are reduced. It is better to use less raw materials, or recycle or reuse it for positive environmental impacts which save money. In addition, if less energy and water are used, less garbage will be produced and less traveling on distance. Hence, huge materials can be saved and operational and administrative costs are reduced. Many companies are already harvesting these savings, which are added to the bottom line [4]. This is a simple, true, and logical idea that can be accepted.

Nowadays, business and all transactions are done globally. Hence, companies will face difficulties in attracting new customers to do business with them and also penetrating the markets. Company should try to enter a global business and compete with other companies. At the same time in online or global market, it is possible to attract regular customers and also make the market as customers' choice [4]. All customers should be appreciated by providing good services for them. Likewise, the same procedures can be applied to suppliers to create good values for business. However, attracting new customers should be prioritized instead of existing customers.

A February 2008 survey conducted by Cone Inc., in conjunction with the Boston College Center for Corporate Citizenship, confirmed this claim [4]. In this survey, around 59% of 1,080 participants are concerned on environmental factors and they are changing their habits of purchasing. From that survey, 66% participants said that they actually know and consider thoroughly when they are buying something. However, 68% of them said that their action to buy product or service depends on the reputation of the company in caring for environment. If the company is applying green continuously, the company is much better in attracting new customers and retaining existing customers. This way is much better if compared with other companies that do not apply green.

Employees need to be taken care of their welfare to earn their loyalty to a company. Hence, employers must focus on related aspects that show sensitivity in terms of hiring, maintaining, and retraining all of them. The global competition needs to take risks on retaining and attracting good employees. Focusing on green or environmental sustainability issues taken up by the organization will help to address this challenge, because employees will stay longer with a company they believe has treated them well and fairly [4]. Of the 28% of respondents from Cone Inc.'s Millennia survey (discussed previously) who worked full time, 79% of them wanted to work for a company that cares about how it impacts or contributes to society [4]. However, 64% are loyal to their company due to the company's social and environmental activities and another 68% said that they do not want to work if the company does not concern on social and environmental responsibilities. In short, if companies have the intention to keep their good employees and at the same time try to attract their employees, it can consider applying green and can avoid being bypassed by their competitors who already embrace that principle.

The next benefit is the ability to grow. A company can grow by leading, developing, and increasing the size of business by following up an increasing demand of something. This can be made more clear by stating which costs to cut subject to community or government price, for example, the petroleum price. A company will pursue and find environment-friendly alternatives if it focuses deeply on environmental impacts of their existing products. Hence, the inevitable increase in material cost for the remainder of the products' life can be eliminated. Another example involves operations that are energy intensive. Focusing continually on energy conservation while supporting alternative clean energy sources will help in leveraging the energy cost. Focusing on improving the company's environmental impact will see energy reduction ideas flow from staff to ensure that the right energy decisions are continually being made.

Harvesting materials or natural resources faster than they are regenerating, or using materials that are not capable of being regenerated, will at some point inhibit growth and curtail production. It is imperative that in order for a business to sustain growth it must start the journey toward environmental sustainability before it is too late, and in its planning all elements of the supply chain must be considered. With a focus on going green, along with lean, companies' continuous improvement will be a constant concern and becomes a part of everyday life.

Innovation will be stressed as the challenge turns to how things can be done differently and better to reduce environmental impacts. By constantly striving to do things better, innovation becomes critical to success. By committing to lean and green practices, tools and thinking process innovation will become a high priority. Lean will be one of the strongest allies and if combined with going green, it will drive innovation. It is no secret that innovation is a key factor in the success of any organization because it allows a company to reduce cost and lead times and increase capacity. All of these elements will be valued by customers. Being innovative refers to develop the ability to do things differently.

This requires the implementation of lean thinking and a no blame work environment that is based on respect and process thinking. These traits combined with open communication, lean leadership, and meaningful employee involvement will lead to cost savings and other benefits of going green. Other advantages include enhanced productivity and capacity, reduced lead times, leading to more savings, benefits, and true value for customers. For example, using less material leads to less processing, resulting in reduced process costs. If less transportation is involved, this too will contribute to reduce lead times, and there is less garbage to move and deal with and more time to build product and expedite the delivery to customers. Focusing on green and lean will make the staff to think outside the box by experiencing firsthand how they can help customers reduce their environmental impact through the use of the company's products and services.

Focusing on lean and green will further encourages innovation that may lead to the development of new technologies the customers want which in turn can help companies to reduce their environmental footprint. It is easy to see how going green drives process improvement and new technology development. The market for new green technologies both on a business-to-business and business-to-consumer levels is growing exponentially and will only continue to grow as the world looks for ways to lessen environmental impact. Focusing on green will drive the development of new and greener technologies, allowing it to offer solutions that customers are looking for and ensuring the ability to meet the customer demand at present and in the future.

This is the last benefit of lean green. However, lean green can be seen as a platform for company to increase more profits and at the same time make huge savings beyond the original projection. By going green, it is not unusual for companies to increase their profits by 35% or more in less than 5 years. A study found that a large enterprise can yield profits for 38% within 5 years. Willard's study also showed that smaller enterprises can yield even higher percentage with some achieving over 50%.

This increase in profit is typically derived from the continuous improvement of cultural growth, commitment to given process and productivity of employees, material use, and operating costs. The amount of success will depend on the degree to which every employee is involved and aligned with the overall vision and business objectives. Another contributor will be the increased revenues from new and loyal existing customers. Success is evident not only from empirical studies, but also from cold, hard financial results that are subject to audit especially from companies listed in the Dow Jones Sustainability Index (DJSI). This index tracks the financial performance of the leading sustainability-driven companies throughout the world which consistently outperform the rest of the market. In the eyes of the shareholders, this is the value that investors are seeking.

18.6 Lean green disadvantages

With the increase of government regulations and stronger public awareness in environmental protection, currently firms simply cannot ignore environmental issues if they want to survive in the global market. In addition to complying with the environmental regulations for selling products in certain countries, firms need to implement strategies to voluntarily reduce the environmental impacts of their products. The integration of environment, economic, and social performances to achieve sustainable development is a major business challenge for the new century.

Four types of indicators, each expressing a different kind of impact on the state of natural resources, are as follows:

(a) *Waste water*. These indices consist of total water consumption and some specific critical water wastes (carbon dioxide, total nitrogen, phosphorous, dissolved salts) due to the supplier's plants.

(b) *Air emissions*. Air pollution occurs when the air contains gases such as SO_2, NH_3, CO, HC1, dust, and fumes in harmful amounts. The amount of air emission could be harmful to human's health or comfort which could cause damage to plants and materials.

(c) *Solid wastes*. Solid wastes are any discarded (abandoned or considered waste-like) materials. Solid wastes can be solid, liquid, and semisolid or containerized gaseous material. Examples of solid wastes are scrapped metal, discarded appliances and vehicles, domestic refuse, and uncontaminated used oil. A material is discarded if it is abandoned by being disposed of or burned.

(d) *Energy consumption*. The use of energy as a source of heat or power or as a raw material input to a manufacturing process such as an indicator point out of the supplier's total amount of energy consumption within the year.

18.7 Elements integrate lean green

Green manufacturing goals are to conserve natural resources for future generations. The benefit of green manufacturing is to create a great reputation to the public, decrease cost, and promote research and design. Green manufacturing has several processes to go through to keep our environment clean. It "involves the smart design of products, processes, systems, organizations, and of smart management strategies implementation, that effectively harness technology and ideas, to avoid environmental problems before they arise." In other words, companies are making "cleaner processes and products" for creating a better environment. The different hazards and wastes that are produced during manufacturing represent money that is lost and liability risks that are unnecessary. The elimination of these two factors will, in turn, increase consumer demand for green products, and the corporation will gain an advantage in the competing market because of the cleaner products.

There are seven elements of lean green integration:

1. *Focus on the customer and eliminate waste*: Waste elimination should be viewed from customers' perspectives and all activities that do not add value to them should be perused thoroughly and eliminated or reduced. In an integration context, customer refers to an internal sponsor or a group within an organization that uses, benefits from, or pays for, the integrated capabilities.
2. *Improve continuously*: A data-driven cycle of hypothesis–validation–implementation should be used to drive innovation and improve the end-to-end process continuously. Adopting and institutionalizing lessons learned and

sustaining integration knowledge are related concepts that ease this principle establishment.

3. *Empower the team*: Creating cross-functional teams and sharing commitments across individuals empower the teams and individuals who have a clear understanding of their roles and needs of their customers. The team is also supported by senior management to innovate and try new ideas without fear of failure.
4. *Optimize the whole*: Adopt a big-picture perspective of the end-to-end process and optimize the whole to maximize the customer value. This may at times require performing individual steps and activities which appear to be suboptimal when viewed in isolation but aiding in the end-to-end process streamlining.
5. *Plan for change*: Application of mass customization techniques such as leveraging automated tools, structured processes, and reusable and parameterized integration elements lead to reduction in cost and time for both the build and run stages of the integration lifecycle. Another key technique is a middleware service layer that presents applications with enduring abstractions of data through standardized interfaces that allow the underlying data structures to change without necessarily impacting the dependent applications.
6. *Automate processes*: Automation of tasks increases the ability to respond to large integration projects as effectively as small changes. In its ultimate form, automation eliminates integration dependencies from the critical implementation path of projects.
7. *Build in quality*: Process excellence is emphasized and quality is built in rather than inspected in. A key metric for this principle is the *first time through percentage*, which is a measure of the number of times an end-to-end process is executed without having to do any rework or repeat any of the steps.

The following are the advantages derived by adopting the green lean integration practices:

1. *Efficiency*: Typical improvements are in the scale of 50% labor productivity improvements and 90% lead-time reduction through continuous efforts to eliminate waste.
2. *Agility*: Reusable components, highly automated processes, and self-service delivery models improve the agility of the organization.
3. *Data quality*: Quality and reliability of data are enhanced and data becomes real asset.
4. *Governance*: Metrics are established, which drive continuous improvement.
5. *Innovation*: Innovation is facilitated using fact-based approach.
6. *Staff morale*: IT staff is kept engaged with high morale, driving bottom-up improvements.

18.8 Lean green tools

There are various lean green tools, namely kaizen, 5S, cellular manufacturing, just-in-time (JIT), total productive maintenance (TPM), and six sigma. These tools have been introduced at the beginning of this chapter. Other than these tools, 3P, also known as preproduction planning, is one of the most powerful and transformative advanced manufacturing tools and typically used only by organizations that have experience in implementing other lean methods. While kaizen and other lean methods adopt a production process and seek to make improvements, the 3P focuses on eliminating waste through product and process design.

18.8.1 Production Preparation Process (3P)

Lean experts typically view 3P as a method to meet customer requirements by starting with a clean product development slate to rapidly creating and testing potential product and process designs that require the least time, material, and capital resources. This method typically involves a diverse group of individuals in a multiday creative process to identify several alternative ways to meet the customer's needs using different product or process designs. 3P typically results in products that are less complex, easier to manufacture (often referred to as design for manufacturability), and easier to use and maintain. 3P can also design production processes that eliminate multiple process steps and that utilize homemade, right-sized equipment that better meet the production needs.

Ultimately, 3P method represents a dramatic shift from the continuous, incremental improvement of existing processes sought with kaizen events. Instead, 3P offers potential to make "quantum leap" design improvements that can improve performance and eliminate waste to a level beyond which can be achieved through the continual improvement of the existing processes.

The teams spend several days (with singular focus on the 3P event) working with 3P to develop multiple alternatives for each process and evaluate each alternative against manufacturing criteria (e.g., designated takt time) and a preferred cost. The goal is typically to develop a process or product design that meets the best customer requirements in the "least waste way." The typical steps in a 3P event are as follows:

1. *Define product or process design objectives/needs*: The team seeks to understand the core customer needs to be met. If a product or product prototype is available, the project team breaks it down into component parts and raw materials to assess the function that each plays.
2. *Diagraming*: A fishbone diagram or another type of illustration is created to demonstrate the flow from raw material to finished product. The project team then analyzes each branch of the diagram (or each illustration) and brainstorms key words (e.g., roll, rotate, form, bend) to describe the change (or "transformation") made at each branch.
3. *Find and analyze examples in nature*: The project team then tries to find examples of each process key word in the natural world. For example, forming can be found in the nature when a heavy animal such as an elephant walks in mud, or when water pressure shapes rocks in a river. Similar examples are grouped, and examples that best exemplify the process key word are analyzed to better understand how the examples occur in nature. Here, team members place a heavy emphasis on how and why nature works in the examples. Once

the unique qualities of the natural process are dissected, team members then discuss how the natural process can be applied to the given manufacturing step.

4. *Sketch and evaluate the process*: Sub-teams are formed, and each sub-team member is required to draw different ways to accomplish the process in question. Each of the sketches is evaluated and the best is chosen (along with any good features from the sketches that are not chosen) for a mock-up.
5. *Build, present, and select process prototypes*: The team prototypes and then evaluates the chosen process, spending several days (if necessary) working with different variations of the mock-up to ensure that it will meet the target criteria.
6. *Hold design review*: Once a concept has been selected for an additional refinement, it is presented to a larger group (including the original product designers) for feedback.
7. *Develop project implementation plan*: If the project is selected to proceed, the team selects a project implementation leader who helps determine the schedule, process, resource requirements, and distribution of responsibilities for completion.

18.9 Lean enterprise supplier networks

Lean manufacturing can entail significant organizational and technological changes to a firm's manufacturing operation. With these changes, and the potential for significant resulting gains, companies may often determine that it will be more productive to work with suppliers who have also adopted lean techniques and who can pass on resulting financial savings and better accommodate their "customer's" lean production schedules. Fundamentally, lean supplier networks involve the application of lean manufacturing techniques across multiple supply chain partners to deliver products of the right design and quantity at the right place and time, resulting in mutual cost and waste reduction benefits for all members of the network. Leaning can require a dramatic change in production thinking and organization among suppliers, hence working closely with existing suppliers to either directly or indirectly facilitate lean is preferable over a threat to switch suppliers if they do not lean on their own [5].

The first step in creating lean supplier networks (and attempting to introduce lean to suppliers not yet exposed to lean) is to convey the obvious benefits of leaning—not only to the network but also to each individual participant. The next step is to take a role in helping suppliers make the "lean transformation" by sending experienced trainers to work with the suppliers (free of charge) and/or by agreeing to share the ultimate savings with them [6]. More specific options are as follows:

- *Network technical assistance*: Often, the most effective approach is for a leaned "customer" within a network to send its own lean production team out to work with suppliers. Experts can conduct both lean instruction and "learning by doing" sessions for suppliers, and they make periodic return visits to ensure thorough understanding and implementation.
- *Network lean exchanges*: Where members of the network have already begun to implement lean but can benefit from the wisdom of their supply chain

partners, companies can "swap" lean technical experts and other employees and learn from each other's experiences.

- *Lean consultants*: Companies who would like suppliers to become lean can hire outside consultants with leaning experience in their industry to assist in the transformation.
- *Supplier associations*: Supplier associations represent another way to accrue benefits from shared lean experiences. Such associations can be organized by "tier" (where all first-tier suppliers are brought together to learn from each other). These associations can then pool its resources to focus on leaning second-tier suppliers. Table 18.1 shows the environmental benefits of lean methods.

A total of eight lean methods with their corresponding potential environmental benefits is shown.

Table 18.1 Environmental benefits of lean methods

Lean method	Potential environmental benefits
Kaizen	• Continual improvement culture focusing on eliminating waste • Uncover and eliminate hidden wastes and waste-generating activities • Quick, sustained results without significant capital investment
5S	• Decrease lighting, energy needs when windows are cleaned and equipment is painted light colors • Notice spills and leaks quickly • Decrease potential for accidents and spills with clearly marked and obstacle-free thoroughfares • Reduce contamination of products resulting in fewer product defects, which reduces energy and resource needs (avoids waste) • Reduce floor space needed for operations and storage, potential decrease in energy needs • Less unneeded consumption of materials and chemicals when equipment, parts, and materials are organized, easy to find, less need for disposal of expired chemicals • Visual cues can raise awareness of waste handling or management procedures, workplace hazards, and emergency response procedures
Cellular manufacturing	• Eliminate overproduction, thereby reducing waste and the use of energy and raw materials • Fewer defects from processing and product changeovers, reducing energy and resource needs (avoids waste) • Notice defects quickly, preventing waste • Less use of materials and energy (per unit of production) with right-sized equipment • Less floor space needed, potential decrease in energy use and less need to construct new facilities • Easier to focus on equipment maintenance, pollution prevention

(Continues)

Table 18.1 (*Continued*)

Lean method	Potential environmental benefits
JIT	• Eliminate overproduction, thereby reducing waste and the use of energy and raw materials • Less in-process and postprocess inventory needed, avoiding potential waste from damaged, spoiled, or deteriorated products • Frequent inventory turns can eliminate the need for degreasing metal parts • Less floor space needed, potential decrease in energy use, and less need to construct new facilities • Can facilitate worker-led process improvements • Less excess inventory, reducing energy use associated with transport and reorganization of unsold inventory
TPM	• Fewer defects, reducing energy and resource needs (avoids waste) • Increase longevity of equipment, decrease needs for replacement equipment and associated environmental impacts (energy, raw materials, etc.) • Decrease number and severity of spills, leaks, and upset conditions—less solid and hazardous waste
Six Sigma	• Fewer defects, reducing energy and resource needs (avoids waste) • Focus attention on reducing the conditions that result in accidents, spills, and malfunctions, thereby reducing solid and hazardous wastes • Improve product durability and reliability, increasing product life span, reducing environmental impact of meeting customer needs
3P	• Eliminate waste at product and process design stage, similar to "Design for Environment" methods • Nature (inherently waste free) is used as a design model • Right-sized equipment lowers material and energy requirements for production • Reduce the complexity of the production process ("design for manufacturability") can eliminate or streamline process steps, environmentally sensitive processes can be targeted for elimination, because they are oftentimes, resource- and capital-intensive • Less complex product designs can use fewer parts and fewer types of materials, increasing the ease of disassembly and recycling
Lean enterprise supplier networks	• Magnify environmental benefits of lean production (reduced waste through fewer defects, less scrap, less energy usage, etc.) across the network • Realize environmental benefits by introducing lean to existing suppliers rather than finding new, already lean suppliers

18.10 Lean green application

The organizations that implement lean green can be seen in Table 18.2. The topic and tools are stated along with the discussion and results obtained from the implementation.

From the table, a total of two companies implement lean successfully and saved cost from material, space, energy, and waste disposal.

18.11 Lean green standards

Green standard is used to standardize environmental management system in the world. It comprises a series of documents in managing the waste without affecting the environment. Thus far, green does not have a complete implementation standard. However, ISO 14000 series are fundamental in applying the environmental management system.

The ISO 14000 series of standards have been designed to help enterprises meet their environmental management system needs, developed by the ISO since 1991. They consist of a set of documents that define the key elements of a management system that will help an organization address the environmental issues. The management system includes the setting of goal and priorities, assignment of responsibility for accomplishing them, measuring and reporting on results, and external verification of claims. Even though the first standards in the series were not published until late 1996, many organizations have been implementing the system using the drafts since mid-1995. There is an intense interest in these standards around the world. However, there is often a lack of clear understanding about what they are and what role they can play.

In addition, the ISO 14000 standards have been designed to help an organization implement or improve its environmental management system. The standards do not set performance values but they provide a way of systematically setting and managing performance commitments instead. That is, they are concerned with establishing "how to" achieve a goal, not "what" the goal should be. In addition to the core management systems standards, there are also a number of guidelines that provide supporting tools. These include documents on environmental auditing, environmental performance evaluation and environmental labeling, and lifecycle assessment.

The President and CEO of International Institute for Sustainable Development have stated that a decade from now we may recognize ISO 14000 standards as one of the most significant international initiatives for sustainable development. ISO 14000 defines a voluntary environmental management system. The standards are used along with appropriate goals and management commitment to improve corporate performance. They will provide an objective basis for verifying a company's claims on its performance. This is particularly important in relation to international trade, where at present almost anyone can make assertions about environmental performance and there are only limited means to address veracity.

Consumers, governments, and companies in the supply chain are all seeking ways to reduce their environmental impact and increase their long-run sustainability.

Table 18.2 Organizations that implement lean green

No.	Year/author	Topic	Tools	Discussion	Results
1.	2009/ON Semiconductor	Divert recyclable material from landfill to local organization for reuse among local recyclers	5S	Implement blue booties recycling: provide area realtors and builders associations for reuse at open house and parade of house	Avoid a disposal of 800 pairs of booties per month
		Create green team to have an ongoing company-wide ownership of green projects	Kaizen	Implement plastic bottle, aluminum can, newspaper, and magazine recycling	Recyclables are now collected and the on-site collection is managed by green team members to prevent an increase in workload for custodial services
		Improve and standardize spray paint techniques to optimize transfer efficiency and reduce overspray	Kaizen	Right-sized nitrogen gas production system	Energy savings and reduce overproduction of nitrogen gas
		Consolidate buildings and like processes	5S	Consolidate office space and duplicate services	Save 3.4 million gallons of water annually Save 12,000 thermos of natural gas annually Save US$32,000 annually in electricity Save US$400,000 annually in rent and other costs
		Reduce energy consumption	5S	Turn off lab incubators when not in use	Reduce incubator energy consumption by 70%
		Reduce landscaping water use	5S	Revegetate south lawn with native plants, create walking trails, and raise beds for employees to grow organic vegetables or plant	Save 4.9 million gallons of water annually, promote employee health and wellbeing, save US$18,000 annually by reducing water use and maintenance
		Reduce maintenance and repair cost associated with ground maintenance	5S	Eliminate redundant lab refrigerators	Discontinued using one refrigerator by consolidating materials, saving energy

(Continues)

Table 18.2 (Continued)

No.	Year/author	Topic	Tools	Discussion	Results
2.	2011/General Motors Corporation	Reusable containers	Kanban	Reusable containers serve as kanban/ signals for when more parts are needed in a particular process area	Eliminate tons of packaging wastes each year and reduce the space, cost, and energy needs of managing such wastes
		Paint booth cleaning	5SKaizen	Reduce cleaning frequency and use purge solvent	Improve process "up-time" and flow and decrease 369 tons per year
		Electronic-based system in RFQ process	Kaizen	Improve procurement efficiency and lower cost by saving time and eliminate waste	Save 2 tons of paper per year

For companies, the key goals are to become more efficient—get more output per unit of input—while earning profits and maintaining the trust of their stakeholder. The ISO 14000 voluntary standards will help. It is important to note that the ISO 14000 standards do not themselves specify environmental performance goals. These must be set by the company itself, considering the effects the standards have on the environment, and the views of the company's stakeholders. Implementation of a management system-based approach will help companies focus attention on environmental issues and bring them into the main stream of corporate decision-making. ISO 14000 is designed to provide customers with a reasonable assurance that the performance claims of a company are accurate. In fact, ISO 14000 will help integrate the environmental management systems of companies that trade with each other globally.

The ISO process has not fully involved all countries or levels of business. Some consumers and environmental organizations may well be skeptical of voluntary standards. There is a large measure of capacity building needed throughout the world in order for this system to work well. Finally, sustainable development requires that issues of human well-being be added to environmental and economic policies. While sustainable development is introduced within the ISO 14000 standards, the detailed documents deal almost exclusively with the environmental issues.

18.12 Chapter summary

Lean green, a term for the utilization of a wide range of lean tools for sustainable development, has brought great benefits to the environment by reducing wastage for both resources and energy. With the reduction of waste, there are greater direct costs saved. Although lean green may not have a firm standard yet, the ISO 14000 series are being used at present. Lean green is an asset to the world as it can provide the guideline for a greener environment toward Industry 4.0.

References

[1] Shahbazi, S., Kurdve, M., Zackrisson, M., Jonsson, C., and Kristinsdottir, A. R. (2019). Comparison of four environmental assessment tools in Swedish manufacturing: A case study.

[2] Pacala, S. (2004). Stabilization wedges: Solving the climate problem for the next 50 years with current technologies. *Science*, 305(5686), 968–972. doi: 10.1126/science.1100103

[3] Thanki, S. J., and Thakkar, J. (2018) Interdependence analysis of lean-green implementation challenges: A case of Indian SMEs. *Journal of Manufacturing Technology Management*, 29(2), 295–328. doi:10.1108/jmtm-04-2017-0067

[4] Wills, B. (2009). *The Business Case for Environmental Sustainability (Green) Achieving Rapid Returns from the Practical Integration of Lean & Green*. Retrieved from http://seedengr.com/documents/TheBusinessCaseforEnvironmentalSustainability.pdf

[5] Womack, J. P., and Jones, D. T. (1996). *Lean Thinking: Banish Waste and Create Wealth in Your Corporation.* New York: Simon & Schuster.
[6] Emiliani, M. L. (2000). Supporting small business in their transition to lean production. *Supply Chain Management: An International Journal*, 5(2), 66–71. doi:10.1108/13598540010319975

Chapter 19

Lean government in improving public sector performance toward Industry 4.0

Puvanasvaran A. Perumal[1]

19.1 People development system

In today's competitive world, the success of social development depends on a competent, well-functioning government and public sector [1]. However, no organization in business and even in public sector in developing countries can afford to reduce and eliminate the waste resources whenever they face the challenge of delivering a wide range of services essential for development—from infrastructure and social services to the functioning of the legal system and enforcement of property rights—all of which pose the challenge of how to get governance "right." In both organizations, the most underutilized resource is their people.

A good governance has always been important in organizations, even before it becomes fashionable, where people are one of the few influential assets. Nevertheless, governance is complex, both as an idea and in the work needed to realize good governance in the midst of its infusion into our collective conscience and pervasion of the way we talk about management within business, community organizations, or public administration. The quality of governance institutions also has a significant impact on economic growth.

Unfortunately, most of the developing countries are finding centrally regulated public service policy a hindrance to effectively deliver public services in modern globally competitive scenarios [2]. One reason is that governments always design public services based on the one-size-fits-all model that eventually leads to a notion that a standard service offers economies of scale [3]. In fact, governments that invest in that policy become more efficient, flexible, and responsive and are more likely to meet their goals in relation to the part of the economy concerned by providing the basic government services [4].

Making the government function better implies not only improving efficiency and cost-effectiveness of public sector functions and operations but also improving all of the public sector effectiveness so that government policies and programs work

[1]Department of Manufacturing Management, Faculty of Manufacturing Engineering, University Teknikal Malaysia Melaka, Malaysia

smoothly, achieve the stated desired objectives, treat recipients with respect and dignity, and positively affect people in which they are designated to minimize any negative distortionary side effects [1]. Moreover, the governments all over the world have been under pressure to reduce the size of public sector, budgets, and expenditures (sometimes especially in the social sector), and at the same time improve their overall performance, then come the challenges to the government, particularly in striking the right balance between accountability and increased flexibility [5].

The range of issues in improving the provision and quality of public sector involves establishing public services where they are needed yet lacking, while in the cases where they do exist, increasing their effectiveness to achieve improved outcomes. The reason is that the public sectors are the portion of society controlled, where they also provide services that non-payers cannot be executed from services that benefit society rather than just the individual who uses the service that encourages equal opportunity as well. The public sectors are those entities owned and/or controlled by the government, as well as the entities and relationships that are funded, regulated, and operated solely or in part by the government.

However, the most important and heavily criticized issues of various public sector limitations are as follows:

- Lack of motivation. Employees of public organizations are partially prepared to take larger workload and think that requirements raised for them are rather too small than too large
- Status officials. Too strictly defined status determines lack of flexibility, and this first of all does not allow optimal use of personnel capabilities, thus preventing public officials from seeking personnel career in the public sector
- Lack of possibilities to pursue career and develop skills
- Automatic position upgrading
- Limited possibilities in personnel selection.

19.2 Lean implementation in public sector

Most research on expectations related to consumers' purchase and consumption experiences in private markets are theoretically highly relevant to people's experiences with public service, although they require modification for use in the context of public sector. Dr. Zoe Radnor and Mr. Paul Walley, Warwick Business School researchers, have found the method employed by Toyota in making their production system "lean" that can be applied to public sector services. Using "lean" in the public sector will have a positive impact on employees' morale, customer satisfaction, and process efficiency because the organization will improve in customer waiting times, service performance, processing times, customer flow, and quality by achieving more for less, generating a better understanding of the process, better joined-up working, improved use of performance data, increased staff satisfaction and confidence, and improved continuous culture [6]. In a nutshell, lean technique takes the premise that all organizations are made up of a number of organizational processes in a search for customer value determinants. Bhatia and Drew [3] commented that a lean approach is crucial for public sector because it breaks the prevailing view that there has to be a trade-off between quality of public services and cost of providing them.

The term "lean" stemmed from a 1990s bestseller *The Machine that Changed the World: The Story of Lean Production*. The book authors, Daniel Jones and James Womack, identified five core principles of lean as follows:

- Specify the value desired by the customer.
- Identify the value stream for each product providing value and challenge of all the wasted steps.
- Make the product or service "flow" continuously along the value stream.
- Introduce "pull" between all steps where continuous flow is impossible.
- Manage toward perfection so that the number of steps and the amount of time and information needed to serve the customer continually falls.

Read [7] commented that the lean manufacturing process can be used by government departments to improve customer focus and avoid some common business pitfalls because lean concept encourages services to consider three main elements as follows:

- Failure demand (what extra work is created because processes fail)
- Remove processes that do not add value (what work can be stripped out because they do not add value for the majority customer)
- Reduce the movement of work between departments (reduce "double" handling or more work is dealt at the point of contact).

Therefore, to succeed, the public sector organizations must find a way to align their growth strategy—providing new and better services at limited costs—with regard to the interests of their workers besides optimizing costs, quality, and customer service constantly. By using lean principles, business companies and government can realign their organizations and invest in the development of team leaders.

Although they differ in the value of the resources (capabilities and environments) and implications for designing and implementing the strategy, such private and public sector organizations are striving to produce value for stakeholders in their environment by deploying resources and capabilities [8]. Bridgman [9] stated that compliance against the cost is a significant and mandatory investment. Even though compliance alone is insufficient for high performance or good governance, understanding the relationship among compliance, performance, and good governance increases the prospect of achieving return on investment. A strategy allows a clear policy deployment and concentration of effort which in turn allow increasing process capability and exploiting new capabilities as shown in Figure 19.1.

To become fully lean, an organization must understand lean as a long-term philosophy where the right processes will produce the right results and value can be added to the organization by continuously developing people and partners, while continuously solving problems to drive organizational learning [10]. Although there is no exact definition for a fully lean organization, it is important that an organization must understand and apply all of the practices and principles. It is also important to understand that lean thinking, which affects the whole business model,

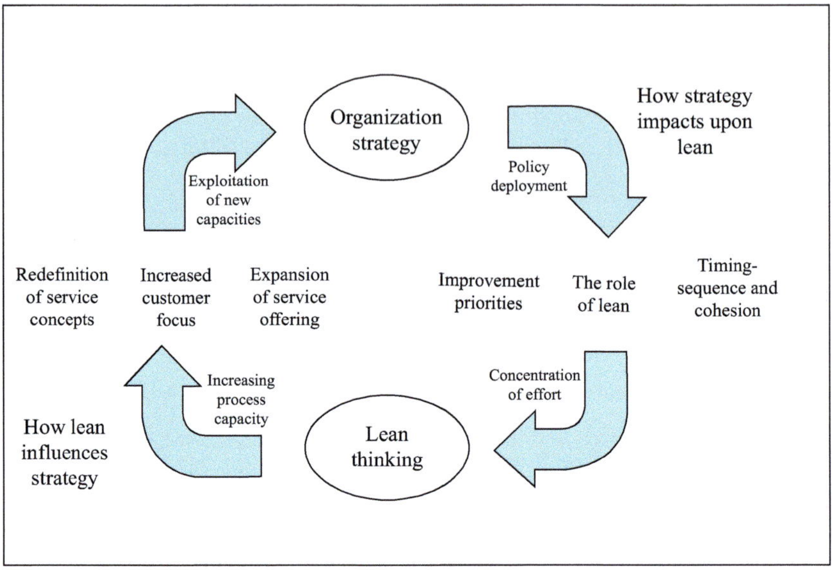

Figure 19.1 Relationship between strategy and lean [6]

is the key and not solely leaner production, where only parts of the whole of lean philosophy are applied.

19.3 Performance measurement in improving public sector

At present, performance measurement is central to a new public management, which is characterized as a global movement reflecting liberation management and market-driven management [11]. The liberation management involves public sector managers who are surrounded by a plethora of cumbersome and unnecessary rules and regulations. Instead of controlling input factors, outcome measures should be the primary focus.

Public sectors are mostly institutional in nature and their effectiveness may vary according to cultural and historical contexts, legislative frameworks and institutions, as well as differences in levels of socioeconomic development among high-, middle-, and low-income countries. Some organizational forms are not consistent with public administrations; besides, the scale of an organization impacts on its capacity to deliver on complex governance activity. Big agencies can afford the organizational costs of strong compliance and performance activity, whereas smaller ones may struggle to meet these overheads. But, if scale makes finding resources easier, it can also make governing harder because most government entities are allocated a budget through a centralized process. This has important implications and points to differences from other sectors. Therefore, performance measurement may provide data on how effectively and efficiently public services are delivered. Within public sectors, performance measurement has fostered a move toward a contract value on various levels. Anyhow, managing and measuring performance have been the key drivers in the reform of the public sector in recent years.

Performance measures are necessary to be competitive with the private sectors. A few examples of activities practiced by the private sectors are as follows:

- *Voice and Accountability* (in China, citizens of the Lioning province participate in the local hearing against major cases of corruption and the diversion of public funds) [12]
- *Simplification of Administration* (communities and advisory bodies of communities, e.g., KGSt or *Kommunale Gemeinschaftsstelle für Verwaltungsvereinfachung* that works with local government board and is an advisory body of German local governments (cities, municipalities, and countries). They come up with the new steering model that uses Tilburg model as performance measurement reference scheme) [13]
- *One Stop Shops* (in Latvia into public administration with objective to change the priorities of public administration, so the intention would be serving the people and anyone attending the competent institution only once could receive the necessary service) [14]
- *Strategic Personnel Management* (in Kaunas-Lithuania in personnel management through new public management) [15]
- *Expectations Disconfirmation, Expectations Anchoring and Delivery Process* (in England against service and households refuse collection) [16]
- *Total Quality Management* (in Malaysia) [17]
- *The Results-Oriented Management Initiative* (in Uganda) [18]
- *Privatization and Downsizing* (in UK and New Zealand in converting service departments into free-standing agencies or enterprise, whether within civil service or outside it altogether) [19].

Market-driven management seeks to create internal and external competition for budgetary resources to decrease X-inefficiency and budget-maximizing behavior. Čiarnienė *et al.* [15] commented that for an organization to become "High Performance," public organization needs to focus on the following aspects of their organization:

- Vision, mission, and goal-directed with continuous improvement
- Preference to multiskilled worker
- A flatter and more flexible one replaces the tall and rigid organization hierarchy
- Job enrichment and dispersed decision-making, resulting from promoting continuous learning at all organization levels
- Managerial control is maintained less by exercise of formal authority.

Critical success factors must be met [6] as follows:

- (Developing) Organizational readiness
- Organizational culture and ownership

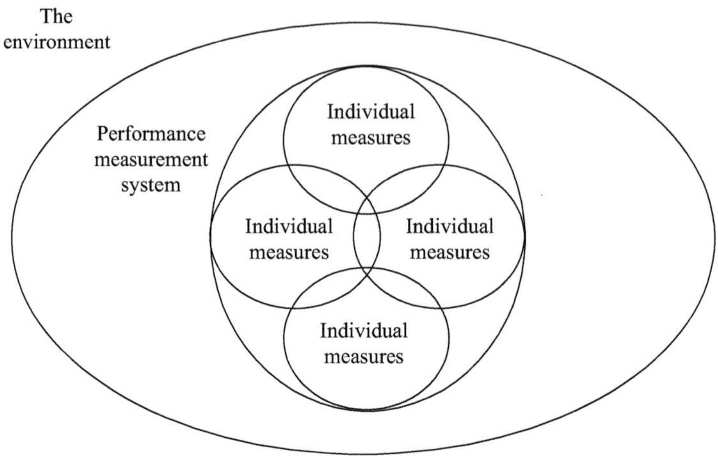

Figure 19.2 Individual measures when integrated will develop a performance measurement system [22]

- Management commitment and capability
- Adequate resources fund initial changes and external expertise that will create ongoing internal skills and competences
- Clear communication process and engagement
- Strategic deployment and management of lean activities (strategic approach)
- External support
- Teamwork
- Timing.

Performance control system can serve two purposes: to measure and to motivate [20]. The firm becomes what it measures [21]. Measurement has become such an accepted approach within organizations that a considerable effort is utilized in identifying *"What" can be measured* and "*How" to measure it.* Every measurement activity incurs costs to both implementation and maintenance realms. A few individual performance measurements will be integrated into a performance measurement system. This is shown in Figure 19.2.

19.4 Problems of lean implementation in public sector

Of course, there are obstacles to overcome. Applying lean is difficult in the private sectors, and more so in the public ones. The biggest challenge faced in implementing lean in organizations is in freeing up resources from existing activities to devote to new initiatives because public sector managers sometimes lack the skills, experience, and mindset to launch this approach [3]. Organizational rigidity and silo structures and thinking tend to inhibit cooperation and communication throughout agencies, where finding the necessary resources becomes much more complicated [4]. Therefore, the developers of a lean system should identify end-to-end processes from a customer's perspective and then design and manage the system to keep information and materials flow smoothly through those processes.

When performance measurement is used primarily to assist administrators to manage their agencies, the task of developing them is more likely to be seen as an investment. However, when the development of performance measurement is imposed from the outside, it may help internal accountability, but it is a mean of assuring external accountability and thus may meet resistance [23]. These problems result from the nature of the metrics or the instruments used for benchmarking and assessing change in performance, the political context of agency operations, and the possible dysfunctions of performance measurement [24–26].

Lean in the public sector is not a quick fix, yielding results steadily over a long implementation of time span. Anyhow, there is a need for many organizations that adopt lean immediately to get the change management experience or the right leadership style to make the transition straightaway [6] because most public organizations do not have agility or frontline empowerment to respond to changing demands of their customers. This is because the customers often have no choice of providers.

The key characteristic of lean organization is its ability to improve itself constantly by bringing problems to the surface and resolving them, then a system masks which underlying problems in many organizations that keep their "water levels" high and deal with problems drive the managers in public sector are often temptation to add something to the system (e.g., installing expensive IT systems), whereby these temptations could be failures. Huge benefits probably would have been more likely even without the new IT systems if government managers manage to tackle the underlying problems.

Successful lean transformations must close the capability gap early in the process, so managers and staff can make the transition to a new way of working. This is because lean requires more than courage to uncover deep-seated organizational problems; it may call for the ability to deal with job losses as well. A lean process requires a performance-tracking system that breaks down top-level objectives into clear, measurable targets that workers at every level must understand, accept, and meet.

The overriding purpose of a lean system is to configure assets, material resources, and workers in a way that improves the process flow for customers' benefit while minimizing losses caused by waste, variability, and inflexibility. To become fully lean, an organization must understand lean as a long-term philosophy where the right processes will produce the right results and value can be added to the organization by continuously developing people and partners while continuously solving problems to drive organizational learning [10]. While there is no exact definition for a fully lean organization, it is important that an organization must understand and apply all the practices and principles. Besides, lean thinking, which affects the whole organization model, is the key and not solely leaner production, where only parts of the whole of lean philosophy are applied.

However, the major difficulties an organization encounter in attempting to apply lean are a lack of direction, planning, and project sequencing [27], since decades ago the lean concepts were viewed as a counterintuitive alternative to traditional manufacturing model proposed [28]. The concept of waste has not yet been effectively extended to the self-defeating behaviors of individuals and groups of people in the workplace [28].

Pullin [29] insisted that to reap the full benefit, we need to view lean not as an abstract philosophy, but one that includes both concepts—a philosophy, and practices, tools, or processes. In addition, Toni and Tonchia [30] argued that in lean production, management by process is an organizational method, aiming to carry out, at the same time, several performances, including their continuous improvement, by means of an organization structure operation flows oriented toward results and flexibility with regard to changes.

Famholtz [31] stated that four key areas are involved in which all organizations must manage their culture or values:

- the treatment of customers
- the treatment of an organization's own people or human capital
- standards of organizational performance
- notions of accountability.

Barnes *et al.* [32] stated that the key shift for organizations in the increasingly competitive operating environment revolves around the heightened emphasis on the concepts of upgrading, and the different conceptualization of what value means for manufacturing firm. In addition, Parker [33] asserted that when lean production is introduced, it is often accompanied by modifications based on local orientations. In addition, lean organization can also vary in terms of implementation. Gates and Cooksey [34] asserted that the conventional approaches that over-rely on single-loop learning process that ignores dynamic complexities in the human condition and in organization system have failed to take into account the more demanding experiential side of human learning and development which may be just as important as the rational approach that so often captures the delivery agenda. Importantly, new skills demands are no longer solely technical skills. There is now a strong emphasis on the development of the "softer" skills that are required to increase team-working with the acceptance of increased responsibility, and new focus on the communication and transmission of knowledge and ideas both within and outside the organization.

Emiliani [28] defined repeated mistakes as another primary type of waste and argued that a business that is unable to learn and change its behavior will, "no doubt, risk the future existence of their entire enterprise as currently governed." In a lean organization, learning continues because "lean is a continuum and not a steady state" [35].

19.5 Conclusion

The development of performance measurement toward public sectors which is imposed as part of lean implementation may face resistance. Therefore, the improvement in performance across a range of business, technical, and human factors in public sectors depends on top managers who lead the lean transformation through the direct participation and consistent application of both principles: "continuous improvement" and either explicitly or implicitly "respect for people" where the managers directly participate in kaizen and other process improvement activities [35]. If the focus is on improving people, a likely outcome is that those people will possess the right skill set to continue improvement activities on other processes even though these varied accounts toward the effective and behavioral aspects of lean are largely as important as its cognitive dimension when it comes to implementing it [36], as the lean implementation in the public sector is not a quick fix and the lean itself is a long-term philosophy to drive organizational learning in problem-solving capability.

References

[1] UN Expert Group Meeting. (2003). *Improving Public Sector Effectiveness*. 42nd Session of the Commission for Social Development Priority, Dublin Ireland, June 16–19.

[2] Matheson, A. (1998) *Managing Public Service Performance: Some Ideas from the Commonwealth*. Paper to the Commonwealth Advanced Seminar on Setting Agenda for Public Service Reforms, Wellington.

[3] Bhatia, N. and J. Drew. (2008). *Applying Lean Production to Public Sector*. http://www.mckinseyquarterly.com/Public_Sector/Applying_lean-production_to_the... (accessed on August 11, 2008).

[4] Jarrar, Y. and G. Schiuma (2007). Measuring performance in the public sector: Challenges and trends. *Measuring Business Excellence*, Vol. 11, No. 4, pp. 4–8.

[5] Halachmi, A. (2002). Performance measurement and government productivity. *Work Study*, Vol. 51, No. 2, pp. 63–73.

[6] Warwick Business School, W. (n.d.). Can The Public Sector Become Lean? Retrieved October 27, 2019, from http://www.wbs.ac.uk/news/releases/2006/06/16/can/The/Public

[7] Read, C. (2008). Applying Lean in the Public Sector: How We Use Lean Manufacturing Techniques in Government Departments. 10 May 2008. Retrieved October 28, 2019, from. http://customer-management.suite101.com/article.cfm/applying_lean_in_ the_public_sector

[8] Ackoff, R. L. (2003). Creating a competitive strategic advantage. *Journal of Innovative Management*, Vol. 9, No. 1, pp. 31–51.

[9] Bridgman, P. (2007, April). Performance, conformance, and good governance in the public sector. *Key Issues: Risk Management—Keep Good Companies*, Decisive Consultants Pty Ltd, pp. 149–157.

[10] Liker, J.K. (2004). *The Toyota Way: 14 Management Principles from the World's Greatest Manufacturer*. New York: McGraw-Hill.

[11] Ginakis, G.A. (2002). The promise of public sector performance measurement: Anodyne or placebo? *Public Administration Quarterly*, Vol. 26, No. 1, pp. 34–64.

[12] Knox, G. and Z. Qun. (2007). Building public service-oriented government in China. *International Journal of Public Sector Management*, Vol. 20, No. 5, pp. 449–464.
[13] Greiling, D. (2005). Performance measurement in the public sector: the German experience. *International Journal of Productivity and Performance Management*, Vol. 54, No. 7, pp. 551–567.
[14] Riga. (2000). Activities for public service quality improvement in public administration institutions of Latvia. *The 8th NISPAcee Annual Conference*. Working group of Better Quality Administration for Public.
[15] Čiarnienė, R., Sakalas, A., and Vienažindienė, M. (2006). Strategy personnel management in public sector: the case study of Kaunas Municipality. *Engineering Economics*, Vol. 47, No. 2, pp. 62–69.
[16] David, F. (2007). *Professional Fellowship: Models Measurement and Inference in Social Research*. Full Research Report, ESR End of Award Report, RES-153-25-0036. Swindon: ESRC, pp. 22–45.
[17] Common, R. (1999). Accounting for administrative change in three Asia-Pacific states: the utility of policy transfer analysis. *Public Management*, Vol. 1, No. 3, pp. 429–438.
[18] Langseth, P. (1995). Civil service reform in Uganda: lessons learned. *Public Administration and Development*, No. 15, pp. 365–390.
[19] Polidano, C. (1999, November). *The New Public Management in Developing Countries*. IDPM Public Policy and Management Working Paper No. 13.
[20] Mintzberg, H. (1978). Patterns in strategy formulation. Management Science, Vol. 24, No. 9, pp. 934–948.
[21] Hauser, J.R. and Katz, G.M., 1998 Metrics: you are what you measure!, *European Management Journal*, Vol. 16, No.5, pp. 517–528.
[22] Neely, A.D., Mills, J., Platts, K., Richards, H., Gregory, M., Bourne, M., and Kennerley, M. (2000). Performance measurement system design: developing and testing a process-based approach. International Journal of Operations & Production Management, Vol. 20, No. 10, pp. 1119–1145.
[23] Townley, B. and Cooper, D. (1998). Performance measures: rationalization & resistance. *A Paper for Performance Measurement: Theory and Practice Conference*. Cambridge University, Cambridge, July 17.
[24] Halachmi, A. (1996). Promises and possible pitfall on the way SEA reporting. In A. Halachmi and G. Bouckaert (Eds.), Organizational Performance and Measurement in the Public Sector. Westport, CT: Quorum Books, pp. 77–100.
[25] Halachmi, A. (1997). Government reform and public productivity: do we have all the answer? *Work Study*, Vol. 46, No. 7, pp. 233–245.
[26] Halachmi, A. and Boorsma, P. B. (1998). *Inter- and Intra-Government Arrangement for Productivity: An Agency Approach*. Boston, MA: Kluwer Publisher.
[27] Bhasin, S. and Burcher, P. (2006). Lean viewed as a philosophy. *International Journal of Manufacturing Technology Management*, Vol. 17, No. 1, pp. 56–72.

[28] Emiliani, M.L. (1998). Continuous personal improvement. *Journal of Workplace Learning*, Vol. 10, No. 1, pp. 29–38.

[29] Pullin, J. (2002). In pursuit of excellence. *Professional Engineering*, Vol. 15, pp. 1–6.

[30] Toni, A.D. and Tonchia, S. (1996). Lean organization, management by process and performance measurement. *International Journal of Operations & Production Management*, Vol. 16, No. 2, pp. 221–236.

[31] Famholtz, E. (2001). Corporate culture and the bottom line. *European Management Journal*, Vol. 19, No. 3, pp. 268–275.

[32] Barnes, J., Bessant, J., Dunne, N., and Morris, M. (2001). Developing manufacturing competitiveness within South African industry: the role of middle management. *Technovation*, Vol. 21, pp. 293–309.

[33] Parker, S.K. (2003). Longitudinal effects of lean production on employee and the mediating role of work characteristics. *Journal of Applied Psychology*, Vol. 88, No. 4, pp. 620–634.

[34] Gates, G.R. and Cooksey, R.W. (1998). Learning to manage and managing to learn. *Journal of Workplace Learning*, Vol. 10, No. 1, pp. 5–14.

[35] Liker, J.K. (1998). *Becoming Lean: Inside Stories of U.S. Manufacturers*. Portland, OR: Productivity Press.

[36] Emiliani, M.L. (2006). Origins of lean management in America. *Journal of Management History*, Vol. 12, No. 2, pp. 167–184.

Chapter 20

Lean dominancy in service Industry 4.0

Puvanasvaran A. Perumal[1]

Lean is a system that is applied in manufacturing industry. It is the systematic elimination of waste, and the implementation of continuous flow concepts and customer pull. Lean is derived from Toyota Production System by Henry Ford and has been applied at Ford automobile. Lean is the best management system for satisfying customers on delivery, quality, and price. In Bowen and Youngdahl [1], the manufacturing sector has typically led the service sector in developing ways to resolve performance trade-offs among low cost, dependability, quality, and flexibility that are assumed to exist in early manufacturing strategy research. Lean is commonly applied by manufacturing company, but nowadays people start applying lean in-service industry.

20.1 Definition

Although in service industry, the definition of lean in service is still the same as lean in manufacturing, lean manufacturing is an operational strategy that orients toward achieving the shortest possible cycle time by eliminating waste. In service industry, lean is also an operational strategy that orients toward achieving the shortest possible service cycle time by eliminating waste and increasing profit.

In service industry, lean has been used to support company growth strategy: financial services companies to put mergers back on track, energy companies to lower costs, telecommunications companies to improve customer service, and retailers to increase efficiency while boosting customer service in the store. Those are examples of advantages of applying lean in service industry. Guarraia *et al.* stated that there are steps in applying lean in service [2].

The steps in applying lean in service are:

1. Enterprise value stream mapping (VSM). Initially, a map of the operation's processes and the costs associated with them are developed. Then, the enterprise is scanned, and its primary processes are mapped to identify the biggest opportunities to reduce cost by reducing wasted time and materials.

[1]Department of Manufacturing Management, Faculty of Manufacturing Engineering, University Teknikal Malaysia Melaka, Malaysia

2. Benchmarking. The performance of processes is measured against internal and external benchmarks to gauge shortcomings and establish improvement targets.
3. Prioritizing. Team determines which process improvements will yield the greatest results when lean is applied. The traditional five steps DMAIC—define, measure, analyze, improve, and control. Process can begin on the targeted area.

20.2 Transformation in lean services

Lean is an approach that is used to manage the operation to reduce, maximize, or minimize in terms of waste, late time and the main purpose is to improve the activities to satisfy the human priority. Long time ago, lean is established in automobile sector as lean operation. Then, lean operation began as lean manufacturing in the mid of 1990s. The main development for lean concept can be seen in the Japanese automobile manufacturer, Toyota. Figure 20.1 illustrates the elements of transformation of the lean.

Bitner *et al.* stated that service now constitutes the majority of employers and source of income to developed economies, accounting for approximately three-quarters of gross domestic product in the USA and the UK [3]. Even though the service constitutes the majority of employers and sources, Dickson *et al.* suggested

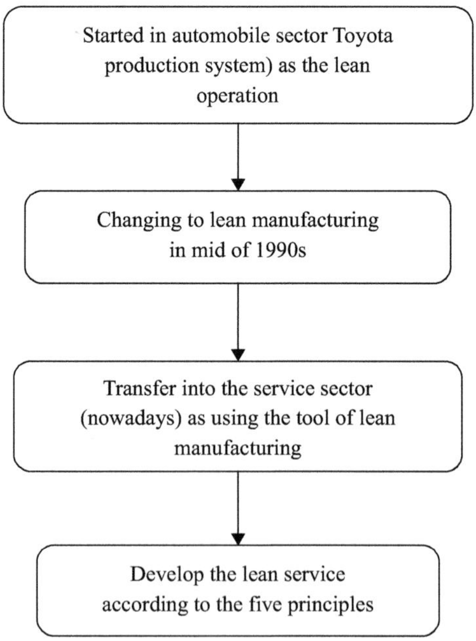

Figure 20.1 Flowchart of lean transformations

that the level of service quality is actually declining, with years on service deteriorated by significant amount [4].

Many researchers and practitioners have echoed their calls for lean adoption in services [5]. Bowen and Youngdahl stated that lean approaches such as work redesign, training increment, and process mapping in retail, airline, and hospital management could generate positive but a more general set of change principles, many of which share commonality with some aspects of lean thinking [1]. The benefits of lean application in services may become a potential competitive priority for all organizations.

Other than that, in lean services, the transformation has focused not on the handling of commercial products but on the handling of patients such as healthcare system. One focus of research is on the purchasing or supplying chain inputs to the healthcare system, applying lean supply partnership and inventory reduction approach to improve responsiveness and cost. According to Womack and Jones, there are five lean principles identified to guide organizations in economic sectors that include services [5] in lean transformation which are:

- Value is determined by what customer values (specifically, what they are prepared to pay for) in the product or service.
- The value stream state map out (with a process or value stream map) how value is delivered. Use this as a basis for eliminating any that does not add value.
- Flow must ensure products and information seamlessly flow from start to finish of the value stream. Then, remove inventory or buffer zones with the use of structural enablers such as modular design, cellular working, general purpose machines, and multiskilled workers.
- Pull-only deliver what is actually demanded (pulled) by the customer rather than serving from stocks or buffers.
- Perfection is continually sought to improve the process and system with the above principles, striving for perfection.

20.3 Eight wastes of lean

Sarkar mentioned that anything that does not add value to customer is a waste [6]. Referred as eight wastes of lean manufacturing by Taichi Ohno, the father of Toyota Production System, it has a universal application, including service industry. Table 20.1 shows the type of waste, its meaning, and examples for service industry.

From the table, eight types of wastes listed are overproduction, defects, inventory, overprocessing, transportation, waiting, unnecessary motion, and unutilized workforce.

Table 20.1 Eight wastes for service industry

Type of waste	Meaning	Examples
Waste of overproduction	Processing too soon or too much than required	• Information sent automatically even when not required • Printing documents before they are required • Processing items before they are required by the next person in the process
Waste of defects	Errors, mistakes, and rework	• Rejections in sourcing applications • Incorrect data entry • Incorrect name printed on a credit card • Surgical errors
Waste of inventory	Holding inventory (material and information) more than required	• Files and documents awaiting to be processed • Excess promotional material sent to the market • Overstocked medicines in a hospital • More servers than required
Waste of overprocessing	Processing more than required wherein a simple approach would have done	• Too many paperwork for a mortgage loan • Same data required in number of places in an application form • Follow-ups and costs associated with coordination • Too many approvals • Multiple MIS reports
Waste of transportation	Movement of items more than required resulting in wasted efforts and energy and adding to cost	• Movement of files and documents from one location to another • Excessive e-mail attachments • Multiple hand-offs

(Continues)

Waste of waiting	Employees and customers waiting	• Customers waiting to be served by a contact center • Queuing in a grocery store • Patients waiting for a doctor at a clinic • System downtime
Waste of motion	Movement of people that does not add value	• Looking for data and information • Looking for surgical instruments • Movement of people to and fro from filing, fax, and Xerox machines
Waste of unutilized people	Employees not leveraged to their own potential	• Limited authority and responsibility • Managers common • Person put on a wrong job

20.4 Lean tools in service industry

Lean tools are processes and strategies which are used to identify issues in the production of goods or services and increase the efficiency of the operation. The multiple tools used to evaluate situations and help minimize waste and clear the way for greater profits. The lean tools that are used in service industry here are Poka Yoke, 5S, Kanban, and VSM. The positive impact of these tools will be stated and discussed alongside with the few examples of successful implementations by few companies.

20.5 Critical elements in lean service

Lean service is the application of the lean manufacturing concept to service operations. Most people understand the value of lean principles relatively easily, especially when presented with some good examples. The critical factors that have been determined as the success of implementing the concept of lean manufacturing within SMEs are leadership, management, finance organizational skills and expertise, among other factors that are classified as the most critical factors for successful lean manufacturing for SMEs environment [7]. The important elements that have been found for success in lean service are described in the following subsections.

20.5.1 Leadership and management

The method, revolutions, and thresholds that must be crossed to compete in the global market cannot be accomplished from the bottom up. They have to start from the top down [8]. The leadership will represent that success or failure for any company. Bill Selby, who retired as a vice president from Boeing in 1996, was involved in so many groundworks for the productivity improvement and employee improvement. Bill Selby said that, “I put managers in three categories.”

- There are people who truly do want to change.
- There are those who are sitting on the fence waiting to see how things go.
- There are people who think that change is wrong and that we should work using traditional way.

Leaders must believe that creating a world-class production system is not just important but vital, non-negotiable, and that it must begin now, even if it takes many years to achieve. Leadership is found as an important characteristic for success in lean service.

20.5.2 Customer focus

Customers are the most important consideration in everything that you do. Customers, like employees, contribute inputs that impact the organization’s productivity through both the quantity and the quality of those inputs and the resulting quality of output generated [9]. For example, in contributing information and effort in the diagnoses of their ailments, patients of a healthcare organization are part of the service production process. If they provide accurate information in a timely fashion, physicians will be more efficient and accurate in their diagnoses. Thus, the quality of the information patients provide can ultimately affect the quality of the outcome. Furthermore, in most cases, if patients follow their physician's advice, they will be less likely to return for follow-up treatment, further increasing the healthcare organization's productivity [10].

20.5.3 Empowering employment

Employee involvement is the foundation. Without it, one cannot build a lean, world-class production system [11]. Without employee, no success will be achieved in any field of company. There are three steps for success in services and manufacturing:

1. Obtaining the top management participation and support. Successes are marked by active management participation and failures by the lack of it.
2. Being careful in research and planning.
3. Giving full commitment to employees' training and involvement. Employees must understand the goals and be part of them.

20.5.4 Quality

Quality is often defined as meeting of standards, specification, or tolerances at an acceptable level of conformance [12]. This acceptable level is variously defined such as zero defects, Six Sigma and Lean Six Sigma, defective parts per million. This quality management is not quality control or quality assurance, but it is actually an appropriate term to convey that the goal of achieving and improving quality is an area of management concern and responsibility. There are three lean business policies developed by quality management systems:

1. Pursuit of quality is a strategic company goal.
2. Top management is committed to and actively involved in achieving quality objectives.
3. A permanent, organized company-wide effort to continuously improve product process quality, including training and measurements are maintained.

The most fundamental definition of a quality product is one that meets the expectations of the customer. However, even this definition is too high level to be considered adequate. To develop a more complete definition of quality, some of the key dimensions of a quality product or service must be considered. There are eight key dimensions of a quality product or service: performance, features, reliability, conformance, durability, serviceability, aesthetics, and perception.

Dimension 1, Performance (Does the product or services do what it is supposed to do within its defined tolerances?) Performance is often a source of contention between customers and suppliers, particularly when deliverables are not adequately defined within specifications. The performance of a product often influences profitability or reputation of the end user. As such, many contracts or specifications include damages related to inadequate performance.

Dimension 2, Features (Does the product or services possess all of the features specified, or required for its intended purpose?) While this dimension may seem obvious, performance specifications rarely define the features required in a product. Thus, it is important that suppliers designing product or services from performance specifications are familiar with its intended uses, and they maintain close relationships with the end users.

Dimension 3, Reliability (Will the product consistently perform within specifications?) Reliability may be closely related to performance. For instance, a product specification may define parameters for uptime, or acceptable failure rates. Reliability is a major contributor to brand or company image, and it is considered a fundamental dimension of quality by most end users.

Dimension 4, Conformance (Does the product or service conforms to the specification?) If it is developed based on a performance specification, does it perform as specified? If it is developed based on a design specification, does it possess all of the features defined?

Dimension 5, Durability (How long will the product perform or last, and under what conditions?) Durability is closely related to warranty. Requirements for product durability are often included within procurement contracts and specifications. For instance, fighter aircraft procures to operate from aircraft carriers including design criteria intended to improve their durability in the demanding naval environment.

Dimension 6, Serviceability (Is the product relatively easy to maintain and repair?) As end users become more focused on total cost of ownership than simple procurement costs, serviceability (as well as reliability) is becoming an increasingly important dimension of quality and criteria for product selection.

Dimension 7, Aesthetics (The way a product looks is important to end users.) The aesthetic properties of a product contribute to a company's or brand's identity. Faults or defects in a product which diminish its aesthetic properties, even those that do not reduce or alter other dimensions of quality, are often causing for rejection.

Dimension 8, Perception (Perception is reality.) The product or service may possess adequate or even superior dimensions of quality, but still fall victim to negative customer or public perceptions. As an example, a high quality product may get the reputation for being low quality based on poor service by installation or field technicians. If the product is not installed or maintained properly, and it fails as a result, the failure is often associated with the quality of the product rather than its quality of the service it receives.

The most fundamental definition of a quality product is one that meets the expectations of the customer. However, even this definition is too high level to be considered adequate. To develop a more complete definition of quality, some of the key dimensions of a quality product or service must be considered.

20.5.5 Challenges in lean services

As we know when talk about management, all systems are involved which include human, operations, and others. The major challenge for lean is waste because waste is a term that starts the lean concept. For example, in making something or arranging some activity in an industry, basically people will not see their work in progress as a waste unless they have good reasons. The implementation of lean service is described as points of the process to improve the challenges illustrated in Table 20.2.

From the table, it can be seen that the challenges in service industry requires plenty of skill set, leadership, and support to overcome them.

All of these challenges often occur when the lean approach is implemented because human mistake is the natural thing that occurs. To implement the lean service, people or human emotion is not increase when they are not fully understanding the lean concept, but they can be handle with great care. However, these challenges can be avoided if lots of skill, leadership, and support from all organizations in the area are improved according to the lean service in lean principle concept.

20.6 Application of lean in services

The application of lean in service industry such as hotel, hospital, construction, and office is shown and discussed subsequently.

20.6.1 Lean hotel

Lean hotel is a service industry in which productivity is highly influenced by the efficiency and attention to details of the people who are working manually with tools or operating equipment (Mekong Capital). The applications of some basic lean principles, which are common in manufacturing, often fail in service industry because of the failure on the part of service organizations by not changing the way

Table 20.2 Challenges in service industry

Item	Description
Need a lot of skill	• Manage the problem well without aggravating another problem. • Understand the lean principle concept to apply into the system when needed. • Identify the major cause for each problem that occurs before managing the process.
Leadership	• Have proper skill in solving especially lean concept when getting some problems. • Take others idea to improve the problem.
Support	• Provide training to increase the skill. • Generate the activity according to the maximum skill.

they think. A change of thinking, as Heidi learned, can take an organization to an entirely different direction in Marriot hotel.

The hotel industry consists of companies within the food services, accommodations, recreation, and entertainment sectors. This industry is a several billion-dollar industry that mostly depends on the availability of leisure time and disposable income. The common unit in hotel's industry such as a restaurant or hotel consists of multiple groups such as:

- Facility maintenance
- Direct operations
- Management
- Marketing
- Human resources.

20.6.1.1 Implementation and tool

1. 5S

At housekeeping room, the notices of the storage locations are marked on the shelves to staff doing friendly job. Figure 20.2 shows storage before and after implementing 5S. They label not only "what" goes where, but also "how many," which is a key point for 5S-ing stock shelves and locations [13].

2. Poka-yoke

In service industry such as hotels, the Poka-yoke is used to prevent damage to equipment or furniture in hotels. Figure 20.3 is an example of the application of Poka-yoke.

Figure 20.3 illustrates the technique of Poka-yoke. Many hotel rooms now have LCD TVs in them. The old heavy tube TVs are not likely to be pulled forward and off of a cabinet or table. But the LCDs might be. The hotel does not hang an ugly sign saying "Caution: Don't pull TV off of table!" So, Poka-yoke is suitable in replacing that sign.

20.6.1.2 Waste in hotel

Waste is a problem. There are many general categories of waste. For example, Taiichi Ohno's seven wastes, including the overproduction, waiting, transportation, processing, inventory, movement, and making defective products.

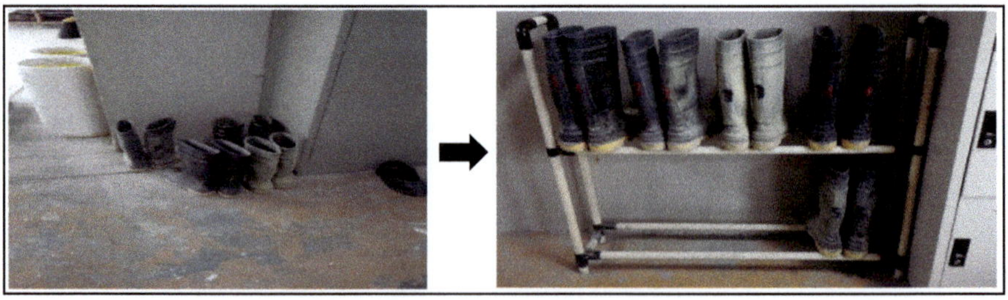

Figure 20.2 Before and after 5S

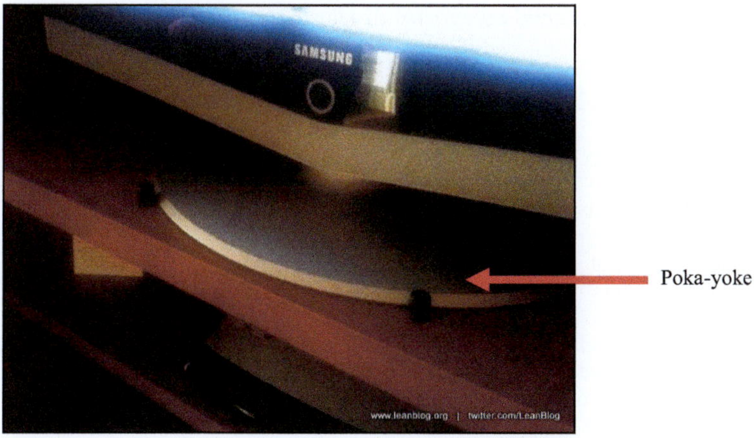

Figure 20.3 Poka-yoke [13]

Table 20.3 Waste in hotel

No.	Waste	Explanation
1.	Overproduction	Overproduction is unnecessarily producing more than demanded or producing it too early before it is needed. Excessive production of food in the cafe or restaurant in the hotel will become the overproduction waste. This increases the management cost of hotel.
2.	Waiting	Waiting is idle time for guest or worker due to bottlenecks or inefficient management flow of hotel. Waiting also includes small delays between processing of units. Waiting results in a significant cost insofar as it increases labor costs and depreciation costs per unit of output.
3.	Inventory	Hotels tend to have many materials' expenses. As this is a 5-star hotel, it is likely that their materials' expenses will be higher than a 3-star or 4-star hotels', as they will provide customers with more amenities which results in more materials and resources being used.

Table 20.3 highlights wastes in hotels. There are three typical sources of waste in hotels: kitchen, guest/staff rooms, and garden. Each of these sources has its own waste composition. For example, waste from the garden contains a lot of leaves and tree trimmings, whereas there are more vegetable and fruit wastes in kitchen waste. Table 20.4 gives more details on items that may cause wastes in hotels.

20.6.1.3 Benefits of lean hotel

The satisfactory levels of customers are very important in the service industry. The application of lean systems can eliminate significant levels of waste or inefficiency. As a result of applying lean into Marriot hotel, in <1 year the hotel's service score

Table 20.4 List of material categories [14]

Material category		Description
Waste from guest room		
Compostable	Fruit waste	Rambutan, watermelon, dragon fruit, litchi, pineapple, mangosteen shells
	Leftover food	Bread, sticky rice
	Flower waste	Daisy, roses
Recyclables	Metal	Aluminum cans
	Glass	Beer bottles, glasses
	Recyclable plastics	Bottles, clear bags
	Recyclable paper	Tissues, cardboard, print paper, newspaper
Miscellaneous	Non-recyclable paper	Damp napkins, tissues
	Non-recyclable plastics	Plastics bags, hard plastics, etc.
	Others	Wood chips, sand, dust, rubber, cloth, etc.
Waste from kitchen and restaurant		
Compostable	Leftover food	Rice, cooked meat, cakes
	Fruit waste	Rambutan, watermelon, dragon fruit, litchi, pineapple
	Vegetable waste	Water-morning glory, potherbs, onions, spinach, carrot
	Egg shells	–
	Seafood waste	Lobster, crab, shells, fish
Recyclables	Recyclable paper	Tissues, cardboard, print paper, newspaper
	Metal	Aluminum cans
	Glass	Beer bottles, kitchen stuff
	Recyclable plastics	Bottles, clear bags
Miscellaneous	Non-recyclable paper	Damp, napkins, tissues
	Non-recyclable plastics	Plastics bags, hard plastics, etc.
	Others	Bone, dirt, sand, cloth, rubber, etc.
Waste from garden		
Compostable	Leaves	Pine needles, brush, leaves, tree trimmings
	Grass clippings	–
Miscellaneous	Others	Dirt, sand, rubber, cloth, foam, etc.

rose 10 points to 87 and went from the bottom half to the top 10% in its peer group. This improved score not only positively impacts hotel guests, but also impacts the hotel's bottom line through repeat customer stays. The other benefits are:

1. Rebuild every work process from the customers' perspective to save labor costs and fund improvements in customer service.
2. Make "upstream" improvements and eliminate or simplify the tasks that are invisible to customers.
3. Strengthen standard benchmarking approaches with rigorous performance management tools to compare operating and financial metrics and achieve performance improvements.

A cleaner workplace by being waste wise has become a routine part of many people's lives, so most staff expects that their workplace will provide programs to reduce, reuse, and recycle waste. Waste wise creates the opportunity for staff to work together to reduce waste through practical tips and activities that they can easily use in their everyday work practices.

20.6.2 Lean in hospital

Today healthcare is very expensive. The costs are rising and people around the world are getting hurt and killed by medical error. This is not real encouraging story to know. Every day, the hospital employees such as nurses, physician, and nurses are frequently annoyed with the same problem they face. Due to this, many of them going home worn out or end up leaving this profession. In the hospital, there are a lot of miracles, and a wonderful thing happens when a life is still going on. However, those things are being blurred by incompetency of the administrator of the hospital due to some problems.

Lean is a tool set, a management system, and a philosophy that can revolutionize the way hospitals are organized and managed. Lean is a methodology that allows hospital to improve the quality of care for patient by reducing error and waiting times. Its approach can support employees and physicians, eliminating obstacle and allowing them to focus on providing care. Lean is a system for strengthening hospital organization for the long term by reducing cost and risks while also facilitating growth and expansion. It also helps to break the barrier between the other departments to cooperate together for the benefits of the patients.

Someone might ask how lean technique can help solve the everyday, nagging problem that so many committees and teams have already tried fixing. Lean is special in that the methodology shows people how to look by the people who do the work, rather than relying on experts to tell them exactly what to do. It also helps leaders to see and understand that it is not the individuals who are broken, but the system itself. This approach also requires the continued learning: continue to improve (Kaizen) and professional development of employees, for their own sake and the organization itself.

20.6.2.1 Tools and implementation

Visual Management (VM), 5S, Kanban (Signboard), Kaizen (Wall of Fame), VSM, and Heijunka Box are the tools that help to reduce the waste. Table 20.5 summarizes lean tools in hospitals.

20.6.2.2 Waste in hospital

Nowadays, hospitals also face severe shortage of key skilled employees, including nurses, pharmacists, and medical technologies. Lean improvement through standardized work, layout, and flow improvement and 5S are reducing waste and non-value-added time, reducing labor requirement or allowing caregivers to spend more times with patients. The typical lean types of waste are seen throughout the hospitals. These wastes are the same with the other sectors, however, within its own department. The ideas are not the same even though the meanings are the same. Table 20.6 shows the seven common wastes of lean in hospitals.

Table 20.5 Tools in lean hospital

Tool	Application and implication
Heijunka Box	This technique is generally employed for scheduling of production activities to control inventory, decreasing lead times, and producing a mix of products and in appropriate volumes as per the customer demand.
7S	First S: Sort—where to sort; five things to remove; red tagging. Second S: Straighten—some simple rules; a place for everything: VM. Third S: Shine—cleaning means inspection. Fourth S: Standardize—prevention. Fifth S: Sustain—how to sustain? Tool to sustain. Sixth S: Safety—ergonomic consideration and rules; safety checklist. Seventh S: Security—impact of changes on hospital alert.
Kanban	To create value, services need to be provided in line demand. All work, material, and information should be pulled toward the task of delivering health services. There is an emphasis on reducing patients' waiting and queuing time.
Kaizen	The goal of Kaizen is to implement a medical supplies replenishment system based on the Kanban methodology. The selected team will identify a target area to be improved within the duration of the event. The Kaizen team will receive all the necessary training and materials required to ensure a complete learning experience.
VM	The goal of VM is to make waste, problem, and abnormal condition readily to employees and manager. The main idea is to expose it so it can be fixed.
VSM	Through this process of developing the Future State, the participants will also identify waste in the current value stream and be guided through the lean methods to improve a process.

Table 20.6 Waste in hospital

Type	Laboratory	Example
Defects	Mislabeled patient specimens	Wrong medication delivered to patient
Overproduction	JUST IN CASE blood tubes drawn from patient, but not used	Patient seen MD faster than can be treated
Transportation	Moving specimen long distances from receiving to testing	Long walks from MD clinic to chemotherapy
Waiting	Specimen waiting in batches for testing	Patient waiting time due to physician lateness or schedule exceeding capacity
Inventory	Expired test reagents	Expired chemotherapy drugs
motion	Technologist walking due to poor layout	Nurses search for missing or poorly located supplies
Overprocessing	Time and date stamps on label that are not used	Time spent creating a schedule that is not followed
Human potential	Employees' ideas not listened	

20.6.2.3 Benefits of lean hospital

Lean benefits patients (through reduced waiting times), nurses, or even a doctor where they are not just saving a life but improving the life (self-awareness). The hospital and the quality of healthcare are improved where patients will have confidence to put their life and maybe a life could be saved. Physician will benefit too through improved productivity that translates into a higher pay by reducing or eliminating the need for multimillion-dollar expansion. Lean hospital does more than implementing simply tool and technical methods. Lean also features a cultural change and a management system, a transformation that takes time, effort, and persistence. Figure 20.4 shows the 7S and VM in a hospital.

Figure 20.4 shows the visual color difference that allows a distinctive separation of rooms, walls, flooring, and hand support by the wall. Figure 20.5 shows the Kanban system for the hospital liquid inventory storage.

Figure 20.5 shows the neat and organized inventory being stored on the shelves.

20.6.3 Lean construction

Lean construction is a production management-based approach to project delivery. It is also a new way to design and build capital facilities. When the perception of lean was first adapted from the manufacturing industry, the concern was in abstracting the underlying lean production principles and then fit them to the construction industry. The challenge was then how to take the lean construction principles and make them operational. A new project delivery system was needed which encompasses all the core principles of lean construction. This was, however, not intended to be a formulaic delivery system as lean construction does not have a standard set of tools. A lean project delivery system (LPDS) is one that is structured, controlled, and improved in pursuit of maximizing value to owners and

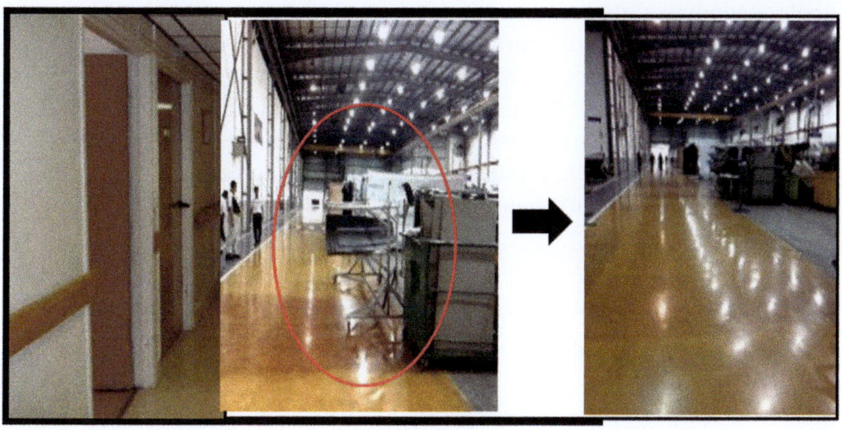

Figure 20.4 7S and VM

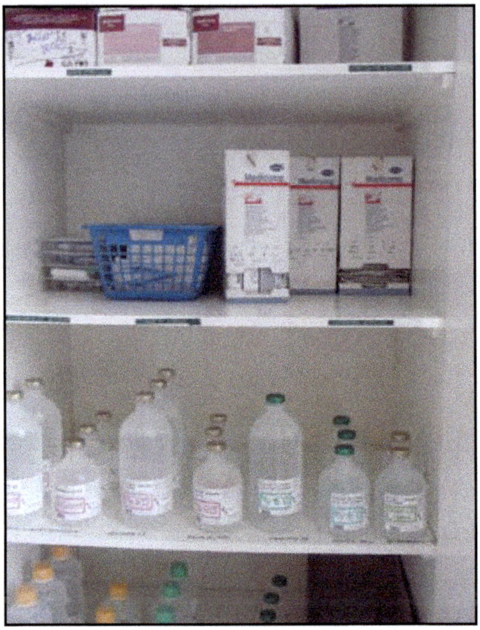

Figure 20.5 Kanban system [15]

workflow reliability on construction sites. Figure 20.6 shows the sample picture for shed construction.

The LPDS is a conceptual framework developed by Ballard and Zabelle to guide the implementation of lean construction on project-based production systems [17], i.e., the structures that build. Figure 20.7 describes LPDS. The LPDS was developed as a set of interdependent functions (the systems level), rules for decision-making, procedures for execution of functions, and implementation aids and tools, including software when appropriate. Lean production management has caused a revolution in manufacturing design, supply, and assembly. When applied to construction, lean changes the way work is done throughout the delivery process. Lean construction extends from the objectives of a lean production system—maximize value and minimize waste—to specific techniques and applies them in a new project delivery process. Figure 20.7 shows the diagram of an LPDS.

From Figure 20.7, the explanatory of the diagram is indicated as follows:

- The project definition phase consists of the modules: needs and values determination, design criteria, and conceptual design.
- Lean design consists of conceptual design, process design, and product design.
- Lean supply consists of product design, detailed engineering, and fabrication/logistics.
- Lean assembly consists of fabrication/logistics, site installation, and testing/turnover.

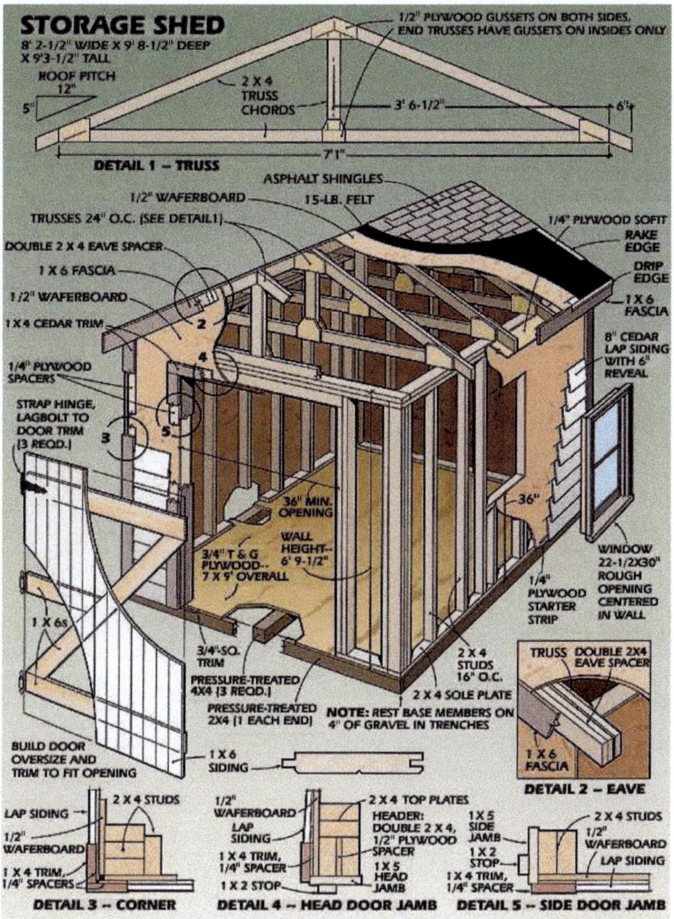

Figure 20.6 Sample picture for shed construction [16]

- Production control consists of work flow control and production unit control.
- Work structuring and post-occupancy evaluation are, thus far, only single modules.

What is lean construction? A simple, not simplistic, definition is that lean construction is a "way to design production systems to minimize waste of materials, time, and effort in order to generate the maximum possible amount of value" [18].

Another definition states that lean construction is a holistic facility design and delivery philosophy with an overarching aim of maximizing values to all stakeholders through systematic, synergistic, and continuous improvements in the contractual arrangements, the product design, the construction process design and methods selection, the supply chain, and the workflow reliability of site operations [19].

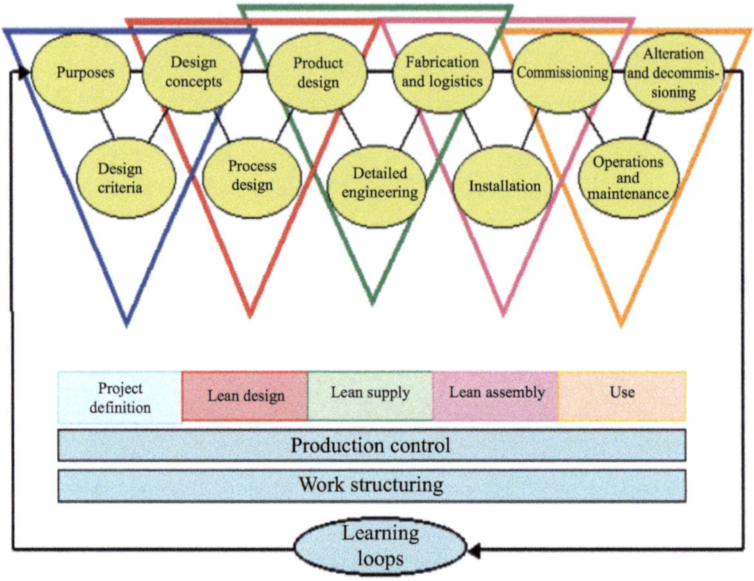

Figure 20.7 Lean project delivery system [17]

20.6.3.1 Implementation

In the last 10 years, lean has been taken up by the construction industry with consistently growing enthusiasm and effectiveness. In the beginning, there was a high level of resistance within the industry with the often-heard comment, "that may work for manufacturing but construction is different." This type of statement has proved to be both true and false. True is that a slightly different lean approach had to be developed for the project-based environment of construction. False is that once the frameworks and methods of lean planning systems (see last planner) and lean logistics have been accomplished, the application of all other lean principles and lean methods are effectively used to produce the dramatic results that have been common in manufacturing for many years. Additionally, a lean construction approach brings effective business management into construction companies that often suffer from casual and undisciplined methods of management (lean construction delivery system).

Construction is possibly the last frontier for lean. Although manufacturing's productivity has improved during the last 40 years, the construction industry has experienced a slight decline. Even though the construction world has embraced high-tech tools, we still manage projects the same way we always have, and we are still getting the same poor results. Less than 30% of projects come in on time, on budget, and within specification. The answers in improving construction productivity are not found in more software or technology.

Waste is everywhere in construction and it has been for hundreds of years. This is not a statement of blame, but just a fact. It is so much a way of life that most construction managers overlook the fact. They accept waste as inevitable and unpreventable and add it into the cost of the job. Thus, the customer pays for it. Implementing lean construction has its benefits reaped by not only contractors, but also the architect and owner, who are set to gain a lot by this practice. However, some construction companies do not view waste as a necessary part of doing business.

20.6.3.2 Tools

The tools and their implementation are listed and described in Table 20.7.

Lean tools are very useful in improving construction management, especially tools such as VSM, Kaizen, and standardized work [21]. Besides, lean tools such as Six Sigma and Kanban enable sustainability in a construction industry [22].

20.6.3.3 Wastes in construction

Waste (also known as rubbish, trash, refuse, garbage, and junk) is unwanted or useless material. Litter is a waste that has been disposed of improperly, in living organisms; waste is the unwanted substances or toxins that are expelled from them. Waste is directly linked to human development, both technologically and socially. Table 20.8 identifies the eight types of wastes readily found in construction.

20.6.3.4 Benefits of lean construction

The reliable release of work between specialists in design, supply, and assembly ensure that value is delivered to customers and waste is reduced. Lean construction is particularly useful on complex, uncertain, and quick projects. It challenges the belief that there must always be a trade between time, cost, and quality. There are few advantages or benefits of applying lean concept in construction:

- The facility and its delivery process are designed together to better reveal and support customer purposes. Positive iteration within the process is supported and negative iteration is reduced.
- The work is structured throughout the process to maximize value and to reduce waste at the project delivery level.
- Efforts to manage and improve performance are aimed at improving the total project performance because it is more important than reducing the cost or increasing the speed of any activity.
- "Control" is redefined from "monitoring results" to "making things happen." The performance of the planning and control systems is measured and improved.

Figure 20.8 shows a photo of a garage before applying lean concept in the layout. The condition was messy and disorganized.

Figure 20.9 shows the condition due to improvements made in the garage after 6 months of applying 5S of lean concept. The condition of the garage has been improved as the tools and equipment in the garage can be identified.

Figure 20.10 shows the overall transformation of the garage after the complete application of lean principles in the garage. The garage now is in an orderly manner and has a distinctive indicator of work area and walkway.

Table 20.7 Tools in lean construction

No.	Tools	Description/implementation
1	5S	The 5S process (sometimes referred to as the Visual Work Place) is about "a place for everything and everything in its place." It has five levels of housekeeping that can help in eliminating wasteful resources (Kobayashi, 1995) [27]. Masaaki Imai said, "If one is not doing **5S**, one is not doing lean." In construction, 5S (traditionally translated as "sort," "set in order," "shine," "standardize," and "sustain") has been applied in the shop with great success.
2	Kanban	A basic **kanban approach** has been used in construction, both in the field and in the shop, to signal the need to refill consumables. As an example, when one bin of bolts or screws is empty, the second bin replaces it. The empty bin becomes a signal to refill it by placing it in a designated location. In the bottom of the bin is a card that has the bolt type, part number, quantity to be ordered, and the supplier's contact information.
3	Kaizen	"Building Quality at Veridian Homes" in the October 2006 issue of Quality Progress by Denis Leonard is detailing on **Kaizen** (i.e., waste reduction) events that have been used successfully in construction fields. Veridian Homes, headquartered in Madison, Wisconsin, uses what it calls Kaizen redline events to review all of its home plans. This helps Veridian to develop plans that are more constructible. The company has 50 different plans and involving the people who design and build these models has led to major improvements. In these Kaizen redline events, the team examines every part of the design, looking for ways to simplify, reduce duplicate efforts, and standardize the installation work. The event consists of two parts. First, a team of representatives from design, construction, and customer service reviews a specific plan. This includes the key subcontractors. Then, the team presents its findings and concerns regarding that plan to all design, construction, management, and customer-service people in a 90-min monthly meeting.
4	VSM	Some construction companies are using VSM to improve how they deliver the design and the project, based on the Peter Dumont, P. E., of Lean Engineering, in a presentation made to the Lean Project Delivery Workshop on April 11, 2007 [20]. The Boldt Co., a construction-services contractor headquartered in Appleton, Wisconsin, used VSM to reduce accounts payable from 87 to 67 days. Tracer Industries Canada Ltd., based in Edmonton, Alberta, specializes in providing turnkey management systems for industrial and commercial projects. The company has applied VSM and decreased total engineering design time from 20 to 1.2 days. Direct labor cost per drawing has decreased and generated a US $1.3 million annual increase in earnings before interest and taxes.

20.6.4 Lean office

The philosophies, tools, and methodologies associated with lean manufacturing are well-documented and have been successfully applied in a broad range of businesses and industries. It is a process that focuses on the elimination of everything that is not beneficial to customers—all types of waste and any elements that prohibit the flow. This process applies equally well to the office.

Table 20.8 Waste in construction

No.	Type of waste	Description
1	Defects	Everyone in construction understands this type of waste. It includes doing the wrong installation defects in fabrication, and errors in punch lists. Not meeting the required code is a waste. Rework in construction is rarely measured.
2	Overproduction	This waste happens when we fabricate material too early or stockpile material in the warehouse or at the job site. Estimating and bidding jobs that are not won is a form of this waste. Printing more blueprints or making more copies of a report than needed is an overproduction.
3	Transportation	This waste occurs when we move material around the shop, load it on the truck or trailer, haul it to the job site, unload it, and move the material from the lay-down or staging area to the installation point.
4	Waiting	Construction is full of this waste, including when a crew waits for instructions or materials at the job site, when a fabrication machine waits for material to be loaded, and even when payroll waits for the late time sheets.
5	Overprocessing	This waste includes over engineering, requiring additional signatures on a requisition, multiple handling of time sheets, duplicating entries on forms, and getting double and triple estimates from suppliers.
6	Motion	These "treasure hunts" happen when material is stored away from the job or when workers look for tools, material, or information. This waste also occurs in the office or job site trailer, when looking for files, reports, reference books, drawings, contracts, or vendor catalogs.
7	Inventory	This includes uncut materials, work in process, and finished fabrications. Some contractors claim that they have no inventory because they job-cost all material. While this may work for accounting, if the material is not yet installed and is not being used by the customer, it is considered a waste. This waste includes spare parts, unused tools, consumables, forms and copies, employee stashes, and personal stockpiles. One could argue that the unfinished facility is considered an inventory and is treated as a waste until operational.
8	Human	This type of waste will occur when the number of persons who handle the job are more than the required for a given job.

Figure 20.11 shows the implementation of lean concept in the office. In the office, the focus is on reducing the cycle time for all processes, especially those between order placement and receipt of payment. The objective is to optimize the flow of information and improve company's ability to collect, move, and store information, utilizing the absolute minimum resources necessary to fulfill customer expectations.

Figure 20.8 Before applying lean concept [23]

Figure 20.9 Months after applying lean concept [23]

20.6.4.1 Implementation

Figure 20.11 shows the flowchart of implementing lean concept in the office which involves six steps.

Figure 20.12 shows the initial office condition before the implementation of lean concept. The condition was messy.

Figure 20.10 Current condition after applying lean in construction [23]

START

Finding Waste in the Office

Eliminating waste allows value to flow with higher speed and quality through the business process. Waste is anything that does not add value to a product or service.

Discovering and Optimizing Value in the Office

Purging the office of wastes hastens the journey toward the lean office: cost-effectiveness, customer satisfaction, and profitability.

Mapping the Value Stream in the Office

Value-stream mapping provides a unifying plan that helps focus lean office efforts on the process issues that most impact organization's strategic intentions

Preparing for a Kaizen in the Office

The five preparatory activities that help enhance the success of your lean office Kaizen event.

- Evaluating the situation
- Developing a charter
- Gathering data
- Updating the charter
- Planning for the Kaizen

Conducting a Kaizen in the Office

From the beginning kickoff to the final report-out, Kaizen workshops can be exciting and productive for everyone involved. No matter what the outcome, it is a valuabel learning experience and an opportunity for team development.

Sustaining Office Kaizen Improvements

Sustaining and building on office Kaizen improvements needs a support system. Because Kaizen stands for continuous incremental improvement, sustaining the changes on a continual basis is at the heart of the Kaizen philosophy—and the Lean office.

Figure 20.11 Implementation of lean in office [24]

Figure 20.12 Before lean concept implementation; office champion training [25]

Figure 20.13 After lean concept implementation; office champion training [25]

After implementing lean concept in the office, the office area becomes neat and orderly as shown in Figure 20.13.

20.6.4.2 Tools

Table 20.9 shows the description of seven tools that can be used in a lean office.

Table 20.9 Tools in lean office

Tool	Description
5S	• A general cleanup of office areas can remove clutter and increase efficiency. • It may also involve changes to the office layout. • The goal is to create specific places for forms, documents, and equipment so that they can be found within 30 s.
Visual controls	• Color coding, status boards, and other communication tools. • It includes whatever visual that would help employees understand better where things go the first time.
Batch reduction/ elimination	• Moving smaller quantities of work though the system will increase throughput, shorten lead times, and improve quality. • Tailored Label Products Inc., a Menomonee Falls manufacturer of custom labels and die-cut adhesives, discovers a bottleneck at their planner's desk, who received work from four or five customer service representatives who are batching their works. • Eliminating batching even out the workload.
Standardized work	• The opportunities to standardize work are often numerous. Forms and checklists can be revised so that they collect all the needed information. • Tailored label updates the form that relays order information to the plant so that the information is organized by task, making it easy for workers to find what they need in one place. • Standardizing processes may also be necessary, because different people do the same job, resulting in people do task using their own method. • The risk is the difference in the outcomes. Staff members who develop a standard procedure will often reach a consensus that reflects the best practices for everyone, making the process even better.
Leveling	• The amount of work in an office is not constant. Lean office provides proactive measures to respond to busy periods. • For example, if a backlog of work reaches a certain point, a floater could be brought in to help. • Leveling can also be used to balance high- and low-priority tasks. • Activities that bottleneck the value stream could be done at designated times to ensure all tasks get their attention.
Pull/Kanban	• Using pull, downstream tasks pull from upstream tasks to even out capacity and demand. • For example, in a customer service call center, there might be a point where the number of calls on hold signals more people to take calls. • When the number of calls falls below that point, those people return to other tasks. • Kanban is a visual signal to indicate when something is needed, such as cards placed in literature racks to indicate when more copies are needed.

(Continues)

Table 20.9 (*Continued*)

Tool	Description
Office layout/work cells/teams	• Sometimes the office layout needs to be rethought, not just in terms of printer and copier locations, but in terms of organizational structure. • Most front offices are organized by function, with purchasing, customer service, and engineering departments. That is not optimal when you have a value stream focus. • Cross-functional teams organized in work cells can greatly increase efficiency, because the team understands the overall process and it is not focused on a single task.

20.6.4.3 Waste

There are eight common office wastes as shown in Table 20.10. They resemble the manufacturing wastes to some extent. These wastes include defects, overproduction, waiting, non-value-added processing, variation, excessive inventory, extra motion/transportation, and underutilized employees.

20.6.4.4 Benefits of lean office

- Lean office creates results
- Reduce lead times
- Increase customer responsiveness
- Improve on-time delivery
- Get more done with the people and equipment available
- Improve communication, boost morale, and productivity
- Improve flow of information.

20.7 Lean in service versus lean in manufacturing

There is a considerable interest in applying the principles of lean in service. The popular thought is that those same lean principles that have transformed manufacturing could do the same for services. Table 20.11 shows the difference between lean in service and manufacturing.

The differences from a manufacturing situation are inventories between each stage of production and a finished goods inventory which can be increased or decreased as demand fluctuates. For example, three stages are involved in a production process. In the first stage, a production process starts with the raw material. Semifinished goods that go into an inventory are the end of stage 1.

The semiprocessed and finished goods might be in inventories for a considerable amount of time (in most operations, one aspect of lean is to minimize inventory). There would be a considerable lag between the time the product is produced and time it is consumed. In service operation, there are no lags as in the

Table 20.10 Waste in office

Type of waste	Description	Examples
Defects	A failure or flaw that may not allow the affected thing to function properly	• Data entry errors • Pricing errors • Lost files or records • Incorrect information on a report or document
Overproduction	Production of something in excess of what is required	• Printing paperwork out before it is needed • Pricing errors • Losing files or records • Writing correct information on a report or document
Waiting	A period of time spent while expecting something to happen or ready for something to happen	• System downtime • Waiting for approvals • (Delays) in receiving information • Waiting for supplies • Waiting for instructions
Non-value-added processing	Effort that adds no value to the product or service from the customers' viewpoint	• Re-entering data • Extra copies • Unnecessary reports • Excessive reviews
Variation	Multiple methods or tools for performing the same activity	• Multiple forms for the same information • The same activity is performed in several different ways • Time to perform activity is difficult to predict

(Continues)

Table 20.10 (Continued)

Type of waste	Description	Examples
Excessive inventory	Stock that an organization has on hand beyond what is considered necessary	• Office supplies • Sales literature • Reports • Letterhead papers
Extra motion/transportation	Movement of people, information, or paperwork that does not add value to the product or service	• Excessive e-mail attachments • Fax machine or copier too far • Multiple handoffs • Ineffective filing • Poor office layout
Underutilized employees	People’s abilities, not their time, not effectively or appropriately used	• Limited employee authority and responsibility for basic tasks • Management command and control • Inadequate/unavailability of business tools

Table 20.11 Lean in service versus lean in manufacturing [26]

Lean in service	Lean in manufacturing
Simultaneous production and consumption (cocreation between producer and consumer)	Consumption and production are at different stages
Many critical aspects are intangible	Many critical aspects are tangible
Concept of inventory may not be material, but can be virtual such as requests and, in healthcare, patients waiting for service can be considered a type of inventory	May have inventory and buffers
Considerable variability in service delivery	Some variations
Open universe in a variety of service cases	A closed set in a variety of product in manufacturing
Substantive and peripheral benefits	Mainly substantive product benefits

manufacturing operation. Indeed, there is almost a cocreation of the product between the manufacturers and customer.

20.7.1 Lean manufacturing

Lean manufacturing principle is used to determine how lean transformation principle can be utilized. This principle can lead to a guideline formation for the lean in services as the transformation in lean services. This principle is divided into five categories [5]. First, the principle value precisely specifies the value of product or family. This principle is focusing on any analysis or improvement only on the product or service that the process customer would specify what she needs as the output from a given process. It is measured by what the customer is willing to pay.

The second principle specifies values stream or value stream for each product or service. This principle is identifying the sequences of tasks that produce the output for which the customer is willing to pay. These principles do not show the sequence of data. Otherwise, the customer is generally not interested in paying for data that support a process. They are interested only unnecessary presence, cost reduction, quality improvement, or speed delivery. Using this second principle, value added can be identified within the process stream by taking the role of the customer. This value-added changing is a change for the process or task that the customer would see as a change that makes the process or output process more valuable to them. It may be a change that a customer requests, that is a legally mandated requirement, or one that is proposed by the process owner, staff, or management personnel.

The third principle is the flow, or it can be described as making value flow without any interruption. This is to develop the uninterrupted movement of a product through the steps in a process and between processes. This is to ensure that all processes are smooth and clear for each product and each service that is involved.

The fourth principle is pull, or customer pull value from the process owner. This principle is a signal to initiate a process that must come from a downstream process and ultimately the customer or services. According the lean performance project, team members will often develop improvement by working backward or pull through for all process. This is to identify the customer requirements in the process stream. Last principle is the perfection. This purpose of this principle is to maintain and continuously improve the element of the four preceding principles, and those elements are called process tasks. When these five principles are implemented into the lean services, the transformation can be realized using the tools of lean such as Kanban and Poka Yoke.

20.8 Chapter summary

From this chapter, we are able to identify or even differentiate how lean is implemented in the service industry. Furthermore, this chapter lends insights into how the tools of lean are beneficial to the service industry. Nonetheless, a wider perspective is engaged on how lean can be implemented and used at different industries such as services. Lean has a great potential and this can be seen in the following chapters.

References

[1] Bowen, D. E and Youngdahl, W. E. (1998). "Lean" service: in defense of a production line approach. *International Journal of Service Industry Management*, 9(3), 207–225.

[2] Guarraia, P., Carey, G., Corbett, A., and Neuhaus, K. (2008). *Lean Six Sigma for the Services Industry*. Retrieved from http://www.bain.com/publications/articles/lean-six-sigma-for-services-industry.aspx (accessed on October 20, 2019).

[3] Bitner, M. J., Zeithaml, V. A., and Gremler, D. D. (2010). Technology's impact on the gaps model of service quality. *Handbook of Service Science Service Science: Research and Innovations in the Service Economy*, 197–218. doi: 10.1007/978-1-4419-1628-0_10

[4] Dickson, D., Ford, R. C., and Laval, B. (2005). Managing real and virtual waits in hospitality and service organizations. *Cornell Hotel and Restaurant Administration Quarterly*, 46(1), 52–68. doi: 10.1177/0010880404271560

[5] Womack, J. P. and Jones, D. T. (1996), *Lean Thinking: Banish Waste and Create Wealth in your Corporation*. New York: Simon & Schuster.

[6] Sarkar, D. (2009). *What Is Lean in a Service Context?* Retrieved from https://www.processexcellencenetwork.com/lean-six-sigma-business-transformation/columns/what-is-lean-in-a-service-context (accessed on November 5, 2019).

[7] Hallowell, R. (1996). The relationship of customer satisfaction, customer loyalty, and profitability: an empirical study. *International Journal of Service Industry*, 7(4), 27–42.

[8] Black, J. and Black, J. R. (2008). *Lean Production: Implementing a World-Class System*. Norwalk: Industrial Press Inc.

[9] Mills, P. K. and Morris, J. H. (1986). Clients as 'partial' employees: role development in client participation. *Academy of Management Review*, 11(4), 726–735.

[10] Bitner, M. J., Faranda, W. T., Hubbert, A. R., and Zeithaml, V. A. (1996). Customer contributions and roles in services delivery. *International Journal of Service Industry Management*, 8(3), 193–205.

[11] Johnson, H. T. (2006). Sustainability and lean operation. *Journal of Cost Management*, 40–45.

[12] Carroll, B. J. (2008). *Lean Performance ERP Project Management: Implementing the Virtual Lean Enterprise*. New York, London: Auerbach Publications.

[13] Graban, M. (2016). *Lean Hospital: Improving Quality, Patient Safety and Employee Engagement*, second edition. London, New York: Productivity Press.

[14] Hoang, P. C. (2005). *Audit of Solid Wastes from Hotels and Composting Trial in Halong City, Vietnam*. Retrieved from http://sites.utoronto.ca/waste-econ/HoangPhuongChi.pdf (accessed on October 20, 2019).

[15] Ballé, M. (2006). *Lean as a Learning System in a Hospital Ward*. Retrieved from http://www.lean.enst.fr/wiki/pub/Lean/LesPublications/HospitalLearning.pdf (accessed on October 20, 2019).

[16] TedsWoodworking. (n.d.). *Type of Projects You Can Start Building Today*. Retrieved from http://www.tedswoodworking.com/new/vsl/?utm_expid=31820540-141.s01b4pRiQHa3Di0kRf_ehg.0&utm_referrer=http%3A%2F%2Fyourplans.de.vu%2Fpdfshedplans%2F2014%2F02%2F17%2Frona-free-shed-plans-pdf-randkey%2F (accessed on August 16, 2019).

[17] Ballard, G., and Zabelle, T. (2000). *Project Definition White Paper #9 Lean Construction Institute*. Retrieved from https://www.leanconstruction.org/media/docs/ktll-add-read/Project_Definition.pdf (accessed on September 18, 2019).

[18] Koskela, L., Ballard, G., Howell, G., and Tommelein, I. (2002). The foundations of lean construction, in: *Design and Construction: Building in Value*. Oxford, UK: Butterworth-Heinemann.

[19] Abdelhamid, T. S. (2004). The self-destruction and renewal of lean construction theory: a prediction from Boyd's theory. *Proceedings of the 12th Annual Conference of the International Group for Lean Construction*. Helsingør, Denmark, August 03–06, 2004.

[20] Sowards, D. (2018). *Lean Construction, New Tools and Old Combine to Reduce Waste*. Retrieved from https://www.qualitydigest.com/nov07/articles/02_article.shtml (accessed on October 20, 2019).

[21] Leksic, Ivan (2018). Lean model for construction project effectiveness measurement. *Annals of the Faculty of Engineering Hunedoara – International Journal of Engineering*, 16(1), 45–51.

[22] Dixit, S., Mandal, S. N., Sawhney, A., and Singh, S. (2017). Area of linkage between lean construction and sustainability in Indian construction industry. *International Journal of Civil Engineering and Technology*, 8(8), 623–636.

[23] Sowards, D. (2016). *Lean Construction Practices*. Retrieved from http://www.pinp.org/wpcontent/uploads/2016/03/Lean_Contracting_Practices.pdf (accessed on September 18, 2019).

[24] Venegas, C. (2007). *Flow in the Office: Implementing and Sustaining Lean Improvement*. New York: Productivity Press.

[25] Michigan Manufacturing Technology Center (MMTC). (n.d.). Retrieved from https://www.the-center.org/ (accessed on September 18, 2019).

[26] Shmula. (2010). *Difference between Manufacturing and Service*. Retrieved from http://www.shmula.com/lean-difference-between-manufacturing-service/ (accessed on September 18, 2019).

[27] Kobayashi, I. (1995). *20 keys to workplace improvement*. Portland, Oregon: Productivity Press.

Chapter 21

Case study: security system for solar panel theft based on system integration of GPS tracking and face recognition using deep learning

Bhakti Yudho Suprapto[1], Meydie Tri Malindo[1], Muhammad Iqbal Agung Tri Putra[1], Suci Dwijayanti[1] and Leong Wai Yie[2]

Security system is important to protect the objects, including solar panel modules. In this study, an integrated system that combines image processing and object tracking is proposed as a security system of solar panel. Face recognition using deep learning is used to detect unknown face. Then, the stolen object can be tracked using Global Positioning System (GPS) that works using General Packet Radio Service and Global System for Mobile communication system. The results show that the integrated security system is able to find the suspect and track the stolen object. Using the combination of FaceNet and deep belief network, unknown face can be recognized with an accuracy of 94.4% and 87.5% for offline and online testing, respectively. Meanwhile, the GPS tracking system is able to track the coordinate data of the stolen object with an error of 2.5 m and the average sending time is 4.64 s. The duration of sending and receiving data is affected by the signal strength. The proposed method works well in real-time manner and they can be monitored through a website for both recorded unknown face and coordinate data location.

21.1 Introduction

Solar power plant is one of the alternative energies which is utilized in many applications in our daily life. Nevertheless, the price of solar panel is relatively high and so it might be stolen. Thus, a security system is needed for the solar panel system.

Most of the security systems considered only one aspect such as the suspect or the position of the stolen object. One of the methods to detect the suspected thief is using image processing approach. Face recognition is often applied for security system because of its high accuracy compared to other biometrics [1]. Thus,

[1]Department of Electrical Engineering, Universitas Sriwijaya, Palembang, Indonesia
[2]Faculty of Engineering and IT, Mahsa University, Kuala Lumpur, Malaysia

surveillance system can be developed using a security system based on face recognition. Nurhopipah and Harjoko [2] developed a monitoring system using closed circuit television (CCTV) that can detect motion and face and then recognize the face. Another research by Wati and Abadianto [3] developed a home security system based on face recognition. However, this system could not work in real time, especially for unknown face that might be the suspected person. In addition, the system did not have any central storage to save the detected face.

Another aspect in a security system is to track the position of the stolen object. Many research have been done to tracking the position using Global Positioning System (GPS). GPS module and Global System for Mobile Communication (GSM) have been used to track the stolen item using the study by Liu *et al.* [4] and Salim and Idrees [5] who proposed security system based on GPS and General Packet Radio Service (GPRS). Singh *et al.* [6] developed an object tracking system using GPS with communication media of GPRS and Short Message Service (SMS), and Shinde and Mane [7] utilized GPRS and GPS to track the location of the object through website or smartphone. However, it did not have backup communication system when there is no GPRS signal [5,7]. On the other hand, no central storage may allow users to save the input data received from GPS manually [4,6].

Based on the problem mentioned earlier, it is important to make one system for securing the object. Hence, this study proposed an integrated system that combines the face recognition for finding the suspect who stole the solar panel module and tracking system to find the current position of the stolen object. This chapter is organized as follows: Section 21.2 describes the proposed method to recognize unknown face and to track the data coordinate of stolen object. Results and discussion are presented in Section 21.3. Finally, this chapter is concluded in Section 21.4.

21.2 Method

In this study, the proposed method is designed as the combination of face recognition and tracking system. CCTV or webcam is attached close to the solar panel. When an unknown person approaches the solar panel, his/her image will be captured by CCTV or webcam and this image will be recognized using deep learning. If the person is unrecognized, then the image will be sent and stored to online database that can be utilized to track the identity of the thief. In the meantime, the position of the stolen solar panel will be tracked using GPS. The detailed process of the proposed method will be described in the following subsections.

21.2.1 Face recognition using deep learning

The face of an unknown person will be a key in the proposed security system in finding a suspected person. The flowchart of the working process of face recognition can be seen in Figure 21.1.

As shown in the flowchart, the first stage is initialization and it is followed by face detection. Here, the system detects the face using Viola Jones algorithm [8]. The next stage is face detection that is utilized whether the face is detected. If the face is

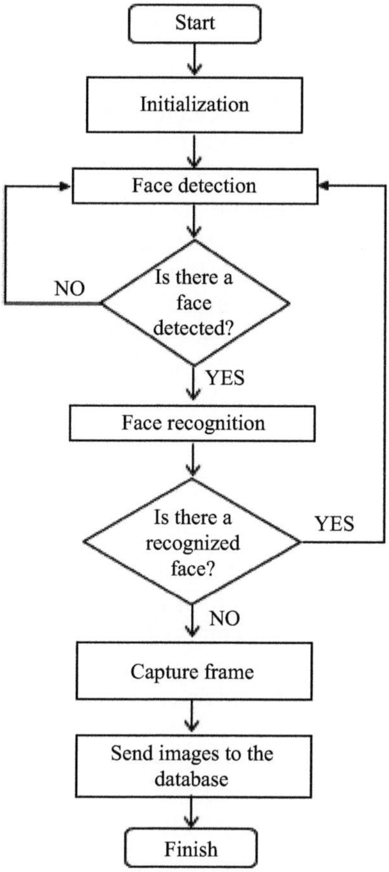

Figure 21.1 Flowchart for face recognition system

detected, the system may continue with the following stage to recognize the face. In this process, the face feature is extracted using embedding Facenet algorithm [9] and the recognition process is performed by deep belief network (DBN). The recognized face will be determined from the confidence value above 80%. The confidence value below 80% is determined as unknown individual who will be captured and sent to the database.

Data used in this research are obtained from different conditions, namely bright and dark. In addition, the object must not cover his or her face. Data used in the training process consist of two people, hereafter called as operator. This training data are obtained from video CCTV. Then, to test the system, eight new faces are utilized.

In obtaining the model, 4,000 images of operators are used where 3,000 images for training and 1,000 images for testing data. FaceNet is used to extract face features to obtain 128 domain points, which is called face embedding. These embedding files are used as the input for DBN. The obtained model is then tested in two conditions, namely offline and online. In the offline test, system is tested for recognized and unrecognized faces coming from images and videos. Nine unrecognized faces are utilized to test the system in both bright and dark. In the online test, the system may recognize the faces in real time and the unknown faces are sent to the database for both bright and dark.

The training process of the proposed method is shown in Figure 21.2. The first stage is to resize the image into 96 × 96 to comply with the input for the purpose of embedding process for FaceNet.

Architecture of FaceNet used in this study is model inception_blocks_v2 in the structure as shown in Table 21.1. Instead of using artificial neural network (ANN) for classification stage, this study utilizes a generative model DBN that consists of restricted Boltzmann machines (RBMs) [10]. DBN does not have any connection in intra layer. Parameters for training network of DBN used in this study can be seen in Table 21.2.

21.2.2 GPS tracking

Face recognition in the previous subsection may be helpful to capture a suspect who stole the solar panel. In the meantime, the stolen object may have been moved to different position. Hence, the security system integrates the face recognition system with GPS tracking to track the position of stolen object.

In this study, the proposed GPS tracking uses the combination of GPRS and GSM as communication system to send the location of the stolen solar panel from Internet network and SMS, respectively. This security system works in parallel with face recognition system. The proposed process of tracking is shown in Figure 21.3.

As shown in Figure 21.3, the system will determine the coordinate location of solar panel using GPS sensor that is aimed as a receiver from GPS satellite to receive and process data to obtain accurate coordinate. When the location is obtained, coordinate data are sent to the database using GPRS or GSM through an SMS. Location and position movement of the stolen solar panel will be tracked in real time and shown in the website. This security system also has website and android application as an interface to track the position and location of the solar panel module. Figures 21.4 and 21.5 represent the website interface and android application, respectively .

While sending the data location, Internet might be interrupted and it will depend on the strength of signal. Hence, this proposed tracking system is also provided to read SMS that is sent to the smartphone.

21.3 Results and discussion

This proposed security system integrates two systems: face recognition to capture the suspect and GPS tracking to track the location of the suspect and the stolen solar panel. In the experiment, both systems will be examined.

21.3.1 Deep learning model for face recognition system

Data used for training consist of two people in two conditions during the day and night as shown in Figure 21.6. Total data for training and testing are 4,000 images.

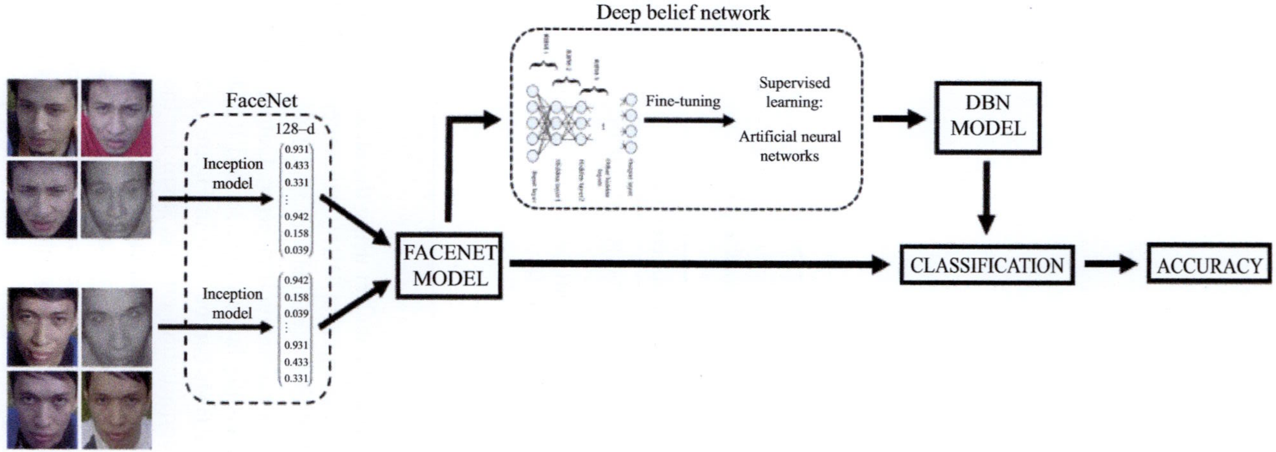

Figure 21.2 Training process

Table 21.1 Inception blocks V2 architecture

ZeroPadding
Conv (7 × 7 × 3.2) + norm
ZeroPadding + max pool
Conv (1 × 1 × 3.1) + norm
ZeroPadding
Conv (3 × 3 × 3.1)+norm
ZeroPadding + max pool
Inception (1a)
Inception (1b)
Inception (1c)
Inception (2a)
Inception (2b)
Inception (3a)
Inception (3b)
Average pool
Flatten
L2

Table 21.2 Parameter training network DBN

Parameter	Parameter values
Hidden layer structure	[3,000, 4,500, 3,000]
Jumlah epoch RBM	10
Learning rate RBM	0.001
Jumlah epoch ANN	3,000
Learning rate ANN	0.01
Batch size	32
Dropout	0.2

After obtaining the face feature through FaceNet using the structure as shown in Table 21.1, the features are used as the input to DBN. Parameter structure for DBN is shown in Table 21.2. As DBN is a combination of RBMs, such RBMs reconstruct the input and minimize the reconstruction error. Output of the first hidden layer will be the input for the visible layer in the second RBM and so on. The graph of reconstruction error for RBMs can be seen in Figure 21.7.

As shown in Figure 21.7, the reconstruction error is small for the last RBM. This may indicate that RBM has formed a model that can detect the pattern of the data. This process is performed in an unsupervised manner. The next stage in the training process is performed in a supervised manner which is the same as ANN [11] and its training loss can be seen in Figure 21.8. This training loss decreases significantly in 356th epoch which may indicate that the training process needs relatively high computation time. However, this model is robust to detect the testing data which is not included in training process coming from the operator.

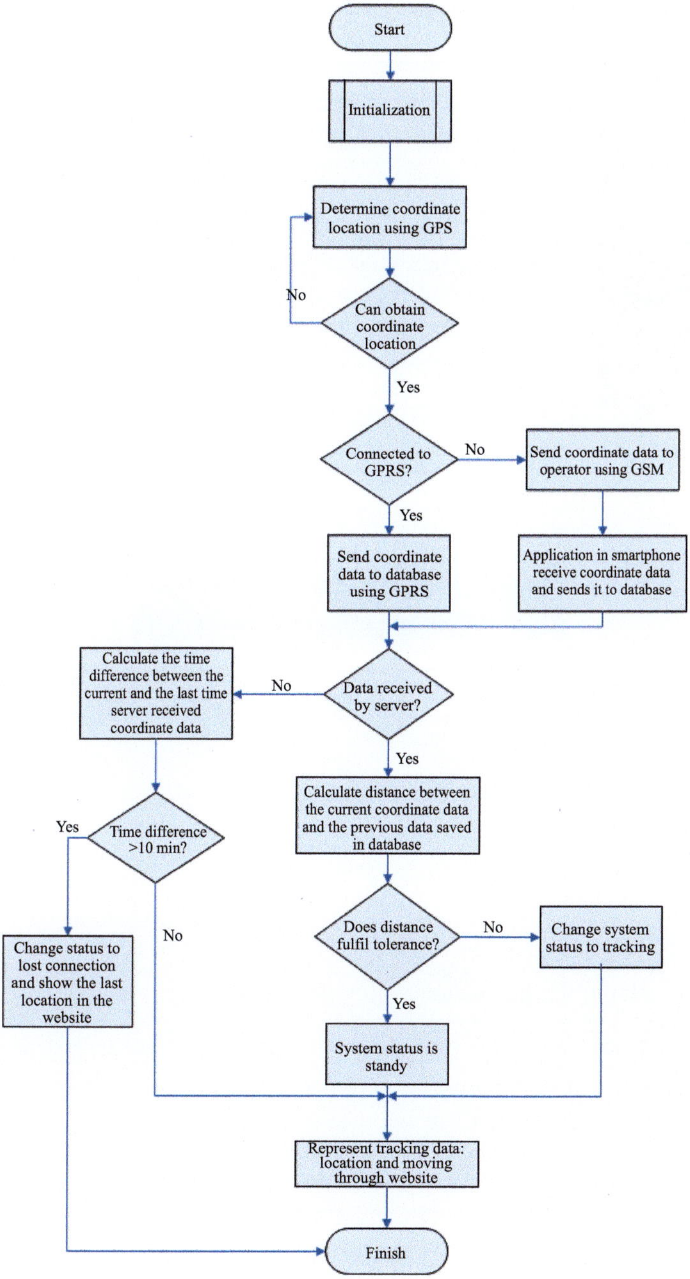

Figure 21.3 Flowchart tracking system using GPS

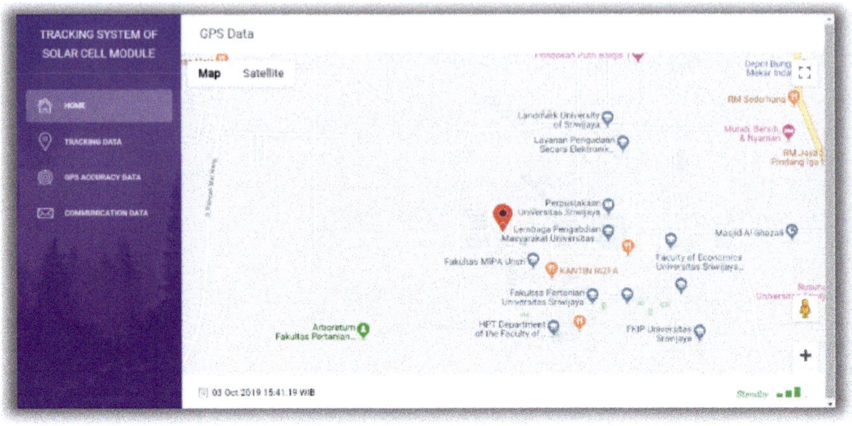

Figure 21.4 Website interface

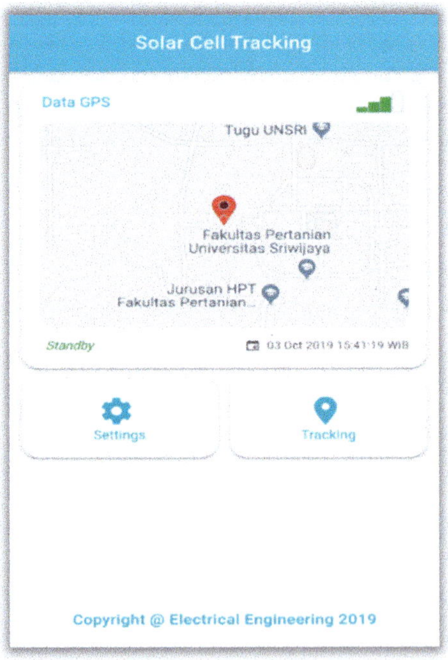

Figure 21.5 Android application interface

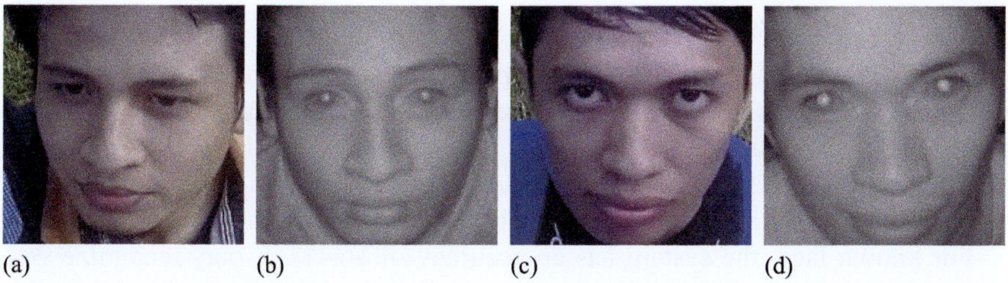

Figure 21.6 Sample of known face (2 operators) used in training, bright 1 (a), dark 1 (b), bright 2 (c), and dark 2 (d)

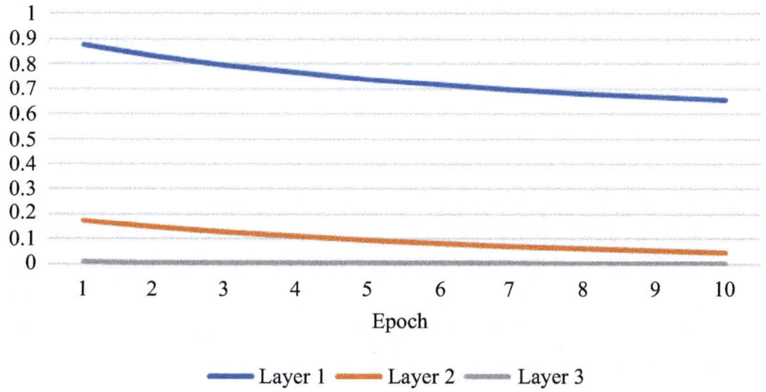

Figure 21.7 RBM reconstruction error

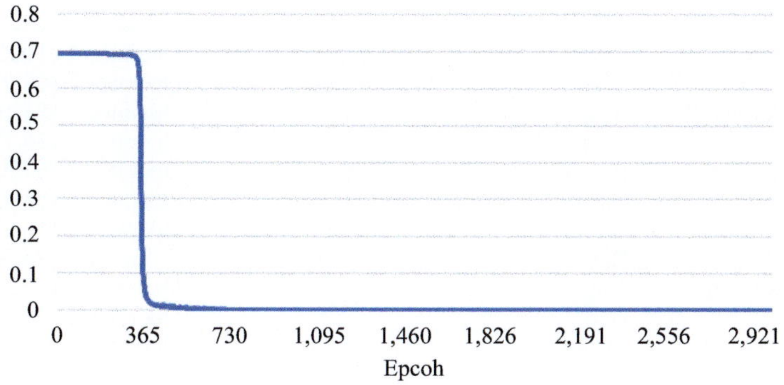

Figure 21.8 Training loss of ANN

21.3.2 Offline test

Offline test is divided into known face (operator) and unknown face (suspect). To detect unknown face, the system must represent the confidence probability based on the model of FaceNet and DBN which have been trained earlier. Here, the confidence probability below 90% may imply unknown face. Higher confidence probability is chosen because the aim of this face recognition is performed as a security system.

For known face, the system has an accuracy of 96.4%. It may recognize 964 images of operators from 1,000 images that are not included in the training process. This result shows that the system is good enough to recognize the known face.

Later, the system is tested using unknown images that are totally new data. These images come from nine different individuals for both image and video. The results of testing using image can been seen in Table 21.3.

As shown in Table 21.3, the proposed system is able to recognize unknown face for each person. The error still occurs in the first unknown image that is recognized as operator. There might be some factors that cause it such as light intensity and the feature of unknown face may have similar characteristics as the operators' faces used in the training.

Next, the test was also performed in the video for unknown faces. Videos for nine unknown faces were recorded using CCTV and webcam to see the performance of the deep learning model in two different cameras. The results can be seen in Table 21.4.

From Table 21.4, the system can work well for moving images (video) using CCTV and webcam. Accuracy of recognizing unknown face is 94.44% for video. However, for the night where lighting is limited, CCTV gives better accuracy than

Table 21.3 Results of offline test for unknown face

No.	Face image	Lighting condition	Recognition	Prediction (confidence)
1	Unknown 1	Bright	False	Operator 1 (100%)
		Dark	False	Operator 1 (99.91%)
2	Unknown 2	Bright	True	Operator 1 (83.16%)
		Dark	True	Operator 2 (56.12%)
3	Unknown 3	Bright	True	Operator 1 (67.55%)
		Dark	True	Operator 1 (84.85%)
4	Unknown 4	Bright	True	Operator 1 (66.23%)
		Dark	True	Operator 2 (75.55%)
5	Unknown 5	Bright	True	Operator 1 (73.38%)
		Dark	True	Operator 2 (52.2%)
6	Unknown 6	Bright	True	Operator 1 (52.94%)
		Dark	True	Operator 1 (61.35%)
7	Unknown 7	Bright	True	Operator 1 (72.43%)
		Dark	True	Operator 2 (78.45%)
8	Unknown 8	Bright	True	Operator 2 (82.94%)
		Dark	True	Operator 2 (79.99%)
9	Unknown 9	Bright	True	Operator 2 (81.95%)
		Dark	True	Operator 1 (50.96%)

Table 21.4 Offline test using video for unknown faces

Face image	Camera	Lighting condition	Capture	Probability (confidence) (%)
Unknown 1	CCTV	Bright	Fail	100
		Dark	Success	77.88
	Webcam	Bright	Fail	100
		Dark	Success	85.27
Unknown 2	CCTV	Bright	Success	70.02
		Dark	Success	64.93
	Webcam	Bright	Success	50.88
		Dark	Success	66.96
Unknown 3	CCTV	Bright	Success	82.77
		Dark	Success	87.84
	Webcam	Bright	Success	81.19
		Dark	Success	74.29
Unknown 4	CCTV	Bright	Success	83.88
		Dark	Success	74.81
	Webcam	Bright	Success	54.65
		Dark	Success	55.39
Unknown 5	CCTV	Bright	Success	73.19
		Dark	Success	79.8
	Webcam	Bright	Success	69.84
		Dark	Success	63.61
Unknown 6	CCTV	Bright	Success	83.74
		Dark	Success	52.9
	Webcam	Bright	Success	73.97
		Dark	Success	73.65
Unknown 7	CCTV	Bright	Success	67.93
		Dark	Success	73.95
	Webcam	Bright	Success	72.79
		Dark	Success	87.74
Unknown 8	CCTV	Bright	Success	50.53
		Dark	Success	63.55
	Webcam	Bright	Success	86.13
		Dark	Success	83.26
Unknown 9	CCTV	Bright	Success	67.65
		Dark	Success	84.11
	Webcam	Bright	Success	55.18
		Dark	Success	85.78

webcam because it has infrared which is helpful in capturing the face compared with webcam as shown in Figure 21.9.

21.3.3 Online test

As the proposed security system will work in real time, the online test should be performed. As shown in Figure 21.9, the face captured using CCTV gives better result than webcam. However, CCTB brand used in the experiment has difficulty in making connection to Python used in face recognition system. Hence, during the online test, webcam is utilized to capture image in video format.

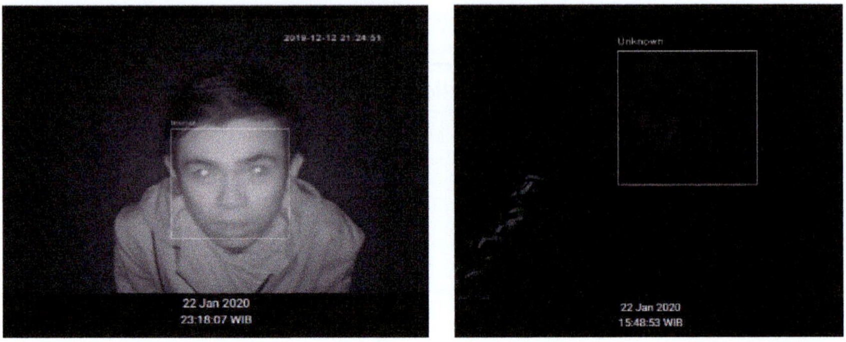

(a) (b)

Figure 21.9 Captured image for unknown face, using CCTV (a) and webcam (b)

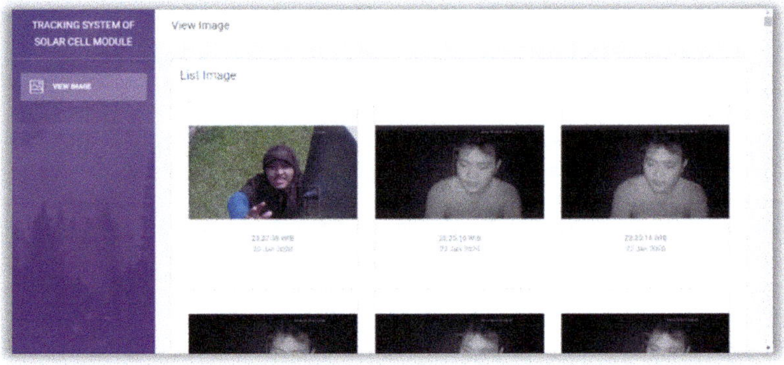

Figure 21.10 Website interface for storing the unknown face captured by the face recognition system

In the online test, using face recognition system that has been trained by deep learning earlier, the unknown detected face will be captured and sent to website as shown in Figure 21.10. Here, website has a role as the database which presents unknown faces captured by the system. Results of online test can be seen in Table 21.5.

As shown in Table 21.5, the system can detect unknown face and send it to the database. This can be helpful for the operators to monitor solar panel, especially when the solar panel is stolen. This security system based on face recognition works well in the day when the light is bright. Nevertheless, in the night, the system experiences difficulty in detecting and recognizing the face because the face image is not clear. Overall, the proposed system is able to recognize unknown face with an accuracy of 87.5% for real-time condition.

21.3.4 GPS tracking test

As an integrated security system, the GPS tracking system may work when the stolen solar panel has moved from its original position. The face recognition may be helpful

Table 21.5 Results of online test

Thief	**Upload image**	
	Bright (day light)	**Dark (night)**
Unknown 1	Success	Success
Unknown 2	Success	Success
Unknown 3	Success	Success
Unknown 4	Success	Success
Unknown 5	Fail	Success
Unknown 6	Success	Success
Unknown 7	Success	Success
Unknown 8	Success	Success

Table 21.6 Initial coordinate data of solar panel

Smartphone data		**Sensor GPS data**	
Latitude	Longitude	Latitude	Longitude
−3.21525°	104.64857°	−3.21521°	104.64859°

for the operator to find the suspect who stole the solar panel. However, the solar panel that has been stolen needs to be tracked so that its position can be found.

In the first experiment for testing the accuracy of GPS system, the solar panel is placed in the initial position and the coordinate for this initial position can be seen in Table 21.6.

As shown in Table 21.6, the coordinate GPS sensor has good accuracy as the difference between the coordinate from smartphone which is obtained from Google Maps is 0.00004 and 0.00003 for latitude and longitude, respectively. Then, the solar panel module box is moved to different locations of 1, 10, 50, 100, 150, 200 m away from the initial location. The results of coordinate changes can be seen in Table 21.7. As shown in the table, the first data have coordinates of −3.21523° and 104.64860° for latitude and longitude, respectively, when the GPS sensor is moved for 1 m. Hence, we can find the distance from two coordinates (see Table 21.6 for initial coordinate and the data in Table 21.7) using Haversine equation [12] as follows.

$$\Delta \text{lat} = \text{lat2} - \text{lat1}$$

$$\Delta \text{lng} = \text{lng2} - \text{lng1}$$

$$a = \sin^2\left(\frac{\Delta \text{lat}}{2}\right) + \cos(\text{lat1}) \times \cos(\text{lat2}) \times \sin^2\left(\frac{\Delta \text{lng}}{2}\right)$$

$$d = 2 \times R \times \text{arc sin}\left(\sqrt{a}\right)$$

Table 21.7 Coordinate data after moving the position of GPS sensor

Smartphone data		Sensor GPS data		Distance (m)		
Latitude	**Longitude**	**Latitude**	**Longitude**	**Smartphone**	**Sensor**	**Actual**
−3.21523°	104.64860°	−3.21523°	104.64860°	4.01	2.49	1
−3.21509°	104.64858°	−3.21515°	104.64862°	17.85	7.47	10
−3.21477°	104.64860°	−3.21477°	104.64860°	53.54	48.99	50
−3.21433°	104.64860°	−3.21432°	104.64860°	102.47	99.08	100
−3.21391°	104.64860°	−3.21387°	104.64860°	149.21	149.17	150
−3.21345°	104.64870°	−3.21342°	104.64860°	200.90	199.26	200

where d is the distance in meter, R is the radius of earth at the equator which is 6,378,137 m, and Δlat, Δlng, lat1, lat2, lng1, and lng2 are in radian.

From Table 21.7, the difference between actual data and GPS sensor as well as smartphone is large enough for the first three data. This difference becomes smaller for the fourth to sixth data. The difference is about 2.53 m which is suitable with the specification of the sensor in this research with the error of about $\pm$2.5 m [13].

21.3.5 GPS tracking: communication system

The proposed tracking system must have good communication system. Thus, the success of sending data and time while sending data is important. In this study, GPRS will be utilized as communication data system for Internet network and GSM is a backup system for SMS.

The initial position is shown by red bullet in Figure 21.11. This figure represents the moving lane of solar panel box. The aim is to get the coordinate data and time for sending the data from initial to the current position.

Results for communicating system using GPRS after moving object 16 times can be seen in Table 21.8. As shown in the table, the time need for sending and receiving the coordinate data is around 2–4 s with the average of 2.38 s. And the time to send the moving object from one location to other is about 17–22 s. Meanwhile, server may receive the data in about 18–21 s. The average time is 19.53 and 19.40 s for sending and receiving, respectively. Table 21.8 also shows the signal strength with the unit of received signal strength indicator (RSSI) [14] where the higher is the better signal strength.

In addition, this study also performed experiment for different speeds while the object is moving from one to another position. The speed varies from 20 to 60 km/h as shown in Tables 21.9 and 21.10.

We can see from Table 21.10 that the duration for sending the data is the same as Table 21.9 which is 2 s. Time delay for sending and receiving the data is 19 and 18.75 s, respectively. This means that the speed of moving object may not influence GPRS in sending and receiving the data. Using GPRS, the tracking system needs 2.38 s for receiving coordinate data and 19.4 s for receiving them. This process of sending and receiving data may depend on the signal strength.

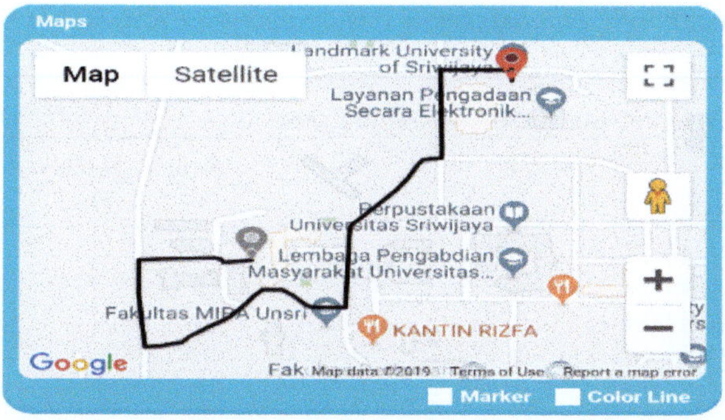

Figure 21.11 Moving lane of solar panel box to test communication system

Table 21.8 Results of communication system using GPRS

Sending time (WIB)	**Receiving time (WIB)**	**Sending time duration (s)**	**Time differences**		**Signal strength**
			Sent (s)	**Received (s)**	
13:38:57	13:38:59	2	0	0	17
13:39:16	13:39:18	2	19	19	17
13:39:35	13:39:37	2	19	19	18
13:39:54	13:39:56	2	19	19	21
13:40:13	13:40:15	2	19	19	19
13:40:32	13:40:34	2	19	19	19
13:40:51	13:40:53	2	19	19	14
13:41:10	13:41:12	2	19	19	16
13:41:28	13:41:31	3	18	19	15
13:41:48	13:41:51	3	20	20	14
13:42:08	13:42:12	4	20	21	14
13:42:30	13:42:33	3	22	21	18
13:42:51	13:42:53	2	21	20	19
13:43:10	13:43:13	3	21	20	23
13:43:30	13:43:32	2	20	19	29
13:43:48	13:43:50	2	18	18	31
Average		2.38	19.53	19.40	19

As communication data also utilizes GSM, the testing of sending and receiving coordinate data is also performed in this study. The results can be seen in Table 21.11.

Table 21.11 shows that the duration of sending data using GSM is longer than using GPRS. The average time is 4.87 s which is twice that of GPRS. It might happen because GSM has two stages in sending data. First, the system sends the coordinate data through SMS to smartphone and later, application in smartphone reads SMS and sends the coordinate data to database using Internet connection. Figure 21.12 shows the SMS sent to the smartphone.

Table 21.9 Results of communication system using GPRS (20 km/h)

Sending time (WIB)	Receiving time (WIB)	Sending time duration (s)	Time differences		Signal strength
			Sent (s)	Received (s)	
11:14:02	11:14:04	2	0	0	18
11:14:21	11:14:23	2	19	19	16
11:14:40	11:14:42	2	19	19	15
11:14:59	11:15:01	2	19	19	16
11:15:18	11:15:20	2	19	19	15
Average		2	19	19	16

Table 21.10 Results of communication system using GPRS (60 km/h)

Sending time (WIB)	Receiving time (WIB)	Sending time duration (s)	Time differences		Signal strength
			Sent (s)	Received (s)	
11:35:05	11:35:07	2	0	0	25
11:35:24	11:35:26	2	19	19	18
11:35:42	11:35:44	2	18	18	21
11:36:01	11:36:03	2	19	19	17
11:36:20	11:36:22	2	19	19	16
Average		2	18.75	18.75	19.4

As shown in Figure 21.12, there are some information sent in this SMS. The sequences information is as follows: the sending time, signal strength, status, type of communication, latitude and longitude. As seen in Table 21.11, GSM needs 18.71 and 18.57 s time delay for sending and receiving the coordinate data, respectively. This may imply that the time delay for sending and receiving may not be influenced by the type of communication data and the speed of moving object. It may be affected by the feedback response given by SIM808. From Tables 21.8 to 21.11, the time delay for sending each data is almost equal to about 19 s. Hence, communication using GSM can be utilized as a backup system when signal strength is low or Internet is unavailable.

21.3.6 GPS tracking: real-time system test

The purpose of GPS tracking in the security system of solar panel is to track the position of the stolen solar panel box in a real-time manner. The initial coordinates for this real-time test is shown in Table 21.12.

When the initial position has been saved in the database, the system will have "standby" status automatically as shown in Figure 21.13. The object will not be moved for 10 min to distinguish whether the position is moving or not. Table 21.12

Table 21.11 Results of communication system using GSM

Sending time (WIB)	**Receiving time (WIB)**	**Sending time duration (s)**	**Time differences**		**Signal strength**
			Sent (s)	**Received (s)**	
14:11:29	14:11:35	6	0	0	18
14:11:51	14:11:56	5	22	21	16
14:12:09	14:12:14	5	18	18	21
14:12:27	14:12:32	5	18	18	18
14:12:45	14:12:50	5	18	18	28
14:13:06	14:13:11	5	21	21	21
14:13:24	14:13:29	5	18	18	18
14:13:42	14:13:46	4	18	17	16
14:13:59	14:14:04	5	17	18	16
14:14:17	14:14:22	5	18	18	19
14:14:35	14:14:40	5	18	18	14
14:14:53	14:14:57	4	18	17	14
14:15:10	14:15:15	5	17	18	20
14:15:30	14:15:35	5	20	20	26
14:15:51	14:15:55	4	21	20	29
Average		4.87	18.71	18.57s	19.6

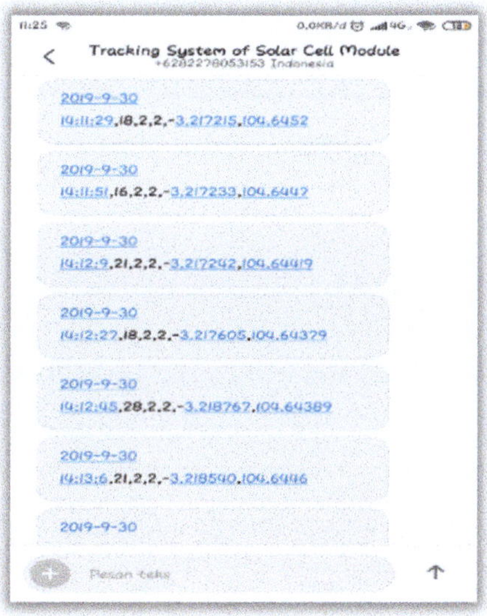

Figure 21.12 Coordinate data from GSM communication system through SMS

Table 21.12 Coordinate data for initial position to test the real-time GPS system

	Coordinate		Signal strength	Status
	Latitude	Longitude		
Initial position	−3.21737°	104.64520°	15	Standby
After 10 min without moving the object	−3.21737°	104.64520°	20	Standby

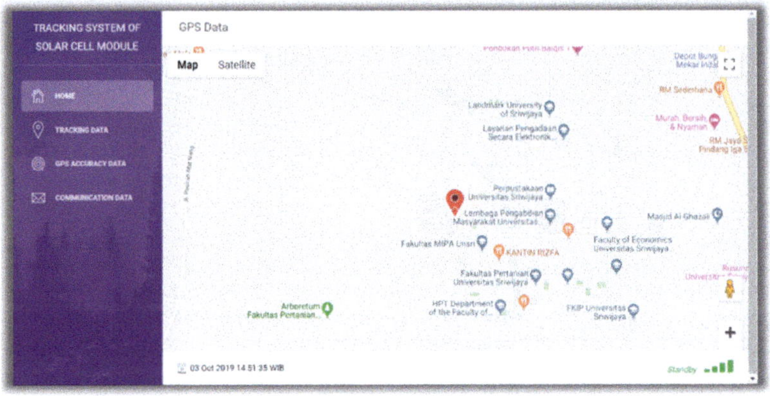

Figure 21.13 System interface in standby mode

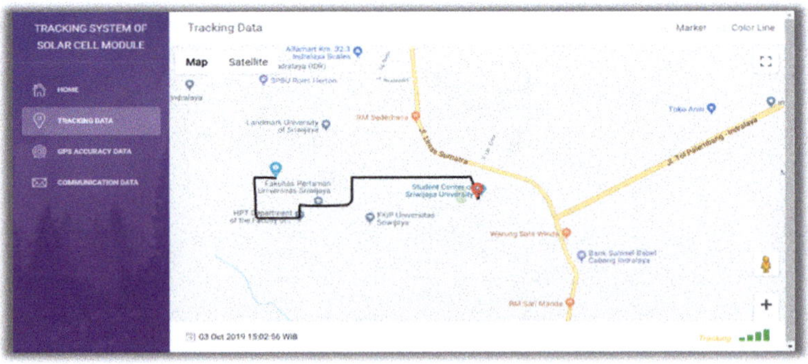

Figure 21.14 System interface in tracking mode

shows that the coordinate data do not change for 10 min when the object is not moving. Thus, the system is able to distinguish the standby mode.

Later, the object is moved to new position as shown in Figure 21.14 and the coordinate data can be seen in Table 21.13. As shown in Figure 21.14, pin location colored blue represents the initial position of solar panel module and the red represents the final location of it. From this figure, we can see that the proposed system

Table 21.13 Coordinate data for real-time GPS tracking system

No.	Sending time (WIB)	Coordinate		Signal strength	Communication type
		Latitude	Longitude		
1	14:54:56	−3.21726°	104.6446°	15	GPRS
2	14:55:12	−3.2173°	104.644°	20	GPRS
3	14:55:28	−3.21784°	104.64388°	18	GPRS
4	14:55:44	−3.21913°	104.64376°	24	GPRS
5	14:56:00	−3.2201°	104.64378°	17	GPRS
6	14:56:15	−3.22041°	104.6442°	15	GPRS
7	14:56:31	−3.22039°	104.6454°	15	GPRS
8	14:56:47	−3.22044°	104.6465°	17	GPRS
9	14:57:03	−3.22001°	104.64682°	13	GPRS
10	14:57:19	−3.21947°	104.6468°	13	GPRS
11	14:57:36	−3.21956°	104.6475°	11	GPRS
12	14:57:53	−3.21952°	104.6483°	10	GPRS
13	14:58:09	−3.21950°	104.64900°	9	GSM
14	14:58:25	−3.21954°	104.64970°	9	GSM
15	14:58:41	−3.21921°	104.6504°	12	GPRS
16	14:58:57	−3.21823°	104.65059°	12	GPRS
17	14:59:35	−3.21729°	104.6508°	21	GPRS
18	14:59:51	−3.21735°	104.6514°	19	GPRS
19	15:00:08	−3.21733°	104.6526°	18	GPRS
20	15:00:24	−3.21729°	104.6534°	23	GPRS
21	15:00:42	−3.21734°	104.6548°	22	GPRS
22	15:00:59	−3.21731°	104.656°	18	GPRS
23	15:01:16	−3.21731°	104.657°	19	GPRS
24	15:01:34	−3.21729°	104.6585°	19	GPRS
25	15:01:52	−3.21808°	104.65902°	19	GPRS
26	15:02:10	−3.21853°	104.65939°	19	GSM
27	15:02:26	−3.21897°	104.6593°	24	GPRS
28	15:02:42	−3.21898°	104.6593°	23	GPRS

may recognize the position of the solar panel from standby to moving. Figure 21.14 also shows that the tracking system is able to draw the lane that is passed by the solar panel box. In Table 21.13, data of 28 coordinates represent different communication systems. It is in GSM mode for three times (13th, 14th, and 26th data). The system will automatically move to GSM when the system cannot send the data using Internet or there is an error as well as disturbance while sending the data which may cause failure. In the 13th and 14th data, the signal strength is 9 which is considered low and hence the system uses GSM directly. Meanwhile, in the 26th data, the signal strength is 19 which is good enough but there might be failure or error while sending the data.

Figure 21.15 shows the visualization of coordinate data as shown in Table 21.13. The red and gray colors represent the data sending by GPRS and GSM, respectively. The details of sending time can be seen in Table 21.14.

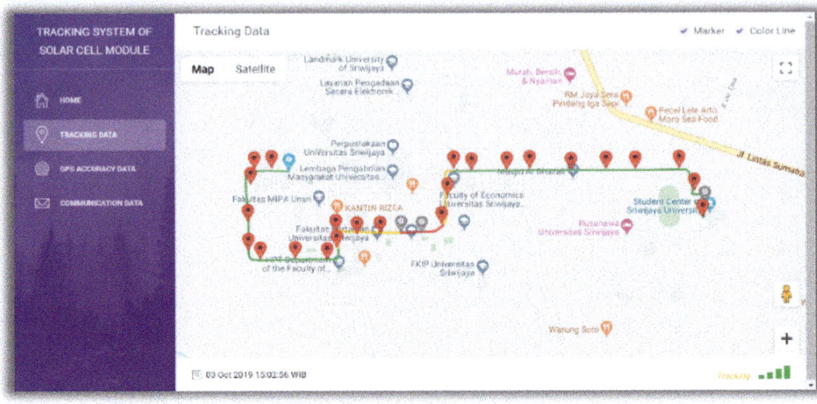

Figure 21.15 Visualization of coordinate data and the signal strength in real-time test of tracking system

From Table 21.14, the average sending time is 4.64 s. This time duration may be affected by the strength of the signal. In this experiment, the signal strength is 16.93 so the system needs lesser time in sending the coordinate data.

Lastly, the experiment is also performed for a condition when the system cannot connect to the satellite or when there is no GPRS or GSM connection. Figure 21.16 shows the interface when the system cannot be connected to the server.

As shown in Figure 21.16, the interface is quite different with Figure 21.14. Status system has changed to a lost connection and there is a circle around the red pin. The radius represented by this condition is set to be 100 m. This may indicate the last location of the solar panel box. This condition may occur after 10 mins that the system cannot send data to the server.

From the results and discussions earlier, we can see that the face recognition system and tracking system work together as integrated security system. The security system is able to capture the unknown face who is suspected to be harmful to the solar panel module. The face recognition developed using deep learning can distinguish the known and unknown faces. On the other hand, the tracking system that is aimed to track the position of the stolen solar panel box works well using GPS technology based on GPRS and GSM. The tracking system can recognize the solar panel module position whether it stays still or move to a new location. Those both face recognition and tracking system can be monitored through an interface based on website. Thus, this security system can be applied in real-time condition. In addition, the tracking system is able to manage the lost connection status by giving the information of the last location in radius of 100 m after 10 min when the system cannot send new coordinate data.

Table 21.14 Data for communication system in real-time test of tracking system

Sending time (WIB)	Receiving time (WIB)	Sending time duration (s)	Time differences		Signal strength
			Sent (s)	Received (s)	
14:54:56	14:55:02	6	0	0	15
14:55:12	14:55:18	6	16	16	20
14:55:28	14:55:34	6	16	16	18
14:55:44	14:55:50	6	16	16	24
14:56:00	14:56:06	6	16	16	17
14:56:15	14:56:21	6	15	15	15
14:56:31	14:56:37	6	16	16	15
14:56:47	14:56:53	6	16	16	17
14:57:03	14:57:06	3	16	13	13
14:57:19	14:57:22	3	16	16	13
14:57:36	14:57:40	4	17	18	11
14:57:53	14:57:56	3	17	16	10
14:58:09	14:58:14	5	16	18	9
14:58:25	14:58:30	5	16	16	9
14:58:41	14:58:44	3	16	14	12
14:58:57	14:59:05	8	16	21	12
14:59:35	14:59:38	3	38	33	21
14:59:51	14:59:55	4	16	17	19
15:00:08	15:00:11	3	17	16	18
15:00:24	15:00:28	4	16	17	23
15:00:42	15:00:45	3	18	17	22
15:00:59	15:01:03	4	17	18	18
15:01:16	15:01:20	4	17	17	19
15:01:34	15:01:39	5	18	19	19
15:01:52	15:01:55	3	18	16	19
15:02:10	15:02:17	7	18	22	19
15:02:26	15:02:31	5	16	14	24
15:02:42	15:02:45	3	16	14	23
Average		4.64	17.26	17.15	16.93

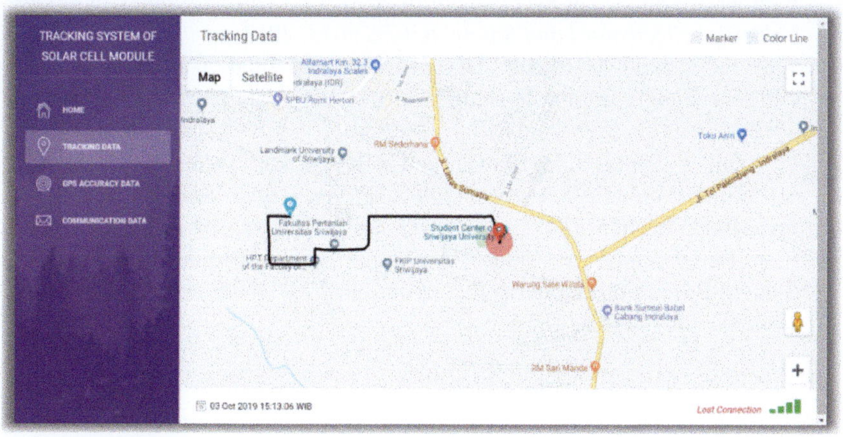

Figure 21.16 Interface for system response when the solar panel box has lost connection to the server

21.4 Conclusion

This study proposed an integrated system for a security system to find and track the stolen good. The first level of security system is face recognition system that may recognize unknown face using FaceNet as the feature extraction and DBN for the classifier. This system is integrated to the database. From the experiment performed in offline and online manner, the system is able to recognize unknown face with an accuracy of 94.4% for offline and 87.5% for online or in real-time condition. The unknown face recognized by the system is captured and sent to the database may be helpful to find the suspect who stole the object. Meanwhile, the stolen object position needs to be tracked. This tracking is performed in the second level of security system once the object has been stolen. The tracking system utilizes GPS integrated to the database using GPRS and GSM as communication system. The error of GPS sensor is about 2.5 m with the sending time duration of 4.64 s. The system can also track the coordinate location well using both GPRS and GSM and this is affected by the strength of the signal.

The proposed security system could be useful for the security system as it combines the technology of face recognition and GPS tracking.

References

[1] P. Kumar, M. Agarwal, and S. Nagar, "A Survey on Face Recognition System—A Challenge," *Int. J. Adv. Res. Comput. Commun. Eng.*, vol. 2, no. 5, pp. 2167–2171, 2013.

[2] A. Nurhopipah and A. Harjoko, "Motion Detection and Face Recognition for CCTV Surveillance System," *IJCCS (Indonesian J. Comput. Cybern. Syst. Yogyakarta, Indones.)*, vol. 12, no. 2, pp. 107–118, 2018.

[3] D. A. R. Wati and D. Abadianto, "Design of Face Detection and Recognition System for Smart Home Security Application," *2017 2nd Int. Conf. Inf. Technol. Inf. Syst. Electr. Eng.*, pp. 342–347, 2017.

[4] Z. Liu, A. Zhang, and S. Li, "Vehicle Anti-Theft Tracking System Based on Internet of Things," in *Proceedings of 2013 IEEE International Conference on Vehicular Electronics and Safety, ICVES 2013*, 2013.

[5] K. A. Salim and I. M. Idrees, "Design and Implementation of Web-Based GPS–GPRS Vehicle Tracking System," *IJCSET Dec*, vol. 3, no. 3, pp. 5343–5345, 2013.

[6] P. Singh, T. Sethi, B. B. Biswal, and S. K. Pattanayak, "A Smart Anti-Theft System for Vehicle Security," *Int. J. Mater. Mech. Manuf.*, vol. 3, no. 4, pp. 249–254, 2015.

[7] P. A. Shinde and Y. B. Mane, "Advanced Vehicle Monitoring and Tracking System Based on Raspberry Pi," *Proc. 2015 IEEE 9th Int. Conf. Intell. Syst. Control. ISCO 2015*, 2015.

[8] P. Viola and M. J. Jones, "Robust Real-time Object Detection," *2nd Int. Work. Stat. Comput. Theor. Vis.—Model. Learn. Comput. Sampling. Vancouver, Canada*, vol. 57, pp. 1–30, 2001.

[9] F. Schroff, D. Kalenichenko, and J. Philbin, "FaceNet: A Unified Embedding for Face Recognition and Clustering," *2015 IEEE Conf. Comput. Vis. Pattern Recognit.*, pp. 815–823, 2015

[10] I. Goodfellow, Y. Bengio, and A. Courville, *Deep Learning*, vol. 1. London, England: Nature Publishing Group, 2016.

[11] G. E. Hinton, "Deep Belief Networks," *Scholarpedia*, vol. 4, no. 5, p. 5947, 2009.

[12] I. Setyorini and D. Ramayanti, "Finding Nearest Mosque Using Haversine Formula on Android Platform," *J. Online Inform.*, vol. 4, no. 1, p. 57, 2019.

[13] S. W. Sun, X. Wang, X. Xiao, L. Teng, X. Zhang, and H. Yang, *SIM808 Hardware Design*. Shanghai: Shanghai SIMCom Wireless Solutions Ltd., 2015.

[14] SIMCom, *SIM800 Series AT Command Manual*, vol. 1. Shanghai: Shanghai SIMCom Wireless Solutions Ltd., 2015.

Chapter 22

Project Dragonfly

Ting Rang Ling[1], Lee Soon Tan[1], Sing Muk Ng[2] and Hong Siang Chua[1]

Project Dragonfly is a two-in-one industrial wastewater and air toxicity monitoring solution that is environmentally friendly, noninvasive, and cost-effective. The name suggests a biomimicry, where the drone in the project is able to fly, change its altitude, move forward and backward at relatively fast and slow speeds, change direction, spontaneously stop and hover, in addition to performing the signature act of dipping into the surface of water, resembling the nature of a dragonfly. The project utilizes Microsoft Azure platform along with Microsoft's proprietary cloud products and services, and Android mobile application for its software components and database: Lolin D32 Pro, Neffos Y5i, multiple sensors and a quadrotor helicopter (quadcopter) are among its main hardware components. The sensors are packed into two functional units: air monitoring unit (AMU) and water monitoring unit (WMU).

22.1 Introduction

22.1.1 Overview

In the recent years, technology has evolved swiftly. This leads to the emergence of new areas of business activities and industrial development as well as expansion, for example transforming toward Industry 4.0. Nonetheless, pollution has become a major concern in modern metropolises due to industrial emissions and increasing urbanization. It would be the responsibility of factories to control their emissions of pollutants into the air and/or water. The project presents remarkable significance when regulation of industrial emissions becomes crucial. Access to otherwise-inaccessible areas would be made possible. Moreover, due to safety concerns, using a drone to monitor levels of pollutants is much more desired. Primarily to the regulatory authority, governing body, and scientific institutions, water quality and air toxicity monitoring can be performed systematically and periodically, with minimal effort by

[1]Faculty of Engineering, Computing and Science, Swinburne University of Technology Sarawak Campus, Sarawak, Malaysia
[2]Sarawak Energy Berhad, Malaysia

humans, should there be a well-developed system in place which also utilizes the famous Internet of Things (IoT) concept. The information obtained could be used by enforcement teams to penalize companies that violated the set standards. The primary objective of the project is to design and create an environmental monitoring system that is reliable, linked to World Wide Web (WWW), user-friendly, environmentally sound, and cost-effective.

22.1.2 Air quality

Gas emissions during industrial activities may be harmful to humans, animals, and/or plants, leading to air toxicity [1]. This project addresses several notable air pollutants as highlighted in World Health Organization (WHO) air quality guidelines. Also, greenhouse gases such as carbon dioxide (CO_2) and methane (CH_4) are considered. Overall, for the purpose of air quality monitoring, the project focuses on the concentrations of nitrogen dioxide (NO_2), ammonia (NH_3), carbon monoxide (CO), carbon dioxide (CO_2), and methane (CH_4), besides supporting parameters such as ambient temperature and humidity. The gases are detected by using metal-oxide semiconductor (MOS)-type gas sensors.

22.1.3 Water quality

Environmental protection laws govern the treatment of industrial wastewater, where the specified organic matter, inorganics, pathogens, and nutrients need to be removed before the water is considered treated and subsequently discharged into water bodies [2]. Typical wastewater treatment methods are solids removal, oils and grease removal, biodegradable organics removal, activated sludge process, treatment of acids and alkalis, and treatment of toxic materials, as listed by the International Water Association [3]. Generally, wastewater would receive pretreatment at the factories, thereby minimizing pollutant load, prior to discharging into water bodies. However, especially in developing countries, sometimes industrial effluent is discharged without any treatment process [4]. Thus, an efficacious system to monitor water quality is crucial. In the project, turbidity and electrical conductivity (EC) values are considered to give a picture of the water quality at a certain location.

22.2 Related research

22.2.1 IoT concept

The IoT refers to a network of physical devices or objects that are connected to the Internet and have the capability to identify themselves (UIDs) as well as communicating with one another over the network. It enables interconnected devices to be accessible anywhere through the Internet and can be remotely monitored and controlled. Based on Cisco's VNI report, it is stated that the production of IoT devices are expected to surge up to 10 billion by 2020 and 22 billion by 2025. Thus, there is no doubt that IoT plays an indispensable role in the future. Majority of business processes and systems are likely to incorporate IoT element in the upcoming years [5]. The applications of smart technology have significantly impacted every aspect of human lives, by bringing a better quality of living as well as keeping us connected with the technologies. Providing a unified access mechanism is the fundamental concept in IoT. With better controllability and configurability of electronics devices under IoT environment, many countries are now seeking for IoT-oriented technology to assist in paving the conformation of related field.

Generally, IoT architectures are comprised of three main components, namely hardware, middleware, and services. Hardware components can be any devices, sensors, or even embedded systems which made up the fundamental parts of a smart system. They function to collect data and transmit it to a specific platform (such as middleware) for processing. The middleware provides the functionality to gather and process the data sent from hardware components and store it whenever necessary. With established network connection, it can transmit these data out to online cloud infrastructures for further computations and storage purposes. These stored resources can be fetched by application software that is substantially devised to meet the project design specifications as well as to depict the necessary insights on the information collected via graphical user interface. The emergence of IoT sensing technology has made it a key component in the monitoring and healthcare applications, especially in agricultural and industrial fields. Such applications include automatic plant health monitoring, soil fertility and erosion monitoring, ceramic tile grade monitoring, and gas pipe leakage detection system.

In most cases, IoT water monitoring systems are having the standardized workflow structure, in which water quality detection sensors are deployed in specific places of interest to inspect the surrounding physiological status. Simultaneously, sensor data measurement will be taken and transmitted to the dedicated cloud platform for processing and storage purposes via a local area network. For better real-time data analysis and event detection, it will be crucial to have some form of feedback paradigm, aimed to deliver comprehensive overview on the environmental condition. With respect to that, a software application that works alongside with the cloud server will be deployed to allow retrieval and dissemination of information to end users. Figure 22.1 depicts the standard block diagram of IoT monitoring system as explained previously.

Although there were many existing researches done on the water quality monitoring system, the methods of realization and deployment of such system may be slightly different with reference to the technologies and software applications used.

22.2.2 Air quality

22.2.2.1 Effects of industrial activities on air quality

Increasing urbanization is necessary to cope with the ever-rising world population number, which is predicted to be 7.81 billion at the time of writing [6].

From Figure 22.2, one could observe that the number of people living in this world had skyrocketed in the recent decades. Thus, it is naturally justifiable that more industrial activities had occurred as compared to the previous centuries.

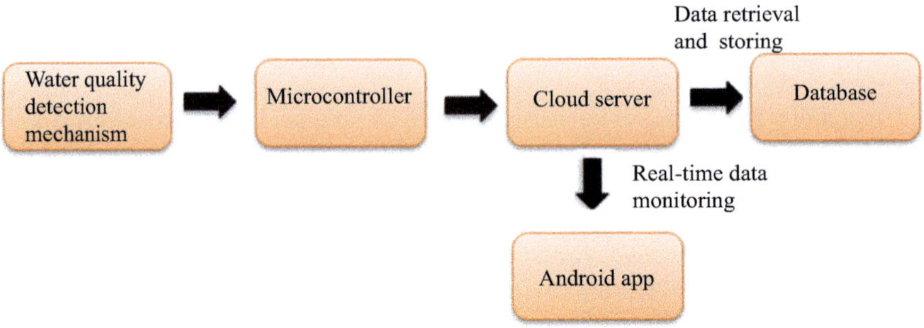

Figure 22.1 Block diagram for IoT based water quality monitoring system

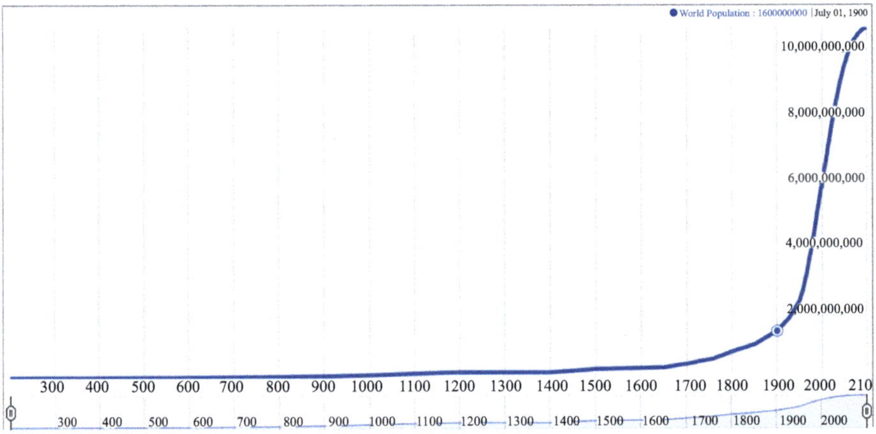

Figure 22.2 World population: past, present, and future

As highlighted by [7], industries can be classified according to the type. There are generally four types of industries: primary, secondary, tertiary, and quaternary.

The obtainment of raw materials, for instance mining, fishing, and farming, fall into primary industry. Secondary industry includes all manufacturing activities, for example, the making of cars, electronic components, electrical appliances, tiles, steel, and so on. Tertiary industry means providing or delivering services. Examples include teaching, coaching, and nursing. Research and development falls under quaternary industry.

Industrial activities that will harm the environment are usually of primary and secondary categories. Quaternary industry rarely leads to noticeable negative effects on the environment, although it does have the potential if carried out on a large scale. Tertiary industry is the most gentle toward the environment.

Fuel combustion has been known to emit gases that pollute the environment, for example, sulfur dioxide (SO_2), nitrogen oxides (NO_x), and carbon dioxide (CO_2). In Asia, intensive emissions had been observed at various locations, especially the Pacific Rim region of Japan, North Korea, and South Korea, and coastal areas of Taiwan and China [8]. In the inland of China, high emissions of SO_2 were measured at several grids. Historical data showed that NO_x emissions in Asia had surpassed those of North America and Europe [9]. Besides, Fenger [10] pointed out that photochemical air pollution is a large-scale phenomenon in the northern part of Europe. Also stated by Fenger [10], rapid urbanization has often led to uncontrolled growth and deterioration of the environment in developing countries. Industrial activities contribute to transboundary pollution, resulting in the increase of greenhouse gas concentrations in the global scope. Mitigation of air pollution requires efforts from developed and developing countries.

22.2.2.2 Air pollutants

WHO air quality guidelines

Designed to serve as a guidance in reducing the health impacts of air pollution, the WHO air quality guidelines were first published in the year 1987 and subsequently updated in the year 1997 [11]. Later, several revisions had been made. The guidelines can be regarded as the most reliable international air quality guidelines. The guidelines are constructed based on expert evaluation, with the availability of latest scientific evidences. From time to time, WHO has undertaken to review the accumulated scientific studies in the field of air quality. The guidelines produced are applicable across all WHO regions, which include African Region, Region of the Americas, South-East Asia Region, European Region, Eastern Mediterranean Region, and Western Pacific Region [12].

The WHO air quality guidelines shown herein are based on the global update in 2005 (Table 22.1).

Notable air pollutants

$PM_{2.5}$ refers to particles smaller than 2.5 μm, while PM_{10} refers to particles smaller than 10 μm. In both short-term and long-term exposures to particulate matter, the adverse effects are evident. Prominently, diseases and health problems of the respiratory and cardiovascular systems are observed. Presently, measurements of

Table 22.1 Overview of WHO air quality guidelines [11]

Pollutant	Value (μg/m^3)	Averaging time
$PM_{2.5}$	10	1 year
	25	24 h
PM_{10}	20	1 year
	50	24 h
O_3	N/A	1 year
	100	8 h
NO_2	40	1 year
	200	1 h
SO_2	20	24 h
	500	10 min

PM_{10} are more routinely conducted. PM_{10} measurements collectively include the fine ($PM_{2.5}$) and coarse (PM_{10}) particles that may enter the respiratory tract.

Ozone is generally formed in the atmosphere through photochemical reactions. Presence of sunlight is mandatory. It also requires precursor pollutants, for instance volatile organic compounds (VOCs) and NO_x. Noticeable negative health effects are expected in the event of 8-h concentrations, with values higher than 240 $\mu g/m^3$.

Indicated by the experimental studies on humans and animals, concentrations of more than 200 $\mu g/m^3$ render the gas toxic with notable health effects in short-term exposures. Adverse effects will occur in long-term exposures. A large portion of NO_2 is originally released as NO, but is expeditiously oxidized by O_3 to become NO_2. As the 200 $\mu g/m^3$ guideline has yet to be challenged by recent studies, it is maintained from the previous revision of the WHO air quality guidelines [13].

Exposure to SO_2 is detrimental to human health. Studies have indicated that short-term exposure as short as 10 min may lead to respiratory symptoms and impact the pulmonary function. Notable decrease in the health effects was noticed when sulfur content was greatly reduced, based on a research done in Hong Kong [14]. In a similar manner, increased mortality is associated with sulfate and fine particulate air pollution at levels commonly found in U.S. cities [15].

Although there are evidences that point to the health effects caused by SO_2, there is still controversy that surrounds the matter. There is a possibility that the adverse health effects are caused by ultrafine (UF) particles, or other factors that are still out of the reach of current studies. Nonetheless, based on the current scientific literature, it is reasonable to regard SO_2 as a deleterious gas.

A spatial assessment was conducted on the air quality patterns in Malaysia, as documented in Dominick *et al.* [16]. The multivariate analysis incorporated several methods, including hierarchical agglomerative cluster analysis (HACA), principal component analysis (PCA), and multiple linear regression (MLR) to determine the spatial patterns, major sources of the pollution, and contributions of the air pollutants, respectively. From the PCA analysis, air pollution cases in Malaysia are largely due to emissions from industries, vehicles, and zones in which the population density is high. This is a literature evidence to support the hypothesis whereby air pollutants are evident in the industrial areas. From the MLR analysis, the variability in the air pollutant index (API) at the air monitoring stations in Malaysia was predominantly due to particulate matter, specifically PM_{10}. It was even shown in further analysis that high concentration of PM_{10}, in the context of the research conducted, was mainly caused by carbon monoxide (CO). In this case, it can be inferred that combustion processes, including those carried out in factories and industrial facilities, as well as the engines of vehicles, will contribute toward air pollution in the local environment.

Ammonia (NH_3) is also one of the toxic gases. It is commonly used in the production of dry fertilizers, food processing, manufacture of plastics, dry fertilizers, textiles, dyes, and so on. Ammonia gas may leak from storage tanks accidentally, as documented in [17], and thereafter polluting the local atmosphere and causing health risks. As monitoring of air quality at industrial areas is a major part of the project, the ability of the system to detect leakage gases like ammonia in the industrial areas will be an added feature.

On the other hand, methane (CH_4) is one of the major heat-trapping gases which contribute to the so-called "greenhouse effect." In other words, it is a greenhouse gas. Some of the sources of methane include anaerobic decomposition in natural wetlands and paddy fields, emission from livestock production systems, biomass burning, and exploration of fossil fuels [18]. Besides, digesting tanks and lagoons of wastewater treatment plants might have emitted great amounts of methane [19]. Excessive release of methane into the atmosphere is definitely undesired.

22.2.2.3 Existing air quality monitoring methods

There are two categories for mainstream air quality monitoring methods, namely manual air sampling and periodic monitoring using air quality monitoring stations (AQMS).

Manual air sampling

On-site quick analysis is perhaps the most primitive of all. Nevertheless, its effectiveness cannot be underestimated, especially for air quality monitoring at places where the air monitoring stations are not installed. It involves a worker wearing the air sampling accessories. The worker will need to enter the monitoring site, which is possibly filled with abundance of toxic air pollutants. A pump is used to sample the air near to the breathing zone, and the wearable device will make quick analysis of the sampled air.

For more comprehensive analysis, the worker needs to collect air samples and bring them back to the laboratory. SanAir Technologies Laboratory [20] described several ways to achieve this.

Whole air sampling is the easiest of its kind, and it involves the collection of air samples using Tedlar bags and Summa canisters. The samples are sealed.

Active air sampling is generally used for collection of volatile and semi-VOCs. It involves usage of a tube that is filled with solid sorbent material. Using a sampling pump, air is passed through the tube and the pollutants are chemically absorbed by the material.

The solid sorbent material is also used in the case of passive air sampling. The difference is that the absorption of the pollutants depends on diffusion process.

Filter sampling is ideal for collecting pollutants in vapor form. Using a pump, the air is passed through a filter cassette. Chemical reaction occurs between the filter and the air containing the pollutants. Then, a stable derivative is formed. The type of filter is specific toward the individual pollutant.

In impinger sampling, air is bubbled through a solution. The solution (in liquid form) is reactive. The pollutants in the air will chemically react with the solution. Impinger sampling is typically used in cases where the source is considerably static, moisture level is high, and ambient temperature is high.

Using AQMS

The air quality monitoring process can be done remotely by constructing permanent air monitoring stations at fixed locations. Typically, the AQMS are capable of detecting at least PM_{10} (particulate matter with diameter less than 10 μm). Newer AQMS are able to detect smaller particulate matter, that is, $PM_{2.5}$. Most AQMS also take into account other pollutants such as ozone (O_3), sulfur dioxide (SO_2), carbon monoxide (CO), and nitrogen dioxide (NO_2).

Nonetheless, an AQMS would only be able to give the air quality readings at the specific point it is installed. The air quality in that city is commonly assumed to be consistent with that measured at the AQMS installation site. A substantial number of AQMS will be needed to quantify air qualities at different points in a particular area. Besides, AQMS are relatively costly and bulky, and they require significant amount of inspection and maintenance works from time to time.

22.2.2.4 Water quality index and pollutants

In general, there are numerous parameters that can influence the quality of water. Department of Environment (DOE) in Malaysia has asserted some of the utmost components that best describe the water quality index (WQI). This includes biochemical oxygen demand (BOD), pH, and chemical oxygen demand (COD). The classification of WQI is stated in Table 22.2.

Table 22.3 shows the suitable utilizations of water for different classes, depending on the range specified. Classes I–III describe the water quality required to sustain aquatic life, whereas water sources from Class V are considered to have minimal beneficial usage due to the presence of high amount of pollutant compounds. Conversely, water sources from Class I have the minimal level of contamination. For safe level of drinking, other classes still require a proper treatment system. WQI is a general definition that signifies the cleanliness of water. It serves to summarize large amount of quality data of a river in simpler term and it is useful in communicating water quality information to the local communities [22].

The water quality parameters can be categorized into three properties, namely chemical, physical, and biological [22]. pH, EC, dissolved oxygen, salinity, and oxidation–reduction potential are considered part of the chemical properties in water quality. The acidity or alkalinity of a solution is determined by the pH value whereas the EC is influenced by the number of ions present in the solution. The ability of water to remove impurities by itself is known as oxidation–reduction potential. Likewise, having a high value for oxidation–reduction potential signifies a good quality of water. On the other hand, the physical factors of water quality include turbidity and

Table 22.2 DOE WQI classification [21]

Parameters	Unit	Class				
		I	II	III	IV	V
Ammoniacal nitrogen	mg/L	<0.1	0.1–0.3	0.3–0.9	0.9–2.7	>2.7
Biochemical oxygen demand	mg/L	<1	1–3	3–6	6–12	>12
Chemical oxygen demand	mg/L	<10	10–25	25–50	50–100	>100
Dissolved oxygen	mg/L	>7	5–7	3–5	1–3	<1
pH	mg/L	>7	6–7	5–6	<5	>5
Total suspended solid	mg/L	<25	25–50	50–150	150–300	>300

Table 22.3 DOE class definitions [21]

Class	Definition
I	• Conservation of natural environment • Water supply I—practically no treatment necessary (except by disinfection or boiling only) • Fishery I—Very sensitive aquatic species
IIA	• Water supply II—Conventional treatment required • Fishery II—Sensitive aquatic species
IIB	• Recreational use with body contact
III	• Water supply III—Extensive treatment required • Fishery III—Common of economic value, and tolerant species; livestock drinking
IV	Irrigation
V	None of the above

temperature. Turbidity represents the cloudiness or opacity of a solution. High value of turbidity indicates a large amount of material dissolved in a solution (less opacity). The amount of bacteria, phytoplankton, virus, and algae form the biological factors of water quality. All these factors are used to evaluate water quality. However, some of them (algae, virus, etc.) are still immeasurable by existing sensing technology and can only be realized through chemical analysis [22].

22.2.2.5 Existing water quality monitoring methods

Several methods have been proposed to monitor and provide real-time data on the quality of water using sensors. In every monitoring-based IoT system, choosing the right water quality sensors is important since different sensors may serve for different purposes [22]. So, to suit the project specifications, the selection of appropriate sensors must be made, depending on the sensor efficiency, cost, and the measurement parameters.

According to Radhakrishnan and Wu [22], most IoT systems employ sensors that are commercially available in the market. Such sensors include DFRobot–Turbidity sensor, gravity–pH sensor, gravity–analog DO sensor, and DFRobot–temperature sensor. These are considerably low-cost sensors with single quality parameter detection mechanism and are commonly utilized in simple IoT projects. Madhavireddy and Koteswarrao [23] had published a proposal for cheap portable water monitoring system. This research ensures a safe supply of drinking water. For the system implementation, simple and yet low-cost sensors are used and installed at water distribution network. These include pH sensor, temperature sensor, and SKU:SEN0219 sensor (for detecting CO_2). Their respective parameters will be measured and stored at web server. For efficient dissemination of pollutant information, Madhavireddy and Koteswarrao [23] have developed a custom webpage to facilitate real-time monitoring of the sensor data as shown in Figure 22.3.

For production purposes or IoT applications in broad-scale economic sectors, a better sensor with multiparameter detection mechanism should be used [22]. An

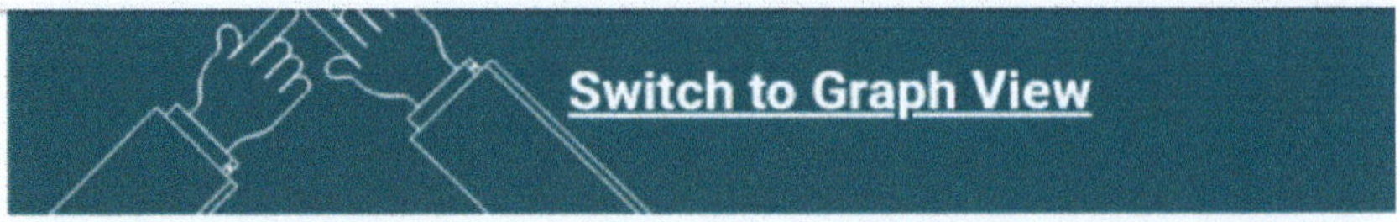

Previous Page 16 of 92 Next

S.No	S1	S2	S3	S4	Date
301	HIGH_PH				2018-04-08 13:07:45
302	HIGH_TEMPRATURE				2018-04-08 13:07:17
303	HIGH_PH				2018-04-08 13:06:49
304	HIGH_TEMPRATURE				2018-04-08 13:06:20
305	HIGH_PH				2018-04-08 13:05:52
306	HIGH_TEMPRATURE				2018-04-08 13:05:24
307	HIGH_PH				2018-04-08 12:57:12
308	HIGH_PH				2018-04-08 12:56:45
309	WATER_LOW				2018-04-08 12:56:18
310	HIGH_PH				2018-04-08 12:55:52
311	HIGH_PH				2018-04-08 12:55:25
312	GAS_DETECTED				2018-04-08 12:54:58
313	HIGH_PH				2018-04-08 12:54:31
314	HIGH_PH				2018-04-08 12:54:04
315	HIGH_PH				2018-04-08 12:51:04
316	HIGH_PH				2018-04-08 12:50:38
317	WATER_LOW				2018-04-08 12:47:27
318	GAS_DETECTED				2018-04-08 12:47:01
319	HIGH_PH				2018-04-08 12:44:02
320	HIGH_PH				2018-04-08 12:43:36

Figure 22.3 Sensor data accessing in the web through IoT [23]

example would be Kapta 3000 AC4 that is designed to measure chlorine level, pressure, temperature, and conductivity of water, and it costs 3,200 GBP. Spectro::lyser is another type of sensor which measures turbidity, color, and water contaminations and it costs around 5,000 GBP. Likewise, these are pragmatically competent sensors but expensive in terms of cost, and hence impractical for deployment especially in a project with broad monitoring sites (where numerous sensors are required) [22]. Table 22.4 lists some of the conventional water quality detection sensors that are readily available in the market and their respective estimated cost.

Based on the research done by Gholizadeh *et al.* [24], the physical collection of water sample is quite a challenge, especially in obtaining the quality indices in a

Table 22.4 Commercially available water quality sensors and its measuring parameters [22]

Sensors	Quality detection parameters	Estimated cost (GBP)
DFRobot–Turbidity sensor	Turbidity	10
Gravity–Analog DO sensor	Dissolved oxygen	130
Kapta 3000 AC4	Conductivity, chlorine, temperature, pH, pressure	3,200
Spectro::lyser	Turbidity, color, dissolved ions, temperature pressure	5,000
SmartCoast (Wireless sensor)	pH, temperature, phosphate, water level conductivity, dissolved oxygen	4,100
Gravity–pH sensor	pH	29

large scale. To pave up these limitations, some researchers have proposed the use of remote sensing technique.

Remote-based system utilizes visible and near-infrared bands of solar to identify the correlation between the water reflection and the pollutants level [24]. This technique makes it possible to have instantaneous temporal view of the surface water quality parameters. According to Wang and Yang [25], remote sensing does support the evaluation of changes in water quality (through observation of color and refractive index) and detection of algal bloom. However, such visual sensing methods are not sufficiently precise and dependable as it could be influenced by optically complex conditions such as soil runoff and sediment. As a result, even until now, this technique is not widely used. Moreover, the algorithm developed for the system is applicable only for a specific area, and the quality data obtained could sometimes be fallacious due to unforeseen environmental changes such as storm [26]. Thus, as far as result accuracy is concerned, it would be advisable to focus on the physical sensors that are readily available in the market.

Sensor selection factor

As mentioned by Radhakrishnan and Wu [22], when it comes to devising an IoT monitoring infrastructure, it may not be a productive way to incorporate every single quality parameter as it would require a high workload and substantial resources (involving many sensors) for realization of such system. So, the parameters of water quality to be considered should be based on the property in which they exhibit and the location it is measured. In most cases, the main constituents of water source at the monitoring site and how it is substantially affected by external factors (such as industrial wastewater, climate change, disposal of waste, etc.) will be taken into considerations in choosing the right quality parameters for an IoT monitoring system. The similar concept is applied in the project done by Hamuna *et al.* [26]. They have launched a smart water wireless platform for ocean and coastal quality monitoring. Since their project is dedicated for measuring seawater pollution, the parameters measured by sensors are mainly concerned with dissolved ions and salinity. These include pH, EC, and dissolved ions (such as fluoride, nitrate, chloride, lithium, etc.). Another example can be discerned in the work proposed by Zakaria and Michael [27]. The devised system is a cloud-based wireless sensor network (WSN) with embedded sensor module and gateway node. It functions to collect water quality data from the discharged industrial wastewater and transmit it to the cloud platform via wireless communication module. As most wastewater discharged by industries contains a high amount of dissolved ions and suspended solid particles, the proposed sensor unit will comprise pH, conductivity, and dissolved oxygen sensors.

Topological factor and sensor deployment location

High operational cost and frequent maintenance are common issues encountered in most water-based IoT system. Such issues are believed to be related to the sensor deployment location and environmental conditions. In that matter, the concept of deploying various sensors in different locations such as river and streams has been widely tested. Based on the project conducted by Kruger *et al.* [28], over 220 units of sensors are installed to monitor water level and pollution content. The deployed system is comprised of sensor module (consisting of ultrasonic, turbidity, and temperature sensors), solar energy system (for powering up the sensors), and a GPS receiver. The ultrasonic sensors in this case are used as distance measuring module that estimates the water level and sends alert notification to end users whenever any abnormalities are detected in the reading which could signifies flooding. As bridge provides a sturdy mounting platform, these sensors will be installed underneath it.

Temperature and turbidity sensors are water-contact sensors which only work when its detection section is in contact with the water. Hence, float gauges will be bind to these sensors so that real-time water quality data can be collected overtime. Additionally, a cell modem is installed to allow integration with cellular network and transmission of sensor data to dedicated cloud platform. When activated, sensor measurement will be taken periodically (every 5 min) and sent across the network. However, after months of operation, it has been found that the system potentially demands a high level of maintenance. Throughout the period, many of the water-contact sensors were damaged due to continuous changes in the environmental conditions [28]. Rains and solid materials carried by stream currents seem to pose serious threat to these hardware components, by inducing salt and other contaminations into the system, which could lead to corrosion of wire and unintended short circuits in long term. Nonetheless, noncontact components such as ultrasonic sensors are not affected in this case as they are not in direct contact with the water. Thus, the work from Kruger *et al.* [28] has proven that the longevity of noncontact hardware system to be much better than those that are directly exposed to the environment.

An image-based flood alarm system was proposed for use in disaster warning and detection [29]. In this system, remote camera would be employed to monitor rivers with high risk of flooding. By using image processing technique, the water edges will be identified and been converted to its respective water-level value. In terms of efficiency and flexibility, Lo *et al.* [29] have asserted that image-based system is still preferable than the traditional water-level estimation techniques that involve installation of sensors at pipes in rivers and walls of drains. The approach of using traditional methods have limits the location, range, and sensor density of the water-level observation. Furthermore, sensors installed in-site can be easily damaged due to flood or even buried by debris. As a result, it has been further clarified that sensors with direct physical contact with the environment will likely to impose lifespan problems during long-term operation. As of now, based on existing research and findings, the only approach to tackle these issues will be implementing noncontact sensing technology. For the proposed solution of this research project, serious measures will be taken in correspond to this matter.

22.3 Methodology

22.3.1 Overview

A human operator would fly the Dragonfly to the desired location. The view captured by the camera on the drone is reflected on the app, along with current coordinates and altitude to guide the pilot in maneuvering. Real-time sensor data are also shown on the app. Then, by activating the monitoring process through the mobile application, data would be captured periodically. Air monitoring and water monitoring functions are available. Data are processed to yield relevant results, presented in graphical forms. Historical data are kept in the database and can be retrieved from the mobile application when necessary.

In-situ environmental monitoring is rendered much efficient as the staff no longer need to be at the exact coordinate or very near to it to gather the samples, as the inspection area is accessed only by the drone. This mitigates the risk of human exposure toward harmful air pollutants and at the same time eliminates the tedious journeys workers have to take to enter the monitoring sites.

Additionally, locations such as middle course of a river and areas above chimneys of great heights are accessible using a drone. The case where multiple sets of sensors are permanently installed at different monitoring stations at the same water body/industrial area will be rather obsolete, as a drone with merely one set of sensors is able to travel to various points within a specified area and have the monitoring process done expeditiously.

Furthermore, maintenance is necessary only for the Dragonfly, in contrast to the demanding regular checks and maintenance required, which again comes with a rather taxing journey taken by workers, should sets of sensors be permanently installed at the sites. Having said that, installation, operation, and maintenance costs are kept to minimum if Project Dragonfly is implemented. Optionally, the system setup can be scaled up to include multiple dragonflies, if the environmental profile of a much wider area is to be mapped.

22.3.2 System setup

Implementing modular design principles, the necessary operations are grouped into individual blocks (units).

The system consists of hardware and software components. The hardware is a Dragonfly that contains the air monitoring unit (AMU), water monitoring unit (WMU), and communication unit (CU). The software components include Microsoft Azure and Microsoft's proprietary cloud products and services, grouped as Cloud, complemented with an Android mobile application (Figures 22.4 and 22.5).

22.3.3 Quadcopter

Essentially, all the sensors will be packaged in a compact manner and mounted to the drone. Indirectly, the takeoff weight will be increased. Hence, the drone must be able to accommodate high payload. The model of the drone chosen is KaiDeng K70C, which is capable of handling up to 0.5 kg payload. Buoyancy aid is attached to the landing gear to enable the drone to float on water.

22.3.4 Hardware modules

22.3.4.1 Air monitoring unit

The parameters for air quality monitoring include nitrogen dioxide (NO_2), carbon monoxide (CO), carbon dioxide equivalent (CO_2-eq), ammonia (NH_3), and methane (CH_4). The concentrations of these gases are monitored using metal-oxide gas sensing

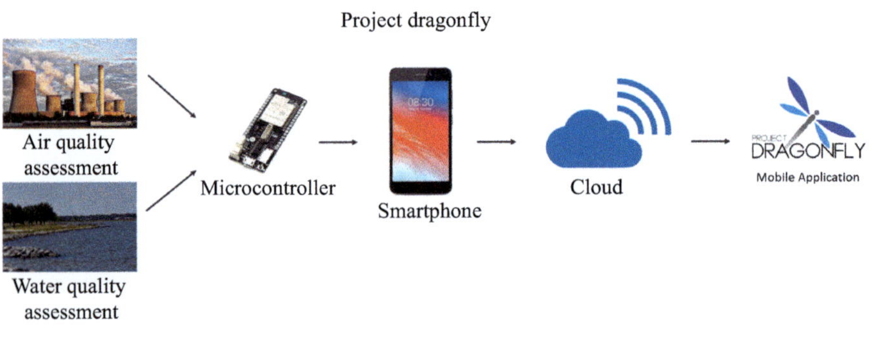

Figure 22.4 Graphical view of Project Dragonfly system setup

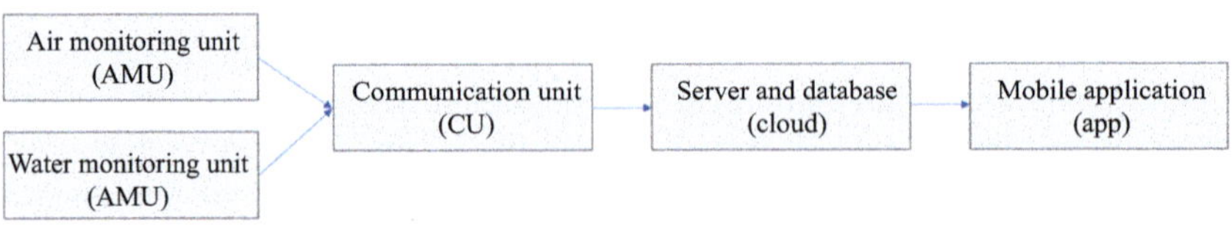

Figure 22.5 Block diagram of the system

Table 22.5 Sensors in AMU and their respective functions

Sensor	Parameter(s) measured
MiCS-6814	NO_2, CO, NH_3, CH_4
CCS811	CO_2-eq
BME280	Temperature, Pressure, Humidity

technology. Besides, there are supporting parameters such as ambient temperature, pressure (to determine altitude), and humidity.

Two components included in AMU are Grove-Multichannel Gas Sensor (MiCS-6814) and SparkFun Environmental Combo Breakout (CCS811/BME280). The sensors utilize inter-integrated circuit (I2C) interface (Table 22.5).

The AMU also houses Lolin D32 Pro microcontroller.

22.3.4.2 Water monitoring unit

The measure of the degree the water loses its transparency due to the presence of suspended particulates is known as turbidity. The turbidity sensor implemented in the project is DFRobot gravity–analog turbidity sensor. Besides, EC is measured using a custom-built EC probe. The two sensors are installed on the "tail" of the Dragonfly, which is lowered automatically by gravity during takeoff and retracted again during solid-surface touchdown. When lowered, the tail is maintained to be 30° off normal to enable the sensing device to be submerged and also minimize the impact on the module during solid-surface touchdown.

22.3.4.3 Communication unit

A Neffos Y5i smartphone is mounted firmly on the drone. The rationale behind installing a smartphone on the Dragonfly is that broadband cellular network connectivity is established (4G technology). Hence, data could be sent to the server speedily and reliably. Besides, the camera on the smartphone is used to provide first-person view (FPV) of the drone on the mobile app, which might be operating quite far away. Moreover, altimeter and GPS antenna on the smartphone are utilized to supply data on altitude, longitude, and latitude. The smartphone battery, on the other hand, powers up AMU and WMU.

22.3.4.4 Azure Cloud Service

Azure Cloud Service provides inextricable link between IoT and Cloud. It offers IoT-based products and services to allow microcontroller devices such as Arduino to develop IoT applications and enable device-to-device communications and management of raw data across their open and flexible cloud platform. Figure 22.6 shows the simplified block diagram of the sensor data processing system across the database and server side.

In the following sections, the Microsoft Azure services that make up the entire cloud database and server system will be elaborated on its usage and functions.

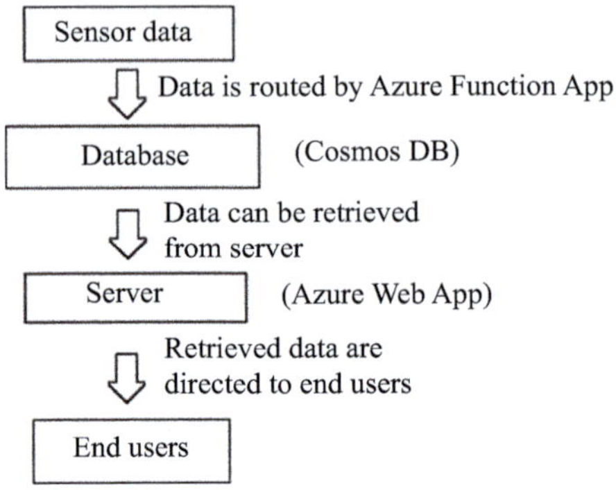

Figure 22.6 Block diagram of the data processing and server communication system

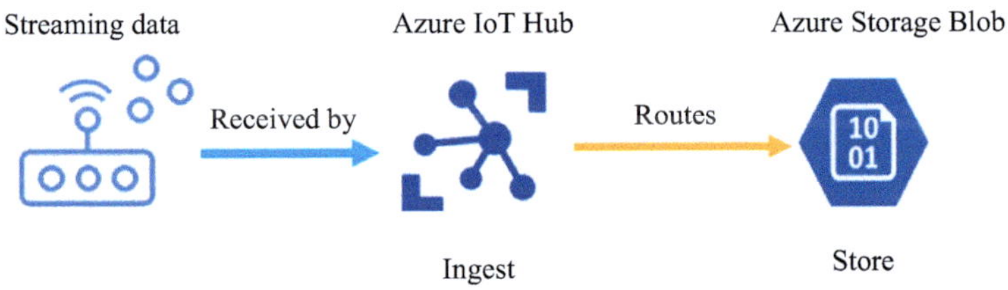

Figure 22.7 Standardized flow of telemetry data across Azure IoT Hub

IoT Hub

Azure provides convenient services such as IoT Hub to simplify device–cloud messaging and data routing. For this project, Azure IoT Hub will be used as the intermediary cloud for the communication between Arduino Nano and database. By default, telemetry data sent from microcontroller to IoT Hub are routed to the Microsoft Azure Storage as the final end point as shown in Figure 22.7.

Due to the difficulties in accessing and modifying the data stored in Azure Storage, the end point will be set to Azure database instead. Hence, no additional workflow needs to be done since data are automatically routed toward the selected database. The routing details are further explained in later section.

Azure Cosmos DB

Cosmos DB will be the telemetry database for the IoT Hub whereas Azure Function App is used as the telemetry data router. Therefore, the actual IoT Hub message routing is shown in Figure 22.8.

Cosmos DB is a NoSQL type of database. Part of the reason for choosing this database architecture is its high scalability and ability to handle and query large amount of raw data at superior speed compared to other databases such as MySQL, PostgreSQL, etc. This is already being proved in Literature Review section. Hence, it will be more

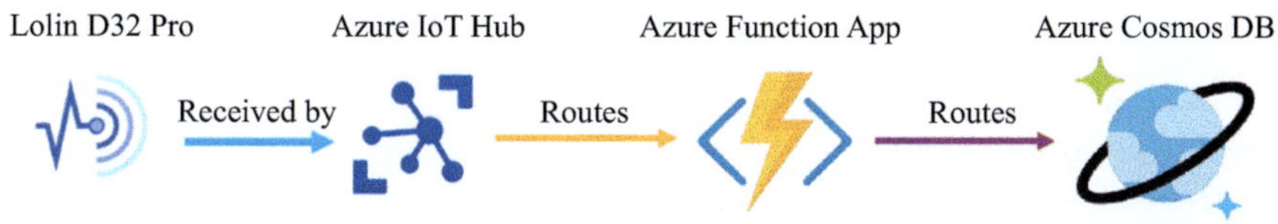

Figure 22.8 Reconfigured routing of data flow toward Azure Cosmos DB

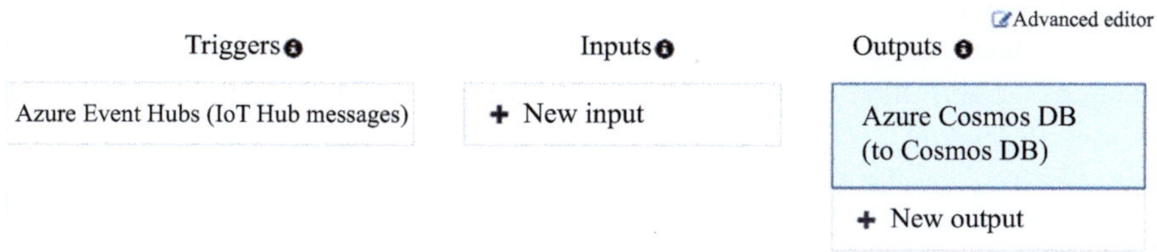

Figure 22.9 Internal configuration of Function App to bind telemetry data to Azure Cosmos DB

practical to use this database's system to manage unstructured data generated from IoT applications/devices.

Azure Function App

Function App is a convenient platform that executes codes in a serverless environment whenever an event has occurred. It makes it easy to integrate with other Azure products and services by just binding the inputs and outputs as shown in Figure 22.9.

For this project, it will be configured to run whenever the IoT Hub receives telemetry data from a registered microcontroller device. Once triggered, it will route the raw data directly to the Azure Cosmos DB. Thus, the combination of IoT Hub together with Function App and Cosmos DB have made up a complete data routing and storage system.

Azure Web App and server communication protocol

Having a plain device-to-cloud communication is not enough as there is still a need for server to allow mobile backend interaction with the data stored in the database. To realize such scheme, Azure Web App will be used as a backend server of this project. The purpose of this server is to allow retrieval and modification of the sensor data stored in the Cosmos DB database. Node JS will be chosen as the runtime environment for the server. An android app will be made to connect to the server via MQTT. MQTT is generally an improved version of WebSocket with fallback options supported and broadcasting functionality. As stated in the Literature Review section, MQTT is a perfect candidate for IoT applications, and hence the server will be built based on this algorithm. Once connected to the server, mobile device will be able to retrieve and monitor sensor data from the server in real time. Besides, the server will also be accountable for handling the user authentication and login management in which users are able to log into their own accounts or even register a new account via the custom-made android app. All these user credentials will be stored in Azure MySQL database.

Azure Logic App

Unlike Azure Function App, which is code-triggered during the occurrence of an event, Logic App is having a workflow-triggered structure. These workflows are made with a simple visual interface, combined with a straightforward workflow definition language in the code view. It offers a lot of accessible flexibility especially in connecting to the Software as a service (SaaS) applications and social media, and providing mailing functionality. For this project, the Logic App will be managing the sending of mails to newly registered users for email ownership verification purposes. The request for sending of the activation mail will be prompted by Web App server on registration of new user by sending a HTTP POST request to the Logic App URL which is shown in Figure 22.10.

A unique custom link will be embedded into the activation mail. Once accessed, the Web App server will be able to decrypt the URL and activate the user account.

22.3.4.5 Mobile application

A user-friendly mobile application is vital toward excellent user experience. In this project, Android Studio was used to develop the mobile application. Android Studio is a professional grade software for app development. It is the official integrated development environment for Google's Android operating system.

The activities that are contained in the mobile application were properly planned prior to the coding process (Figure 22.11).

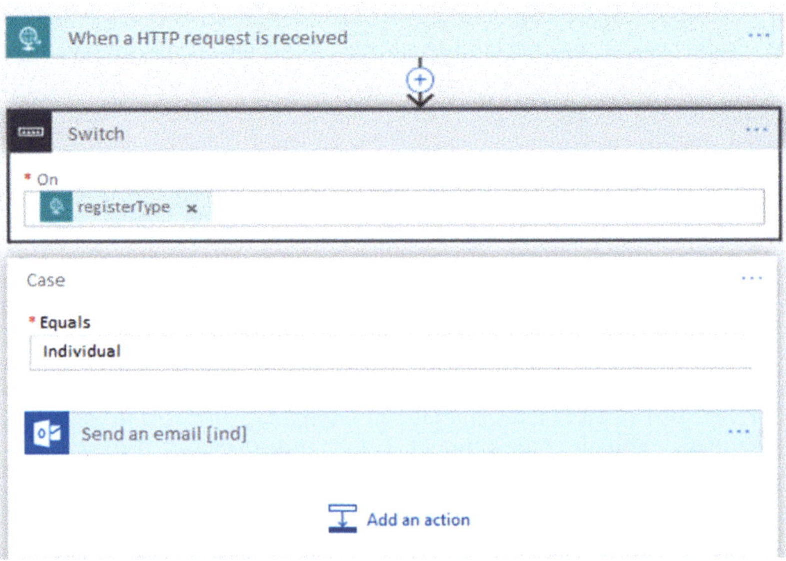

Figure 22.10 Automated emailing on receiving request of new registration

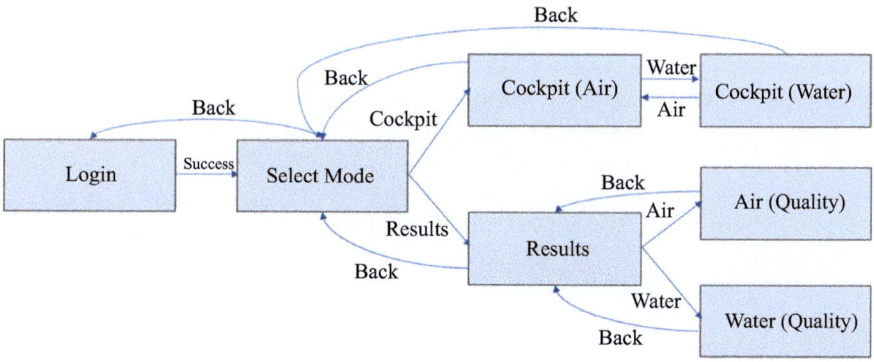

Figure 22.11 Block diagram showing the included activities

The mobile application communicates with the server through Node.js. JavaScript Object Notation (JSON) format is used for the sending of data and receiving of data.

To achieve FPV, WebRTC is used. WebRTC is commonly used for video streaming application between devices. In this case, the FPV from the camera on the drone will be displayed on the mobile application.

22.4 Results

The final product is an environment monitoring system, which at minimum consists of a Dragonfly and a Project Dragonfly software. It can be extended into a network of Dragonflies to effectively achieve wide-area monitoring (Figure 22.12).

22.4.1 Dragonfly

22.4.2 Project Dragonfly Software

The mobile application is connected to Microsoft Azure, and it features user verification, QR code scanning, live data, FPV, network reconnection, historical data, and map integration. Usage of the mobile application is described as follows.

22.4.2.1 User verification

To use the mobile app, user login is required. For first time user, an account can be registered by clicking "Register Here." Data validation will be performed on all the fields in the app. For company registration, company name is required. Additionally, a pop-out dialog will appear to ask for company passcode to verify the identity of the user. Should a company name not exist in the database, the dialog will ask for a new company passcode. After registration, a toast message notifies the user that the process is successful. Then, the user can proceed to log in. If incorrect credentials are entered, the user will be notified through a toast message. Once successfully logged in, a greeting

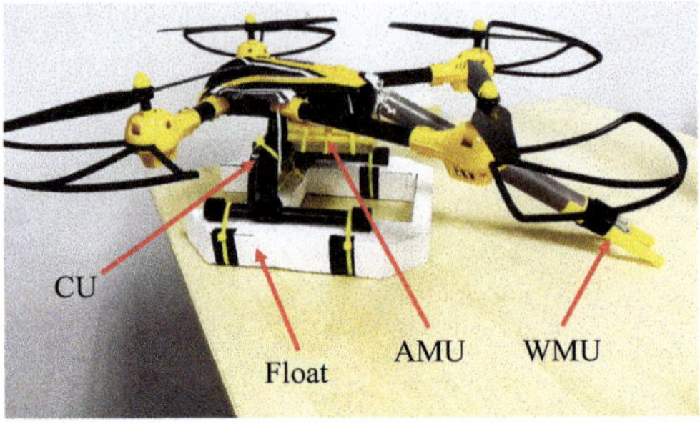

Figure 22.12 Installation of the modules

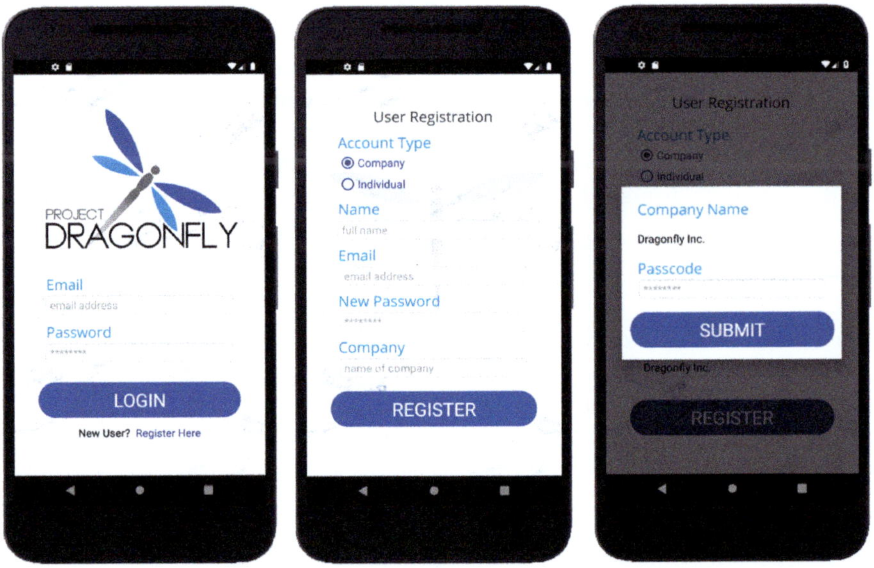

Figure 22.13 Login screen, user registration (personal), user registration (company)

message appears, and the user can choose either of the two modes: Cockpit and Results (Figure 22.13).

22.4.2.2 Cockpit

This mode is used to pilot the drone and get real-time data. The FPV from onboard camera is livestreamed. Flight information includes current longitude, latitude, and altitude. The real-time concentrations of the air pollutants are displayed on the app. If a certain parameter exceeds the defined threshold, it would appear in red to signify alert. Whenever "Activate" button is pressed, all the data, including data

Figure 22.14 Air monitoring sub-mode

Figure 22.15 Water monitoring sub-mode

and time, will be sent to the server at a fixed rate. For example, the fixed rate can be set to be one entry per 5 s (Figure 22.14).

If the user presses the button showing "Air," the sub-mode will be changed, that is, to water monitoring sub-mode in this case (Figure 22.15).

In water monitoring sub-mode, the EC and water turbidity values are displayed.

22.4.2.3 Monitoring results

If Results mode is chosen, the user will be prompted to choose between air monitoring results and water monitoring results (Figure 22.16).

In both cases, interactive time-varying graphs will be shown, with the pollution severities graphically presented with respect to a color scale. There are also map views that show the geographical points where the readings, along with the respective readings, were taken. The user may delete a result by tapping on the red delete button (Figure 22.17).

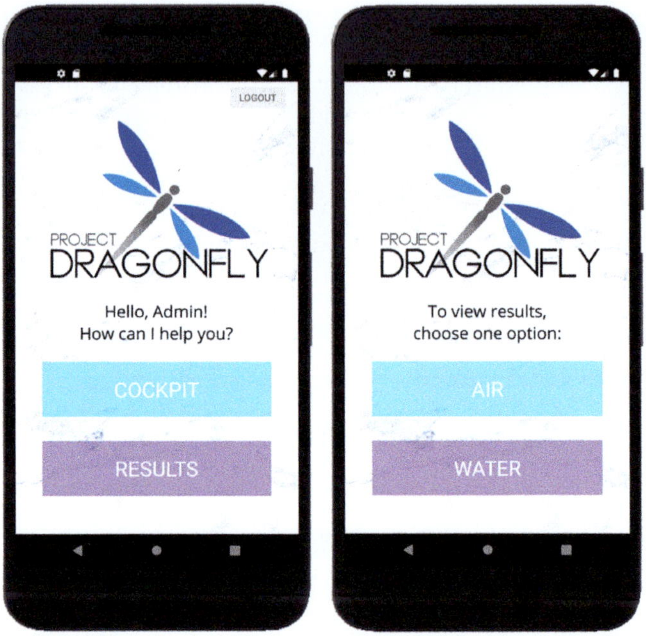

Figure 22.16 User is prompted to select one of the options

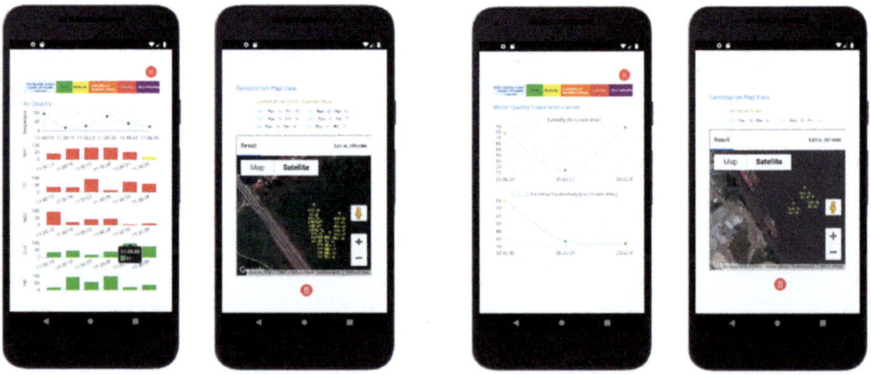

Figure 22.17 Air (left): graphs and map view; water (right): graphs and map view

22.5 Conclusion

Project Dragonfly is a boon toward environmental monitoring works. It offers a low-cost, safe, and convenient method to monitor air quality and water quality.

References

[1] Australian Government Department of the Environment and Heritage 2005, Department of the Environment and Energy. Accessed on: October 2, 2018. [Online]. Available: http://www.environment.gov.au/protection/publications/air-toxics

[2] Evoqua Water Technologies LLC, What is industrial wastewater, ADI systems, Accessed on: September 7, 2018. [Online]. Available: http://www.evoqua.com/en/brands/adi-systems/Pages/industrial-wastewater.aspx

[3] Industrial wastewater treatment, The International Water Association. Accessed on: September 13, 2018. [Online]. Available: https://www.iwa-publishing.com/news/industrial-wastewater-treatment

[4] P.K. Nair, Wastewater treatment issues must be taken seriously, Malay Mail, October 2017. Accessed on: August 27, 2018. [Online]. Available: https://www.malaymail.com/s/1486523/wastewater-treatment-issues-must-be-taken-seriously-prem-kumar-nair

[5] E. Gasiorowski, How the Internet of Things will change our lives, 2016. Accessed on: May 25, 2019. [Online]. Available: https://www.iso.org/news/2016/09/Ref2112.html

[6] Worldometers. (2019). Current world population. Accessed on: August 27, 2020. [Online]. Available: https://www.worldometers.info/world-population/

[7] I. Geography. (2009). Industry theory. Accessed on: May 25, 2019. [Online]. Available: http://www.geography.learnontheinternet.co.uk/topics/industrytheory.html

[8] H. Akimoto and H. Narita, Distribution of SO_2, NO_x and CO_2 emissions from fuel combustion and industrial activities in Asia with $1^\circ \times 1^\circ$ resolution, *Atmospheric Environment*, vol. 28, no. 2, pp. 213–225, 1994.

[9] H. Akimoto, Global air quality and pollution, *Science*, vol. 302, no. 5651, pp. 1716–1719, 2003.

[10] J. Fenger, Urban air quality, *Atmospheric Environment*, vol. 33, no. 29, pp. 4877–4900, 1999.

[11] World Health Organization, *Air Quality Guidelines: Global Update 2005: Particulate Matter, Ozone, Nitrogen Dioxide, and Sulfur Dioxide*. World Health Organization, 2006.

[12] World Health Organization. (2019). Definition of regional groupings. Accessed on: May 17, 2019. [Online]. Available: https://www.who.int/healthinfo/global_burden_disease/definition_regions/en/

[13] World Health Organization, *Air quality guidelines for Europe*, 2nd ed. Copenhagen: World Health Organization Regional Office for Europe, 2000.

[14] A. J. Hedley, C.-M. Wong, T. Q. Thach, S. Ma, T.-H. Lam, and H. R. Anderson, Cardiorespiratory and all-cause mortality after restrictions on sulphur content of fuel in Hong Kong: an intervention study, *The Lancet*, vol. 360, no. 9346, pp. 1646–1652, 2002.

[15] C. A. Pope, M. J. Thun, M. M. Namboodiri, *et al.*, Particulate air pollution as a predictor of mortality in a prospective study of US adults, *American Journal of Respiratory and Critical Care Medicine*, vol. 151, no. 3, pp. 669–674, 1995.

[16] D. Dominick, H. Juahir, M. T. Latif, S. M. Zain, and A. Z. Aris, Spatial assessment of air quality patterns in Malaysia using multivariate analysis, *Atmospheric Environment*, vol. 60, pp. 172–181, 2012.

[17] N. Anjana, M. H. Nair, K. Sajith, A. Amarnath, and I. Indu, Accidental release of ammonia from a storage tank and the effects of atmosphere on the affected area using ALOHA, *Indian Journal of Scientific Research*, vol. 21, pp. 1–7, 2018.

[18] G. K. Heilig, The greenhouse gas methane (CH_4): Sources and sinks, the impact of population growth, possible interventions, *Population and Environment*, vol. 16, no. 2, pp. 109–137, 1994.

[19] Y. Shirai, M. Wakisaka, S. Yacob, M. A. Hassan, and S. i. Suzuki, Reduction of methane released from palm oil mill lagoon in Malaysia and its counter-measures, *Mitigation and Adaptation Strategies for Global Change*, vol. 8, no. 3, pp. 237–252, 2003.

[20] I. SanAir Technologies Laboratory. (2016). How are air samples gathered and sent to a lab for testing? Accessed on: May 23, 2019. [Online]. Available: https://www.sanair.com/how-are-air-samples-gathered-and-sent-to-a-lab-for-testing/

[21] Interim National Water Quality Standards for Malaysia. Accessed on: May 20, 2019. [Online]. Available: http://www.wepa-db.net/policies/law/malaysia/eq_surface.htm

[22] V. Radhakrishnan and W. Wu, IoT technology for smart water system, in *2018 IEEE 20th International Conference on High Performance Computing and Communications; IEEE 16th International Conference on Smart City; IEEE 4th International Conference on Data Science and Systems (HPCC/SmartCity/DSS)*, 2018, pp. 1491–1496.

[23] V. Madhavireddy and B. Koteswarrao, Smart water quality monitoring system using IoT technology, *International Journal of Engineering & Technology*, vol. 7, no. 4.36, p. 636, 2018.

[24] M. H. Gholizadeh, A. M. Melesse, and L. Reddi, A comprehensive review on water quality parameters estimation using remote sensing techniques, *Sensors (Switzerland)*, vol. 16, no. 8. MDPI AG, August 16, 2016.

[25] X. Wang and W. Yang, Water quality monitoring and evaluation using remote-sensing techniques in China: a systematic review, *Ecosystem Health and Sustainability*, vol. 5, pp. 47–56, 2019.

[26] B. Hamuna, R. Tanjung, and A. Alianto, Assessment of water quality and pollution index in Coastal Waters of Mimika, Indonesia, *Journal of Ecological Engineering*, vol. 20, no. 2, pp. 87–94, 2019.

[27] Y. Zakaria and K. Michael, An integrated cloud-based wireless sensor network for monitoring industrial wastewater discharged into water sources, *Wireless Sensor Networks*, vol. 9, pp. 290–301, 2017.

[28] A. Kruger, W. F. Krajewski, J. J. Niemeier, D. L. Ceynar, and R. Goska, Bridge-mounted river stage sensors (BMRSS), *IEEE Access*, vol. 4, pp. 8948–8966, 2016.
[29] S. W. Lo, J. H. Wu, F. P. Lin, and C. H. Hsu, Cyber surveillance for flood disasters, *Sensors*, vol. 15, no. 2, pp. 2369–2387, 2015.

Chapter 23

Improving round-robin through load adjusted–load informed algorithm in parallel database server application

Nseabasi Peter Essien[1], *Uduakobong-Aniebiat Okon*[1] *and Peace Asuquo Frank*[1]

Industry 4.0 has recently become an important topic in the software development context. This standard-based strategy integrates physical systems, the Internet of Things, and the Internet of Services with the aim of extending the capacity of software development process. Although many software development experts have presented the advantages of different software development models and approach, software development refers to an architecture that allows the correct implementation of Industry 4.0 applications using the load-balancing approach model (LBAM 4.0). This study exposes the essential characteristics that allow software to be retrofitted to become Industry 4.0 applications. Specifically, an intelligent software system based on a load-balancing approach was developed and implemented using equal and unequal clustering processing capabilities. To evaluate the performance of LBAM 4.0, implementation was carried out on a cluster with equal and unequal nodes using round-robin algorithm. It was discovered that the performance of the algorithm is quite good when the nodes are of equal capacities, but very poor when the capacities of the nodes are not equal. Load adjusted–load informed algorithm was used to improve the worst case of the round-robin algorithm to prevent the worst situations of using nodes of different capacities of which the results showed remarkable improvement.

23.1 Introduction

Industry 4.0 is the information-intensive transformation of manufacturing (and related industries) in a connected environment of data, people, processes, services, systems, Internet of Things (IoT)-enabled industrial assets with the generation, leverage, and utilization of actionable data, and information as a way and means to realize smart industry, ecosystems of industrial innovation and collaboration [1].

[1]Department of Vocational Education, University of Uyo, Uyo, Nigeria

To understand Industry 4.0, it is essential to see the full value chain that includes suppliers and the origins of the materials and components needed for various forms of software development. Industry 4.0 has now been considered as a core element of the value chain and a crucial component in the technological development, job creation, and economic stability of a country [2]. The traditional industrialized countries have assumed a leading role in the fourth industrial revolution and have adopted Industry 4.0 as a strategy to confront new requirements in the global market and position themselves more competitively against emerging countries. This strategy has been planned to offer new potential to the manufacturing industry such as meeting individual customer requirements, optimizing decision-making, and adding new product capacities.

A well-supported definition of Industry 4.0 is presented by Juan [1] as "Industries 4.0 is a collective term for technologies and concepts of value chain organization. Within the modular structured Smart Factories of Industries 4.0, Cyber-Physical Systems (CPS) monitor physical processes, create a virtual copy of the physical world and make decentralized decisions. Over the IoT, CPS communicates and cooperates with each other and humans in real time. Through IoS (Internet of Services), both internal and cross-organizational services are offered and utilized by participants of the value chain." According to the definition presented, it is against this background that the fundamental technologies in Industry 4.0 were conceived and load-balancing approach model (LBAM 4.0) was developed to realize new services and efficient business models in software development.

23.2 Application of LBAM 4.0

One of the problems with the clustering capabilities in MySQL is that it provides no means to distribute the query/transaction load in a cluster. This leads to poor response time to multiple users' queries. Therefore, the current databases have to handle high loads while providing high availability. The solution often used to cope with these requirements is a cluster. Load balancing is a technique (usually performed by load balancers) to spread work between two or more computers, network links, CPU, hard drives, or other resources to get optimal resource utilization, throughput, or response time [3]. Using multiple components with load balancing, instead of a single component, may increase reliability. The balancing service in recent days is usually provided by a dedicated hardware device (such as a multilayer switch). The technique is commonly used to mediate internal communications in computer clusters, especially highly available clusters [14].

Schneider and Dewitt [4] stipulated that distributing load is usually handled by a load balancer. A load balancer selects a particular node for which the job is going to be executed. The goal for the load balancer is to equalize the load on all nodes to avoid bottlenecks and achieve maximum throughput. Some load balancers blindly follow a specific pattern to reach this goal, and others evaluate information obtained from the nodes and base their decision on that. This study focuses on developing a load balancer to distribute loads. The goal of our load-balancing algorithm is to decrease the total number of heavy nodes in the system by distributing loads among nodes. To evaluate the performance of our load balancer, we did an implementation on an equal and unequal nodes using round-robin algorithm. It was discovered that the performance of the algorithm is quite good when the nodes are of equal capacities, but very poor in an unequal capacities. To improve the worst case of the round-robin, load adjusted–load informed algorithm was combined as a means to prevent the worst situations of using nodes of different capacities. From the results, this goal was largely achieved. Linux kernel exposes ways of detecting current CPU load on a node using the file/proc/loadavg. When using LAMP, data are captured through screen. When using WAMP through MySQL, data are retrieved through "Show Process List Command." In this work, the two platforms were used interchangeably.

23.3 Cluster database system

A cluster is a group of highly interconnected hardware servers housing databases [5]. Historically, a server in a cluster is called a node. The clusters provide high availability in allowing transparent fail-over in case of failure on a node. The clusters also provide a means to distribute load, as requests can be handled on any of the cluster nodes.

Shivaratri *et al.* [6] argued that the job is the work that should be executed on the cluster. In the context of databases, it is a query/transaction. A job is generated on a client machine and sent across the network to a node that processes the job. The node then returns the result to the client, when the job has been executed. In databases, communication between the node and the client is usually via TCP connection. TCP is a protocol where messages are guaranteed to be delivered in order and without loss of data. To make this happen the two ends of the communication must go through a setup phase. For this reason, a TCP connection is always between machines (called unicast). Another detail is that TCP is not packet oriented [7].

Casavant [8] stated that some scheduling algorithms require information about the individual nodes. The information must ultimately come from each node in the cluster. Also the information must be transported from the nodes to the scheduling algorithm [9]. One can assume that the cluster and the clients communicate through an IP network. Furthermore, one can also assume that the nodes are on the same physical network, but that it is not necessarily true for the clients.

23.3.1 Methodology

For the purpose of implementing LBAM 4.0 through the concept of Industry 4.0 in software development process, courseware was designed. Designing a web-based courseware is not merely writing a series of pages, linking them together, and presenting them as a course. A good and efficient design strategy had to be developed so that the participants get the fullest benefit possible from the instructions with the help of technology. Before the actual development of the module, the overall structure of the module was first created. Key topics were identified bearing in mind the requirements of the course, the learner, and the teacher [10]. The design of virtual learning system courseware was based on the internal design of courseware. Keane and Monaghan [11] suggested that internal design of courseware consist of

1. Selection of an instructional strategy to instruct the learner in acquiring specific knowledge and skills.
2. Allowing maintenance, extension, and reuse of the courseware.

This section uses software design techniques to design the proposed courseware. The designed technique used was based on Object Oriented Analysis and Design as well as Rational Unified Process (RUP), which are road maps to creating application using Unified Modeling Language (UML) tools [12,15]. UML tools were used to capture user requirement of the courseware, logical and conceptual design, and implementation of the courseware [13].

The class diagram in Figure 23.1 shows the basic components and their interactions for the implementation of virtual learning system. These basic components are

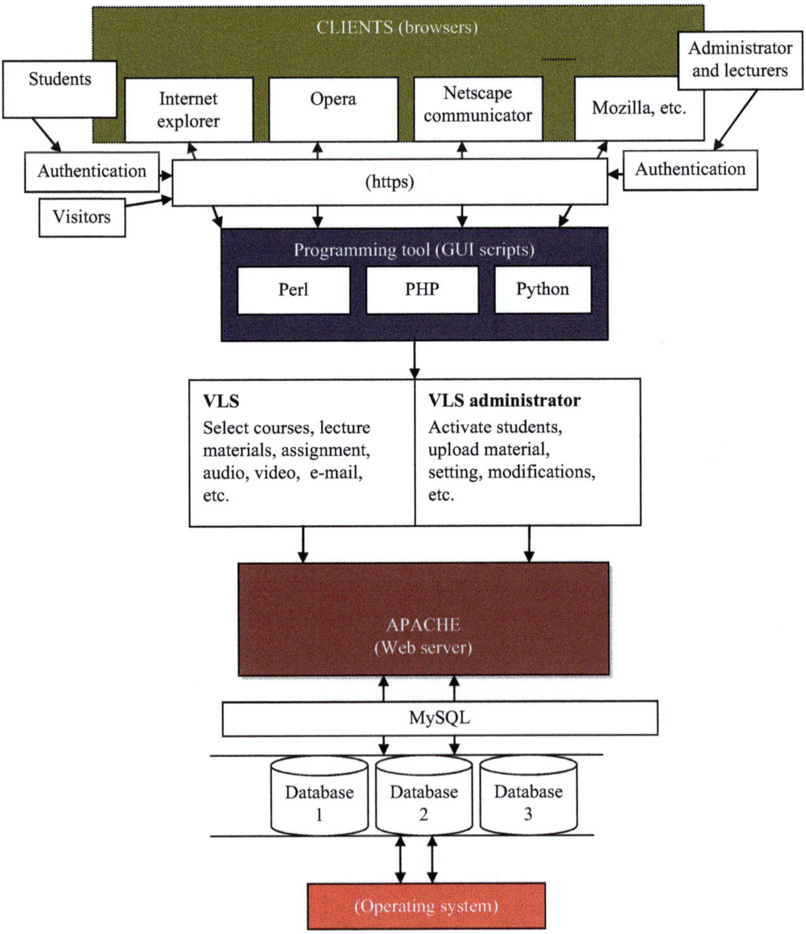

Figure 23.1 Class model diagram architecture of virtual learning system

- Client layer
- PHP scripts implementing virtual learning system (middleware) layer
- The web server layer using Apache
- Relational database layer using MySQL
- Operating system layer using Linux.

Client layer: The application developed with MySQL and PHP make use of a web browser as interface for the client. The clients have the flexibility of using different browsers to access the system. In virtual learning system, all requests are rendered in the browser by the clients at the client's layer. When such requests are rendered, it passes into the next layer of processing in the virtual learning system through the interception of the PHP scripts.

PHP scripts implementing virtual learning system (middleware) layer: PHP scripts belong to a class of languages known as middleware. These scripts work closely with the browser (client layer) and are processed in the web server (layer) to establish connection between the two. They interpret the requests made from the clients through Local Area Network (LAN) or World Wide Web (WWW), process these requests, and then indicate to the web server layer exactly what to serve to the client's browser. In virtual learning system application, once the request is rendered to the browser, the middleware running the application (PHP scripts) performs some level of authentication as shown in Figure 23.5 (activity diagram) to determine whether the client is an administrator or an ordinary user before such requests are allowed to access the web server.

The web server layer using Apache: The web server has what seems to be fairly straightforward job. It sits there, running on top of operating system and established connection with relational database layer, listening for requests that somebody on the web browser might make via middleware layer, responds to those requests, and serves the appropriate digitalized materials that have been stored in the rational database (MySQL) through appropriate web pages. Apache in this case serves as the server-based application development tool for virtual learning system. It does the work of browser's communications and the responds to those requests needed from the database (MySQL) application tool.

Relational database layer using MySQL: A relational database stores whatever information the application requires. MySQL provides a great way to store and access complex information in a relational database. In virtual learning system, digitalized learning materials are stored in MySQL relational database which interact with the application web server to enable the server responds to those requests and serves out the appropriate web pages to client's request.

Operating system layer using Linux: In virtual learning system, all the applications runs on top of operating system (Linux) to handle and perform certain amount of integration between the programming language and the web server.

Figure 23.2 depicts the setup block diagram of virtual learning system. The system structure is summarized in the following description layers:

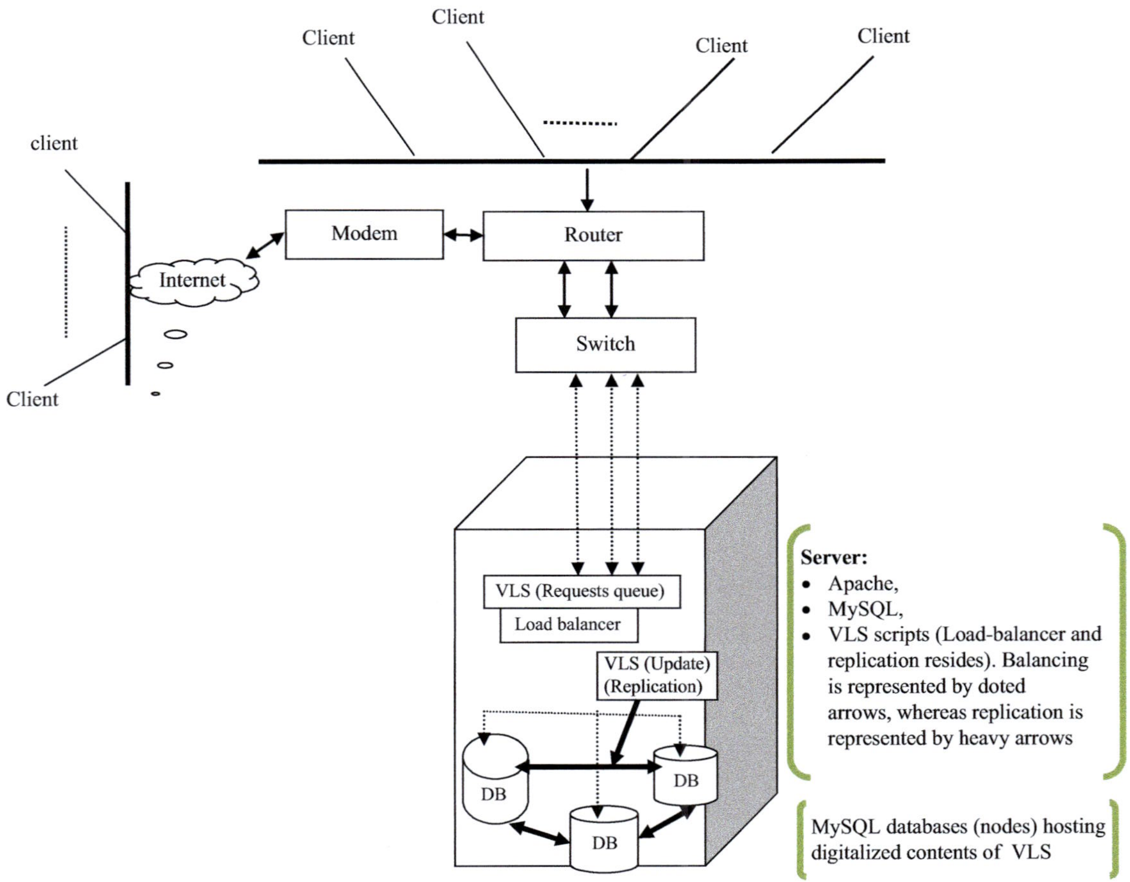

Figure 23.2 Block diagram of virtual learning system

1. *Client layer*: This can be referred to as user's interface layer. Clients access virtual learning system through this layer by using their respective browsers. All clients render their query transaction requests that can be read into the system application in this level either through LAN or through WWW.
2. *Transmission layer*: The transmission layer is responsible for communication. It communicates various transaction requests made by clients at client's layer to the web server machine layer. All clients transaction request are intercepted by the application and communicated to the web server machine layer to be processed by scripts implementing the virtual learning system and also providing feedback on the requests to the clients. Transmission layer components include modem, router, switch, and the cabling system.
3. *Web server machine layer*: This can be referred to as domain application layer. This serves as a logic box where all clients' requests transmitted are given rational decision. This layer is the domain on which virtual learning system run and for the persistent data management, global control flow, the access control policy, and the handling of conditions. The web server machine hosts Apache, MySQL, VLS scripts (load balancer and replication resides). Balancing is represented by doted arrows whereas replication is represented by heavy arrows.

23.4 Experimentation

23.4.1 Hardware and software requirement for load-balance approach model

There are many possible environments the system could be tested. The system was tested in two different environments. The system was tested in an environment where the capabilities of the servers are

1. Equal nodes or closely related in capabilities
2. Unequal nodes capabilities.

1. Equal nodes capabilities

   Hardware requirement: The first is where two nodes are closely related in capabilities. The setup has two 400 GHz Pentium III and 126 MB RAM memory. Each client used a dual Pentium IV with 2.0 GHz and 1 G memory with a 1 Gbit network interface. The entire setup is connected through an 8-port 10/100 mbps switch.

   Software requirement: We also have Apache/2.2.4 (win32)—5.2.3 web server, MySQL 5.0.41, Phpmyadmin 2.10.1. Proxy server, HP Intel Pentium 4 Processor with 3.4 MHz—running Linux.
2. Unequal nodes capabilities

The second setup was combined with the first setup during the experimentation. The node is Pentium III 700 MHz with 256 Mb memory.

23.5 Algorithm description

Each scheduling algorithm has its pros and cons. Which of them works best depends on a lot of factors. The round-robin and the weighted round-robin are great when the queries are alike in complexity [7]. Dennis and Klaus used C++ in the demonstration of their algorithm and developed assumptions. In this study, we experimented round-robin under web application environment using PHP algorithm translation and implementation. We observed the performance of the virtual learning system in terms of balancing request on the cluster as it transverses around the cluster.

23.5.1 Round-robin algorithm

Load Balance in C++

1: **Input:** A list of nodes (nodes)
2: Iterator into nodes (next_node)
3: **Output:** The node to use
4: GETNEXTNODE()
5: Node *result = next_node;
6: next_node++;
7: **if** (next_node == nodes.end()) {
8: next_node = nodes.begin();
9: }
10: Return result;

Source: Load balancing for MySQL by [7]

23.5.2 Translation of round-robin algorithm into PHP algorithm

<?php

1: **Input:** A list of nodes (nodes) //Lines 1, 2, 3 are comments or remark statements that could be used to explain certain concepts in the body of the program. These could be variable descriptions, housekeeping, what the program is doing etc. It is very optional

2: Iterator into nodes (next_node)

3: **Output:** The node to use

4: function getNextNode() // this is function that establish connection to available node

5: $next_node=$node * $result; // the variable "result" is a pointer declaration of type "Node," being initialized to "next_node"

6: $next_node++; // The node is incremented as the pointer is advancing to the next node

7: if ($next_node==$node.end()) // This is a test to see if there is available node to take or whether it has reached the end

8: $next_node==$node.reset(); // This mean from the interpretation of the condition above, if it has reached the end, set the pointer to the beginning{

```
}
echo $result; // show or print the final number of all the nodes used in the operation it is optional
?>
```

23.5.2.1 Implementation (PHP which is a web server base application—program extract)

- **Transverse all the nodes with the set parameters within the timeout specified**
- <?php
- /*
- Function to Handle connection
- */
- function Connect_server()
- {
- $node1 = “122.10.1.2”; //Node 1 ip address
- $username1 = “username”;
- $password1 = “password”;
- //Node 2
- $node2 = “122.10.1.3”; //Node 2 ip address
- $username2 = “username2”;
- $password2 = “password2”;
- //Node 3
- $node3 = “122.10.1.4”; //Node 3 ip address
- $username3 = “username3”;
- $password3 = “password3”;
- // connection timeout
- $timeout = 5; // timeout in seconds
- }
- .
- .
- .
- .
- ?>

This algorithm is very simple. However, as we can see in the performance below, the algorithm can only be efficient when:

1. The nodes are identical in capacity. Otherwise performance will degrade to the speed of slowest node in the cluster.
2. Two or more client connections must not start at the same time. Should they, the node chosen will be the same, because the order of nodes retrieved from the cluster is the same every time.
3. The jobs must be similar to achieve optimum load distribution among the nodes. If a single node is more loaded than the others it will become a bottleneck to the system. This goes a long way to confirm the assumption of Dennis (2003).

We now think of balancing unequal capacity node by introducing load adjusted–load informed algorithm that we have been able to develop a round-robin algorithm for distributing loads among nodes but except the nodes is of equal capacity. A way to solve the problem would be to choose randomly between the nodes with the least or lowest load. This way the jobs will be evenly distributed between the least loaded nodes. The implementation of this work is that the balancer periodically probe the nodes as queries arrives or perform lookup operation to find the node that is heavy or light, then transfer may take place between the two nodes.

23.6 Load adjusted–load informed algorithm

Load informed in C++ adopted from Job Informed

```
1: Input: A list of nodes (orig_nodes)
2: ONLOADUPDATE()
3: token_tot = 0;
4: active_jobs_tot = 0;
5: for (iterator it = orig_nodes.begin();
It != orig_nodes.end(); ++it) {
6: token_tot += it.original_token_count;
7: active_jobs_tot += it.active_jobs;
8: }
9: nodes_expired.clear();
10: nodes.clear();
11: for (iterator it = orig_nodes.begin();
it != orig_nodes.end(); ++it) {
12: it.tokens = floor((it.original_token_count/token_tot -
it.active_jobs/active_jobs_tot) * active_jobs_tot);
13: if (it.tokens <= 0) {
14: it.tokens += it.original_token_count;
15: if (it.tokens <= 0) {
16: it.tokens = 1
17: }
18: nodes_expired.push_back(it);
19:       } else {
20:             nodes.push_back(it);
21:        }
22:    }
23: GETNEXTNODE()
24: Return weightedRoundRobin.GetNextNode();
```

Source: Load balancing for MySQL by Dennis and Klaus [7].

23.6.1 Translation of load adjusted–load informed algorithm into PHP algorithm (adopted from Job Informed)

<?php

1: Input: A list of nodes (orig_nodes) // List of nodes in a cluster

2: ONLOADUPDATE() // function onloadUpdate()

3: $token_tot=0; // queries are taken as input and separated them into pieces and store in this variable

4: $active_jobs_tot=0; // active queries are taken as input and separated them into pieces and store in this variable. At first instance zero is initialized

5: for ($It=$orig_nodes.reset(); $It!=$orig_nodes.end(); $It++){// a loop body setting counter “it” of type iterator to start from orig_nodes.begin. // Continue execution as long as it is not equal to orig_nodes.end() and step of 1. // The iterator declared as ït” is like a pointer, locating one item in the container (node) as it traverses

6: $token_tot += $original_token_count; // equivalent to incrementing the original variable with original or current content. // This statement accumulate all the tokens executed in each node

7: $active_jobs_tot += $active_jobs; // equivalent to incrementing the original variable with original or current content. // this statement accumulate all the active token currently under execution in each node

8: }

9: unset($node_expired); // a call to method “clear” the node expire. // a call to method “clear” the node object. Clear or remove all the memory content(s) in readiness

10: unset($nodes); // a call to method “clear” the node expire. // a call to method “clear” the node object. Clear the memory space in ready for action

11: for ($It=orig_nodes.reset(); $It!=orig_nodes.end(); $It++){ // a loop body, setting counter "it" of type iterator to start from orig_nodes.begin(). // Continue execution as long as it is not equal to orig_nodes.end() and step of 1. The iterator transverses through the nodes starting from the first node till the end of the cluster. It must be declared in a “for “loop

12: $tokens=$floor(($original_token_count/$token_tot – $active_jobs/$active_jobs_tot) * $active_jobs_tot); //. This is arithmetic expression computing the job capacity as the iterator is visiting each node, it has the capacity to determine whether there are actually items at the iterator’s current node position and then decide on the status of the node using the arithmetic expression to know whether the node is over loaded. The “floor” statement means take the lowest returned value from the computation and store in it.token variable

13: if ($tokens<=0) { // conditional statement, testing if “token”< = 0. // Test to see if iterator’s token value is less or equal to 0

14: $tokens +=original_token_count; // add original_token_count to tokens accumulatively with all the queries currently running in the mode to the Iterator’s variable

```
if ($tokens<=0){// test again to see if "token" is < = 0. Test again to see if iterator's token value is still less or equal to zero 0
it.token = 1; : nodes.reset();}//Maintain connection and pass request
}
$nodes_expired.reset(); // this means the node in the cluster is over loaded or busy. Kill the connection or connection die, if it is true; it means the node is overloaded, try the next connection
        } else {
$nodes_push_back() // connection continue
                }
            }
GETNEXTNODE() // GetNextConnection in the cluster. This is the next connection that is call if any node is expired or overloaded
?>
```

Also in C++ algorithm, the calculation ***it.original_token_count/token_tot - it. active_jobs/active_jobs_tot***
is a node's percentage of capacity in the cluster minus the percentage of node's load. The result is the distance from the optimum load for the node. Multiplying the distance with the amount of active jobs, the distance in jobs is obtained. If the resulting number is negative, the node is more loaded than the average node (Dennis, 2003).

23.6.2 Implementation (program extract PHP)

This function will handle the connections as PHP cannot "talk" to processors directly like C/C++. We have to check connections to each of the MySQL NODES, and if no connection is returned, then the server is either busy, down, or encountering a problem.

<?php

- /*
- Function to Handle connection
- */
- function Connect_server()
- {
- $node1 = "122.10.1.2"; //Node 1 ip address
- $username1 = "username";
- $password1 = "password";
- //Node 2
- $node2 = "122.10.1.3"; //Node 2 ip address
- $username2 = "username2";
- $password2 = "password2";
- //Node 3
- $node3 = "122.10.1.4"; //Node 3 ip address
- $username3 = "username3";
- $password3 = "password3";
- // connection timeout

```
$timeout = 15; // timeout in seconds
if($fp = fsockopen($node1, 3306, &$errno, &$errstr, $timeout)){
fclose($fp);
return $link = mysql_connect($node1, $username1, $password1);
}
//check server 2
if($fp = fsockopen($node2, 3306, &$errno, &$errstr, $timeout)){
fclose($fp);
return $link = mysql_connect($snode2, $username2, $password2);
}
//check server 3
if($fp = fsockopen($node3, 3306, &$errno, &$errstr, $timeout)){
fclose($fp);
return $link = mysql_connect($node3, $username3, $password3);
}
}
?>
```

Sample input data to round-robin load balancer: The sample data to round-robin load balancer were the number of queries submitted by clients as request to the database. The implementation was strictly based on clients input queries. The number of queries at interval of time was used in appraising the performance of the round-robin load balancer. The goal of our entire load-balancing algorithm is to decrease the total number of heavy nodes in the system by distributing loads among nodes.

Load-balancer output using round-robin: To evaluate the performance of our load balancer, we did an implementation on an equal and unequal nodes using round-robin algorithm. We discovered that the performance of the algorithm is quite good when the nodes are of equal capacities, but very poor in an unequal capacities. To improve the worst case of the round-robin algorithm, load adjusted–load informed algorithm was combined as a means to prevent the worst situations of using nodes of different capacities (Table 23.1).

Part of the objective of this study was improving round-robin algorithm such that it will balance unequal nodes capacities. Therefore to achieve this, load adjusted–load informed algorithm was integrated with the round-robin algorithm. That is, the balancer periodically probes the nodes as queries arrive or perform lookup operation to find the

Table 23.1. MySQL process commitment of varying server capabilities used in the experimenting round-robin algorithm

Time/s	**8**	**16**	**32**	**64**	**128**	**256**	**512**
CPU1	20	345	310	310	300	320	250
CPU2	0	100	330	330	345	335	340
CPU3	0	100	400	550	510	550	560

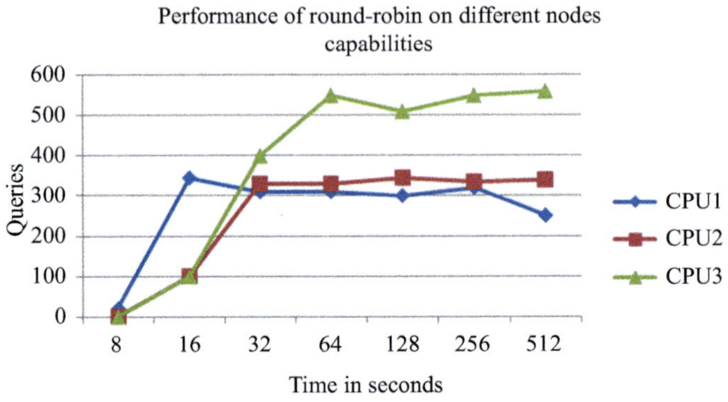

Figure 23.3 Performance of round-robin algorithm on different nodes capabilities. From the graph, it can be seen that CPU1 and CUP2 seem to be closely related in capabilities, whereas CPU3 has the lowest configuration. This may lead to worst-case situation

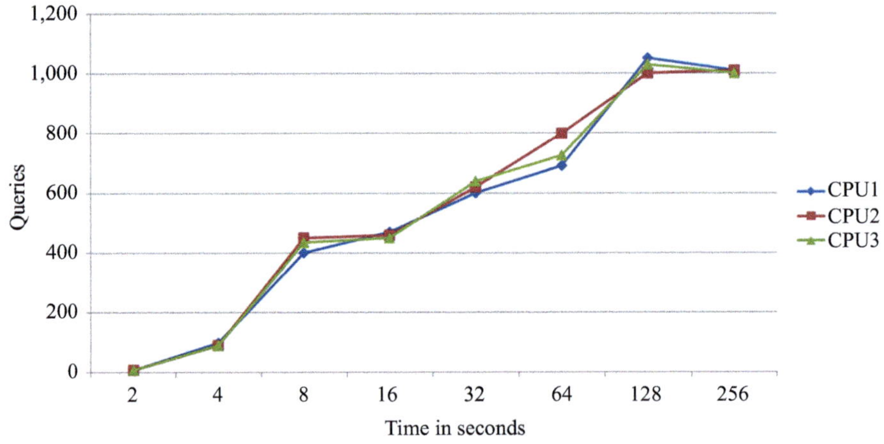

Figure 23.4 Performance of load adjusted–load informed algorithm on different nodes capabilities. From the graph, it can be seen that there are significant improvement on all CPU capabilities irrespective of the volume of queries allocated to the CPU

node that is heavy or light, and then transfer may take place between the two nodes or more in the cluster. The problem can be solved by viewing the number of active jobs on a node's load. The result is a simpler algorithm (Figures 23.3–23.6).

Sample data to load balancer: The sample data to load balancer are the number of queries submitted by clients as request to the database. The goal of entire load-balancing algorithm was to decrease the total number of heavy nodes in the system by distributing loads among nodes (Tables 23.2–23.4).

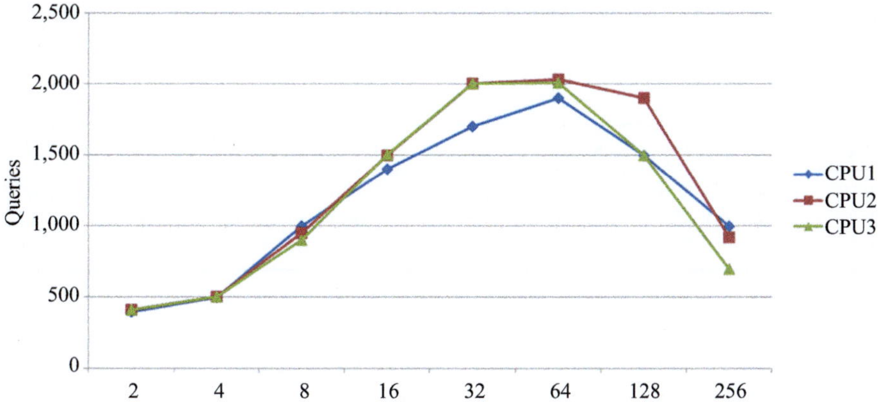

Figure 23.5 Performance of round-robin + load adjusted–load informed on different nodes capabilities

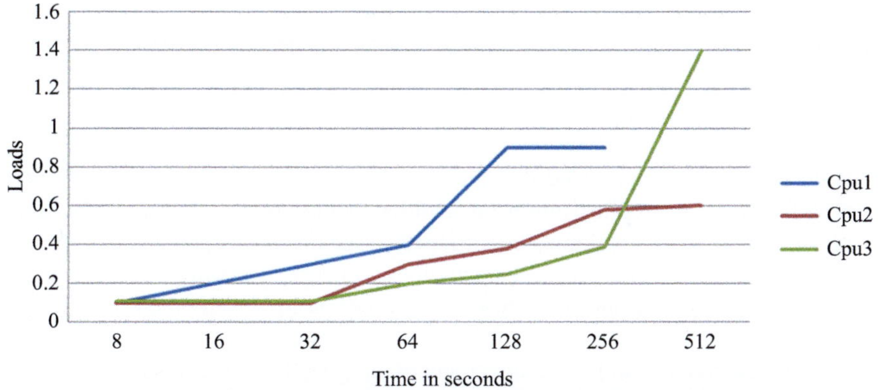

Figure 23.6 Performance of round-robin algorithm on different nodes capabilities and initial average load. From the graph, it can be seen that CPU1 and CPU2 seem to be closely related in capabilities, whereas CPU3 has the lowest configuration. This may lead to worst-case situation as reported earlier

Table 23.2 MySQL process commitment of varying server capabilities used in the experimenting round-robin + load adjusted–load informed algorithm

Time	**2**	**4**	**8**	**16**	**32**	**64**	**128**	**256**
CPU1	9	100	400	470	600	690	1,050	1,010
CPU2	8	90	450	460	620	800	1,000	1,010
CPU3	8	90	435	450	640	726	1,030	1,000

Table 23.3 MySQL process commitment of varying server capabilities used in the experimenting round-robin + load adjusted–load informed algorithm with high-populated queries

Time	**2**	**4**	**8**	**16**	**32**	**64**	**128**	**256**
CPU1	400	500	1,000	1,400	1,700	1,900	1,500	1,000
CPU2	410	502	950	1,500	2,000	2,030	1,900	920
CPU3	410	505	900	1,502	2,001	2,005	1,500	700

Table 23.4 The initial average load of the three CPUs used in the experiment (all other readings were based on this)

Time	**8**	**16**	**32**	**64**	**128**	**256**	**512**
CPU1	0.1	0.2	0.3	0.4	0.9	0.9	
CPU2	0.1	0.1	0.1	0.3	0.38	0.58	0.6
CPU3	0.11	0.11	0.11	0.2	0.25	0.39	1.4

23.7 Recommendations

1. It is therefore recommended that the idea of depending on higher system capabilities or heavy sophistication on load-balancing equipment should be completely rolled out in developing countries if we can invest on open-source software initiatives through Industry 4.0 technology concept. Software developers and researcher should embrace Industry 4.0 concept in development process to enhance flexibility, robustness, reliability, and compatibility of software.
2. The implementation and the achievement of this goal through Industry 4.0 concept have gone a long way in achieving the goal of reducing cost in our ICT hardware need particularly countries with low income. Therefore, developing countries should invest in Industry 4.0 technology in all sectors of their economy.

23.8 Conclusion

One major problem with clustering capabilities in MySQL is that it provides no means to distribute the transaction loads to the nodes. At the moment, it is very expensive to obtain balancing software; in most cases, organizations use hardware that is very costly. This work explored existing load-balancing algorithms and translates them into software code through the concept of Industry 4.0 technology that was integrated into the virtual learning system to support distribution of transactions to various nodes during operations.

References

[1] Juan, R. (2018). Developing of Industry 4.0 applications. *International Journal of Online Engineering (iJOE)*, *13*(10), 30. How we measure 'reads' A 'read' is counted each time someone views a publication summary (such as the title, abstract, and list of authors), clicks on a figure, or views or downloads the full-text. DOI: 10.3991/ijoe.v13i10.7331

[2] David, M. D. (2018). Developing of Industry 4.0 applications. Industry 4.0: the essence explained in a nutshell. *International Journal of Online Engineering (iJOE),* https://doi.org/10.3991/ijoe.v13i10.7331

[3] Brown, H. (2015). *Resource Allocation and Scheduling for Mixed Database Workloads*. Madison, WI: University of Wisconsin Press Limited.

[4] Schneider, D. and Dewitt, D. J. (2001). A performance evaluation of four parallel joint algorithms in a shared-nothing multiprocessor environment. In *Proceedings ACM SIGMOD Conference*, *18*(2). Retrieved May 12, 2019 from https://www.sitepoint.com/mysql-master-slave-replication-setting-up/

[5] Dewitt, D. J. and Schneider, D. (1990): Tradeoffs in processing complex join queries via hashing in multiprocessor database machines. In *Proceedings of 16th International Conference on Very Large Databases* (pp. 469–480). Brisbane, Australia.

[6] Shivaratri, N., Krueger, P., and Singhal, M. (1993). Load distributing for locally distributed systems. *IEEE Computer*, *25*(12), 33–34.

[7] Dennis H. and Klaus, S. (2003). Load balancing for MySql. Retrieved June 24, 2019 from http://www.mysql.org/support/

[8] Casavant, T. (1988). Taxonomy of scheduling in general-purpose distributed computing systems. *Journal of IEEE Transactions in Software Engineering*, *14*(2), 141–154.

[9] Yen, S. (2016). *Designing an English Reading Web-Based Instruction in Engineering Education*. Kao-Yuan Institute of Technology. Retrieved February 17, 2019 from http://www.kyit.edu.tw

[10] Hanisch, F. and Klein, R. (2000). Challenges in the development of web-based courseware using virtual experiments. In *11th Annual Conference on Innovations Education for Electrical and Information Engineering (EAEEIE) for Data Engineering (ICDE)*. Graphisch—Interaktive System Universität Tübingen, Germany.

[11] Keane, M. M. and Monaghan, P. F. (2015). *Courseware Relating to Heat Loss Calculations through Walls in Buildings*. Computer Aided Thermal Engineering Research Unit (CATERU), University College Galway, Ireland. Retrieved February 19, 2008 from http://zuse.ucc.ie/iruse/papers/keane-heatloss.pdf

[12] National Institute of Information Technology (NIIT). (2016). *Unified Modeling Language (UML) Tools. Guide to Software Development Process*. NITDA, Nigeria, Abuja.

[13] Larman, C. (2015). *Applying UML and Patterns: An Introduction to Object-Oriented Analysis, Design and Iterative Development* (3rd ed.). USA: Addison-Wesley.

[14] Dahlin, M. (2015). Interpreting state load information. *IEEE Transaction on Parallel and Distributed Systems*, 11(10), 33. Retrieved April 04, 2019 from http://citeseer.nj.nec.com/article/dahlin98interpreting.html

[15] Selic, B. (2017). UML 2.0: Designing accessible web based courseware with authoring tools. The Joint Information Systems Committee. *Zaria Journal of Education Studies*, 5(1), 65–72.

Chapter 24

5G network review and challenges

Wei Siang Hoh[1], Bi-Lynn Ong[2], Wai Yie Leong[1] and Si-Kee Yoon[2]

5G network becomes a reality and it is expected to bring a wealth of new opportunities in various vertical fields. Operators demand for 5G network technologies. The reasons are that 4G network services become saturated and revenue is flattening or declining every year. 5G is expected to enable further economic growth and pervasive digitalization of a hyperconnected society with all people and devices/things are connected to the network virtually. 5G will provide enterprises and consumers with plenty of new use cases, including Virtual Reality (VR) application, telemedicine by using ultra-reliable communications and large-scale Internet of Things (IoT) such as smart cities. Through 5G technology revolution, it is able to provide low latency, large bandwidth, high reliability and capacity and it also brings better user experience. The operator needs to develop new network infrastructure (5G New Radio) or restructure the current 4G network infrastructure to adopt the 5G system. Hence, this chapter reveals the foundation of 5G network in terms of key capabilities, use cases and network structure in detail. In addition, the 5G network deployment is also considered in this chapter which reveals the deployment option for non-standalone (NSA) and standalone (SA) network architectures, spectrum specification and 5G key technologies such as massive multiple input multiple output (MIMO) and uplink and downlink (UL/DL) decoupling mechanisms. The key challenges of 5G network which are faced by worldwide operators are also included in this chapter. The future aim of 5G mobile network communication while moving is made possible, which brings digital to every person, home and organization for a fully connected, intelligent world.

24.1 5G network overview

In the recent decades, the total number of Internet users increased dramatically and reached 4.54 billion over 7.75 billion of human population in 2020 [1]. However, 4G wireless technologies are not viable to provide support in the future. The Internet data traffic has been explosive. Besides that, existing 4G network also cannot deliver instant

[1]Faculty of Engineering and Built Environment, Mahsa University, Selangor, Malaysia
[2]School of Computer and Communication Engineering, Universiti Malaysia Perlis, Perlis, Malaysia

cloud services, Internet of Things (IoT) [2], enhanced Vehicle-to-Everything (eV2X) and tactile Internet [3,4], considering the quality of experience (QoE) to mobile users. These characteristics have become vital necessities for future services. Fourth Industrial Revolution (4IR) based on digital revolution has been launched and the fifth generation of mobile communications (5G) has become a central point for research and development to overcome the 4G issues. International Telecommunication Union-Radio communication Sector (ITU-R), the Next Generation Mobile Networks (NGMN) Alliance and the 3rd Generation Partnership Project (3GPP) have focused their research on 5G services and they standardize the technical requirement. In the ITU-R Working Party (WP) 5D, 5G is defined by the name of International Mobile Telecommunications 2020 (IMT-2020) [5]. The 5G network which is able to provide huge improvement in speed, coverage, energy efficiency and reliability will totally replace the current 4G technology. The overall objectives of IMT-2020 were set via ITU-R M.2083. Table 24.1 shows the eight parameters that are considered as key capabilities of IMT-2020 [6].

The peak rate of IMT-2020 for enhanced mobile broadband is expected to reach 10 Gbit/s, and under certain conditions and scenarios it would support up to 20 Gbit/s, as shown in Figure 24.1 [6]. IMT-2020 would support different user experienced data rates, 100 Mbit/s as minimum requirement for wide area coverage cases, and for hotspot cases the user experience data rate could reach higher values of about 1 Gbit/s. For spectrum efficiency, IMT-2020 is expected to be three times higher compared to IMT-advanced. When spectrum efficiency becomes higher, the area traffic capacity is expected to achieve 10 Mbit/s/m^2. IMT-2020 would be able to provide 1 ms over-the-air latency and high mobility up to 500 km/h with acceptable Quality of Service (QoS) and support a connection density of up to 106 per km^2. The network energy efficiency is able to improve 100 times as great as the envisaged traffic capacity increases.

24.1.1 5G network use cases

A broad variety of capabilities would be tightly coupled with intended different usage scenarios and applications for IMT-2020. Figure 24.2 shows three types of 5G application scenarios. Different key capabilities are included based on the 5G application scenarios and influence the performance. The usage scenarios for IMT-2020 can be classified into three categories which are:

- Enhanced Mobile Broadband (eMBB)
  - In eMBB scenario, peak data rate, user experienced data rate, energy efficiency, area traffic capacity, mobility and spectrum efficiency have high importance on the improvement of a user's perceived experience.
- Ultra-reliable and Low Latency Communications (uRLLC)
  - Latency and reliability have highest importance, with everything highly interconnected, IoT and other vertical industries will bring about a large number of Wireless Sensor Networks (WSNs), posing high requirements on network access quantities and power-consumption efficiency.
  - Massive Machine Type Communications (mMTC)

Table 24.1 Key capabilities of IMT-2020 [6]

No.	Key capability	Description	Unit measure
1	Peak data rate	The maximum data rate can be achieved under ideal conditions per user/device	Gbit/s
2	User experienced data rate	Achievable data rate across the coverage area to a mobile user/device in Mbit/s or Gbit/s	Mbit/s or Gbit/s
3	Latency	The time delay between the sending and receiving of information	ms
4	Mobility	Maximum speed at which a define QoS by internet engineering task force (IETF) and seamless transfer between radio nodes which may belong to different layer and/or radio access technologies	km/h
5	Connection density	Total number of connected devices per unit area	Per km^2
6	Energy efficiency	Network side refers to the quantity of information bits transmitted to/received from users, per unit of energy consumption of the radio access network (RAN) device side refers to the quantity of information bits per unit of energy consumption of the communication module	bit/joule
7	Spectrum efficiency	Average data throughput per unit of spectrum resource per cell	bit/s/Hz
8	Area traffic capacity	Total traffic throughput served per geographical area	$Mbit/s/m^2$

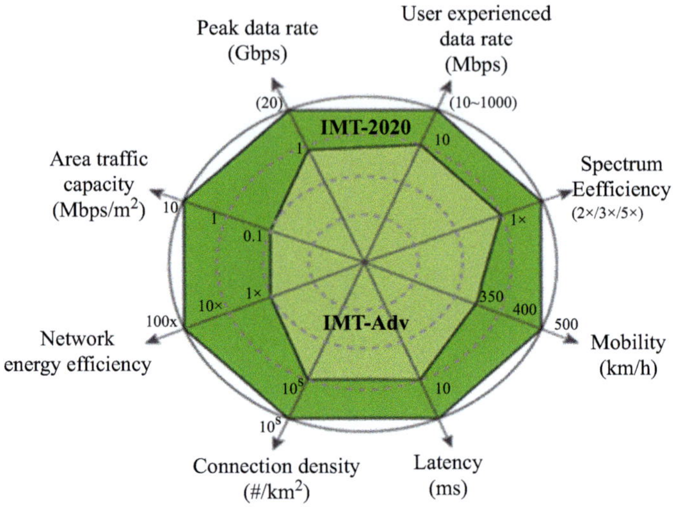

Requirement		Value
Data rate	Peak	Downlink: 20 Gbit/s Uplink: 10 Gbit/s
	User experienced	Downlink: 100 Mbit/s Uplink: 50 Mbit/s
Spectral efficiency	Peak	Downlink: 30 bits/s/Hz Uplink: 15 bit/s/Hz
	5th percentile user	Downlink: 1.12~0.3 bits/s/Hz Uplink: 0.045~0.21 bit/s/Hz
	Average	Downlink: 3.3~9 bits/s/Hz Uplink: 1.6~6.75 bit/s/Hz
Mobility		500 km/h
Latency	User plane	1~4 ms
	Control place	20 ms

Requirement	Value
Connection density	1,000,000 devices per km^2
Network energy efficiency	Loaded: depends on spectral efficiency
Area traffic capacity	10 Mbit/s/m^2
Reliability	1–10^5 success probability of transmitting a layer 2 Protocol Data Unit (PDU) of 32 bytes within 1 ms
Bandwidth	100 MHz
Mobility interruption time	0 ms

Figure 24.1 Enhancement of key capabilities from IMT-advanced to IMT-2020 [6]

- ○ High connection density and network energy efficiency are needed to support tremendous number of devices in the network.
- ○ Automated driving, telemedicine, smart grid and other vertical industries require high reliability and low latency.

24.1.2 5G network architecture

The global mobile communications standards organization 3GPP defines new core network, referred as 5G Core (5GC), and also new radio access technology called 5G 'New Radio' (NR). 3GPP classified the 5G network into two tracks that are non-standalone (NSA) and standalone (SA). Telecommunication operator can

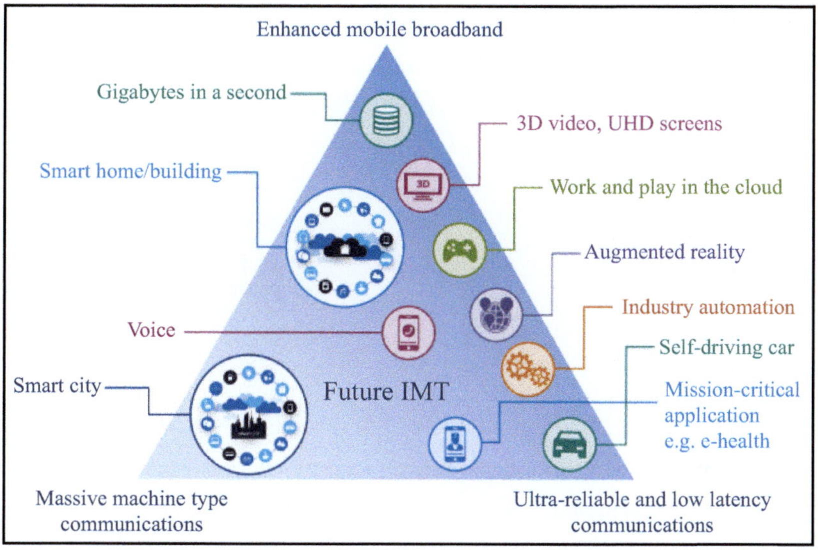

Figure 24.2 Three types of 5G application scenarios [5]

choose to deploy a full 5G SA that provides end-to-end (E2E) 5G experience or deploy a 5G NSA network that can be complemented and supported by 4G network.

The initial phase of 5G NSA deploys 5G cells with support of existing 4G network. The 5G NSA architecture works in master–slave structure, where the 4G access node is the master and 5G access node is the slave. The NR cells are combined with LTE radio cells using dual connectivity (DC) to provide radio access and the core network may be either 5GC or Evolved Packet Core (EPC) depending on the choice of operator. The operator that wish to leverage existing 4G deployment will target for this scenario, which is able to combine the NR and LTE radio resources with existing EPC and/or that enables new 5GC to deliver 5G mobile services. NSA scenario requires tight interworking with the LTE radio access network (RAN) and the end user experience will be dependent on the radio access technologies used. Figure 24.3 shows the 5G NSA network architecture [7].

Due to master–slave structure, NSA is more economical compared to 5G SA network. However, in the final network state, all operator will aim for 5G SA network. In other words, operator will eventually migrate to a fully 5G SA network although they choose the NSA option as a starting point. 5G SA network refers to have an independent 5G network with new 5G air interface, NR and 5GC. This means that the NR or evolved LTE (eLTE) radio cells are used for both control and user planes. Nevertheless, SA network will still interoperate with the existing 4G/LTE network to provide service continuity between the two network generations.

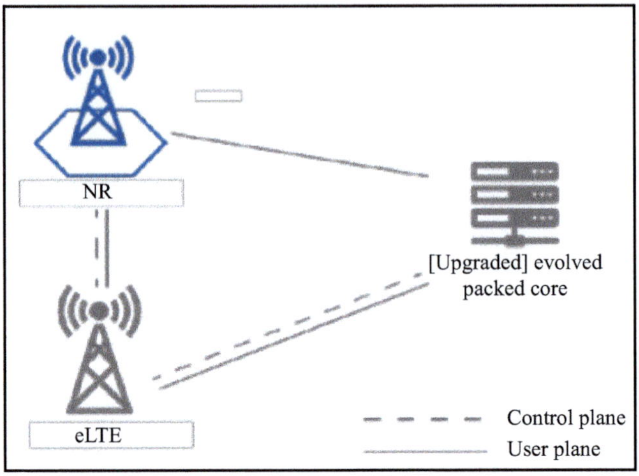

Figure 24.3 5G NSA network architecture [7]

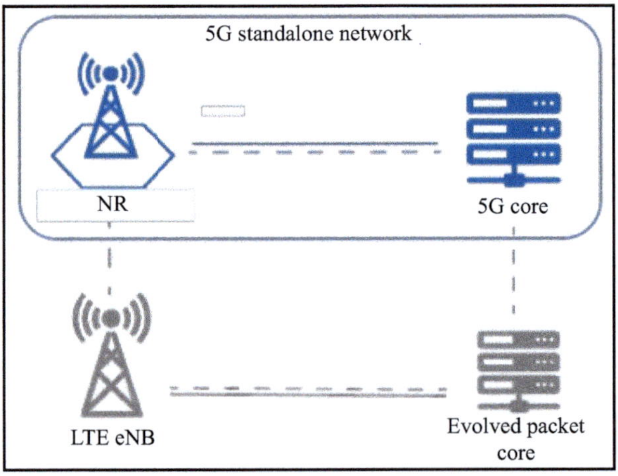

Figure 24.4 5G SA network architecture [7]

The SA option deploys an independent network using normal inter-generation handover between 4G and 5G for service continuity. Figure 24.4 presented the 5G SA network architecture [7].

Table 24.2 illustrated the comparison of 5G NSA and SA network architecture under four main perspectives that include investment cost, development time, use cases support and network key capability performance.

Table 24.2 5G SA vs NSA architectures

	NSA	SA
Description	Combines NR radio cells and LTE radio cells using dual connectivity to provide radio access and the core network may be either EPC or 5GC	Uses only one radio access technology (5G NR or the eLTE radio cells) and the core networks are operated alone
Investment cost		
Short-term	Low to medium	High
Long-term	High	N/A
Development time	Short	Long
Use cases support	Support only eMBB	Covers eMBB, uRLLC and mMTC
Network key capability performance		
Data rate (downlink/uplink)	20 Gbps/10 Gbps	20 Gbps/10 Gbps
Latency	4 ms	1 ms
Network density	1 million devices/km^2	1 million devices/km^2
Advantages	Leverage existing LTE deployments	Simple to manage Inter-generation handover between 4G and 5G

24.2 5G network deployment

24.2.1 Deployment options for NSA and SA

For SA and NSA scenarios, three variations for each scenario are defined by 3GPP as shown in Figure 24.5. The EPC and 5GC networks have been introduced and defined in 3GPP for 5G deployment options for either NSA or SA. The SA includes Option 1 using EPC and LTE eNB access, Option 2 using 5GC and NR gNB access and Option 5 using 5GC and LTE ng-eNB access. The NSA includes Option 3 using EPC and an LTE eNB acting as master and NR en-gNB acting as secondary, Option 4 using 5GC and an NR gNB acting as master and LTE ng-eNB acting as secondary and Option 7 using 5GC and an LTE ng-eNB acting as master and an NR gNB acting as secondary. The early deployment options of 5G will be adopting either NSA Option 3 or SA Option 2 as the standardization of these two options have been completed [8].

The direct path to 5G is to deploy 5G NR using 5GC in SA mode, which is shown as Option 2 in Figure 24.5. The RAN consists of only gNBs (gNode Bs) and connects to 5GC. SA Option 2 has no impact on LTE radio, the capability of 5G can fully exert which is able to support in different usage scenarios for eMBB, uRLLC and mMTC by enabling network slicing via cloud native Service-Based Architecture (SBA). The SBA is proposed in 5GC, capable of providing unprecedented efficiency and flexibility for the network. SBA is an architectural for building system based on fine-grained, interaction of loosely coupled and autonomous components called services. A key Network Function (NF) of SBA is the Network Repository Function (NRF), which provides NF service registration and discovery, enabling NFs to identify appropriate services in one another. SBA principles apply to interfaces between Control Plane (CP) functions within 5GC only. The separation of the control plane and user plane besides adding flexibility in connecting the users allows an easier way to support a multitude of

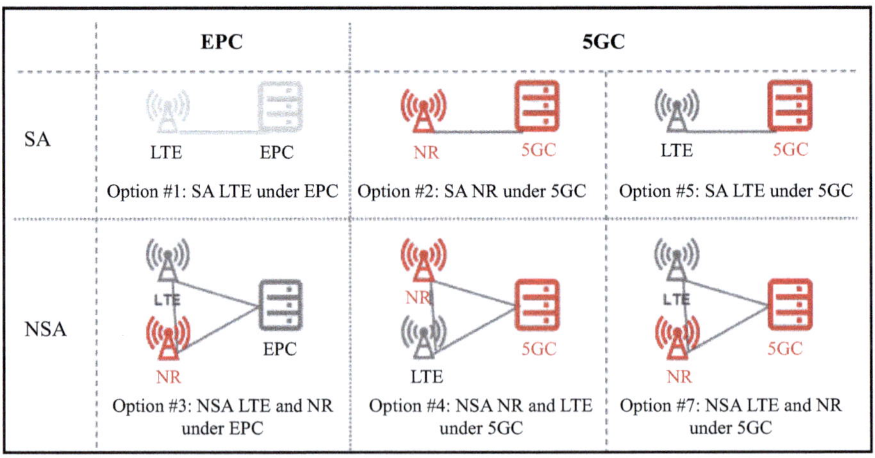

Figure 24.5 Overview of 5G SA and NSA options [8]

access technologies, better support for network slicing and edge computing. 5GC also offers superior E2E network slicing and QoS features.

Another path to 5G in NSA mode is using EPC, in which NR radio cells are combined with LTE radio cells using DC to provide radio access, which is shown as Option 3 in Figure 24.5. EPC is the core network element of 4G network that consists of Mobility Management Entity (MME), Home Subscriber Server (HSC), Serving Gateway (S-GW), PDN Gateway (P-GW) and Policy and Charging Rules Function (PCRF). S-GW and P-GW handle both control and user planes whereas MME handles only the control plane. PCRF provides rules for policy and charging. Control and User Plane Separation (CUPS), a new feature, is introduced in 3GPP Release 14. EPC is able to address latency limits if CUPS is deployed but not otherwise. CUPS support flexible user plane deployment without expanding or upgrading the control plane. In cloud native case, the simpler configuration and maintenance work of user plane nodes can be achieved.

After the introduction of Option 2 in SA mode using 5GC and Option 3 in NSA mode using EPC, the comparison between 5G NSA and SA options is summarized in Table 24.3.

24.2.2 Spectrum of 5G network

More spectrum bandwidth will be mandatory to deploy 5G network (than 4G) to the high capacity requirements, which is able to provide faster data rates and better user experience. The ITU's Standardization Sector (ITU-T) is playing a key role in producing the standards for the technologies and architectures of the wireline elements of 5G systems. New wider 5G bands were purposed, and they support channel sizes ranging from 5 to 100 MHz in bands below 6 GHz, and channel sizes from 50 to 400 MHz in bands above 24 GHz [9]. The full capabilities of 5G will be realized through the wider channel sizes in new 5G bands [10].

Two types of frequency range are defined in 3GPP, whereas sub-6 GHz range is called FR1 and millimetre wave (mmWave) is called FR2. Table 24.4 shows the exact frequency range for FR1 and FR2 [10].

There are two separate approaches to deploy hundreds of MHz of new spectrum for 5G network. The first approach, referred as sub-6 GHz band, focuses on the electromagnetic (EM) spectrum below 6 GHz, which is primarily in the 3 and 4 GHz bands. The second approach, referred as mmWave, focuses on the part of the spectrum between 24 and 52.6 GHz. The spectrum allocation is critical for 5G success, because the spectrum band of choice, whether sub-6 GHz or mmWave, impacts nearly every other aspect of 5G development. Figure 24.6 presents the 5G full-spectrum access.

Based on Figure 24.6, 5G full-spectrum access can be classified into 5G primary bands and 5G complementary bands. The 5G primary bands that allocate below 6 GHz are crucial to support most 5G use scenarios in a wide area. The 5G primary band can be subdivided into two macro categories: low frequencies (below 2 GHz) and medium frequencies (2–6 GHz). To offer good 5G NR coverage, operators will need to have a spectrum band in the sub-6 GHz range to make use of its propagation properties of the signal at these frequencies to enable 5G to create very large coverage areas and deep in-building penetration.

Table 24.3 Comparison between 5G NSA Option 3 and SA Option 2

	NSA Option 3	**SA Option 2**
3GPP 5G specification	Rel-15 (17Q4)	Rel-15 (18Q2)
5G spectrum	Sub-6 GHz and mmWave band are feasible	Sub-6 GHz band is desirable
CN	EPC	5GC
CN interworking	Not required	Required
Network slicing and 5G QoS	Not supported	Supported
Leverage of LTE	Full	Partial (reattach)
LTE upgrade	Required (eNB and EPC)	None/minor
Forward compatibility with Release-16 onwards	Low	High
Service readiness	eMBB	eMBB/uRLLC
NR coverage quality	Low	High (sub-6 GHz + mmWave) Medium (sub-6 GHz or mmWave)
Advantages	Leverage existing EPC deployment	Cloud native Easier to support multiple access
Disadvantages	Cloud support is optional Tight interworking between LTE and NR required	High cost as new deployment is required

Table 24.4 Frequency range of FR1 and FR2 [10]

Frequency range designation	Corresponding frequency range
FR1 (sub-6 GHz band)	450–6,000 MHz
FR2 (mmWave)	24,250–52,600 MHz

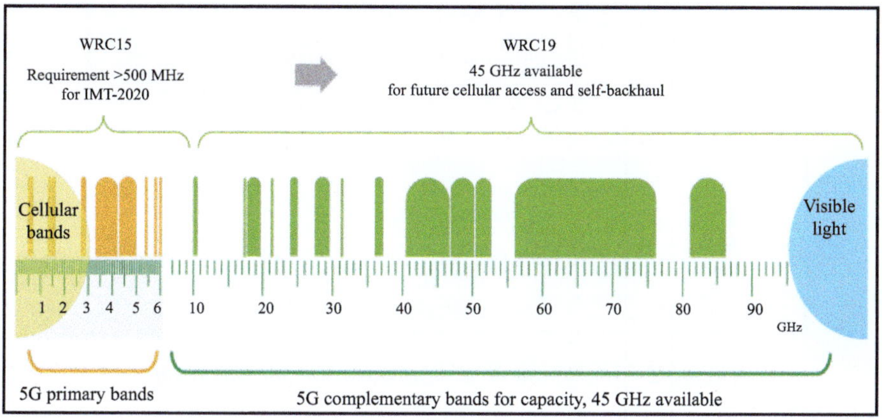

Figure 24.6 5G full-spectrum access [10]

Low frequencies (below 2 GHz) will continue to be essential to extend the 5G mobile broadband experience to wide areas and in deep indoor environments; mMTC and URLLC usage scenarios will also greatly benefit from the low frequencies' extended coverage. The available low frequency bands (e.g. 700, 800, 900, 1,800 and 2,100 MHz) may be exploited for LTE/NR uplink (UL) spectrum sharing in combination with NR on the C-band to allow operators to ensure faster and cost-effective deployment of C-band.

Medium frequencies with ranges 3,300–4,200 and 4,400–5,000 MHz (referred as C-band) are suitable to deliver the best compromise between wide-area coverage and good capacity. In line with the release plans from many countries, the 3,300–3,800 MHz band will be the primary 5G band with highest potential for global harmonization over time. In early deployment of 5G, at least 100 MHz contiguous spectrum bandwidth should be assigned to each 5G network to support user experienced data rate of 100 Mbps.

If sub-6 GHz band is unavailable, operators will have to depend on 5G complementary bands of above 6 GHz. High frequencies (above 6 GHz) will prove indispensable for providing additional capacity and delivering the tremendously high data rates to enable 5G eMBB applications. The 24.25–27.5 and 37–43.5 GHz bands (referred as mmWave band) are prioritized within the ongoing ITU-R work in preparation for World Radiocommunication Conference 2019 (WRC-19). All countries

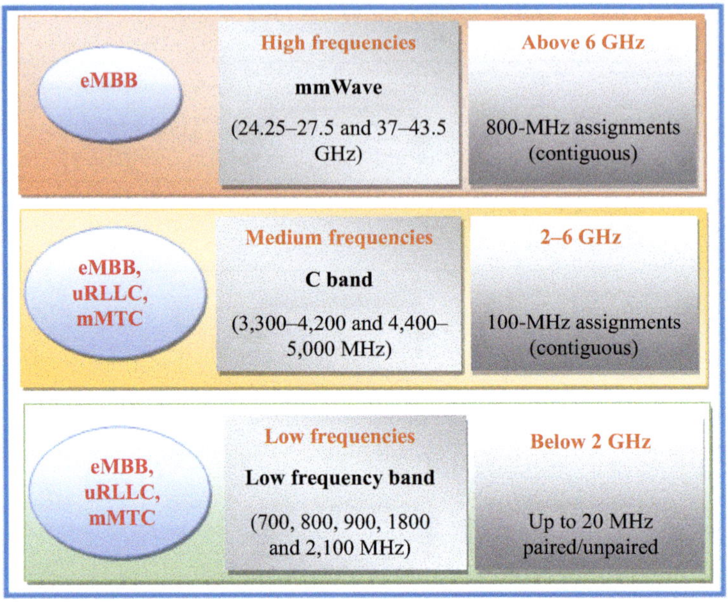

Figure 24.7 Frequencies approach for 5G spectrum

and regions are recommended to support the identification of these two bands for IMT during WRC-19. It should aim to harmonize technical conditions for using these frequencies in 5G. For example, 28 GHz is the band used for many of the Fixed Wireless Access (FWA) trials and commercial launches. The radio cells operating in mmWave are suitable for creating a thick capacity layer where needed (hotspots) as well as for numerous enterprise scenarios. In early deployment of 5G, at least 800 MHz per network of contiguous spectrum bandwidth from high frequencies is recommended. However, mmWave bands are not suitable for wide coverage due to weak propagation characteristics. Somehow, it is a good option for hotspot and local 5G NR coverage, and it offers large network capacity. Figure 24.7 shows the details of frequencies approach for 5G spectrum.

24.2.3 5G technology and technical regulations

5G networks and devices will embrace many new features, exploiting the latest technical innovations. The massive multiple input multiple output (MIMO) and UL/downlink (DL) decoupling mechanisms are able to provide an opportunity for regulators to adjust regulations for more efficient and flexible spectrum utilization.

24.2.3.1 Massive MIMO

Massive MIMO is an extension of MIMO. It expands beyond the legacy systems by adding numerous numbers of antennas on the base station (BS). Massive MIMO is a key 5G feature that implies MIMO and is used by combinations of

Figure 24.8 3D beamforming technique

multiple transmitters/receivers or antennas at both sides of digital communication systems [11]. Along with the increased number of antennas, both the network and mobile devices implement more complex designs to coordinate MIMO operations. The MIMO systems capitalize on three key concepts that include spatial diversity, spatial multiplexing and beamforming. One of the fundamental benefits of MIMO technology is spatial diversity. Diversity aims at improving the reliability of the system by sending the same data across different propagation, either spatial or paths. Spatial multiplexing aims to transmit multiple messages simultaneously without interfacing with one another.

The massive MIMO uses the beamforming technique to focus signals on each user, thereby increasing data rates and decreasing latency. Beamforming is another key wireless technique that utilizes advanced antenna technologies on both mobile devices and networks' BS to focus a wireless signal in a specific direction, rather than broadcasting to a wide area. 3D beamforming can be formed with the massive number of antenna elements in a massive MIMO system. 3D beamforming adopts large-scale antenna arrays that can control the width and tilt, both vertical and horizontal beams towards users, increasing data rates and capacity for all users. Figure 24.8 shows the 3D beamforming technique.

Massive MIMO is a key enabler of 5G's extremely fast data rates and it promises to raise 5G's potential to a new level. The primary benefits of massive MIMO to the network and end users can be summed up as:

- Increased network capacity
  - Massive MIMO contributes to increase capacity first by enabling 5G NR deployment in the higher frequency range in sub-6 GHz (e.g. 3.5 GHz) and second by employing MU-MIMO where multiple users are served with the same time and frequency resources.

- Improved coverage
  - With massive MIMO, users enjoy a more uniform experience across the network, even at the cell's edge. Hence, users can expect high data rate service almost everywhere. Moreover, 3D beamforming enables dynamic coverage required for moving users (e.g. users travelling in cars or connected cars) and adjusts the coverage to suit user location, even in locations that have relatively weak network coverage.

Although the massive MIMO provides many benefits for 5G network, the deployment of Massive MIMO still has to consider from aspects of performance requirements, installation requirement and total cost of ownership (TCO) saving. The most cost-effective configuration, 64T, 32T or 16T, may be different for different deployment scenarios due to the different performance benefits in the scenarios, respectively. In dense-urban high-rise scenarios, 64T or 32T may have performance advantages as the 2D beamforming will provide benefits compared to 1D beamforming and hence may be more cost-efficient. In urban and suburban areas, the user spread in the vertical dimension is normally so small that vertical beamforming will not provide any substantial advantages. Therefore, 16T will have similar performance as 64T or 32T in many such scenarios. Apart from this, 2T2R/4T4R/8T8R solution can also be considered as a basic configuration for 5G deployment. Deploying 16T16R/32T32R/64T64R to meet capacity requirements in those sites that demand for capacity upgrades are heavy loaded. Figure 24.9 illustrates the comparison between 64T/32T and 16T [12].

24.2.3.2 Coverage enhancement mechanisms

Currently, C-band is the primary band for 5G network, where one network operator could obtain 100 MHz or more spectrum, and time division duplex (TDD) is used to facilitate the use of advanced massive MIMO technologies. Compared with the lower frequency bands that are widely used for LTE systems, C-band suffers larger path loss and penetration loss. This issue is even more severe for the UL due to the higher frequency and smaller portion of UL resource allocation. Beamforming and

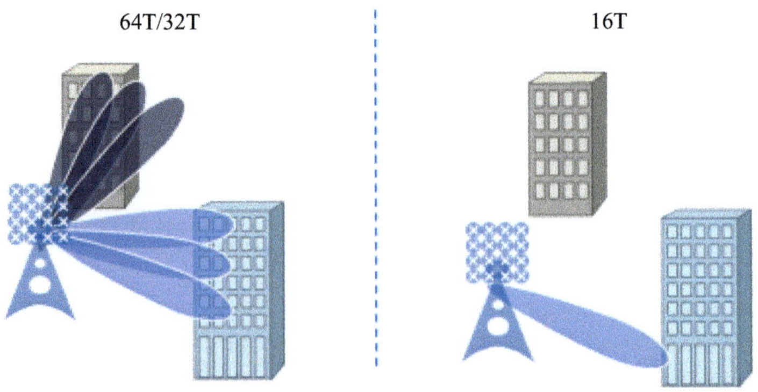

Figure 24.9 Comparison between 64T/32T and 16T [12]

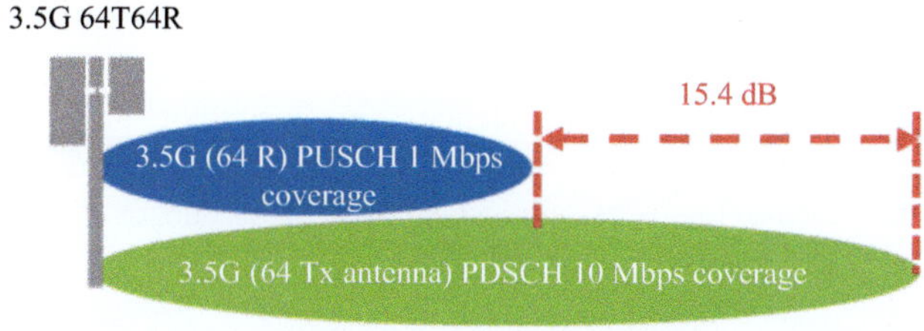

Figure 24.10 Coverage of UL and DL for C-band [13]

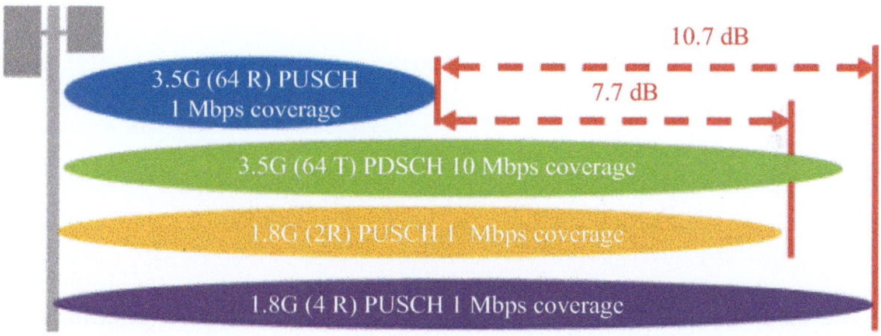

Figure 24.11 Coverage of UL and DL for C-band [13]

cell-specific reference signal (CRS) reduce DL interference and increase the difference between C-band UL and DL coverage. As an example, taking the DL 10 Mbps and the UL 1 Mbps, the C-band UL coverage is up to 15.4 dB smaller than that of the DL, as shown in Figure 24.10 [13].

The C-band DL can achieve the similar coverage as the LTE 1,800 MHz, but there is limitation in the UL coverage and it becomes 5G deployment bottleneck which will affect the user experience. As shown in Figure 24.11, the difference between C-band and LTE 1,800 MHz UL coverage is 7.7 dB for 2R and 10.7 dB for 4R [13].

This UL–DL coverage gap will largely limit the cell coverage. This will increase the number of BS sites to cover the same geographical area and cause higher deployment costs for operators. Therefore, 3GPP Release 15 introduces two mechanisms, namely NR carrier aggregation (CA) and supplementary uplink (SUL), to handle the limited UL coverage on the higher bands. By implementing these mechanisms, these can effectively utilize idle sub-3 GHz band resources, improve the UL coverage of C-band, and enable the provisioning of 5G services in a wider area. For both mechanisms, NR CA and SUL offer transport of UL user data using sub-3 GHz band NR radio resources. NR CA provides the additional benefit and also provides sub-3 GHz DL user data support. Based on the SUL, the feature UL and DL decoupling defines new paired spectrums, where C-band is used for the DL and a sub-3 GHz band (e.g. 1.8 GHz) for the UL. With UL/DL decoupling, these are able to

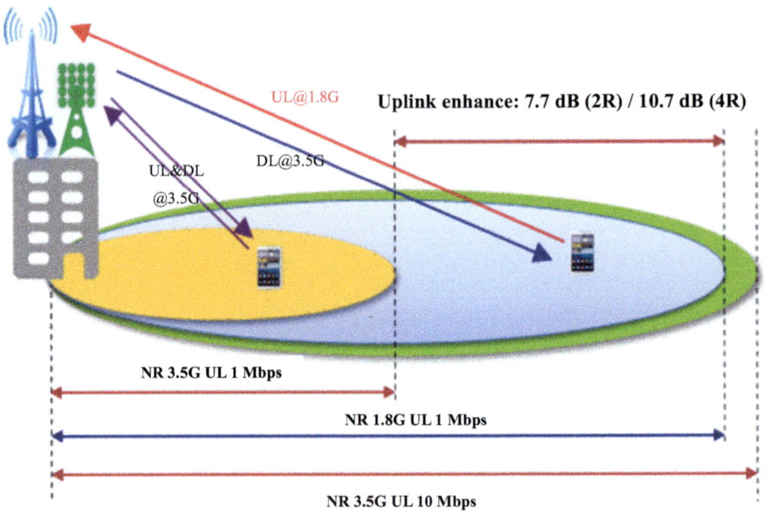

Figure 24.12 UL and DL decoupling

address the issue of UL–DL coverage gap and at the same time enable operators to reuse their existing LTE BS sites, thereby improving UL coverage. Figure 24.12 shows the operation of UL and DL decoupling.

Traditional cell included only one UL carrier and one DL carrier, while UL/DL decoupling for 5G-NR proposes to use an additional 5G-NR UL carrier to supplement a 5G-NR TDD carrier [14]. The additional UL carrier and the TDD carrier are operated in different frequency bands. For example, the additional UL carrier is operating in 1.8 GHz band and the TDD carrier is operating in 3.5 GHz band. Hence, there will be one DL carrier and two UL carriers in the same cell where either of the UL carriers can be used with the DL carrier. Normal TDD carrier will operate when a user equipment (UE) is at the cell centre where there is good channel quality. Additional UL, which is called 9SUL, will operate when the UE is at the cell edge where UL coverage is not guaranteed. Through this method, the UL/DL decoupling can enhance the UL coverage of a 5G-NR cell. Table 24.5 shows the SUL frequency band and band combination definition [15].

24.3 Key challenges of 5G network

After a discussion about 5G architecture, deployment option, use cases, key technologies and technical regulation, numerous key challenges have been found out by the operator to meet the requirement of eMBB, uRLLC and mMTC use cases. The key challenges faced by telecom operator are small cell deployment due to the country regulation and local authority policy, fibre backhaul network deployment and security and privacy.

Table 24.5 SUL frequency band and band combination definition [15]

5G-NR bands			5G-NR band combinations	
			SA	NSA
Band No.	Frequency	Duplex	Band combination number	
n77	3.3–4.2 GHz	TDD	SUL_n78–n80	DC_1-SUL_n78–n84
n78	3.3–3.8 GHz	TDD	SUL_n78–n81	DC_3-SUL_n78–n80
n79	4.4–5.0 GHz	TDD	SUL_n78–n82	DC_3-SUL_n78–n82
n80	1,710–1,785 MHz	SUL	SUL_n78–n83	DC_8-SUL_n78–n81
n81	880–915 MHz	SUL	SUL_n78–n84	DC_20-SUL_n78–n82
n82	832–862 MHz	SUL		DC_28-SUL_n78–n83
n83	703–748 MHz	SUL		
n84	1,920–1,980 MHz	SUL		

24.3.1 Small cell deployment challenges

The local authority policy and regulation have slowed down the development of small cells through excessive administrative and financial obligations on operators. Hence, 5G investment in some countries has been blocked due to these issues. These issues are described in more detail as follows [16]:

1. Local permitting and planning processes
   - Prolonged permitting processes by approving planning application can take 18–24 months to complete from start to finish, largely due to the need to obtain local permits for the installation of the devices.
2. Long engagement duration and procurement
   - Local authorities have use prolonged procurement processes lasting 6–18 months to award wireless providers with exclusive rights to deploy small cell equipment onto street furniture.
3. High fees and charges to access street furniture
   - Local authorities currently charge high fees to access street furniture. For instance, according to U.S. Consumer Institute, US$30,000 application fee is charged to attach small cell equipment onto a utility pole with another locality imposed a US$45,000 fee. Thus, sum up to US$75,000 needs to be invested by telecom operator for small cell deployment in one city.

The issues that are mentioned earlier are prohibiting the rapid and cost-effective roll-out of small cells in city centres where 5G is initially expected to be most in demand. Policymakers who provide simplified and flexible regulatory processes will benefit most from the innovation and economic growth that 5G will bring.

With the purpose of solving the small cell deployment challenges, it is suggested that governments need to work with local community authorities to ensure that reasonable fees are charged during deployment of small cell radio equipment on street furniture. Local authorities suggested to consider improving access by implementing 5G on government-owned street furniture and streamlining engagement processes as an alternative to lengthy procurement exercises. New legislation can be enforced to standardize the small cell deployment across the nation or state. The bills should include:

- Granting providers that are right and non-discriminatory to public property
- Allowing local governments to charge permit fees that are reasonable, fair and cost-based
- Confining the costs charged by the local governments on installing or attaching equipment
- Stopping local governments that impose an unreasonable time on the duration of obtaining permit for the telecommunication service providers.

24.3.2 Fibre backhaul network deployment challenges

On the other hand, the introduction of small cell deployment has an impact on the part of the network referred to as backhaul. Backhaul is responsible for collecting data traffic from the BSs and sending it to the aggregation network mostly in terms of number of required backhaul connections. In fact, a large number of links are required to aggregate the data traffic originating at each small cell, each one working at a peak rate of tens of Mbps. Legacy copper-based infrastructures cannot guarantee the peak rate thus forcing telecom operators to upgrade their backhaul networks to avoid potential bottlenecks.

Therefore, deploy fibre backhaul networks for small cells can support high data rates and low latency to meet most of the 5G use cases requirements. However, some city that is lacked of fibre backhaul network has become one of the largest challenges that is faced by telecom operators. Operators should consider wireless backhaul technologies including mmWave and satellite for fibre backhaul deployment.

Complex wayleave process is one of the barriers to deploy fibre backhaul network. Wayleave agreements allow operators to install telecoms infrastructure on public or private land. Using bespoke wayleaves are expensive, and landowner who is using procurement process to grant wayleaves added risk, time and expenses to the process. Furthermore, lack of early engagement between operators and local authorities can result in planning permission being refused. Local authority policy on the siting and aesthetics of street cabinets can also increase costs and delays while alternative solutions are sought.

Fibre backhaul issues could be solved if policymakers or government consider to remove the tax burdens, thereby helping to reduce the cost of investing. This can facilitate the deployment of 5G networks. On the other hand, local authorities may consider agreeing on standardized wayleave agreements to reduce the cost and time of deploying fibre networks. Governments need to consider stimulating investment in fibre networks and passive assets through setting up public–private partnerships (PPPs), investment funds and offering grant funds and many others.

In setting up an fibre industries, these may need huge investment. Unfortunately, the well-established companies may not have interest to involve in new industries or perhaps there is a monopoly industry where prices are very high. Thus, it is suggested that government can provide loan with low interest in setting up fibre industries. By doing these, it is expected that many company will be setting up which may have competitors with each other. Indirectly, these can lower the cost of fibre.

In addition, device availability is also one of the concerns for fibre backhaul deployment. A device that needs to work within a network implementing NSA/SA option will support all protocols requested by NSA/SA implementation and could support all protocols requested by SA/NSA implementation. Different frequency bands like sub-6 GHz and mmWave require different sets of combination devices. However, sub-6 GHz band is not optimized for new 5G use cases except the mobile broadband and mmWave which may cause fast attenuation of the radio signal.

The device availability issues can be solved since the manufacturers are currently developing technology that embeds 5G, 4G, 3G and 2G onto a single chip and is expected to become available to sell on 2020 for globally harmonized standards.

24.3.3 5G user privacy challenges

In traditional mobile communication networks, users communicate by using SMS or voice/video calls, surf Internet or access app services using smartphones. Nevertheless, 5G network is no longer confined to individual customers but it also serves vertical industries, from which a variety of new services are going to stem. Therefore, user privacy information must be securely protected in the 5G network.

The 5G mobile networks carry data and signalling which contain many personal information such as identity, position and private content. The network-based hop-by-hop security approach that used by traditional mobile communications network may not be efficient enough to build differentiated E2E security for different services. With the purpose of offer differentiated quality of service, networks may need to sense which type of service a user is using. The service type sensing may involve user privacy. Alternatively, data mining can also easily retrieve the user privacy information with its advanced technologies. Hence, the privacy protection in 5G is more challenging.

With an aim to protect the user privacy, service providers have to come out with their 5G security mechanisms. This can prevent the privacy information leakage when users and vertical industries utilized the 5G network. The 5G security mechanism includes [17]:

- Usage management of privacy information

  The 5G network provides customized network services for users by sensing their service features. Yet, privacy information, such as user location, health information and personal information, may be utilized in the service-type sensing process. A service sensing rule must be clearly defined for the 5G network to address users' concerns about privacy. The method must stipulate how the privacy information is used and how it is handled after being used. For example, the privacy policy and user agreement must be issued by user before the third party starts to proceed or access any user private data.

- 5G E2E encryption

  With data mining technologies, a third party or hacker may be able to derive detailed user privacy information. The information leakage risk decreases when telecommunication networks provide 5G E2E encryption that includes:

  1. Mutual authentication

     The end users of the 5G system are authenticated to support charging for network access, accountability (e.g. which user had which IP address and when), and lawful intercept. The network is also authenticated towards the end users so that the end users know that they are connected to a legitimate network.
  2. Security algorithm and protocols implementation

     5G system needs to acquire support from security algorithm expert group of European Telecommunications Standards Institute (ETSI), specifically Security Algorithms Group of Experts (SAGE). For IP layer and earlier, it is suggested to implement well-proven IETF security protocols. As a result, the 5G system is able to provide reliability and robustness against non-malicious unavailability situations, for instance, errors that appear due to unusual but expected bad radio conditions and broken links.

24.4 Summary

5G network introduces revolutionary new technology that will create new business opportunities and change market dynamics. The 5G platform will expand the network ecosystem, bringing new stakeholders, investors and partners into the service provider marketplace. Alongside the standard bodies and others that are helping to define and design 5G, service providers or operator have working on a variety of approaches and technologies to ease the evolution from 4G LTE to 5G. The full 5G system includes eMBB, uRLLC and mMTC which provides greater data bandwidth, 1 ms E2E ultra-reliable latency with massive number of IoT devices. However, high capital expenditures (CAPEX) and operating expenses (OPEX), poor backhaul network and user privacy unprotected issues become key challenges for 5G network development. Few solutions have been proposed to overcome the 5G challenges to enhance the 5G network deployment.

References

[1] Hootsuite, Digital 2020 Global Digital Overview, January 2020, 1–247.

[2] Zikria, Y.B., Yu, H., Afzal, M.K., Rehmani, M.H., and Hahm, O. Internet of Things (IoT): Operating system, applications and protocols design, and validation techniques. *Future Gener. Comput. Syst.* 2018, 88, 699–706.

[3] Fettweis, G.P. The tactile internet: Applications and challenges. *IEEE Veh. Technol. Mag.* 2014, 9, 64–70.

[4] Simsek, M., Aijaz, A., Dohler, M., Sachs, J., and Fettweis, G.P. 5G-enabled tactile internet. *IEEE J. Sel. Areas Commun.* 2016, 34, 460–473.

[5] International Telecommunication Union (ITU). *IMT Vision—Framework and Overall Objectives of the Future Development of IMT for 2020 and Beyond M Series Mobile, Radiodetermination, Amateur and Related Satellite Services*. International Telecommunication Union (ITU), Geneva, Switzerland, 2015.

[6] ITU-R M.2083: IMT Vision – Framework and overall objectives of the future development of IMT for 2020 and beyond, September 2015.

[7] IHS Markit, 5G best choice architecture: White Paper.

[8] GSMA, Road to 5G: Introduction and migration, April 2018.

[9] 3GPP, RP-172115, Revised WID on new radio access technology; 2017. https://www.3gpp.org/ftp/TSG_RAN/TSG_RAN/TSGR_77/Docs/RP-172115.zip

[10] 3GPP specification series: 38series. 3GPP. Retrieved 22 February 2019. https://www.3gpp.org/DynaReport/38-series.htm

[11] Choudhury, R.R., A network overview of massive MIMO for 5G wireless cellular: System model and potentials, *Int. J. Eng, Res, General Sci*, 2014, 2 (4), 338–347.

[12] GSMA, 5G implementation guidelines, July 2019.

[13] Wan, L., Guo, Z., and Chen X., Enabling efficient 5G NR and 4G LTE coexistence, *IEEE Wireless Commun*. 2019, 26(1), 6–8.

[14] 3GPP, R1-1706905. Overview of NR UL for LTE-NR coexistence, 3GPP TSGRAN WG4-NR, June 2017; http://www.3gpp.org/ftp/TSG_RAN/WG1_RL1/TSGR1_89/Docs/R1-1706905.zip

[15] 3GPP Tech. Spec. 38.101-1, User equipment (UE) radio transmission and reception; https://www.etsi.org/deliver/etsi_ts/138100_138199/13810101/15.03.00_60/ts_13810101v150300p.pdf

[16] ITU, Setting the scene for 5G: Opportunities and challenges, 2018. http://www.itu.int/dms_pub/itu-d/opb/pref/D-PREF-BB.5G_01-2018-PDF-E.pdf

[17] Huawei, 5G security: Forward thinking Huawei White Paper, 2015.

Chapter 25
Industry 4.0 and SMEs

Habtom Mebrahtu[1]

Small- and medium-sized enterprises (SMEs) are the most significant contributors to the manufacturing economy in the UK and Europe. However, unlike the big multinational companies, they typically have limited resources and usually lack the capability to invest in new and emerging technologies. Studies indicate that UK industries, especially SMEs, face challenges and also opportunities in their quest to adopting technologies leading to the fourth industrial revolution, Industry 4.0.

For the UK manufacturing industry to remain competitive, it is essential that the potential of Industry 4.0 for growth and increased productivity is seriously embraced by SMEs. As the rate of adoption of new technologies accelerates, the UK SMEs cannot afford to lose their competitive advantage to the more advanced competitors. The UK already has high-performing manufacturing sectors in the application of digitization that it has the potential to be a leader in Europe in digital manufacturing. As an example, the UK has the strongest artificial intelligence and machine learning market in Europe, with over 200 SMEs in the field [Made Smarter Review (2017), Department for business, energy & industrial strategy, available at https://www.gov.uk/government/publications/made-smarter-review]. Therefore, there is a strong need for developing smart methods and tools that could support SMEs in their transformation towards Industry 4.0.

This chapter looks at the opportunities and challenges that SMEs face in adopting Industry 4.0 and their readiness for this fourth-generation industry. It also highlights some potential tools that could help SMEs on their move towards Industry 4.0.

25.1 Introduction

The manufacturing industry has gone through numerous changes in the past two centuries, each of which radically changed the nature and capabilities of the manufacturing industry, while generally yielding superior products and leading to more efficient manufacturing processes and more productive manufacturing [1]. After the first three revolutions involving steam power, mass production with electricity and computer and automation, respectively, the fourth revolution involves the widespread use of data and machine learning to help further automate and digitize processes.

[1]School of Engineering and the Built Environment, Anglia Ruskin University, Chelmsford, UK

Figure 25.1 The four industrial revolutions: (1) mechanization through water and steam power, (2) mass production and assembly lines powered by electricity, (3) computerization and automation and (4) smart factories and CPS (Source: Ian Wright, www.engineering.com)

Industry 4.0 capitalizes on the decades of digitization and adoption of automated manufacturing processes in industry, enhances their capabilities with the integration of cyber-physical systems (CPSs) such as Internet of Things (IoT), Big Data, simulation and digital system integration and improves processes with data-driven machine learning [2].

The adoption of these technologies in the manufacturing sector is already underway and large corporations are utilizing the power of data-driven analytics and machine learning to improve their production efficiency. Organizations not utilizing these technologies will be prone to losing their competitive advantage to the more advanced competitors. Unlike large corporations, SMEs often lack funds and the capability to invest in immerging technologies and are therefore at risk of losing their market share [3].

Manufacturing contributes over £6.7 trillion to the global economy and makes a significant contribution to UK PLC. UK manufacturing is 9.8% of the UK economy (£162 billion gross value added in 2015) and the UK is still in the top 10 largest manufacturing nations (ninth in 2016/2017) and fourth largest in the EU [4]. It employs 2.6 million people and more likely 5.1 million people across the manufacturing value chain. UK exports of goods produced by the manufacturing sector totalled £257 billion in 2015 and were 50% of all UK exports. It accounts for 70% of business R&D and 14% of business investment.

According to Cordes and Stacey [5] and UK Industrial Digitisation Review [4], it has been recognized that by leading industrial digitization, the UK has the potential to realize industrial efficiency gains of up to 25%, manufacturing sector growth of 1.5–3% and delivering annual growth of ~0.5% of GDP. SMEs make the vast majority of businesses in the UK and are significant contributors to the UK economy. It is forecasted that the contribution of SMEs could reach £217 billion by 2020 [6], and amount to 70% of all employment in the UK.

Considering their significance to the UK economy, it is imperative that SMEs have work on adopting Industry 4.0 sooner than later if they are to maintain their competitive advantage nationally and internationally.

25.2 Opportunities and challenges of adopting Industry 4.0 by SMEs

Adopting Industry 4.0 brings opportunities and challenges to SMEs in the UK that could be internal or external. This section looks at some of the common challenges and opportunities.

25.2.1 Opportunities

The UK Industrial Digitisation Review Interim Report [4] emphasized that the faster adoption of industrial digital technologies (IDTs) can become a key driver of improved productivity, global competitiveness and increased exports. A study identified that by fully applying these technologies over the next decade, UK industrial production would be up to 30% faster and 25% more efficient. The strength in areas such as artificial intelligence, blockchain, virtual reality and machine learning creates the opportunity for the UK to be a leader in the development of the digital technologies.

Owing to the fast and significant advancement in IT-related technologies, the *cost of components* such as sensors and cloud infrastructure has significantly dropped in the past few years. This has created an environment where advanced technologies could be integrated to a manufacturing industry at reasonable cost. In recent years, it has been possible to miniaturize embedded systems and put them in a chip [7]. Their performance has significantly improved and they can be equipped with an IP address and modern communication interfaces, integrated to the internet and become part of the CPS.

The vast amount of data generated by production facilities and market analytics everyday make it difficult for SME leaders to keep track of them accurately, and use the data for making most *profitable decisions*. By adopting Industry 4.0, SMEs can make their production processes more efficient and contribute positively to their profit margins. Furthermore, utilizing Industry 4.0 technologies would make it easier for SMEs to monitor their production line more closely and maintain their quality [3].

Looking at the changing market needs, there is a perceived growing demand for the *personalization/customization of products*. Technologies such as additive manufacturing fundamentally change the supply chain, where advantages afforded by high volumes and low labour costs are replaced by proximity to market and the opportunities for products to be unique to each customer. Furthermore, smart automation can achieve this through digitally connected factories that can deliver great variation for customers still at reasonably high volumes, with simultaneously increased speed and reduced waste.

Industry 4.0 technologies could provide *performance optimization* that can be massively beneficial to SMEs. With digital connection all the processes are constantly monitored and apparent problems and redundancies could be quickly removed from the manufacturing process. Additionally, the use of IoT to communicate sections of the manufacturing firm is particularly helpful to accelerate decision-making in the production process.

A significant *support* is forthcoming *from government and UK organizations* with particular focus on SMEs. For instance, Innovate UK and its partners that include the Knowledge Transfer Network (KTN), European Enterprise Network (EEN), Research Councils (RC) and Digital Catapult (DC) are among the forefront UK organizations that support SMEs to adopt Industry 4.0. In July 2019, the government has launched a £30 million R&D competition for projects aimed at radically boosting the productivity and agility of UK manufacturing through digital technologies [8]. In 3 years leading to 2017, Innovate UK has invested over £200 million in transformative digital technologies such as additive manufacturing, robotics and autonomous systems, modelling and simulation, augmented and virtual reality, data analytics and artificial intelligence (AI) [9].

25.2.2 Challenges

The SMEs operating in the manufacturing sector comprise of the largest industry in the UK. However, only around 10% of these firms were utilizing Industry 4.0 standard facilities in their plant machinery and engineering systems [10]. Studies showed that there is a considerable relationship between the size of a firm and their chances of adopting Industry 4.0 technologies. SMEs are less likely to integrate emerging technologies in sophisticated manner in their production facilities and IT systems. Some of the reasons are discussed as follows.

Lack of appropriate *digital strategy* for the adoption of Industry 4.0 by SMEs is one of the leading challenges. SMEs would need to come up with strategies to work out the cost–benefit analysis of the required technologies in carrying out the feasibility in their organization.

There is no appropriate landscape of *business support* with a route to access help and guidance on best practices to adopt Industry 4.0 technologies. The UK is a leader in research and innovation and has started to establish favourable support infrastructure to develop and commercialize technology. However, these innovation assets are under-leveraged and not focused enough on supporting IDT start-ups, meaning the UK is falling behind in creating new innovative companies and industries[4].

Despite all the advancement the Industry 4.0 technologies offer, SMEs are still relatively *cautious* about implementing them in their facilities. This can be attributed to several concerns among the stakeholders, the main being the lack of security standards underlying the technology that could consequentially compromise a large amount of data.

There is lack of *digital standard*s underlying the digital technology and networking. Currently, SMEs mostly adapt to the set of protocols and standards used by their larger companies to which they supply to increase the compatibility of their systems to help exchange data and information smoothly. However, a lack of common standards and norms makes it difficult for SMEs to adopt technologies that may not be compatible with other companies that use different sets of standards.

In addition to setting standard protocols and practices, there are no adequate *cyber security* measures to guard data used within an Industry 4.0 system. Vital company data such as operational data, procedures for crafting patented products and confidential company policies could all be vulnerable if proper security system is not installed. Therefore, the need for developing a common set of standards and security protocols to be used by organizations adopting Industry 4.0 could not be overemphasized.

SMEs usually employ fewer people and they are likely to be affected by *skills gap*. There is shortage of access to people with the right skills that could sustain the fast-moving technological changes. This is particularly more apparent in SMEs that have lesser technical expertise and knowledge as most of them lack a proper dedicated IT department.

Business model is one of the key essentials for business processes to ensure that adequate amount of inspiration is drawn in representing the creative notion of a firm. To sustain a competitive nature, it is vital for firms to consider business model innovation (BMI) as part of their key activities [11].

Other challenges include the high *cost of investments* involved in adopting the technologies and the lack of finance to invest in R&D related to adoption of Industry 4.0.

25.3 Readiness of SMEs to adopt Industry 4.0

A report from the Engineering Employers' Federation (now known as 'Make UK'), on the fourth Industrial Revolution – a primer for manufacturing found that while 42% of UK manufacturers have a good handle on what Industry 4.0 will entail, only 11% think that the UK is geared up for this crucial next industrial age. This suggests that in terms of being 'Industry 4.0 ready' there is still some way to go [12]. It adds, given that SMEs make up half of the UK's manufacturing capacity, it is critical that they embrace Industry 4.0 if it is to be successfully implemented. There are several tools available that could help firms to assess their readiness for Industry 4.0. This section will discuss three of the prominent tools: PwC Industry 4.0 self-assessment tool, Region of Smart Factory (RoSF) and WMG Industry 4.0 readiness assessment tool. One more assessment tool that is specifically geared for SMEs will be discussed in the next section.

25.3.1 PwC Industry 4.0 self-assessment tool

This self-assessment tool is developed by Price waterhouse Coopers (PwC) and is available online at i40-self-assessment.pwc.de [13]. It is based on a survey of over 2,000 companies from 26 countries and is designed to provide companies with an understanding of their position with regards to Industry 4.0 by measuring their maturity along seven dimensions, thereby identifying needs for action and classifying the current maturity. The maturity level will enable companies to benchmark their position against competitors in their sector. The online assessment tool, based on Figure 25.2, provides a guided set of questions and finally outputs the level of maturity of the company in a form of charts and comments. The maturity levels are briefly described as follows.

	1 *Digital novice*	2 *Vertical integrator*	3 *Horizontal collaborator*	4 *Digital champion*
Digital business models and customer access	First digital solutions and isolated applications	Digital product and service portfolio with software, network (M2M) and data as key differentiator	Integrated customer solutions across supply chain boundaries, collaboration with external partners	Development of new disruptive business models with innovative product and service portfolio, lot size 1
Digitization of product and service offerings	Online presence is separated from offline channels, product focus instead of customer focus	Multichannel distribution with integrated use of online and offline channels; data analytics deployed, e. g. for personalization	Individualized customer approach and interaction together with value-chain partners. Shared, integrated interfaces.	Integrated customer journey management across all digital marketing and sales channels with customer empathy and CRM
Digitization and integration of vertical and horizontal value chains	Digitized and automated subprocesses. Partial integration including production or with internal and external partners. Standard processes for collaboration partly in place	Vertical digitization and standardized and harmonized internal processes and data flows within the company; limited integration with external partners	Horizontal integration of processes and data flows with customers and external partners, intensive data use through full integration across the network	Fully digitized, integrated partner ecosystem with self-optimized, virtualized processes, focus on core competency, decentralized autonomy. Near real-time access to extended set of operative information
Data and analytics as core capability	Analytical capabilities mainly based on semi-manual data extracts; selected monitoring and data processing, no event management	Analytical capabilities supported by central business intelligence (B) system; isolated, not standardized decision support systems	Central BI system consolidating all relevant internal and external information sources, some predictive analytics. Specific decision support and event management systems	Central use of predictive analytics for real-time optimization and automated event handling with intelligent database and self-learning algorithm enabling impact analysis and decision support
Agile IT architecture	Fragmented IT architecture in-house	Homogeneous IT architecture in-house. Connection between different data cubes developing	Common IT architectures in partner network. Interconnected single data lake with high-performance architecture	Single data lake with external data integration functionalities and flexible organization. Partner service bus, secure data exchange
Compliance, security, legal and tax	Traditional structures, digitization not in focus	Digital challenges recognized but not comprehensively addressed	Legal risk consistently addressed with collaboration partners	Optimizing the value-chain network for compliance, security, legal and tax
Organization, employees and digital culture	Functional focus in 'silos'	Cross-functional collaboration but not structured and consistently performed	Collaboration across company boundaries, culture and encouragement of sharing	Collaboration as a key value driver

Figure 25.2 Industry 4.0 development dimensions (Source: [13])

Digital novice: At this maturity level, some initial digitization has been achieved in department as well as in products and services but the activities are not well coordinated and geared for the future. The portfolio is typically dominated by physical products and there is limited integration within the vertical and horizontal value chains. Digital risks are not systematically recorded and compliance is not guaranteed in all areas.

Vertical integrator: Companies at this maturity level have already given their product and service portfolio digital functions. The operative processes and some administrative processes are digitized. For example, data from product development are available for physical production, logistics and systems in the company.

Horizontal collaborator: Here collaborators integrate their value chains with customers and partners. The product and service portfolio is connected with external value-chain partners to the extent that customers are offered end-to-end solutions across several steps of the value chain. Innovative concepts optimize customer communication, and customer information is saved and analysed for optimal communication. Digital risks are managed with standardized and efficient method and compliance is maintained for all functions at the company.

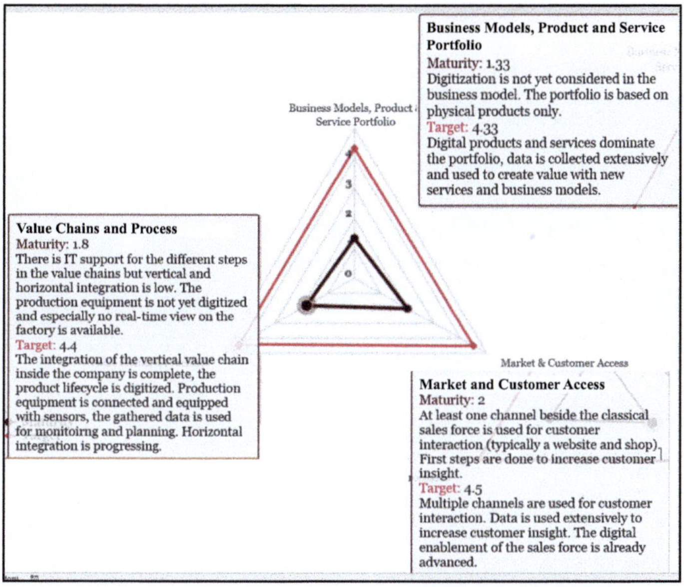

Figure 25.3 Summary output of PwC tool assessment

Digital champion: The digital champion has connected its operational and administrative processes on global scale and will also have virtualized the processes in the main areas. It focuses on the key areas and works with global network of partners. The key administrative processes are digitized and globally optimized according to cost and control criteria.

The assessment will produce a summary output result similar to Figure 25.3 with the tooltips describing the results for each category. This could be followed up by benchmarking the company against others in their sector.

25.3.2 RoSF Smart Factory assessment tool

This tool is developed by the RoSFs launched by the University of Groningen in the Netherlands and is available online at http://www.smartfactoryassessment.com/assessment/#/en [14]. The Smart Factory assessment tool is designed to enable manufacturing firms to assess their own position on how smart their factory is. Although it is a general framework that needs to be customized for individual companies, it is broadly recognized as a very useful guidance to SMEs in their transformation towards a smart factory. It is based on the competencies as shown in Figure 25.4.

This online tool gives five questions/statements on each competency for the company to choose two from (where they currently are and where they want to be). When all the competencies have been chosen, the company is given the opportunity to indicate how relevant each competency is for the company. The system will then

NORTHERN NETHERLANDS REGION OF SMARTFACTORIES

SMART FACTORY ASSESSMENT 2.0

MAKE YOUR FACTORY A SMART FACTORY

Phases ▶ / Competences ▼	Traditional	Advanced	Advanced	Smart Factory now	Smart Factory of the future
Revenue model What is your revenue model?	*Sell & leave.* Customer purchases your product. Any add-on services (such as maintenance or modifications) will be billed separately.	*Sell & care.* Customer purchases your product including a Service Level Agreement, thereby guaranteeing availability.	*Sell & share.* Customer purchases your product together with a Service Level Agreement based on shared actual usage data.	*Product as a service.* Customer pays to use your product, which is managed by you on the basis of real-time usage data (e.g. operational lease).	*Function as a service.* Customer pays for the functionality, which you supply based on the best available technology.
Product intelligence How 'smart' are your products and/or services?	*Modifiable product:* The product can only be modified or (re-)set locally by specialists, within its defined functionality.	*Configurable product:* The product can be modified or reconfigured locally by the user to the specific functionality requirements.	*Connected product:* The product can be monitored and modified online via the IoT (remote support), but this process is not automated.	*Intelligent product:* The product itself detects deviations, mainly based on central data and intelligence via IoT. Intervention is still manual.	*Autonomous product:* The product continuously adapts itself to changing functionality, partly based on central data and intelligence via the IoT.
Product development process How is the development of products and manufacturing processes organised?	Ad-hoc, triggered by incidents (complaints, good ideas), not standardised, based on Trial & Error. Without use of tools.	Plan-based, based on Trial & Error, but professionally executed. So, standardised, documented, utilising product-models, etc.	Plan-based, through systematic empirical research based on structured data. Supported by advanced tools, e.g. Design of Experiments (DoE).	Continuous and digitalised process based on model-based system engineering. Validation by prototypes.	Continuous process organised around self-validating models, self-learning products and Digital Twins. So, first time right products and processes.
Connectivity How well are different functions connected and aligned internally/ externally?	Internal processes are only partially (digitally) connected. So, no integrated control.	At company level, all processes (planning, shop floor, etc.) are digitally connected, but without integrated control.	At company level: 100% internal connectivity, integrated operations and paperless.	Connected Supply Chain. All actors (such as suppliers and products) are digitally linked, directed by the leading company. So, hierarchal.	All players are optimally integrated in the Cyber Physical System, which is decentrally (organically) managed.
Flexibility How flexible is your organisation?	Processes are set in stone. Modifications are not realistic due to high costs.	Rigid. Modifications are complex and must be manually applied by external specialists.	Flexibly automated. Modifications are relatively simple to apply by own people.	Flexibly automated. Modifications are autonomously applied.	No limits. Self-learning, organic production system.
Production technology How 'smart' are the production tools which are used?	Production tools are manually set-up and programmed. Optimisations can only be applied in an ad-hoc manner or in case of long-running series/articles. Requires expert help.	Programs are manually loaded, after which the machine takes over the configuration. Process optimisation is plan-based, but analysis/insights from experts are leading.	Production tools set themselves up, programming is fully automatic, based on available digital product information. Process optimisation is carried out by experts, based on available process data.	Production tools set themselves up, programming is fully automatic based on the available digital product information. Process optimisation is reactive but systematically via standard built-in tools.	Autonomous machines and means of transport are able to recognise products themselves, set themselves up based on the available digital product information and optimise themselves continuously and in real-time.
Operations How are operations managed and controlled?	*Fire & forget:* planning is periodically updated. Major deviations are a daily occurrence. Process is unpredictable.	Operations based on feedback, but not in real-time. Unacceptably large deviations are prevented manually, as far as that is possible.	*Realtime operations management,* so, fast, reliable and flexible. But analysis and adjustment has to be performed manually by specialists.	Real-time and adaptive via human intervention, deviations are automatically signalled.	Real-time, autonomous and self-learning operations across the entire supply chain.
Data management How is data managed and what role does data play within your business?	*Transient data.* Basic process data is gathered locally in an ad-hoc way, and only used for the adjustment of processes. Work for specialists.	*Data based.* Basic process data is retrieved locally on a project-based (ad-hoc) basis and is analysed to solve specific problems (reactive). Primarily to support insights from specialists.	*Data driven.* Data is retrieved structurally per process and centrally stored. Data analyses takes place periodically, on an ad-hoc (trigger based) basis, or proactively, with the intention of strengthening the organisation. So it is not only about solving problems.	*Intelligence based.* Fully automated collection and analysis of high quality data. Central intelligence (AI) reports on any deviations and the causes behind them. Corrections are agreed and applied by specialists on a per machine basis. People make decisions based on Artificial Intelligence.	*Intelligence driven.* Autonomous and connected. Data acquisition, analysis and feedback (deviations from the norm) is performed autonomously for all business processes, and linked to other echelons in the supply chain. (AI @Cyber Physical System is next phase).
Quality How is quality checked and guaranteed?	Subsequent offline or end of line quality control.	Sample-based, inline quality control.	100% inline control, Corrections are applied to following product/batch.	Full inline control. Corrections are applied to running product/batch.	Feed forward control, model-based and online. So, zero defect.
Maintenance How is maintenance carried out?	*Corrective and preventative maintenance.* Repairs rather than maintenance, preventative servicing at fixed intervals for critical process components.	*Preventative maintenance.* Over all sub-processes, based on actual use and Service Level Agreements.	*Condition based maintenance,* on the most critical process components.	*Real-time condition-based maintenance.* Based on real-time condition monitoring over all sub-processes.	*Predictive maintenance.* Based on full-scale real-time condition monitoring and central intelligence.

Figure 25.4 Smart Factory assessment competencies (Source: [14])

generate a profile ranked according to relevance to the company in a form of marked table similar to that shown in Figure 25.5.

25.3.3 WMG Industry 4.0 assessment tool

This Industry 4.0 readiness assessment tool is developed by the Warwick Manufacturing Group (WMG) in conjunction with their industrial collaborators. It is based on a survey

Figure 25.5 Summary results of Smart Factory assessment

with 53 responses from 22 countries and is available online at https://i4ready.co.uk/current-readiness.

Its purpose is to provide a simple and intuitive way for companies to start to assess their readiness and future ambition to harness the potential of the cyber-physical age. It is more comprehensive that the PwC tool and Smart Factory, and it looks at six core dimensions with 37 subdivisions that would assess Industry 4.0 readiness. The core dimensions include

- Products and services
- Manufacturing and operations
- Strategy and organization
- Supply chain
- Business model
- Legal considerations.

The assessment in each dimension is designed around four maturity levels: beginner, intermediate, experienced and expert. Similar to the former tools, each core dimension comprises a set of competencies. For illustration purposes, two of the core dimensions with their competencies, as extracted from [15], are shown in Figures 25.6 and 25.7.

The outcome of assessment, as personalized readiness report, will then be shown in a chart clearly showing where the company's state is compared with the state of Industry 4.0. As an example, a typical report for the Products and Services dimension would include overall benchmark bar chart and an individual position against the

Readiness level	Level 1 Beginner	Level 2 Intermediate	Level 3 Experienced	Level 4 Expert	Don't Know	Not Relevant
Product customization	Product allows for no individualization, standardized mass production	Majority of products are made in large batch sizes with limited late differentiation	Products can be largely customized but still have standardized base	Late differentiation available for most make-to-order products (batch size 1)	Don't know	Not relevant
Digital features of products	Products show only physical value	Products show value only from intellectual property licensing	Products exhibit some digital features and value from intellectual property licensing	Products exhibit high digital features and value from intellectual property licensing	Don't know	Not relevant
Data-driven services	Data-driven services are offered without customer integration	Data-driven services are offered with little customer integration	Data-driven services are offered with customer integration	Data-driven services are fully integrated with the customer	Don't know	Not relevant
Level of product data usage	Data is not used	0–20% of collected data is used	20–50% of collected data is used	More than 50% of collected data is used	Don't know	Not relevant
Share of revenue	Data-driven services account for an initial share of revenue (<2.5%)	Data-driven services account for a moderate share of revenue (2.5–7.5%)	Data-driven services account for a significant share of revenue (7.5–10%)	Data-driven services play an important role in revenue (>10%)	Don't know	Not relevant

Figure 25.6 Competencies for 'products and services'

Readiness level	Level 1 Beginner	Level 5 Intermediate	Level 3 Experienced	Level 4 Expert	Don't Know	Not Relevant
Automation	Few machines can be controlled through automation	Some machines and system infrastructures can be controlled through automation	Most machines and system infrastructures can be controlled through automation	Machines and systems can be controlled completely through automation	Don't know	Not relevant
Machine and operation system integration (M2M)	Machines and systems have no M2M capability	Machines and systems are to some extent interoperable	Machines and systems are partially integrated	Machines and systems are fully integrated	Don't know	Not relevant
Equipment readiness for Industry 4	Significant overhaul required to meet Industry 4.0 model	Some machines and systems can be upgraded	Machines already meet some of the requirements and can be upgraded where required	Machines and systems already meet all futrue requirements	Don't know	Not relevant
Autonomously guided work pieces	Autonomously guided work pieces are not in use	Autonomously guided workpieces are not in use, but there are pilots underway	Autonomously guided workpieces used in select-ed areas	Autonomously guided workpieces are widely adopted	Don't know	Not relevant
Self-optimising processes	Self-optimization processes are not in use	Self-optimizing processes are not in use, but there are pilots in more advanced areas of the business	Self-optimizing processes are used in selected areas	Self-optimizing processes are widely used	Don't know	Not relevant
Digital modelling	No digital modelling	Some processes use digital modelling	Most processes use digital modelling	Complete digital modelling used for all relevant processes	Don't know	Not relevant
Operations data collection	Data is collected manually when required, e.g. sampling for quality control	Required data is collected digitally in certain areas	Comprehensive digital data collection in multiple areas	Comprehensive automated digital data collection across the entire process	Don't know	Not relevant
Operations data usage	Data is only used for quality and regulatory purposes	Some data is used to control processes	Some data are used to control and optimize processes, e.g. predictive maintenance	All data is used not only to optimize processes, but also for decision-making	Don't know	Not relevant
Cloud solution usage	Cloud solutions not in use	Initial solutions planned for cloud-based software, data storage and data analysis	Pilot solutions implemented in some areas of the business	Multiple solutions implemented across the business	Don't know	Not relevant
IT and data security	IT security solutions are planned	IT security solutions have been partially implemented	Comprehensive IT security solutions have been implemented with plans developed to close any gaps	IT security solutions have been implemented for all relevant areas and are reviewed frequently to ensure compliance	Don't know	Not relevant

Figure 25.7 Competencies for 'manufacturing and operations'

benchmark spider web chart as shown in Figure 25.8. The full detail of the WMG assessment tool along with instructions could be found at https://i4ready.co.uk/ [15].

The three Industry 4.0 readiness assessment tools discussed earlier could only help companies to identify their position compared with the Industry 4.0 standard, but they do not assist companies on how to close the gap. Additionally, as important as they are, they may be designed and written based on primarily large industries that their main focus may not have been on SMEs.

25.4 Helping SMEs to adopt Industry 4.0

Industry 4.0 has the potential to create impressive, new and sometimes unimaginable business opportunities for those who are innovative and agile but SMEs need significant support from government organizations and knowledge institutions to realize it. In addition to the support available from organizations such as Innovate UK and its partners, a number of national and European projects are also seen to

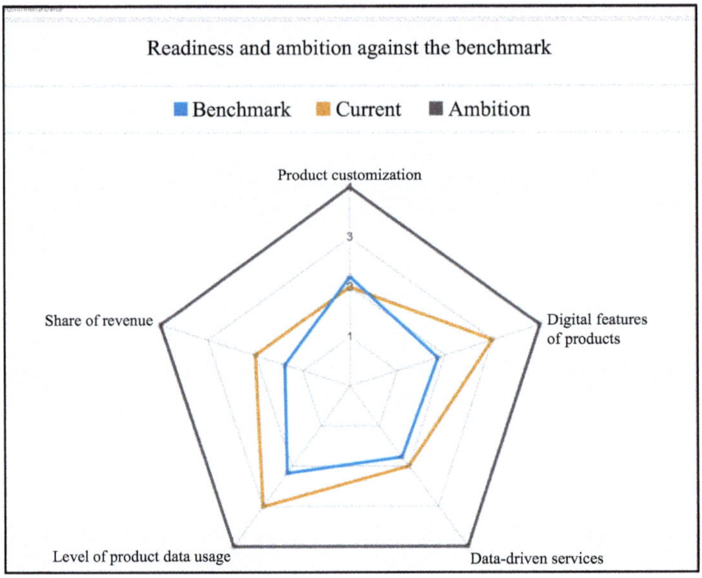

Figure 25.8 Example summary readiness chart from WMG assessment tool

work in supporting SMEs on their move towards Industry 4.0. One notable inter-regional European project dedicated to supporting SMEs in the North Sea Region (NSR) is GrowIn 4.0. The aim of GrowIn 4.0 is specifically to assist SMEs in their transformation towards Industry 4.0 [16].

25.4.1 GrowIn 4.0

The Interreg NSR project, GrowIn 4.0, aims to build strong competences and tools in the participating North Sea regions for the benefit of manufacturing SMEs. The overall objective is to raise the level of innovation and to create more growth within manufacturing SMEs who are heading for Industry 4.0. The EU Commission fears that the European SMEs are falling behind their American counterparts in the implementation of Industry 4.0 [17,18]. The GrowIn project is working within the following three work packages:

New business models and strategy development: The focus of this work package is to strengthen the business strategies and business models of manufacturing SMEs in the context of the opportunities and challenges of Industry 4.0. This is an important precursor that will enable SMEs to assess, improve and adapt their innovation strategy. It will also help them understand how to align their capabilities to their market environment and which aspects of Industry 4.0 will be important for sustaining their competitive advantage.

Better use of technology and development of products: The focus of this work package is in particular to transfer knowledge with Industry 4.0 technologies between knowledge institutions and SMEs. This is necessary in many respects. For

SMEs, the risk is particularly high to use new technologies without having any experience with them. On the other hand, it is indispensable for the competitor to remain technologically at a progressive level.

Training, education and recruitment of Industry 4.0 qualified staff. The focus of this work package is to increase the level of skills and knowledge needed for implementation of Industry 4.0 in SMEs. Technology associated with Industry 4.0 such as Big Data, IoT, Robotics, Additive Manufacturing offers many opportunities for manufacturing SMEs. There is a special focus on how to help the SMEs to develop the required human capital: workers with the right knowledge and competences to apply Industry 4.0 technologies.

To sum up, the NSRs cooperating in this project want to develop a regional approach to help their individual SMEs, their owners, their workers and the entire regional economy to respond to changing opportunities and jointly reap their economic and employment benefits. Around 14 tools are being developed and designed specifically to support SMEs in their move towards Industry 4.0. Most of the tools are still in development and will be made available to SMEs at the end of the project in 2021. A brief look at two of the tools that are still in the final stages of development – the Industry 4.0 readiness/awareness tool and the benefit identification tool – will be highlighted here. The readiness/awareness is expected to assess the SME readiness to Industry 4.0 and identify the main areas to work on whereas the Benefits Identification tool would help the SME to work out how to realize the benefits of the investment in new technologies.

25.4.2 Industry 4.0 readiness/awareness tool

This tool is similar to the three readiness assessment tools discussed in Section 25.3 and uses the PwC and WMG tools as the main sources. However, this is not as comprehensive as some of them and is designed to suit SMEs in its simplicity, time needed to complete the questionnaire and relevance. It is still under development but it has gone through the first phase test. It covers aspects of digitization, regulatory considerations and readiness of a company. It is grouped into five categories (dimensions): business models and products, market and customer access, value chains and processes, legal consideration and cyber/IT security and strategy and experience on Industry 4.0. It comprises a set of questions for each category (total 23 questions) with three possible answers for each question and also the possibility of choosing a button in between when their position is somewhere in between the two. The company would also indicate how relevant the question is to the firm by selecting from the relevance pull down list. As an example, the 'value chain and processes' page with the selected answers in green looks similar to the one shown in Figure 25.9.

When all the questions have been answered, a summary output for all the categories would then be shown in chart forms. For the earlier example, it would look like the one shown in Figure 25.10.

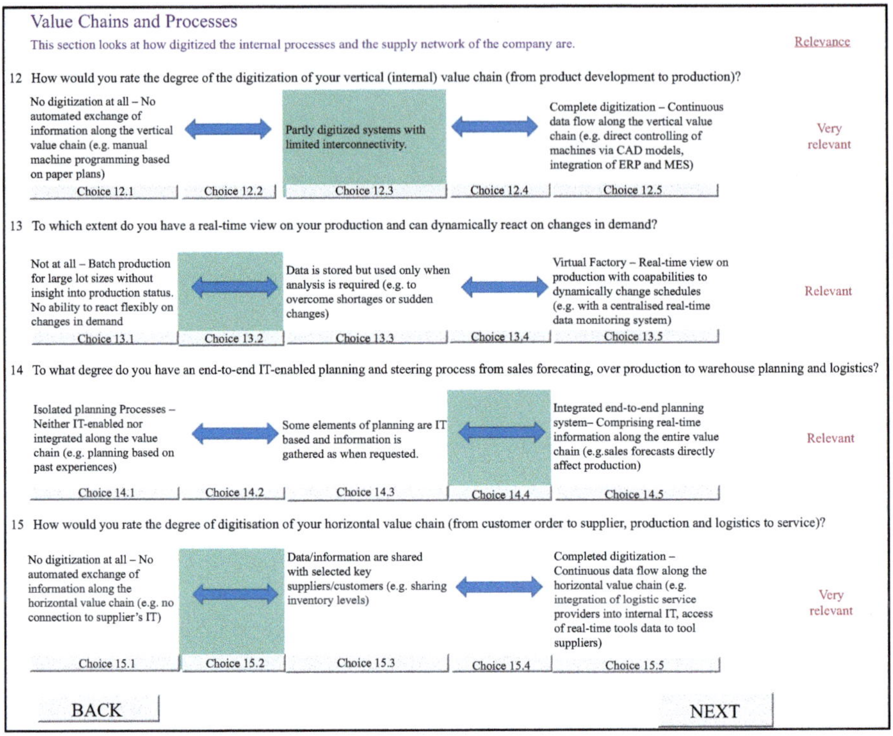

Figure 25.9 Example category page of the readiness/awareness tool

The first phase testing on the suitability of the tool is carried out on ten SMEs along with face-to-face discussions. The results, based on 1 as least favourable and 10 as most favourable, are shown in Figure 25.11.

Figure 25.11 shows that the awareness and understanding of Industry 4.0 by SMEs and their business partners are low and that the interaction with the tool helped to improve their awareness on aspects of Industry 4.0.

There were also free text comments from the companies that highlighted some supporting feedback and areas to work on. In general, all of the participant SMEs indicated that they are happy with the overall content of the questionnaire, its size and simplicity, and they found the tool very useful.

The tool will be developed further following the first test and again tested by a larger number of SMEs in the near future. The final version will then be made available to SMEs at the end of the GrowIn project in 2021.

25.4.3 Benefits identification

The readiness/assessment tool could only help SMEs to understand where they currently are relative to the Industry 4.0 and also to identify the areas they need to work on to transform to Industry 4.0. But they do not give answers on how to get there. Among the

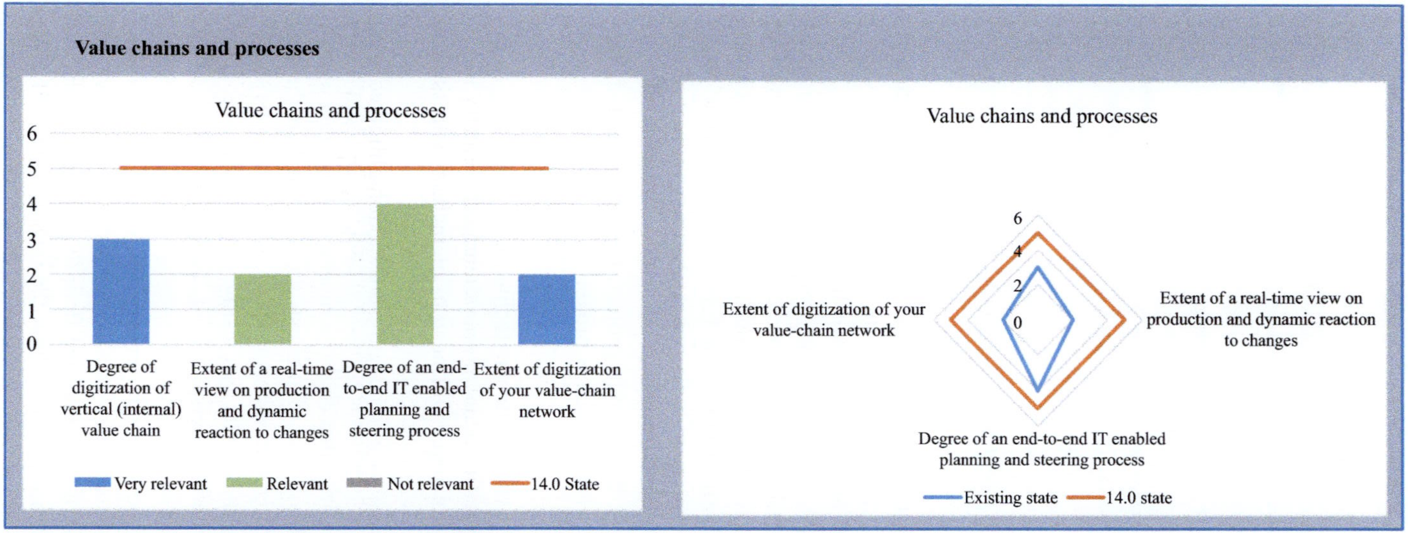

Figure 25.10 Example summary output for value chain and processes category

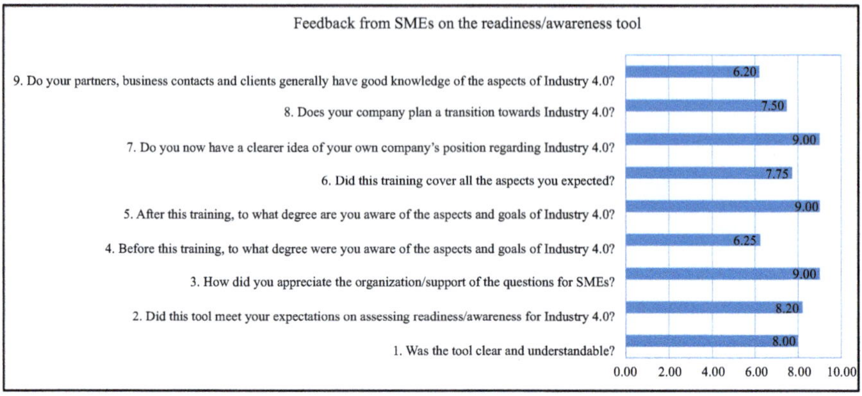

Figure 25.11 How SMEs found the readiness/awareness tool

set of tools that are being developed by the GrowIn project is Benefits Identification. As it is still under development, only a brief account will be highlighted here. It will be made available for the public in full at the end of the GrowIn project.

This tool is expected to help SMEs in their technology adoption decisions by looking at benefits and potential risks. It will require workshops, face-to-face engagement, mapping out benefits and disbenefits, and linking technology choices to strategic objectives. Benefit identification, as a formal process, is a recent addition to the project management battery of tools and can be a central part of both decision to adopt new technologies and managing the implementation of those technologies in the company [19]. In the GrowIn project, this is adapted specifically for SME manufacturing firms. Its approach is broken down into two discrete steps:

- Identifying the specific benefits that particular technologies can bring to the operation of the firm, with a specific focus on operational efficiency.
- Identifying what organizational changes will need to be made to access these potential benefits.

When organizing the process, the benefit identification (and realization) process could be viewed as three distinct activities that could be repeated in different phases. Following identification of the technologies of interest, the benefits that the firm should be exploiting would be identified through workshops. The workshops should ensure sufficient input and visualize the results as a visible map in real time. Later, these can be translated into benefits registers and benefits tracking tools (Figure 25.12).

When running a productive workshop, it is critical that senior managers and professionals are fully involved to ensure that projects are needs-driven. It is advisable that the workshop sessions involve staff from the shop floor to senior managers who could offer strategic or work-related insights into how the technologies will bring benefits to the company. Such diversity of views is important as a plethora of related,

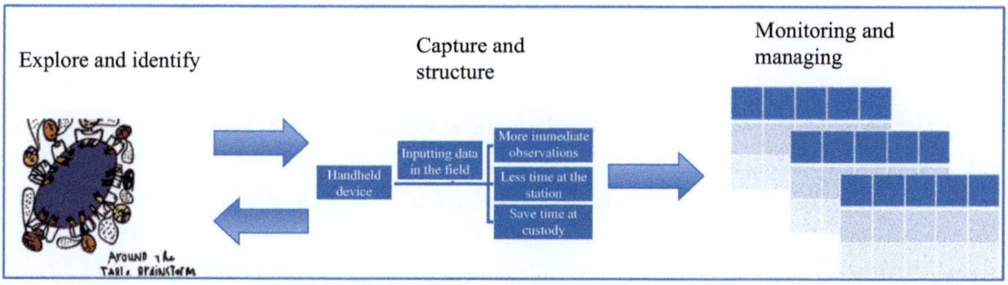

Figure 25.12 Benefits identification activities (Source: Ivory, C., Benefits identification, Unpublished)

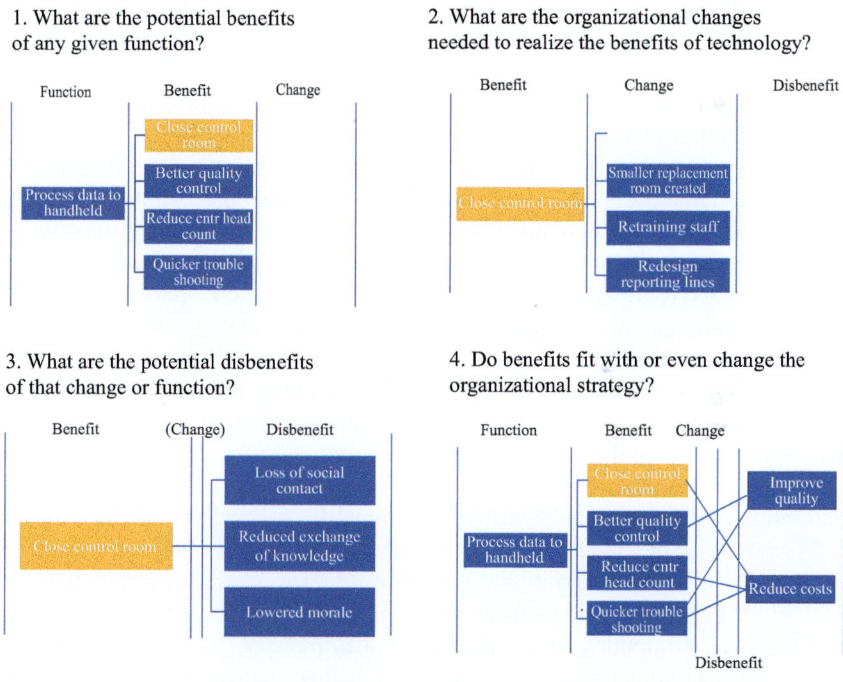

Figure 25.13 Example of mapping structure (Source: Ivory, C., Benefits identification, Unpublished)

yet intangible, factors often determine the level and type of benefits new technologies can bring.

The information garnered in each phase of the benefits identification/realization process will be mapped, so that any relationships between the information at different phases can be more easily identified. The mapping process would follow the basic linear structure – *technology functions*, *potential benefits*, *business changes needed*, *disbenefits* and *strategic benefits*. Each of these is populated and the emerging relationship and interactions are discussed in phases. As the map develops through the different phases, it should stimulate further discussion and contribution. Towards the end, the path should ultimately connect the outputs with the strategic objectives of the company. Examples of linkages that come out during the workshop would look like that shown in Figure 25.13.

It is to be noted that only a brief highlight of the Benefits Identification tool is discussed earlier. As previously stated, the full detail of the tool as adapted to SMEs transforming to Industry 4.0 will be made available at the end of the GrowIn project in 2021.

25.5 Summary

The significance of manufacturing SMEs' contribution to the UK economy is highly recognized but merely 10% of them are using Industry 4.0 standard facilities. Adopting Industry 4.0 brings opportunities and challenges to SMEs, but to maintain their competitive advantage they need to show stronger readiness and work closely with supporting government bodies and knowledge institutions.

The opportunities for SMEs to embrace Industry 4.0 include reduced cost of IT-related technologies, ease of accessing data and making profitable decisions, addressing the growing demand from customers for personalized products, good environment for performance optimization, and the emerging support from the UK government-linked organizations such as Innovate UK and its partners. SMEs also face some challenges that include lack of appropriate strategy and business plan, shortage of support, skills gap, cybersecurity concerns and lack of finance. Studies have also identified that there is a need for developing protocol frameworks and network infrastructures to help in creating a common but secure environment for compatible vertical and horizontal digital integration.

For SMEs to embrace the opportunities and overcome the challenges, governmental organizations, technology developers, professional bodies and knowledge institutions should work together and create an environment for secure, standardized technological systems and narrow the skills gap as required for Industry 4.0 transformation. It would be useful for SMEs to have good guidance and demonstration environments nationally, and easier access to information to stimulate technology transfer.

There are a number of self-assessment tools that companies could use to assess their readiness for Industry 4.0. Three of the existing online tools are the PwC Industry 4.0 assessment tool, the RoSF Smart factory and the WMG Industry 4.0 assessment tool. These tools are developed based on surveys from a number of companies and countries, and they are relevant for all types of manufacturing industries. Although their focus may not have been primarily on SMEs, they are still useful tools for a quick readiness test.

GrowIn 4.0 is one of the European projects aimed at building strong competencies and tools in the Nordic regions for the benefit of SMEs heading towards Industry 4.0. In this context, it focuses on new business models and strategy development, use of Industry 4.0 standard technology and development of products, and overcoming the skills gap through training and education. While trying to achieve this, it is working on developing several tools specifically intended to help SMEs in their transformation towards Industry 4.0. Most of the tools are still under development but, as examples, two tools – the Industry 4.0 readiness/awareness tool and the benefit identification tool – were discussed in this chapter. The readiness/awareness tool would assess the readiness of the SME and would give an indication of the areas the SME needs to work on to achieve Industry 4.0. Tools such as the Benefits Identification tool would then take over and help the SME to realize the implementation of new technologies for Industry 4.0.

References

[1] Popkova, E. G., Ragulina, Y. V., and Bogoviz, A. V. (2019). Fundamental differences of transition to Industry 4.0 from previous industrial revolutions. In *Industry 4.0: Industrial Revolution of the 21st Century* (pp. 21–29). Springer, Cham.

[2] Lee, J., Davari, H., Singh, J., and Pandhare, V. (2018). Industrial artificial intelligence for Industry 4.0-based manufacturing systems. *Manufacturing Letters*, 18, pp. 20–23.

[3] Stentoft, J., Jensen, K. W., Philipsen, K., and Haug, A. (2019). Drivers and barriers for Industry 4.0 Readiness and Practice: A SME perspective with empirical evidence. In *Proceedings of the 52nd International Conference on System Sciences*, January 2019, Hawaii.

[4] UK Industrial Digitisation Review (2017). [Online], available at https://www.tpdegrees.com/globalassets/pdfs/research-2017/idr_interimreport_july17.pdf (accessed on 1 October 2018).

[5] Cordes and Stacey (2017). Is UK Industry ready for the Fourth Industrial Revolution, BCG, [online], Available at https://media-publications.bcg.com/Is-UK-Industry-Ready-for-the-Fourth-Industrial-Revolution.pdf (accessed on 20 March 2020).

[6] Small Businesses (2019). [Online], available at https://smallbusiness.co.uk/sme-contributions-forecast-217billion-2534943/ (accessed on 17 July 2019).

[7] Schroder, C. (2017). The challenges of Industry 4.0 for small and medium-sized enterprises, [Online], available at https://www.researchgate.net/publication/305789672_The_Challenges_of_Industry_40_for_Small_and_Medium-sized_Enterprises (accessed on 25 July 2019).

[8] Department for Business, Energy & Industry, UK research Innovation, and The Rt Hon Greg Clark MP (2019). UK advanced manufacturing gets boost with new investment in digital tech competition, [Online], available at https://www.gov.uk/government/news/uk-advanced-manufacturing-gets-boost-with-new-investment-in-digital-tech-competition?utm_source=e8786557-1c20-46ef-8ade-1c46b119bab6&utm_medium=email&utm_campaign=govuk-notifications&utm_content=daily (accessed on 29 July 2019).

[9] McGregor, L. What does the fourth industrial revolution mean for UK business?, Innovate UK, [Online], available at https://innovateuk.blog.gov.uk/2017/03/28/what-does-the-fourth-industrial-revolution-4ir-mean-for-uk-business/ (accessed on 25 July 2019).

[10] Schumacher, A., Erol, S., and Sihn, W. (2016). A Maturity model for assessing industry 4.0 readiness and maturity of manufacturing enterprises, *Procedia CIRP*, 52, pp. 161–166.

[11] Pucihar, A., Lenart, G., Kljajic Borstnar, M., Vidmar, D., and Marolt, M. (2018). Drivers and outcomes of business model innovation-micro, small and medium-sized enterprises perspective. *Sustainability*, 11(2), pp.1–17.

[12] Eureka (2017). [Online], available at http://www.eurekamagazine.co.uk/design-engineering-features/technology/innovate-uk-wants-to-help-smes-take-more-immediate-action-on-industry-4-0-strategies/154425/ (accessed on 17 July 2019).

[13] PwC (2016). Industry 4.0: Building the digital enterprise, [Online] available at https://www.pwc.com/gx/en/industries/industries-4.0/landing-page/industry-4.0-building-your-digital-enterprise-april-2016.pdf (accessed on 28 March 2018).

[14] RoSF, Smart Factory assessment, [Online], available at http://www.smart-factoryassessment.com/assessment/#/en (accessed on 20 July 2019).

[15] WMG. Industry 4.0 readiness assessment, [Online], available at https://i4ready.co.uk/tool (accessed on 20 July 2019).

[16] GrowIn 4.0 (2019). [Online], available at https://northsearegion.eu/growin4/about-the-growin-40-project/ (accessed on 17 July 2019).

[17] European Commission (September 2016). An analysis of drivers, barriers and readiness factors of EU companies for adopting advanced manufacturing products and technologies.

[18] McKinsey Digital (2015). Industry 4.0 – How to navigate digitization of the manufacturing sector, [Online], available at https://www.mckinsey.de/files/mck_industry_40_report.pdf (accessed on 17 January 2017).

[19] Ivory, C. (n.d.). Benefits identification on the move to Industry 4.0, Unpublished.

Index